Detection of Fire Source in Underground Coal Mine and its Applications

Yang Xiao · Chi-Min Shu · Shuai-Jing Ren

Detection of Fire Source in Underground Coal Mine and its Applications

Springer

Yang Xiao
School of Safety Science and Engineering
Xi'an University of Science
and Technology
Xi'an, Shaanxi, China

Shuai-Jing Ren
School of Safety Science and Engineering
Xi'an University of Science
and Technology
Xi'an, Shaanxi, China

Chi-Min Shu
Department of Safety, Health,
and Environmental Engineering
School of Engineering
National Yunlin University of Science
and Technology
Yunlin, Taiwan, ROC

ISBN 978-981-95-7063-8 ISBN 978-981-95-7064-5 (eBook)
https://doi.org/10.1007/978-981-95-7064-5

This Springer imprint is published by the registered company Springer Nature Singapore Pte Ltd.
The registered company address is: 152 Beach Road, #21-01/04 Gateway East, Singapore 189721, Singapore

Preface

In recent years, with the progress of science and technology and the deepening of understanding of the mechanism of coal fire disasters, underground coal fire detection technology has made great progress. In the field of underground coal fire detection, scholars at home and abroad have carried out a lot of research work, involving geological exploration, groundwater dynamics, seismic monitoring, remote sensing technology and other fields. A series of important research results have been achieved. These achievements not only deepen our understanding of the formation mechanism and evolution law of underground coal fires, but also provide important technical support for the monitoring, prevention and control of coalfield fires.

Underground coal fires are a serious hidden danger to coalmine safety, posing a huge threat to human life, ecological environment and resource protection. About 1,000,000,000 tons of coal are burned by underground coal fires every year, accounting for about 1/8 of the global coal consumption. On this issue, our country is in a particularly serious position. By the end of 2019, there were a large number of fire areas in Xinjiang, China, with a total area of 4,777,300 m^2, and the annual combustion loss was as high as 4,561,600 tons of coal. In Ningxia, there are 37 coalfield fire areas with an area of more than 3,940,000 m^2. Similarly, Inner Mongolia is facing a dire situation, with 26 coalfield fire zones covering an area of 19,030,000 m^2. Accurate detection of underground coal fire sources is the prerequisite for the prevention and control of coal fire disasters, and is the key to accurately detecting and delineating the location and scope of coal fires. This book systematically expounds the detection mechanism of different detection technologies, compares the advantages and disadvantages of different detection technologies, and analyzes the problems existing in current detection technologies.

This book aims to systematically introduce the basic theory and application technology of underground coal fire detection, and deeply explore the formation mechanism, detection methods and fire extinguishing technology of underground coal fire. Through the comprehensive introduction and analysis of underground coal fire detection technology, it is hoped that it can provide more scientific theoretical guidance and practical support for the prevention and control of coal field fires, and make positive contributions to the safe production and ecological environmental protection of coalmines.

The innovation team of coal fire prevention and control, where the author of this book belongs has gradually grown and expanded under the establishment and development of the well-known scholar Prof. Jingcai Xu. The team has been committed to the research of the mechanism and prevention and control technology of underground spontaneous combustion fires in coalmines and surface coalfield fires, and has successively undertaken the National Outstanding Youth Fund project of China, the 973 plan preliminary research special project, the National Natural Science Foundation of China key project, the National Natural Science Foundation of China project, the national "Eighth Five-Year Plan", "Ninth Five-Year Plan", "Tenth Five-Year Plan" and "Eleventh Five-Year Plan" science and technology support plan projects, and China–Australia, China–Germany, China–US international cooperation projects. And go deep into the front line of the mining area, with Shenhua Enterprise Group, Zhongmei Enterprise Group, Shandong Energy Enterprise Group, Shaanxi Coal Chemical Enterprise Group, Chuanmei Enterprise Group, Longmei Enterprise Group, Datong Coal Mining Enterprise Group, Yangquan Coal Enterprise Group, Shanxi Coking Coal Enterprise Group, Jincheng Anthracite Coal Enterprise Group, Lu'an Mining Enterprise Group, Huainan Mining Enterprise Group, Yaojie Coal and Power Enterprise Group, Jingyuan Coal Enterprise Group, Huating Coal Enterprise Group, Xinjiang Production and Construction Corps, Yitai Group and other more than 100 mining areas across the country to carry out technical cooperation and emergency disaster response, more than 160 horizontal scientific research projects, participate in fire extinguishing, More than 50 disaster relief efforts have been carried out, and coal fire disaster prevention and control projects have been successfully carried out in Bangladesh, Australia, Russia and other countries. For the surface coalfield fire area, the fire mechanism, fire area detection and prevention and control technology were mainly carried out in the fire area of Rujigou coalfield in Ningxia, the 4# coal seam outcrop fire in Panzhihua Baoding mining area in Sichuan, the fire area of Gulaben coalfield in Inner Mongolia, the 4 coal seam fire area of Bozidun in Wensu, Xinjiang, the upper coal fire area of the No. 1 Pingshuo Coalmine in Shanxi, the fire area of Dahuangshan coalfield in Xinjiang, and the fire area of Aiweiergou coalfield, and accumulated rich in-site experiences.

Based on theory and field practice, this book introduces the background and research significance of underground coal fire detection, and outlines the domestic and foreign dynamics and development trends in this field. The formation mechanism and evolution law of underground coal fire are discussed in depth, and the multi-field coupling kinetics of underground coal fire are introduced. Traditional detection technology and emerging detection technology, traditional detection technology includes gas detection, borehole detection, magnetic method detection, resistivity detection, isotope detection, etc., and emerging detection technology includes acoustic wave detection, microwave detection, lidar detection, quantum detection, etc. For each technology, this book explains in detail its principles, advantages and disadvantages and scope of application, introduces the corresponding detection equipment, and shows the application effect of coal fire detection technology in different scenarios through actual cases. The future development direction and trend of coal fire detection technology are prospected.

The research of this book has been funded and supported by the National Natural Science Foundation of China (National Natural Science Foundation of China) "Research on Temperature Sensing Method of Coal Fire Micro-Nanoscale Thermal Carriers" (Project No.: 52174204), and we would like to express our heartfelt gratitude. Furthermore, thanks to all experts, scholars and researchers for their attention and support to this book.

Due to the limited level of the authors, there are inevitably shortcomings in the book, readers from all walks of life are kindly invited to criticize and correct, jointly promote the development and application of underground coal fire detection technology, and contribute to the safe production and ecological environmental protection of coalmines.

Xi'an, China — Yang Xiao
Yunlin, Taiwan, ROC — Chi-Min Shu
Xi'an, China — Shuai-Jing Ren
October 2025

Acknowledgments This book is a 20-year achievement of our academic career in the Innovation Team of the Ministry of Education China entitled "Research and Application of New Technologies for Identification and Control of Coal Fires". The following research work supported by the National Natural Science Foundation of China (No. 52174204).

We definitely appreciate the efforts of Prof. Jun Deng and Prof. Hu Wen for generously offering their valuable comments, guidance, and suggestions throughout the preparation of this book.

We wish to extend our grateful acknowledgment to the staff members and graduate students for their valuable contributions to the numerical simulations, experimental work, and data processing, which were instrumental to the progress of this research.

Competing Interests The authors have no competing interests to declare that are relevant to the content of this manuscript.

Contents

Chapter 1
Origin and Development of Coal Fire Disasters

Abstract Coal fire refers to the complex dynamic disaster of low temperature oxidation and eventually spontaneous combustion of coal seams under the influence of natural or human factors, and its evolution process covers the multiphysics coupling of thermodynamics, seepage mechanics, rock mechanics and chemical reactions.

Coal fire refers to the complex dynamic disaster of low temperature oxidation and eventually spontaneous combustion of coal seams under the influence of natural or human factors, and its evolution process covers the multiphysics coupling of thermodynamics, seepage mechanics, rock mechanics and chemical reactions. Such disasters are widely distributed in major coal occurrence areas around the world, especially in mining areas between 25° and 55° north latitude, such as Xinjiang and Inner Mongolia in China, and their formation and spread are regulated by coal characteristics, geological structure, mining activities and ventilation conditions. With the increase of mining depth and intensity, coal fires show new characteristics such as deep development, strong concealment, and frequent recurrence, especially in high-gas mining areas, which is easy to cause thermodynamic coupling disasters, which significantly increases the complexity and technical difficulty of disaster prevention and control.

Coal fires not only cause huge losses of coal resources, but also continue to release pollutants such as CO_2, CO, SO_2, NO_x heavy metals and polycyclic aromatic hydrocarbons, leading to serious pollution of the atmosphere, water and soil, inducing geological disasters, such as surface subsidence and collapse, and posing a long-term threat to ecosystems and public health. At present, international research focuses on the multi-field coupling mechanism, monitoring and early warning technology, and efficient treatment methods of coal fires, especially in the construction of Heat–Flow–Solidification–Chemical (THMC) coupling models, identification of deep fire areas, and heat storage control. It is of great significance to carry out systematic research on the occurrence and evolution mechanism of coal fires and develop precise prevention and control technologies to ensure energy security, promote the construction of green mines and achieve sustainable development.

Y. Xiao et al., *Detection of Fire Source in Underground Coal Mine and its Applications*,
https://doi.org/10.1007/978-981-95-7064-5_1

1.1 Current Situation of Coal Fires

1.1.1 Distribution and Characteristics of Coal Fires

(1) Distribution of coal fires

Underground coal fire refers to the coal seam combustion phenomenon under natural conditions or under the influence of human activities, from low temperature oxidation and spontaneous combustion to intense combustion, forming a certain scale, and producing a series of environmental and ecological impacts [1]. As a global continuous ecological disaster, underground coal fires are common in China, the United States, Australia, India, Indonesia, Venezuela, Germany and other countries, and the overall distribution is more in the northern hemisphere and less in the southern hemisphere, and the coal fire area is mainly distributed between 25° and 55° north latitude. In China, ten provinces (autonomous regions) including Xinjiang, Gansu, Inner Mongolia, Shaanxi, and Shanxi have coal fire burning, as shown in Fig. 1.1, with a burning area of up to 720 km2 and a burning area of 17–20 km^2, of which Xinjiang coalfield fire area is the largest in the world [2, 3]. According to the statistical analysis of existing data [4, 5], the distribution of coal fires in the world is positively correlated with coal resource reserves and actual coal mining intensity, and regional climatic conditions are also important factors affecting the distribution of coal fires. Tables 1.1 and 1.2 are obtained from the 2024 BP World Energy Statistical Yearbook and the collection of relevant data [6]. Table 1.1 shows the proven coal reserves of major countries in the world in 2023, Table 1.2 shows the global coal mining scale in 2024, which can be found that the global coal resources are mainly distributed in three major regions: Asia–Pacific, Eurasia and North America, and show the distribution law of "rich in the north and poor in the south", and countries with a large degree of exploitation (such as the United States, Australia, India) are also facing serious coal fires.

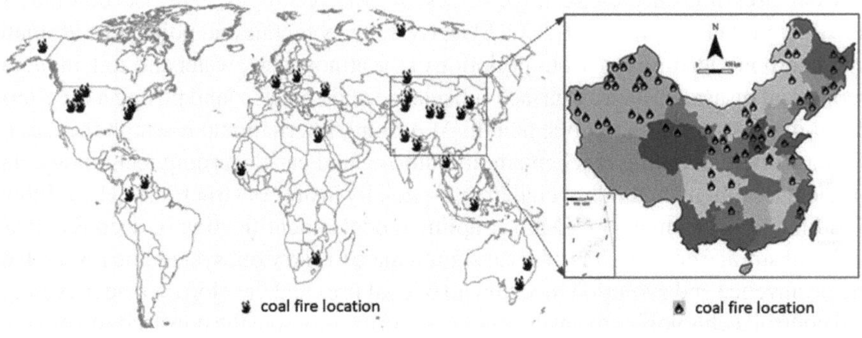

Fig. 1.1 Global underground coal fire distribution [7]

Table 1.1 Coal proven reserves of major countries in the world in 2023 [6]

Ranking	Country	Proven reserves/million tons
1	United States	273,244
2	Russia	178,757
3	China	173,108
4	Australia	164,764
5	India	140,795
6	Germany	39,022
7	Indonesia	38,641
8	Ukraine	37,892
9	Kyrgyzstan	31,415
10	Poland	30,598
11	Kazakhstan	28,225
12	Türkiye	12,098
13	North Korea	11,684
14	South Africa	10,905

Table 1.2 Global coal mining volume in 2024 [6]

Ranking	Country	Mining volume/million tons
1	China	4780.0
2	India	1085.1
3	Indonesia	836.1
4	United States	464.6
5	Australia	462.9
6	Russia	427.2
7	South Africa	235.0
8	Kazakhstan	112.6
9	Mongolia	106.5
10	Germany	91.9
11	Türkiye	87.0
12	Poland	85.2
13	Colombia	52.7

Xinjiang is rich in coal resources, ranking first in the country in terms of predicted reserves, and Xinjiang is also the province with the most serious coalfield fires in China [8]. The implementation of the national strategy for the large-scale development of the western region has made Xinjiang more and more important as a strategic base for the utilization and development of Chinese coal reserves. Most of the coalfields in Xinjiang were formed in the Jurassic period, which led to a short period of spontaneous combustion and ignition due to the late coal formation period, and due to the influence of the Himalayan orogeny, the coal seam was mostly steeply tilted and there were many coal seam outcrops. The combined effect of special geological conditions, natural conditions, climate and human factors has led to large-scale spontaneous combustion of coalfields. The long duration of coal fires, the intensity of fires, and the wide distribution of fire areas have made Xinjiang the most typical and serious region of coal fires in China and the world [5, 9]. According to reports, 46 coal-fired areas in Xinjiang lose 4,420,000 tons of coal resources every year, emitting 13,200,000 tons of carbon dioxide, 103,000 tons of carbon monoxide, 20,500 tons of total hydrocarbons, 44,100 tons of sulfur dioxide, and 10,500 tons of soot, while releasing a large amount of heat, causing millions of square meters of vegetation damage [10]. The high temperature generated during the coal combustion process will change the stress balance of the overlying rock of the coal seam, causing a series of cracks, cracks and surface settlement, which seriously affects the safety of some local infrastructure and people's lives and property. In addition, it can lead to problems such as vegetation degradation, land desertification, water pollution, and ecological environment damage, further deteriorating the fragile ecological environment in mining areas [11, 12]. Under the direction of Premier Enlai Zhou, the Xinjiang Coalfield Fire Extinguishing Engineering Bureau was established in 1958, and then, with the support of the state, it vigorously carried out a series of detections and field surveys of coal fire areas, such as thermal exploration, chemical exploration, drilling, geophysical exploration and remote sensing. In 1995, the state put coal fires in northern China and the ecological and environmental problems they caused on the agenda in the "China Development Plan for the 21st Century". At present, Xinjiang has deployed wireless temperature sensors in some fire areas and connected to remote network monitoring systems, and developed a coalfield fire area information system based on the geographic information system. However, due to the influence of many factors, coal fires in Xinjiang are constantly extinguished, and there are often extinguishing fire zones, and new fire zones appear hundreds of kilometers away [13]. According to the fifth census of coalfield fire areas in Xinjiang in 2019, 18 of the 40 fire areas were newly created after the fourth census of coalfield fire areas. On the other hand, due to the closure of wells and production safety rectification, more than 1,500 mines have been closed in Xinjiang. Some closed coalmines have a lot of coal left underground, a high tendency to spontaneous combustion, and relevant measures are not in place, which is at risk of new coal fires. The continuous emergence of new fire areas, and the rekindling of some ancient fire areas and treated fire areas have brought great challenges to fire extinguishing work.

From a global perspective, the long-term sustainability and governance dilemma of coal fire disasters are common challenges in the field of international research, and a number of typical cases reveal the core contradictions of coal fire problems in different scenarios. The causes of more than 70 underground fire points that have been burning since 1916 in the Galia coalfield in India are closely related to the abandoned roadways formed by private mining, which provide a channel for continuous contact between coal and oxygen, eventually triggering spontaneous combustion of shallow surface coal seams. This case not only reflects the damage caused by disorderly mining to the prevention and control of coal fires, but also highlights the cumulative impact of long-term coal fire combustion on regional ecology and public health, which is highly consistent with the current international research pain points on the coupling mechanism of "coal fire–human activities–environmental risk". The Centralia fire in Pennsylvania, USA, shows the extreme consequences of coalfield fires after the failure of treatment. As one of the more than 260 continuously burning coal wells in the United States, its fire is difficult to control and eventually leads to the migration of residents in the town, this case reveals the spread risk of deep coal fires in the case of insufficient monitoring technology and limited treatment methods, and also confirms the urgent need for "long-term evolution law of coal fire and regional scale prevention and control strategies" in current international research. The "9.24" major fire accident at Sanjiaoshu Coalmine in Panjiang Jingmei Co., Ltd. [14], Guizhou reported by the State Coalmine Safety Supervision Bureau caused 16 deaths, 3 injuries, and direct economic losses of RMB 42,338,200. The cause of the accident was the serious deformation of the roadway of the second belt conveyor section of the 1200 transport flat lane, and the friction between the idler and the conveyor belt bracket heated up, igniting the pulverized coal and non-flame retardant conveyor belt, producing toxic and harmful gases, causing poisoning and death. The accident reflects the lack of solid safety development concept, chaotic safety management, and the failure of the safety monitoring system, as well as the current lagging response of the current fire source identification technology for complex roadway environments. Similarly, on May 9, 2023, the "5.9" major fire accident at Gengcun Coalmine of Dayou Energy Co., Ltd. in Sanmenxia, Henan, China [15], killed 5 people and caused direct economic losses of RMB 14,832,600. The cause of the accident was the spontaneous combustion of the broken coal around the regulating wind window of the return air connection lane of the water silo pumping room in the middle of the 13th mining area, igniting the wooden back plate and the broken coal body in the roadway, reversing the wind flow, and blocking the escape route with high temperature flue gas and high concentration of CO. This case highlights the lack of research on the law of toxic gas diffusion during coal spontaneous combustion and the limitations of existing monitoring systems under complex ventilation conditions. The above cases jointly show that whether it is natural coal seam spontaneous combustion or coal fire accidents in industrial scenarios, their prevention and control faces common problems such as complex causes, difficult governance, and significant multi-dimensional impact, which is also the direction that needs to be focused on breakthroughs in international coal fire disaster research.

(2) Characteristics of coal fires

Since 2011, with the shift of the focus of global coal resource development to high-reserve areas, such as western China, northeastern India, the Appalachian Mountains of the United States, and the promotion of high-yield and efficient mining technology, mine coal spontaneous combustion fires and coalfield fires have shown new characteristics around the world.

① New characteristics of spontaneous combustion of underground coal

The spontaneous ignition of the goaf of the high-inclined working face of large-scale mining is serious. With the development of coal mining technology, the construction of safe, high-yield and efficient mines has become a global trend, and the core is to break through the mining of extra-thick coal seam groups of more than 6 m. Represented by large coal bases such as the Ningdong mining area in China, the Jharkhand coalfield in India, and the Powder River Basin in the United States [14], the coal reserves with a thickness of more than 6 m in the main mining seam account for about 40%. Due to the complex geological conditions, the large range of spontaneous combustion hazard areas, the complex air leakage law in the goaf, and the strong concealment of the ignition source are such as Huafeng Coalmine in Shandong, [15], and Zhangshuanglou Coalmine in Jiangsu, China [16], as well as some mines in Siberia, Russia.

The connection roadway is naturally ignited driving double tunnels of the working face. With the improvement of the mechanization level of coal mining in the world, the continuous innovation of tunneling and mining processes, the length of the underground working face has increased, and there are a large number of connecting roadways between the two roadways in the double-lane tunneling mode, which is difficult to seal and the effect of the sealing wall is poor [17].

Mining of the close coal seam group ignites in the adjacent mined goaf. Globally, most of the fully integrated coal seams mining in close coal seams are thick or extra-thick coal seams, and the lower coal pillars are prone to crushing to produce cracks and form air leakage channels. Spontaneous combustion prediction models, hazardous area determination, gas migration laws [18], coordinated prevention and control of spontaneous combustion and gas [19], and comprehensive fire prevention and extinguishing technology [20] in the Shendong mining area of China are common research hotspots and difficulties faced by Shangwan Coalmine, Paner Coalmine in Huainan, and some mines in the Bowen Basin in Australia.

Integrate coalmine goaf spontaneous ignition. To curb accidents caused by private mining, small coalmines have been integrated or closed in many places around the world, but the remaining large-scale goaf has become a high-incidence area of spontaneous combustion fires, bringing major hidden dangers to fire prevention and control. For instance, in the Shendong mining area, there are six small coalmines in the Huojitu well of the Daliuta Coalmine, Yulin, China, where have as many as 19 smoking spots formed by spontaneous combustion, and 5 open flames. In addition, there are small mine fire areas in Pingshuo No. 2 Coalmine [21] and Anjialing Open-pit Coalmine, Shuozhou, China [22].

The number of coal seam fire accidents caused by external causes has increased. The global mining depth is increasing at a rate of 10–30 m per year, the consumption of mining electrical equipment and cables has increased sharply, the fire load has increased, and the heat generation of some roadway filling materials is too high, resulting in the number of external fires such as cable fires, belt conveyor fires, and spontaneous combustion of filling materials increasing year by year. For example, in 2014, the Huashan Coalmine of Sichuan Coal Group, China caused a fire due to the reaction of polymer filling materials to release heat, and some mines in the

Appalachian region of the United States also experienced frequent cable fire accidents due to aging equipment.

Fire and gas thermal power disasters are prominent. The heat release and transfer associated with fire and gas explosion are the main thermodynamic disaster phenomena common to coalmines around the world [23]. With the deep extension of the eastern mine and the large-scale mining of the huge coal seam in the west, the coupling mechanism of mine fire and gas and the disaster energy transfer and utilization mechanism are unclear, and thermodynamic disasters such as fire and gas explosion are particularly prominent [24].

② New characteristics of coalfield fire development

Impacted by mine mining, coalfield fires develop deeper. Globally, mine mining has broken the original crustal balance, coal seam mining disturbances have led to the development of formation cracks, and the spontaneous combustion of outcropping coals has gradually developed into a fully combustible ignition source under the coupling effect of heat–flow–solidification–chemical, and has expanded to the depths under the condition of continuous oxygen supply [25]. The Zhundong coalfield in Xinjiang, China, the Centralia fire in the United States, and the Galia coalfield in India all have the phenomenon of exacerbating the spread of the fire area due to mining disturbances.

The development of coalfield fires is more hidden. More than 90% of the global coalfield fire area is in an underground smoldering state, which has the characteristics of large range, scattered ignition points, long distance, and hidden fire sources, making real-time dynamic monitoring and fire area management extremely difficult. Due to the hidden location of the fire source, the targeted treatment is generally insufficient [26], and such problems exist in the Helanshan coalfield in China, the Siberian coalfield in Russia, and some fire areas in South Africa.

The difficulty of fire control in residual coalfields has increased. Most of the global coalfield fire areas are distributed in arid or semi-arid areas, where water resources are scarce and loess cover cannot dissipate heat from the fire area in time [27]. Due to long-term continuous combustion in the fire area of the remaining coalfield, the surface cover layer is suspended or collapsed after the deep coal seam is burned out, forming pit cracks, the fire area is easy to reignite, and new fire areas are constantly generated. The precise positioning of the fire source center and the removal of heat energy in the fire area of the remaining coalfield are also the main problems that lead to the current management difficulties [28].

There is a re-ignition phenomenon in the coalfield fire area. The coalfield fire area has large heat storage and significant oxygen-poor oxidation characteristics, and the traditional treatment methods, such as stripping, leveling, grouting, and loess mulching cannot replace and dissipate the internal heat storage in time. Over time, ground covering projects are prone to the formation of new oxygen supply channels under the action of collapse cracks, formation subsidence, and rainwater scouring, which are prone to rekindling [29], and there are records of re-ignition in the Wuda coalfield in Inner Mongolia, China, the Centralia earthfire in the United States, and the Galia coalfield in India.

The coal industry is sluggish, and the speed of fire area management is slow. Globally, the demand for coal continues to be weak, and the acceleration of new energy substorion and the deepening of energy structure transformation have further squeezed the living space of coal, resulting in the problem of overcapacity. In this context, coal prices have been hovering at a low level for a long time, and many coal companies have fallen into business difficulties, and some areas have even experienced continuous losses, and the capital chain is under serious pressure. Coalfield fire control requires a lot of capital investment, from the early fire exploration and plan formulation, to the implementation of the fire extinguishing project in the middle stage, and then to the monitoring and maintenance in the later stage, every link is inseparable from stable financial support. However, due to the overall downturn in the industry, the profitability of enterprises has plummeted, and the support of local finances for governance projects has also weakened. The shortage of funds directly leads to the lag in the update of fire extinguishing equipment, the difficulty of maintaining professional governance teams, and the obstruction of the application of key technologies, which makes the already complex fire area management work slow, which not only threatens the ecological environment and surrounding safety of the mining area, but also forms a vicious circle of "industry downturn, shortage of funds, and lagging governance".

1.1.2 Hazards of Coal Fires

Coal fires not only burn a large amount of coal resources, endanger the safety and economic production of coalmines, but also seriously cause a series of homeland security and health environmental problems [30]. Figure 1.2 visually depicts the formation and development process of the combustion system of a typical coalfield fire disaster and its comprehensive impact on the surrounding environment. According to a report by Beijing Remote Sensing Corporation (BRSC) and China Coal Aviation Remote Sensing Bureau (ARSC), about 10,000,000–20,000,000 tons of coal are directly burned due to coal fires in Inner Mongolia alone every year, and the loss caused by indirect impact on coal resource mining is about 200,000,000 tons [31]. The large amount of toxic and harmful substances produced by incomplete combustion of coal fires also poses different degrees and levels of threats to the global atmosphere, geosphere, hydrosphere, biosphere, and ecosystem [32].

(1) Influence of toxic and harmful gases and aerosols

① Impact of SOx and NOx emissions on the environment

While studying greenhouse gas emissions, other gases are also measured using the same method. For example, Bwesaje Kurban [34] calculated the gas emissions in the

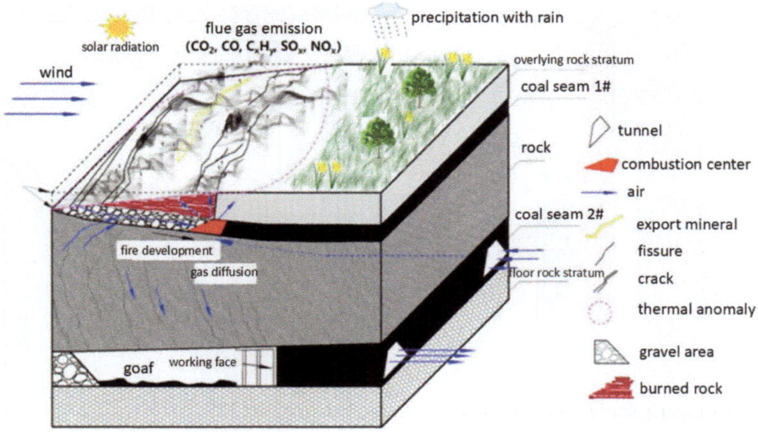

Fig. 1.2 Formation and development of underground coal fires and related environmental impacts [33]

southern fire area of the Shuixigou fire area based on the field test results, of which the CO_2 emission was about 3.7 × 108 m^3/a, the CO was about 1.4 × 104 m^3/a, the NO was about 4.55 × 103 m^3/a, and the SO_2 was about 3.07 × 103 m^3/a. Gas products, such as C_2H_2, C_2H_4, C_2H_6, and C_3H_8 have also been obtained experimentally, and most studies on these gases have been conducted from the perspective of temperature on gas production concentration [34]. Through the monitoring of the spontaneous combustion area of coal in Shanxi, it was found that the emission of pollutants in the spontaneous combustion area was large, especially the emission of CO and SO_2 gas (Table 1.3), which posed a great threat to the surrounding environment [35]. Similarly, a similar situation occurred in the Uda area (Fig. 1.3), where all three gases (SO_2, NO_2, and CO) were above the standard values, i.e., spontaneous combustion zones > residential areas > norms [36].

② Hazards of organic pollutants

Organic pollutants pose a serious hazard to human health, including some known carcinogens, such as benzene, which is closely related to diseases, such as leukemia, Hodgkin's disease, multiple myeloma, and non-Hodgkin's lymphoma. A large number of organic pollutants are produced during the spontaneous combustion of coal in the coalfield fire area of the United States, such as carbon and sulfur compounds

Table 1.3 Pollutant emissions from coal spontaneous combustion observation points [27, 29]

Observation point	Emissions/10,000 tons			
	Soot/dust	SO2	NO2/NO	CO
8 counties and cities in Shanxi, China	2.52	1.30	0.55	9.34
Shanxi Datong Fengzijian fire area	0.13	0.17	0.42	0.55

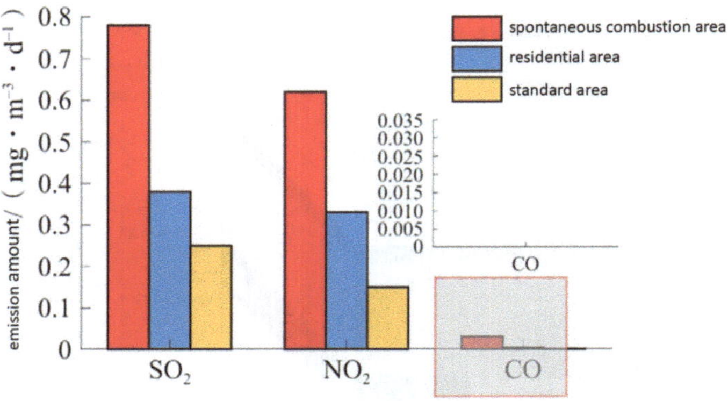

Fig. 1.3 Comparison chart of gas emissions in the Uda region [37]

(OCS, $S(CH_3)2$, CS_2), aliphatic compounds (from CH_4 to C_9H_20), and aromatic compounds (C_6H_6, C_7H_8, C_8H_{10}), even if the emission of these organic pollutants is below the standard limit of the National Institute for Occupational Safety and Health (NIOSH), the gases released by coal fires are still potentially dangerous and may be harmful to people for a long time [38]. The main sources of polycyclic aromatic hydrocarbons (PAHs) are rock-forming sources and pyrolysis sources, resulting in differences in emission components [39].

③ Emissions of Hg and HF and their impact on the environment

As a toxic heavy metal, Hg is released in large quantities during the coal burning process, directly affecting the local environment and the health of residents. Coal fire surveys at Skyline in Kentucky, USA, in May 2008 and January 2009, showed that the Hg concentrations at the site were 500 $\mu g/m^3$ and 2100 $\mu g/m^3$ in May and January, respectively, far exceeding the OSHA 8-h exposure limit (50 $\mu g/m^3$) [38]. The measurement results of underground coal fire and gangue spontaneous combustion area in Wuda area, Inner Mongolia, showed that the mercury emission from surface cracks and vent flue gas in the fire area was 464 ng/m^3, while the mercury emission from the spontaneous combustion of the gangue pile was 12.7 times higher than that of the underground coal fire, as high as 5908 ng/m^3, and the spontaneous combustion fire area and gangue pile were 221 times and 2813 times the background value, respectively, which had a certain impact on the local environment [40]. Fluorine pollution is also closely related to coal mining and spontaneous combustion, leading to the emergence of regional fluorosis [41]. Through on-site measurement of HF emitted from Wuda coalfield, it was found that underground coal fires not only caused local fluorine anomalies in near-coal mining areas, but also caused HF content above the mining areas where underground coal fires occurred due to the high volatility of HF that was higher than the daily values in other cities [40].

④ Impact of aerosol emissions on the environment

Regarding particulate matter research, most scholars take coal-fired power plants as the main research object, and few scholars pay attention to the emission characteristics of particulate matter in spontaneous combustion fire areas [42]. By analyzing the distribution characteristics of organic carbon and elemental carbon in PM10 in the spontaneous combustion areas of coal seam in Shanxi, China (Datong, Ningwu and Hequ), the results show that the total carbon aerosol contributes the most to particulate matter, and there are spatial differences in the distribution characteristics of particulate matter and carbon pollution. The average mass concentration of total suspended solids (TSP) in the ambient air was as high as 14,367μg/m^3, and the enrichment factors of S, Cd, and Cr in the combustion area were higher than those in the surrounding area, but due to the complexity of field testing, the understanding of coal fire aerosols was insufficient [40].

(2) The destruction of land resources and surface vegetation by coal fire

① The impact of coal fire on soil

Spontaneous combustion of coal will lead to changes in the environment such as land resources and surface vegetation in the combustion area. The gaseous substances released during spontaneous combustion destroy the mineral composition and pH of the soil. The pH value is found to be weakly alkaline in the range of 5.61–8.34, and the coefficient of variation (spatial dispersion of soil characteristics, CV) is used as the level of N, P, K, SOM (organic matter) and other parameters in the soil, and the variability of SOM is found to be greater than that of N, P, K, and K [41]. In addition, other scholars have pointed out that the conditions after combustion in the coalfield fire area show sparse vegetation, loose soil, and reduced organic matter content, and the surface soil is compacted during the treatment process, resulting in poor water retention and air permeability [42]. Hu Wen et al. [43] analyzed the physical and chemical characteristics of soil at different combustion stages in the coalfield fire area, among which the soil organic matter content was high during the spontaneous combustion stage and the combustion period of mirabilite was high, while the content of organic matter was very low during the extinguishing stage. In addition, field geological surveys have found that crystal products such as mirabilite, sulfur, and tar are all gaseous condensation products formed by spontaneous combustion of coal (Fig. 1.4), which may be due to the destruction of the surrounding rock under the action of temperature, and some components are free and condensed in the form of gaseous to form simple salts and sulfides, and the salinity changes significantly before and after spontaneous combustion, and soluble ions may make the soil highly acidic [44], which in turn affects the surface vegetation.

Coal fire can also lead to the enrichment or exceedance of toxic and harmful substances in the soil within the burning area. The average pollution index of heavy metals in typical coalfield fire areas is Hg > As > Cu > Ni > Zn > Cr, and Merlot's comprehensive pollution index is heavily polluted, with extremely strong ecological and environmental risks, and the residual arsenic content in the soil is the highest, and the temperature, topography, climatic conditions, and soil properties of the fire area all affect the migration and enrichment of heavy metals [36, 46, 47]. For example,

(a) Mirabilite (b) Sulfur (c) Coal Tar

Fig. 1.4 Gaseous condensation products formed by spontaneous combustion of coal [45]

the study of the content and distribution characteristics of PAHs in gangue samples in the spontaneous combustion area found that PAHs emissions were affected by the degree of combustion, and the higher the degree of combustion, the less toxicity of PAHs, which may be due to the escape of some PAHs during the combustion process [47–51]. At the same time, the PAHs released at different combustion levels are different, mainly benzo[a]anthracene (BaA) when the combustion degree is low, and AcPy is mainly used when the combustion degree is high [51], and some scholars have reached similar conclusions [51], that is, the release ratio of high-cyclic PAHs is greater than that of low-cyclic PAHs at low temperature, and the release ratio of low-cyclic PAHs is greater under high temperature conditions, and the PAHs with different toxicity show a decreasing trend with the increase of temperature, when the temperature reaches 500 °C. Basically, very few PAHs with different toxicities were detected.

② Damage to surface vegetation by coal fire

According to the above discussion, it is not difficult to find that factors such as toxic and harmful gases and soil fertility will affect the survival of vegetation to varying degrees. The thermal effect of coal burning not only seriously threatens the tolerance of plants to high temperatures, but also causes water loss in the soil, which eventually leads to vegetation withering and even death, and may also spread to surrounding forests, grasslands, or farmland [52]. Li [53] used remote sensing technology to investigate the land resources in the Datong mining area, and after extracting the thermal anomaly area, it was found that the impact of coal fire on the surrounding environment was forest land, grassland, residential land, and cultivated land, and this situation was similar in the coal gangue field in Fushun, Liaoning, China [44]. Different combustion stages (oxidative self-heating stage, coal seam combustion period, and extinguishing period) also affect the vegetation in the coalfield fire area [54]: the surface vegetation can basically grow normally during its oxidative self-heating period, which may be due to the fact that temperature, toxic and harmful gases, and soil physical and chemical properties change less during this period [44]. During the burning period, both the characteristics of the product and the soil and surrounding environment underwent great changes, resulting in the branches, leaves and rhizomes of vegetation being damaged to varying degrees until they withered and died. Through the investigation of plant species in the Wuda mining area, it was found that there were few species of plants, most of them were herbaceous, and some low shrub plants were basically no arbor plants, and they were distributed in clusters [36, 55, 56], which may lead to the destruction of biological chains and the stability of the ecosystem [13].

③ Characteristics of pollutants in the corpus cavernosum

The sponge body of coal fire refers to the upward protrusion of the surface layer of the soil due to the upward spread of flue gas, which is mushroom-shaped, highly acidic, and much higher than the background value of the area, which may be due to the acid gas (SO_2 and HF) released during the coal combustion process, resulting in sparse vegetation or no grass on the surface [40]. At the same time, Du et al. [57] also found that the cavernous body of coal fire is also rich in S (HSO_4–, SO_4–, and S +), Fe (Fe +, FeO +, and FeSO +), Ca +, and Cd + plasma, and positive and negative ion interactions lead to the formation of some sulfates, such as gypsum ($CaSO_4$), alum ($Fe_2(SO_4)3$), cadmium sulfate ($CdSO_4$), and ammonium bisulfate (NH_4HSO_4). In addition to acidic ions, some heavy metals and PAHs were detected in the cavernous body of coal fire, and it was found that mercury enrichment occurred in the cavernous body, and the intensity of arsenic content in the cavernous body was 4.6 of the background value, which was relatively small compared with the mercury enrichment [40]. In the analysis of cavernous samples in the fire area of Wuda coalfield, it was found that 15 out of 18 samples had severe environmental pollution and the remaining 3 samples were moderately polluted [58], while the highly enriched PAHs in the coal fire cavernous body led to severe surface soil pollution of about 8,000 m^2, and the emission characteristics of various PAHs were consistent, and the PAHs with low ring structure were dominant, such as naphthalene and phenylene [40].

(3) Pollution of water bodies by coal fire

Some scholars believe that the use of water in the process of coal spontaneous combustion treatment has basically no impact on the surface water and groundwater environment [59, 60]. However, the dissolved oxygen, chemical oxygen demand, pH, and sulfide of surface water measured by Jingjing Tian [35] did not meet the standards in the Ningwu, Pingyao, and Taiyuan sections of the 8 Class III. functional areas, and the Shuozhou section of the 7 Class IV functional areas did not meet the standards. The water quality of other groundwater monitoring data (pH, fluoride, heavy metals, hardness and other parameters) meets the Class III standard of the "Groundwater Quality Standard". Li [60] studied the impact of water fire extinguishing on groundwater in the treatment of coal gangue, mainly evaluating the impact of heavy metals on the environment, and took the secondary standard of GB 15,618−1995 "Soil Environmental Quality Standard" as the evaluation standard, and found that except for the abnormality of suspended solids, other indicators were lower than the standard value. Cao et al. [59] argued that the spontaneous combustion area of Daquanhu coalfield had no effect on groundwater in the area due to simple geological conditions, faults far away from the spontaneous combustion area, poor water permeability of the aquifer, and poor hydraulic connection in the aquifer. PAHs in water bodies affect the water environment. Some scholars have pointed out that there are two main sources of PAHs in the water in coal mining areas: One is generated during leaching, Another is produced during the combustion of coal or coal gangue [61], which is mainly produced by leaching [62], but combined with the literature [35], it is speculated that some PAHs may be derived from spontaneous combustion of coal or gangue. In terms of characteristics, polycyclic aromatic hydrocarbons are not easily soluble in water and mostly exist in water and soil in adsorption state. Chen [53] compared the characteristics of PAHs in surface water and gangue and found that the same components as coal gangue appeared in the water body, suggesting that PAHs had the ability to migrate into the water body and caused pollution of the water body. PAHs are also often adsorbed on solid suspended particles and transported and migrated in water bodies, which leads to higher PAHs content in bottom sediments than in water bodies [47, 49].

(4) Impact of coal fire on geological conditions

① Geological disasters related to coal fire

With the deepening of the degree of combustion, a series of geological disasters are induced, namely "cracks/fractures–burned areas–surface subsidence–collapse". In the early stage of combustion, due to thermal effects, some cracks are formed, and then the thickness of the coal seam will gradually decrease, and even part of the coal will be burned out, resulting in empty shells in the mountain, or due to the loss of support of the overlying rock layer, the surface subsidence, collapse, and surrounding rock collapse [63, 64], which can be seen everywhere in the coalfield fire area (Fig. 1.5). In addition, if the remaining coal pillars in some underground goafs are burned, it will eventually lead to a large area of collapse. Cao et al. [65] systematically classified the fractures in the Wuda coalfield fire area, namely tectonic fractures, mining collapse fractures, and combustion fractures, among which combustion fractures are one of the most common fractures in the Wuda coalfield fire area, which are usually developed on the basis of the first two.

② Influence of burnt rock on geological conditions

A new rock formed by the change of physical and chemical properties of rocks caused by high-temperature baking during the spontaneous combustion of coal is

(a) Combustion Fracture (b) Collapse Fracture (c) Burning Hole

Fig. 1.5 Fissures and voids in the coalfield fire area

called burning rock [66–70]. The shear strength of the exerted rock increases with the increase of depth, and the shear strength of different lithology is different, with the strongest burned sandstone and the weakest incendiary mudstone [71]. The shear resistance index gradually decreases with the increase of water content [71], which may lead to the disintegration of the rock after encountering water, and in severe cases, it may cause a large area of rock properties to change in the burned rock area, and the safety of the slope is greatly threatened [78]. The mechanical properties of the incendiary rock in the Beskuduk open-pit coalmine are quite different from the original rock, and the crack development, frost resistance, and disintegration are worse than those of the original rock, and the slope stability is seriously reduced [73].

The permeability properties of exerted rocks are affected by the properties of the original rock, and it is found that the seepage network is formed earlier than that of incinerated sandstone or calcined argillaceous sandstone [74]. On the one hand, it is conducive to the formation of a water-rich zone to provide water supply sources for coalmine construction, and on the other hand, it brings hidden dangers to the safe mining of coalmines [75–77]. The water of the Poplar River replenishes a large amount of water to the burned rocks in the mining area, with a leakage of 0.300–0.578 m^3/s, and a large spring flow rate of 12.54 × 104 m^3/d) in the burned rock area [78]. Du et al. [82] proposed that the burned rock area of the northern Shaanxi energy and chemical base can be used as a favorable water supply source. With a stable flow rate of 1,000 m^3/d in the Daliuta Coalmine, the Huojitu Coalmine makes full use of the water resources of 4 × 106 m^3 in the pyrotechnic rock area [79]. It has been reported that the Xigou Coalmine dug into the calcined rock area during the mining process, which led to the occurrence of water inrush accidents [80] (Fig. 1.6).

Fig. 1.6 Porous structure of the crested rock [45]

1.2 Conditions for Coal Fires

1.2.1 Conditions for Spontaneous Combustion of Coal

Coal is mainly composed of inorganic and organic matter, and the combustible elements of coal include roughly four kinds, namely carbon, hydrogen, oxygen and sulfur. The above four combustible elements make up combustible compounds, which can be regarded as combustible materials of coal. In addition to combustible compounds, there are also non-combustible substances in coal, such as moisture, mineral ash, and non-combustible substances in coal are regarded as inert. The main reason for spontaneous combustion of coal piles is that coal is oxidized by oxygen in the air, and hydrogen, carbon, and sulfur in coal can react at room temperature to form alkanes [81]. If the heat cannot be dissipated in time, the temperature of the coal pile will gradually rise, and the oxidation of coal will become faster and faster, thereby releasing more heat and combustible materials. When the combustible matter and heat reach a certain level, spontaneous combustion will occur.

The spontaneous combustion of coal piles needs to have the following contents:

(1) It has a certain tendency to spontaneous combustion. The spontaneous combustion tendency of coal belongs to the spontaneous combustion attribute of coal, which can effectively show the deterioration of coal. When spontaneous combustion occurs in coal piles, it needs to have thermal conductivity, particle size, moisture, ash and other conditions. If coal can oxidize at room temperature, it is mainly due to the volatile content. If the volatile content is high, then the spontaneous combustion tendency of coal is in a strong state at this time, and it is very easy to cause spontaneous combustion problems. On the contrary, if the volatile content is low, the spontaneous combustion tendency of coal is in a weak state at this time, and it is not easy to cause spontaneous combustion problems.
(2) Oxygen supply conditions. Coal piles are usually exposed to air for a long time, so the surface of the coal will come into contact with air. The air will flow into the coal pile through the gaps, causing oxidation inside the coal pile. The larger the coal block, the larger the gap between the coal piles, which will create oxidation conditions for the spontaneous combustion of the coal pile, and the coal pile is very prone to spontaneous combustion problems.
(3) Oxidation time. There will be a period of time from the initial oxidation to the final spontaneous combustion of the coal pile, and only when a certain degree of oxidation and oxidation time are reached, the coal pile will produce spontaneous combustion. For example, long-flame coal, which takes one to three months from initial oxidation to eventual spontaneous combustion, while gas coal takes four to six months from initial oxidation to final spontaneous combustion.
(4) Heat storage conditions. The heat generated by the oxidation reaction must not be lost in time. Heat accumulation is prone to heat accumulation in closed environments, thick coal seams or areas with complex geological structures.

The temperature needs to reach a critical point, so that the oxidation reaction enters the self-accelerating stage, and finally triggers an open flame.

1.2.2 Factors Affecting Coal Spontaneous Combustion

(1) Tendency of coal spontaneous combustion

The tendency of coal spontaneous combustion refers to the difficulty undergoing the process of its spontaneous combustion, which is mainly affected by the following factors:

① Degree of coal deterioration

The degree of coal metamorphosis [82] is a decisive factor affecting the tendency of coal to spontaneously combust, and 90% of the 55 coalfields or mining areas in northern China are composed of medium and low metamorphic coal. With the deepening of coalification, the physical and chemical properties of coal change, thus reducing the possibility of oxidation, the oxygen consumption rate of coal decreases, the ignition point increases, and the tendency to spontaneous combustion decreases. Various grades of coal have the possibility of spontaneous combustion, and coal seams with a low degree of deterioration (such as lignite, long-flame coal and gas coal) are more likely to ignite spontaneously, and the risk of spontaneous combustion is greater. Spontaneous combustion of coal seams (such as anthracite) with a high degree of metamorphosis is relatively rare, and the risk of spontaneous combustion is small. However, the degree of coalification must not be used as the only marker to judge the size of the natural tendency, because production practice has proved that some coals with the same degree of coalization have natural characteristics, and some do not spontaneously combust on the whole, the natural tendency of coal decreases with the increase of the degree of deterioration of coal, that is, the natural tendency gradually decreases from lignite, long-flame coal, bituminous coal, coking coal to anthracite. Locally, there is a complex relationship between the natural tendency of coal seams and the degree of coal metamorphosis, that is, the natural tendency of coal with the same degree of coalification may vary greatly in different regions and different mines.

② Coal and rock composition

Different coal and rock components have different oxygen adsorption capabilities and different oxidation properties. Silk charcoal has large porosity and the strongest ability to adsorb oxygen, but the oxidation is the lowest, while mirror coal has the highest oxidation. Therefore, when thin-layered silk charcoal coexists with mirror coal strips, the coal seam is very prone to spontaneous combustion, because silk charcoal can absorb enough oxygen and transfer it to neighboring mirror coal, increasing the spontaneous combustion tendency of coal [83].

③ Sulfur content of coal

The sulfur and other catalysts contained in coal will accelerate the oxidation process of coal. In the same grade of coal, the more sulfur-containing minerals (such as pyrite), the more prone it is to spontaneous combustion. This is due to the formation of iron sulfate and ferrous sulfate when the pyrite contained in the coal is oxidized at low temperature, which increases the volume and makes the coal body expand and become loose, increasing the oxidation surface area, and the heat released during pyrite oxidation also promotes the spontaneous combustion of coal. Therefore, the existence of pyrite will accelerate the spontaneous combustion of coal, and the higher its content, the greater the tendency of coal to spontaneously combust. Statistics show that coal seams with sulfur content greater than 3% are spontaneously ignited coal seams, including anthracite.

④ Moisture in coal

The moisture in coal is an important factor affecting its oxidation process [84–87]. The external and internal moisture of coal and water vapor in the air have a certain effect on lignite and bituminous coal in the low temperature oxidation stage, which not only accelerates oxidation, but also hinders oxidation. In the self-heating stage of coal, due to the generation and evaporation of moisture, a lot of heat must be consumed, and the temperature is difficult to rise to 100 °C before all the moisture in the coal body evaporates, which is why coal with large moisture is difficult to spontaneously combust. However, the moisture in the coal can be filled in the tiny voids of the coal body, removing nitrogen, carbon dioxide, methane and other gases, and activating the adsorption capacity of the coal when dried. Therefore, when the moisture of the coal is large enough, it will inhibit the spontaneous combustion of the coal, but after losing the moisture, the risk of spontaneous combustion will increase. In addition, for coal seams containing pyrite, moisture is an indispensable condition for the decomposition of pyrite. From this point of view, moisture is conducive to the occurrence of spontaneous combustion of coal. Table 1.4 shows the relationship between the moisture content of coal seam and spontaneous ignition in a mine.

Among the 158 spontaneous ignition accidents, 117 occurred in coal seams with a moisture content of 1–4%, accounting for 74.05% of the total number of ignitions. while only 25 spontaneous ignitions occurred in coal seams with a moisture content of less than 1% or more than 4%, accounting for 15.82% of the total number of ignitions. Therefore, keeping the moisture content in the coal seam greater than 4% during the mining process is beneficial to reduce the spontaneous combustion of coal,

Table 1.4 Relationship between coal seam moisture content and spontaneous combustion [90–93]

Water content indicator	Total	Coal body moisture/%										
		< 1	1–2	2–3	3–4	4–5	5–6	6–7	7–8	8–9	9–10	Unknown
Number of fires	158	7	42	52	23	2	5	3	0	3	5	15
Specific gravity	100	4.7	26.6	32.6	14.6	1.2	3.2	1.9	0	1.9	3.2	10.1

reduce the amount of coal dust generated, and reduce the frequency and intensity of ground pressure.

⑤ Porosity and brittleness of coal

The porosity and brittleness of coal also have a greater impact on the self-bias of coal. The larger the porosity of coal, the more likely it is to spontaneously ignite. This is because the porosity is large, and the more oxygen penetrates into the coal. The more brittle the coal with the same degree of deterioration, the more prone it is to spontaneous combustion. Because the brittleness of coal is related to whether the coal is easy to break and form pulverized coal. Once the coal body is broken, the oxygen absorption surface area of the coal will increase, the ignition point will be significantly reduced, and its spontaneous combustion will be significantly improved.

⑥ Coal seam gas content

Gas usually exists in the voids, cracks, cracks and microporous surfaces of coal in the free state and adsorption state, these two gases exist in the pressure state, under the influence of objective conditions such as coal seam pressure relief and temperature rise, the adsorbed gas will undergo desorption phenomenon and transform into free gas. Therefore, gas in its original state or gas under pressure will have an inhibitory effect on air immersion into the coal body. Because the adsorbed gas occupies the surface of the coal molecules and prevents the contact between coal and oxygen, the gas in the coal seam has a good inhibitory effect, which is a favorable factor to prevent spontaneous combustion. On the other hand, when the gas in the coal seam gushes out at its own pressure or gas is pumped, the gas phase pressure in the coal seam decreases, and the gas content decreases, which increases the probability of coal contact with air and enhances the spontaneous combustion of coal. Other factors such as oil content, thermal conductivity, fixed carbon and ash content in coal also affect the natural tendency of coal.

The above factors play an important role in the process of spontaneous combustion of coal. However, the spontaneous combustion of coal does not mean that the coal is exposed to the air and spontaneously ignites immediately, but needs to go through a certain amount of time and suitable external conditions.

(2) Coal seam geological deposits

The geological factors affecting the spontaneous combustion of coal seams mainly include: thickness, dip angle, joints, fractures, roof properties and geological structures (such as faults, folds, magmatic rock intrusion) of coal seams. The larger the thickness of the coal seam, the greater the risk of spontaneous combustion. This is due to the low coal recovery rate, easy damage to coal pillars, difficult closure of goafs, and large air leakage in the process of post-coal seam mining. There are currently two methods of mining thick coal seams: layered mining and secondary mining at full height. During layered mining, the tunneling and mining operations of the lower layered mining roadway are carried out under the false roof, and the coal mining and tunneling processes will be leaked with the upper layered goaf, and the spontaneous ignition is serious. When topping coal is mined, the goaf is large, there is a lot of coal, the working face is slow, and the ignition is more serious, but it is smaller than that of layered mining. Statistics show that the number of spontaneous combustion fires in the goaf of thick coal seams and extra-thick coal seams accounts for about 84% of the number of fires in the goaf. The larger the dip angle of the coal seam, the more serious the ignition. This is due to the fact that when mining coal seams with large inclination angles, the goaf is not easy to fill, and the bottom plate of the coal seam will also slide, especially the steeply inclined coal seam coal pillars are also difficult to retain, and the air leakage is serious. The properties of the roof of the coal seam affect the spontaneous combustion process of coal to a certain extent. The hard and hard-to-collapse roof, the mine pressure on the coal seam and coal pillar is concentrated, easy to be crushed, the goaf is not filled, the air leakage is large, and the closure is not tight, which is conducive to the occurrence of spontaneous combustion, The soft and easy roof plate is filled with goafs, and the air leakage is small, and the natural danger is small. Joints and cracks not only increase the contact area between coal and oxygen, but also increase the air leakage channel. Therefore, coal seams with joints and cracks are more likely to ignite naturally. Coal seams damaged by geological structure are soft, broken, cracked, oxidative, and have good air leakage and oxygen supply conditions, so spontaneous ignitions are much more frequent than areas with normal coal seams. In addition, due to the hydraulic connection at the fault fracture zone, the coal becomes wet, and the oxidation ability of the coal is enhanced after drying. In the magma intrusion zone, the coal seam is distilled, the porosity of the coal increases, the strength decreases, and the risk of spontaneous combustion increases. Statistics show that coal seams with large geological changes are more common than areas with small changes, and in addition to determining the characteristics of coal in geological change areas, the more important factors depend on the active activity of mine pressure.

(3) Conditions for development and exploitation

① Development method

Practical experience shows that the development layout has a great impact on the mines of spontaneously ignited coal seams. It is required that the roadway system is simple, and the use of stone gates and rock roadways to open up the roadway, leave less coal pillars, and reduce the cutting of coal bodies, which is positive and beneficial to eliminate the hidden danger of fire.

The centralized transportation lane and the main return air lane are the main roadways of the mine, and the service time is long. The air volume is large, the air pressure is high, and it often changes due to the adjustment of the air volume of the mining face. There are no transportation tapes and mechanical and electrical equipment in the transportation lane, which create conditions for fire hazards. Therefore, these main roadways should be arranged in rock formations or in coal seam bureaus that are not prone to spontaneous combustion. For extra-thick coal seams that are prone to spontaneous combustion, the selection of development methods is extremely important to inhibit spontaneous combustion. When mining inclined or steeply inclined extra-thick coal seams, the main roadway should be arranged in the bottom rock or the coal seam that is not easy to spontaneously ignite, and the stone door in the mining area should be used to enter the coal seam. Gently inclined and horizontal coal seams can be arranged in the cross-hole panel, but there are different opinions on whether the centralized transportation roadway of a single coal seam is arranged in the rock of the bottom of the coal seam or in the coal seam. The centralized transportation lane is arranged in the rock layer of the bottom of the coal seam, which can not only reduce the maintenance cost of the roadway, but also reduce the chance of spontaneous combustion, but the cost of excavating the roadway in the coal seam is high, the cost is high, and it is not necessarily economical, and the amount of gangue discharged from the mine is large, which affects the surface environment. Excavating centralized transportation roadways in coal seams, scientifically and reasonably determining the size of the roadway protection coal pillars, reducing the impact of mine pressure on coal pillars, so that the coal pillars are not crushed, and then closed in time, is still effective in inhibiting the spontaneous ignition of coal.

② Coal mining methods

The influence of coal mining method on ignition is mainly manifested in the level of recovery rate, the speed of advancement, the length of mining time, roof management and coal seam cutting. For example, high-drop type, knife pillar type, coal skin false roof and hydraulic coal mining with low recovery rate, in addition to losing more coal and low recovery rate, these coal mining methods only partially fall or do not fall at all, and the falling roof rocks cannot fill the goaf, resulting in a large amount of air leakage in the goaf, which is not conducive to preventing the spontaneous combustion of coal. For the mining of shallow and close coal seam groups, when these mining methods are selected, due to the influence of mining, the ground is prone to cracks and interlayer cracks, which can form air leakage channels, and the floating coal discarded in the goaf can obtain the conditions of oxidation and self-heating, so it will accelerate spontaneous combustion, which is not good for fire prevention. Forward coal mining is not conducive to preventing spontaneous ignition due to serious air leakage in the section roadway. Both sides of the backward coal mining inlet and return air lanes are solid coal, which is conducive to fire prevention. In recent years, the widespread application of topping coal mining method in mining extra-thick coal seams has worsened the situation of spontaneous combustion.

Different development systems in coal mining methods have different risks of spontaneous ignition of coal seams. Therefore, it is very important to choose a reasonable development system and coal mining method to prevent spontaneous ignition. A reasonable development system should ensure that there are few coal seam cuts, few coal pillars left, and the goaf can be closed in time. Reasonable coal mining methods should make the roadway layout simple, high coal recovery rate, fast advancement speed, and small air leakage in the goaf.

(4) Ventilation conditions and ventilation management

Air leakage provides necessary oxygen for coal spontaneous combustion, and the strength of air leakage directly affects the heat dissipation of coal. The effect of ventilation on the spontaneous combustion of coal is mainly manifested in the air leakage in the goaf, coal pillar and coal wall cracks. Air leakage oxidizes the coal at these locations to generate heat, and the amount of heat generated and whether it can accumulate depends on the size of the wind speed. When the wind speed is too small, the oxygen supply from the air leakage is very small, and the heat generated by oxidation is less, and it is not easy to self-heat and spontaneous combustion. When the air leakage wind speed is too large, although the oxygen supply is sufficient, the heat generated by oxidation is easy to be taken away, and the heat accumulation cannot be formed, and the coal cannot spontaneously ignite. Some scholars believe that spontaneous combustion fire will not occur when the air leakage intensity in the goaf is greater than 1.2 $m^3/(min \cdot m^2)$ or less than 0.06 $m^3/(min \cdot m^2)$. The most dangerous air leakage intensity is 0.4–0.8 $m^3/(min \cdot m^2)$. In fire prevention work, air leakage must be minimized.

Ventilation management is also an important factor affecting spontaneous ignition, including the rationality of the wind network structure, the distribution of air volume in the mining area, the reasonable determination of wind pressure, the closure of the goaf, the selection of the location of ventilation control facilities, and the engineering quality of ventilation facilities. Therefore, mine ventilation management technology must carefully study the relationship between wind and fire, and choose a reasonable wind network structure and scientific management methods. It is not the mining area to adjust the wind network structure, expand the roadway section, and eliminate the resistance to solve the ventilation problem, but to increase the total wind pressure in the mine to force air supply, which is very unscientific. The increase of the total air pressure in the mine will inevitably lead to an increase in air leakage, which provides sufficient oxygen exchange conditions for the area that may spontaneously ignite or the fire area that is suffocated due to lack of oxygen, and accelerates oxidation and leads to spontaneous ignition or rekindling of the fire area.

1.3 Development and Evolution Process of Coal Fires

The coalfield fire area is formed by the evolution of high-temperature ignition sources, which is mainly the result of the coupling of the temperature field of coal heat storage, the stress field of coal–rock mass cracks, the seepage field formed by aerodynamics and combustion gas products, and the chemical field of coal–oxygen compound combustion [88]. The porous medium characteristics of coal and rock masses and their environment affect the generation of high temperature points and the migration of gases. Therefore, studying the development and evolution process of coalfield fires and mastering their development and evolution laws are important issues in the study of coalfield fire areas.

1.3.1 Multi-field Coupling Mechanism of Development and Evolution in Coalfield Fire Area

As a direct carrier of fluid and heat transport, the coal–rock mass in the coalfield fire area can be simplified into a fracture medium model composed of matrix coal–rock blocks and fractures [89]. The central temperature of the fire area is as high as 1000 °C or more [90], and the heat is released to the surrounding coal and rock mass in a conductive manner. The coal body is heated to produce pyrolysis gas or gas, and the rock mass is heated to rupture to form cracks, providing a channel for air and pyrolysis gas, further promoting the development of the burned area forward, resulting in the collapse of the overlying rock mass due to fracture and weightlessness. It constitutes a complete heat–flow–solidification–chemical coupling process [91], as shown in Fig. 1.7.

(1) Thermal–solid coupling

The coal and rock mass in the fire area expands and deforms due to high temperature, and eventually destroys and produces cracks. The different intrinsic structure and characteristic parameters of coal and rock body lead to uneven temperature field distribution, which constitutes a complete thermo–solid coupling problem. The complete mathematical model of thermo–solid coupling is formed by giving all the conditions for the fixed solution of heat conduction, solid elastic mechanics and coupling equations. Wessling et al. [93] explained the coupling mechanism of coal combustion thermo–solid coupling in view of the abnormal change of surface temperature field in the coalfield fire area, and performed numerical simulation analysis. Drebenstedt et al. [94] established a mathematical model of propagation in the coalfield fire area, analyzed and studied the fractures and cracks of the overlying rock layer in the coalfield fire area, and showed that the large cracks in the overlying rock

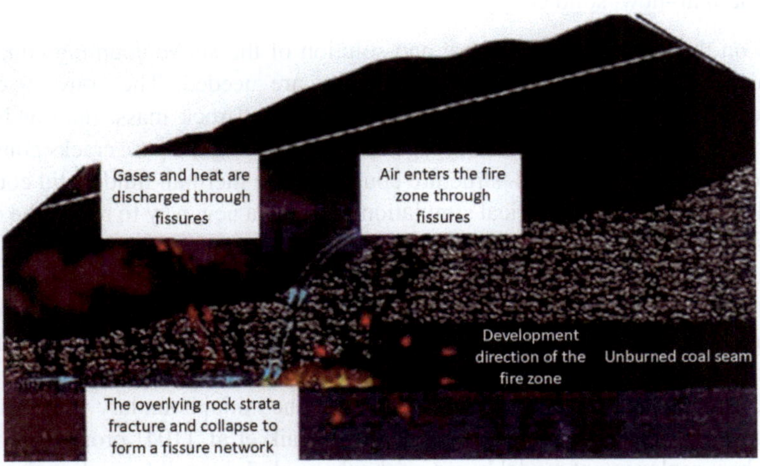

Fig. 1.7 Development and evolution process of coalfield fire areas [92]

layer are the direct oxygen supply channels in the fire area, and the effect of fire wind pressure leads to the penetration of micro-cracks. Eleck [95] studied a large number of cracks formed by surface subsidence caused by the Pennsylvania coal fires in the United States. Heffern and Coates [96] studied the geological history of coal fire burning in the Powder River basin in the United States and demonstrated the formation of numerous fissures after combustion. Rosema et al. [97] established and solved the kinetic model of coal spontaneous combustion by studying the spontaneous combustion process caused by coal oxidation.

(2) Thermal–flow coupling

The heat flow relationship in the fire area is different from other engineering problems, due to the inflow of air through cracks or channels, the coal–oxygen compound heat generation heats up, and at the same time, the thermodynamic product gas outflow and air convection circulate heat exchange. The temperature gradient between the high temperature point and the surface and the gas seepage wind pressure constitute a coupling problem. Zeng et al. [98, 99] established a model describing the flow characteristics of flue gas based on the cracks, temperature, and flue gas flow characteristics in the fire area, and proposed a calculation method for oxygen consumption per unit fire area at the unit fire area. Through the analysis of the physical process of fire wind pressure in the fire area, the fire wind pressure theory is proposed, which is mainly caused by the density difference formed by temperature difference. Li et al. [100] analyzed the influencing factors of fire wind pressure, and concluded that factors such as the depth of the spontaneous ignition point, the occurrence inclination angle of coal seam, the permeability, topography, and wind speed all had a certain influence on fire wind pressure. Wang et al. [101] established a coupling model of outcrop spontaneous combustion seepage and thermodynamic coupling model, and numerically simulated the fire area of a coalfield in Xinjiang, China.

(3) Thermal–flow–solid coupling

Based on the gradual improvement and solution of the above coupling equations, engineering models that are closer to reality are needed. The convective heat transfer of the fluid, the stress–strain of the coal and rock mass, the conduction heat transfer of the fluid, and the seepage of the fluid along the pore cracks constitute the problem of thermal–fluid–structure coupling. The thermal–fluid–solid coupling model combined with numerical simulation provides a new way to reveal the development and evolution mechanism of coal fire. Zhang et al. [102] established a fluid–structure coupling model based on the basic principle of fluid–structure coupling and studied the effects of seepage on the stress field, surrounding rock deformation, and excavation unloading of surrounding rock. Nowak et al. [103] conducted numerical simulations of the thermal–fluid–solid coupling process of underground engineering and proposed a new method to solve the initial boundary value problem of coupled nonlinear finite element schemes. Shaik et al. [104] proposed an estimated heat replacement model based on the thermal–fluid–solid coupling theory by using the fluid medium (water) to extract and replace underground heat from the flow of natural surface fractures. Zhu et al. [105] and Zhang et al. [106] established the heat–fluid–solids model and obtained the variation law of coal permeability under the influence of stress, temperature, and gas desorption. Xia et al. [107] used the thermal–flow–solid total coupling theory to study the stress field, oxygen concentration field, and thermal conductivity of porous medium induced by coal–oxygen compounding reaction in underground coal seams, and performed numerical simulations on actual coal seams.

1.3.2 Evolution of Characteristic Parameters in the Process of Coal Spontaneous Combustion

Coal spontaneous combustion is a series of complex physical and chemical reaction processes, which will lead to significant changes in parameters such as characteristic temperature, heat release, and activation energy in different gas component environments, and microscopic characteristic parameters such as functional groups, free radicals and coal microcrystalline structure will also show regular evolution. Thermogravimetric analysis (TGA), differential scanning calorimetry (DSC), and differential thermal analysis (DTA) were used to study the macroscopic thermal effects of coal spontaneous combustion process.

(1) Characteristic temperature

Hu [108] used TG–DSC synchronous thermal analysis technology to find that the dry cracking temperature increases slowly with the decrease of O2 volume fraction in the oxidizing atmosphere, while CH_4 gas weakens the difference between the active temperature and the dry cracking temperature. Liu et al. [109] studied the effects of CH_4 on coal spontaneous combustion at different temperature stages through thermogravimetric analysis and gas chromatography, and pointed out that CH_4 gas can delay the formation of oxidation products CO and CO_2 in the low oxidation stage, and the oxygen-poor environment caused by CH_4 gas will slow down the speed and ignition temperature of coal–oxygen reaction in the accelerated oxidation stage, and a large amount of O_2 will be consumed due to the intense combustion of CH_4 and the combustion of C and H_2 produced by CH_4 decomposition in the combustion stage. This will inhibit or even terminate the process of coal–oxygen compound reaction. Liu [110] prepared coal samples with different CH_4 contents (1.9–3.8m^3/t), and found that the critical temperature and dry cracking temperature of the adsorbed CH_4 gas coal samples increased by program heating test. Zhou [111] pointed out that with the increase of CH_4 volume fraction in the oxidizing atmosphere, the intersection temperature and critical temperature increased. Deng et al. [112] pointed out that when the temperature of the coal body is lower than the critical temperature, the CH_4 gas adsorbed by the coal body will inhibit the low-temperature oxidation process of the coal, and the higher the gas content, the more obvious the inhibition effect on coal oxidation. This is mainly due to the fact that with the increase of CH_4 volume fraction, CH_4 gas can occupy more active sites on the coal surface under the competitive adsorption effect, resulting in a decrease in O_2 participating in the oxidation reaction on the coal surface and a significant decrease in the intensity of the coal oxidation reaction at low temperature. When the temperature is higher than the critical temperature, the oxidation capacity of coal is released to a certain extent.

(2) Heat release

Lou et al. [113] found that with the decrease of CH_4 gas volume fraction, the TG and DSC curves shifted, and the weight loss rate of coal spontaneous combustion under low CH_4 gas volume fraction was higher than that with high CH_4 volume fraction. Zhang et al. [114] pointed out that CH_4 gas in the oxidizing atmosphere can reduce the heat absorbed and released during low-temperature oxidation, and the higher the volume fraction of CH_4, the less heat release. Hu [108] pointed out that CH_4 gas has a great influence on the heat production of coal in the rapid oxidation stage, and when the CH_4 volume fraction is 14.3–57.1%, it will increase the heat production in the rapid oxidation stage of coal, which means that CH_4 gas reduces the difficulty of coal–oxygen compound reaction, while inhibits coal–oxygen compounding in other volume fraction ranges. Zhang et al. [115] used a C80 microcalorimeter to study the exothermic characteristics of coal in a low volume fraction CH_4 atmosphere, and pointed out that with the increase of CH_4 volume fraction, the heat absorption decrease of coal is less than 10 J, while the heat release decrease is close to 300 J. This is mainly due to the increase of coal surface active groups, the increase of oxygen consumption, and the increase of dependence on oxygen supply increase, resulting in the influence of oxygen content on heat release more and more obviously, while the increase of CH_4 volume fraction will inhibit the effective supply of oxygen under high temperature conditions. As a result, the heat release of coal is significantly reduced.

(3) Activation energy

Activation energy is an important indicator of the kinetics of coal oxidation, which determines the ease of oxidation reaction in coal [116]. The lower the activation energy, the easier it is to carry out the coal oxidation reaction, and the other is not easy to carry out the oxidation reaction [117]. Based on the test data of TG–DSC synchronous thermal analysis test, programmed heating test, C80 heat flow thermal test, etc., the coal oxidation kinetic parameter can be calculated by the Arrhenius equation or the Coats–Redfern integration method, i.e., the activation energy. Deng et al. [112] pointed out that the volume fraction of CH_4 has a significant effect on the activation energy of coal–oxygen compounding, and the activation energy required for low-temperature oxidation of high-metamorphic coal increased by 10% due to CH_4 gas, but had little effect on the activation energy of low-metamorphic coal. Lou et al. [113] showed that when the volume fraction of CH_4 was less than 24%, the activation energy in the coal weight gain stage decreased slightly, and the activation energy in the coal combustion stage was basically stable at 99.5 kJ/mol. When the volume coefficient of CH_4 was 24–72%, the activation energy in the weight gain stage increased and the activation energy in the combustion stage gradually decreased to

62.4 kJ/mol. Hao et al. [118] analyzed the thermogravimetric test data using the Coats–Redfern integration method and pointed out that with the increase of CH_4 volume fraction, the low-temperature oxidation activation energy of coal increases and the pre-index factor decreases, which means that CH_4 gas has a significant inhibitory effect on coal spontaneous combustion. Xincheng Hu [108] pointed out that the activation energy of the slow oxidation stage did not change much in the CH_4-containing atmosphere, and the activation energy in the rapid oxidation stage changed slightly with the volume fraction of CH_4, while the activation energy in the combustion stage was significantly higher than that in the slow and rapid oxidation stages, and there was a slight decrease with the increase of CH_4 volume fraction. With the increase of CH_4 volume fraction or content, the results of coal spontaneous combustion thermal effect and activation energy are shown in Table 1.5. However, it is generally believed that the activation energy of coal is low and the influence of CH_4 gas is small under the low volume fraction CH_4 atmosphere, which is mainly due to the high oxygen fraction, the coal is in the oxygen-rich oxidation state, and the influence of CH_4 gas on the coal oxygen reaction is small. In the atmosphere with high volume fraction CH_4, the oxygen gas fraction is low, the coal is in a fuel-rich state, and the adsorption of CH_4 gas is stronger, resulting in CH_4 gas occupying more active adsorption positions on the coal surface, which is manifested as the inhibitory effect of CH_4 gas on the coal oxygen reaction process, that is, the activation energy increases. However, Ma et al. [119] pointed out that the apparent activation energy of coal in a low CH_4 volume fraction (1%–3%) atmosphere is smaller than that in an atmosphere without CH_4, which means that CH4 gas has a positive effect on coal spontaneous combustion, but no corresponding explanation is given.

(4) Microscopic characteristics of coal spontaneous combustion process under the influence of gas

① Functional group

The spontaneous combustion process of coal is a self-accelerating heating process in which different functional groups are distributed and gradually activated and react with oxygen, which is related to the oxidative activity of the main functional groups in coal, the reaction conditions, and the ambient temperature [120]. The influence of air composition changes on coal spontaneous combustion has attracted attention, and Lv et al. [121] found that the increase in the proportion of nitrogen components in coal spontaneous combustion atmosphere can lead to changes in the evolution characteristics of functional groups in coal. Correspondingly, CH_4, as a hydrocarbon with very low carbon content, has a complex effect on coal spontaneous combustion. With the increase of CH_4 volume fraction, the changes of the main functional groups of the oxidized coal samples are shown in Table 1.6. Ma et al. [119] analyzed that in an atmosphere with a low CH_4 volume fraction (<3%), aromatics had a negative effect on coal spontaneous combustion, while methyl and methylene groups had a positive effect. The inhibitory effect of alkyl ethers on coal spontaneous combustion is weakened, while the inhibitory effect of aromatic rings is enhanced. Deng et al. [112] pointed out that CH_4 gas can delay the low-temperature oxidation process of coal by

Table 1.5 Results of spontaneous combustion thermal effect and activation energy of coal under gas-containing conditions [108, 109, 112–115, 118]

Methane volume fraction/%	Heating	Activation energy		Test method
		Activation energy changes	Calculation method	
0–3	—	Apparent activation energy (↘)	Arrhenius Equation	Procedure temperature test
0–4	heat absorption (↘limited), Heat release (↘significant)	Apparent activation energy (↗)	Arrhenius Equation	C80 calorimetry test
0–4	Low degree of metamorphism coal heat release (not obvious)	Low degree of metamorphism coal (not obvious)	Arrhenius Equation	C80 calorimetry test
	High degree of metamorphism coal heat release (↘)	High degree of metamorphism coal activation energy (↗)		
0–4	Heat absorption (↘not noticeable)	—	—	C80 calorimetry test
	Heat absorption (↘not noticeable)			
0–72	weightless rate (↘), Heat release (↘)	Coal weight gain stage: activation energy (whole↗)	Coats–Redfern Equation	TG–DSC thermal analysis test
		Coal combustion stage: activation energy (whole↘)		
0–85.7	Slow oxidation stage: Weight loss rate (↘) Rapid oxidation stage: Heat release rate (↘) Thermal decomposition and combustion stage: Heat release rate (↘)	Rapid oxidation stage	Coats–Redfern Equation	TG–DSC thermal analysis test
		< 28.6% activation energy (weak↘)		
		28.6%–57.1% activation energy (weak↘)		
		> 57.1%: activation energy (↗)		
29–86	Low temperature oxidation stage: reaction rate (↘)	—	—	TG–S/DTG–S thermal analysis test

(continued)

Table 1.5 (continued)

Methane volume fraction/%	Heating	Activation energy		Test method
		Activation energy changes	Calculation method	
	Accelerated oxidation stage: ignition temperature (↗)			
	Combustion Phase: Weight Loss Rate (↘)			
0–100	Heat release (↘)	Apparent activation energy (↗)	Coats–Redfern Equation	TG–DSC thermal analysis test
0–100	—	Low-temperature oxidative activation energy (↗)	Coats–Redfern Equation	TG–DSC thermal analysis test

inhibiting the content of key active groups in the low-temperature oxidation process of coal, and when the volume fraction is 0–4%, the inhibitory effect of CH_4 gas on low-temperature oxidation of coal becomes significant with the increase of CH_4 volume fraction, and the degree of inhibition of CH_4 on the active groups is oxygen-containing functional groups, aromatic hydrocarbons, aliphatic hydrocarbons, and hydroxyl groups in descending order. Zhang et al. [115] pointed out that the relative contents of aromatic hydrocarbons, aliphatic hydrocarbons, and carboxyl groups decreased with the increase of CH_4 volume fraction in the oxidizing atmosphere. The relative contents of hydroxyl and ether bonds and carbonyl groups showed an increasing trend, which means that CH_4 gas inhibited the breakage of aliphatic hydrocarbons and the oxidation of carboxyl groups in coal. Liu et al. [120] divided the oxidation reaction of–COOH functional group in the spontaneous combustion process of coal under CH_4-containing atmosphere into three stages: (1) the low volume fraction stage (<6.0%), the CH_4 gas had a significant promoting effect on the oxidation consumption of–COOH functional groups, (2) the excitation effect of CH_4 gas on the oxidation of–COOH functional groups was weakened in the middle volume fraction stage (6.0–13.6%), (3) the inhibitory effect of CH_4 gas was significant in the high volume index stage (>13.6%). And the inhibitory effect intensified with the increase of CH_4 volume fraction. Xu et al. [123] analyzed that CH_4 gas had a strong inhibitory effect on–OH oxidation under low volume fraction conditions when the oxidation temperature was 80 °C, while the inhibitory effect on–COOH oxidation mainly occurred under high volume fraction conditions.

② free radical

The free radical theory of coal spontaneous combustion believes that coal–oxygen compounding is the reaction between oxygen molecules and coal surface active

Table 1.6 Changes in the main functional groups during the oxidation of gas-containing coal samples [108, 112, 115, 122–124]

Methane volume fraction/%	Relative changes in the major functional groups
0–4	CH_4 2%: C – C&C – H (↘), COO – (↘), C – O – (↗), C = O (↗) CH_4 4%: C – C&C – H (↘), COO – (↘), C – O – (↗), C = O (↗)
0–4	– OH (↘), C = O&C – O – C (↘), – CH3& – CH2 (↘), Ar – CH&C = C (↘)
0–40	CH_4 < 6.0%: – COOH (↗) CH_4 (6.0%–13.6%): – COOH (↗) CH_4 > 13.6%: – COOH (↘)
0–45, 80 °C	CH_4 < 15%: – CH3& – CH2 (↗) CH_4 (15%–25%): – CH3& – CH2 (—) CH_4 > 25%: – CH3& – CH2 (↗)
0–45	CH_4 < 25%: – CH3& – CH2 (↘) CH_4 > 35%: – OH (↘↗↘) CH_4 > 25%: – C – O (↗), COOH – (—)
5–55 (Experimental temperature 190, 230°C)	C – H&C – C (↗↘), C – O (↘↗), C = O (↗↘)

groups (free radicals) [120]. During the process of spontaneous combustion of coal, the oxidizing atmosphere will change the formation of coal free radicals, which will affect the process of coal–oxygen compounding reaction [124]. Hu et al. [125] studied the radical properties of coal oxidation in gas atmosphere using electron paramagnetic resonance spectroscopy and characterized the radical activity by Lander factor g. The changes of free radical volume fraction and g factor with CH_4 volume fraction of coal under different oxidation temperatures are shown in Fig. 1.8. The effects of CH_4 gas on g factor and free radicals can be roughly divided into three stages: ① in the low volume fraction stage (<25%), with the increase of CH_4 volume fraction, the g factor increases slowly and the free radical volume fraction decreases rapidly, ② in the middle volume fraction stage (25%–35%), the increase of CH_4 volume fraction leads to a rapid increase in g factor, while the decrease rate of free radical volume fraction decreases, ③ high volume fraction stage (>35%), The g-factor remained basically unchanged, while the decrease in the volume fraction of free radicals tended to flatten. This indicates that the presence of CH_4 gas in the oxidizing atmosphere affects the generation and transport of free radicals during the coal–oxygen compound reaction, and the orbital–spin coupling of coal, as well as the types and contents of coal free radicals, are changed. The analysis shows that with the increase of the volume fraction of CH_4 gas, the active site on the coal surface will be occupied by dilutive gas molecules that do not participate in the oxidation reaction, which weakens the adsorption of oxygen molecules by coal and thus affects the process of coal–oxygen compound reaction.

③ Microcrystalline structure

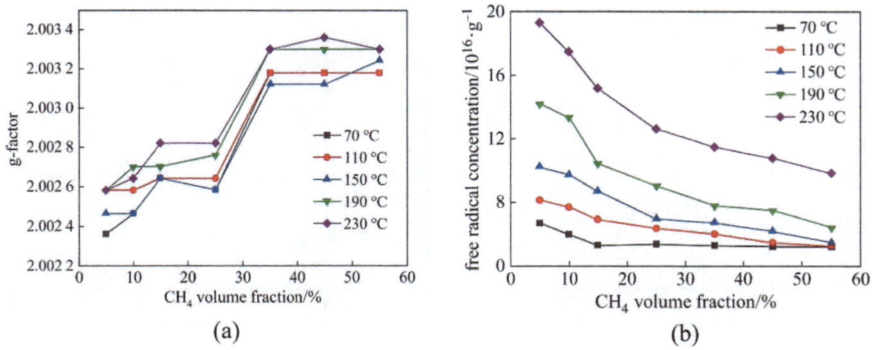

Fig. 1.8 Free radical parameters of coal changed with the volume fraction of CH_4 at different oxidation temperatures [125]

The microcrystalline structure parameters of coal mainly include the stacking height of aromatic layers Lc, the effective stacking number of layers N, the extensibility of the lamellar La, and the degree of graphitization P [126]. Based on X-ray diffraction (XRD) tests, the relationship between the microcrystalline structure parameters of gaseous coal and coking coal samples at constant oxidation temperature (190 °C) and the volume fraction of CH_4 is shown in Fig. 1.9 [127]. The stacking height and effective stacking number of coal aromatic layers decrease rapidly and then tend to flatten with the increase of CH_4 volume fraction, while the ductility shows a relatively stable decrease and the graphitization degree of coal decreases. This reflects that with the increase of CH_4 volume fraction in the oxidizing atmosphere, the volume of coal microcrystalline structural units decreases, the aromatization and condensation degree of coal macromolecular structural units decreases, and the directional arrangement degree becomes worse. Under the same CH_4 volume fraction, the stacking height, effective stacking number and graphitization degree parameters of coking coal are larger than those of gas coal, and the three parameters of the two coal samples change basically the same with the increase of CH_4 volume fraction. As for the extensibility parameters, when the CH_4 volume index is low (less than 10%), the extensibility of the two coal samples does not change significantly with the volume score of CH_4, and with the continuous increase of CH_4 volume fraction, the extensibility of gas coal shows a more rapid and stable decrease trend than that of coking coal, which may be due to the influence of high volume fraction methane gas dilution and competitive adsorption, and the horizontal oxidation reaction of aromatic microcrystalline layers of coking coal is faster than that of gas coal. Hu [108] further studied the effect of oxidation temperature on the aromatic microcrystalline structure of coal under constant CH_4 volume fraction, and found that with the increase of constant oxidation temperature, the number of stacked layers of aromatic lamination gradually increased, the stacking height increased, and the graphitization degree of coal gradually deepened, and this phenomenon became more significant with the decrease of CH_4 volume fraction under constant oxidation temperature atmosphere.

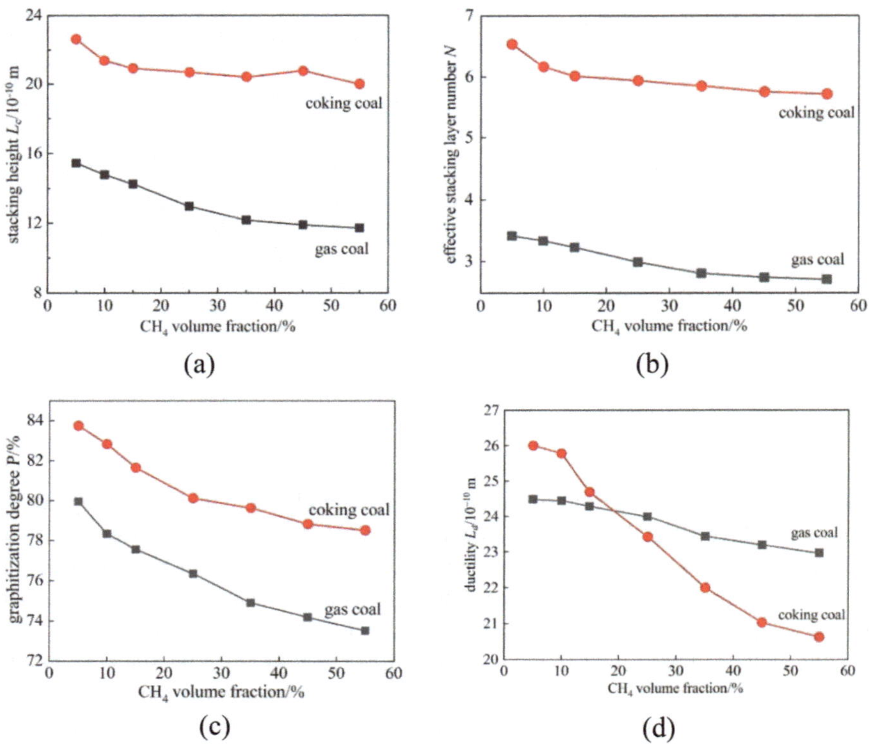

Fig. 1.9 Microcrystalline structure parameters of coal changed with the volume fraction of CH_4 [127]

This indicates that the temperature rise promotes the condensation and polymerization of the aromatic layers of oxidized coal samples and enhances the spatial order, which also reflects the influence of CH_4 gas on the spatial orientation of coal microcrystalline structures in coal oxidation reaction and the oxidation temperature.

④ Pore properties

The pore characteristics of coal (pore shape, specific surface area, pore volume, etc.) are important factors closely related to the coal oxidation process, and they can have a significant effect on the coal–oxygen compound reaction by affecting the physical oxygen absorption capacity during the coal oxidation process [128]. The evolution characteristics of coal pore structure with coal oxidation under different gas atmospheres have attracted attention. Tang et al. [129] conducted a mercury intrusion test on coal samples (bituminous coal) after 12 h of constant temperature treatment at 200 °C under N_2, CO_2, and 10% O_2 atmospheres, and pointed out that the pore volume of pretreated coal samples above 10 nm increased significantly, and the average pore size, porosity, and permeability were larger than those of raw coal samples, reflecting that the thermal effect in low oxygen and inert gas atmosphere promotes more uniform coal pore distribution and weakens the complexity of pore structure, which is conducive to the migration of gas in coal pores. Fractal dimension is an important parameter for characterizing the pore surface roughness and structural complexity of porous materials [130], and Cai et al. [131] conducted low-temperature liquid nitrogen adsorption tests on coal samples (anthracite) after 2 h of constant temperature treatment at 150 °C and 230 °C under different CH_4 volume fraction atmospheres, and the changes of the specific surface area, average pore size, and FHH (Frenkel–Halsey–Hill) fractal dimension with CH_4 volume fraction are shown in Fig. 1.10. With the increase of CH_4 volume fraction in the oxidation atmosphere, the average pore size of BET in the oxidized coal samples at constant temperature 150 °C and 230 °C shows a trend of increasing first and then decreasing, and reaches the maximum value when the CH_4 volume fraction is 25%. The specific surface area and total specific surface area of the oxidized coal samples decreased rapidly and then increased slightly with the increase of CH_4 volume fraction, and reached the minimum value when the CH_4 volume fraction was 35%. At the same time, the minimum specific surface area of coal samples oxidized at a constant temperature of 230 °C was greater than that of the constant temperature of 150 °C, while the average pore size of BET of the former was smaller than that of the latter, reflecting that the higher oxidation temperature was conducive to the development of micropores in coal. Figure 1.5c shows that with the increase of CH_4 volume fraction, there is a difference in fractal dimension between constant temperature 150 °C and 230 °C oxidation treated coal samples, but the overall trend is to decrease first and then increase slightly, and reaches the minimum value when the CH_4 volume fraction is 25%. The influence of CH_4 gas on coal micropores during low-temperature oxidation is related to the volume fraction of CH_4 and oxidation temperature, and CH_4 gas tends to complicate the micropore structure at higher oxidation temperatures. The experimental phenomenon may be due to the limited inhibitory effect of CH_4 gas on coal oxidation when the volume fraction of CH_4 is lower than a certain value (~25%), and the micropores in the coal body under the action of high-temperature oxidation expand and connect to develop into larger pores, resulting in an increase in the average size of micropores and a decrease in specific surface area and an increase in the uniformity of micropore distribution. However, with the continuous increase of CH_4 volume fraction, the inhibitory effect of CH_4 gas on coal oxidation reaction is enhanced, and the expansion and development of coal micropores during the oxidation process are hindered, and the complexity of microporous structure is weakened no longer significantly.

Mercury intrusion and low-temperature liquid nitrogen adsorption methods have advantages in analyzing mesopores and larger pores with pore sizes greater than 2 nm [132], but they cannot well characterize the micropores of coal (<2 nm) characteristics. In future research, it is necessary to combine advanced testing technologies such as low-temperature CO_2 adsorption method and small-angle neutron scattering method to carry out full-scale pore characterization of coal, and then explore the mechanism of coal pore characteristics on coal oxygen reaction under gas-containing atmosphere.

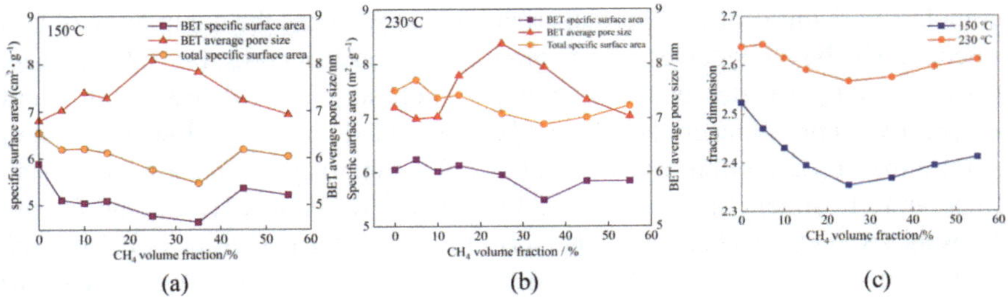

Fig. 1.10 Pore structure parameters of coal changed with the volume fraction of CH_4 [131]

1.3.3 Thermal–Flow–Solidification–Chemical Coupling Process

The development and evolution of coalfield fire areas is an extremely complex and dynamically evolving process of physical and chemical fields. At first, the thin coal seam or outcropping coal seam that was abandoned due to fertilizer mining occurred under the condition of air leakage and oxygen supply, and physical and chemical adsorption released heat. With the accumulation of heat, the coal oxidation reaction further accumulates heat and heats up, forming a unit temperature field inside the coal body. Due to thermal stress, the coal body as a double-gap structure is deformed by thermal expansion, and the unit stress field changes to produce micro-cracks, which in turn provides a channel for oxygen diffusion. After the chemical reaction of the coal body, the stiffness decreases, which makes the rock mass unstable and the shear failure cracks penetrate, providing a convection channel for the product gas and air. Due to the temperature difference between the central ignition source and the outside, the density difference between the internal gas and the outside air forms a seepage field of thermal wind pressure, which promotes a large amount of air to flow into the fire area and accelerates the coal oxidation reaction. The combustion of the coal seam extends from the shallow part to the deep part, and the overlying rock layer forms a large fracture network due to instability and collapse, providing a channel for gas and accelerating the development of the fire area. At this time, the temperature field formed by heat storage and heating due to the chemical field of coal oxidation reaction causes thermal damage to the stress field of coal and rock mass, and further forms a gas seepage field. The whole process is accompanied by chemical reactions at different stages. Therefore, the reaction rate of the chemical field affects the interaction of other fields as the dominant factor, which constitutes the THMC fully coupled cycle system. Figure 1.11 explains the Heat–Flow–Solidification–Chemical coupling mechanism.

The combustion rate of the whole process is a time-dependent variable. The main influencing factor is the reaction rate of coal–oxygen compound. The reaction rate is a certain amount in the dt time. The reaction rate at different stages directly affects the exothermic rate and temperature field T change of the coal body, thereby changing the stress σ, strain ε and stiffness k of the coal–rock mass, and then changing the permeability coefficient Ki. The thermo–solid coupling and chemical–solid coupling

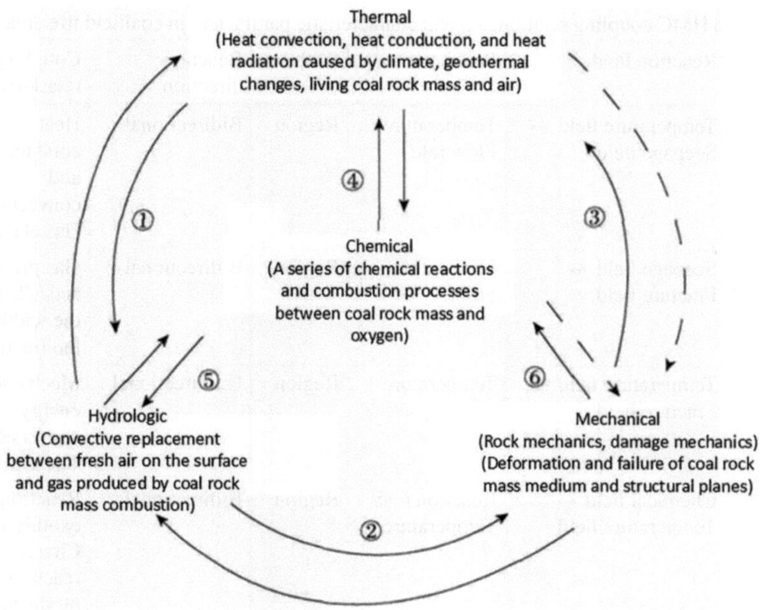

Fig. 1.11 Coupling mechanism of THMC in the development and evolution of coalfield fire area [92]

at high temperature in coal–rock mass are one-way weak coupling, and the thermalization, heat flow and fluidization are two-way strong coupling. The specific characterization parameter analysis is shown in Table 1.7, and the relationship of the coupling mechanism in Fig. 1.11 is further analyzed.

Table 1.7 THMC coupling relationship and characteristic parameters in coalfield fire area [92]

Serial number	Reaction field	Parameter	Scope	Reaction direction	Coupling relationship
1	Temperature field ↔ Seepage field	Temperature/ Flow rate	Region	Bidirectional	Heat conduction and convection/ Gas elasticity
2	Seepage field ↔ Fracture field	Flow rate/ Stress	Region	Bidirectional	Gas pressure acts/Changes the width of the fracture
3	Temperature field → Fracture field	Temperature	Region	Unidirectional	Mechanical energy conversion/ Thermal stress
4	Chemical field ↔ Temperature field	Reaction rate/ Temperature	Region	Bidirectional	Reaction exothermic/ Change reaction products, rate
5	Chemical field ↔ Seepage field	Reaction rate/ Flow rate	Region	Bidirectional	Solid current conversion/ Modification of reactants and products
6	Chemical field → Fracture field	Reaction rate	Region	Unidirectional	Combustion phase transition causes weightless fracture

References

1. Zhu, H. Q., Yuan, J., Zhao, J. L., & Zhu, X. M. (2019). Research on the status of underground coal fire distribution and development of the detection technology. *Industrial Safety and Environmental Protection, 45*(12), 28–32. (朱红青, 袁杰, 赵金龙, 朱晓曼. 地下煤火分布及探测技术现状研究. 工业安全与环保, 2019, 45(12): 28–32.)
2. Deng, J., Li, B., Wang, K., & Wang, C. P. (2016). Research status and outlook on prevention and control technology of coal fire disaster in China. *Coal Science and Technology, 44*(10), 1–7. (邓军, 李贝, 王凯, 王彩萍. 我国煤火灾害防治技术研究现状及展望. 煤炭科学技术, 2016, 44(10): 1–7.)
3. Zhang, J. M. (2008). *Underground coal fires in China: origin, detection, fire-fighting, and prevention*. China Coal Industry Publishing House. (张建民. 中国地下煤火研究与治理. 煤炭工业出版社, 2008.)
4. Song, Z., & Kuenzer, C. (2014). Coal fires in China over the last decade: A comprehensive review. *International Journal of Coal Geology, 133*, 72–99.

5. Wu, J. J., Liu, X. C., Jiang, W. G., Gu, L., & Cai, Z. Y. (2010). Spatial analysis of risk for underground coal fire in Xinjiang, China. *Journal of China Coal Society*, *35*(07), 1147–1154. (武建军, 刘晓晨, 蒋卫国, 顾磊, 蔡忠勇. 新疆地下煤火风险分布格局探析. 煤炭学报, 2010, 35(07):1147–1154.)
6. BPPLC. (2025).*BP World Energy Statistics Yearbook 2024*.
7. Deng, J. C. (2023). *Thermal infrared remote sensing inversion and the characteristics of typical pollutant emissions from coalfield fires*. Ph.D. dissertation, China University of Mining and Technology. (邓进昌. 煤田火灾热红外遥感反演与典型污染物排放特征分析研究. 中国矿业大学, 2023.)
8. Jia, X. Y. (2017). The hazards of coalfield fires in Xinjiang and the work ideas for their control. *Global market*, (14), 80–81. (贾新勇. 新疆煤田火灾危害及治理工作思路. 环球市场, 2017 (14): 80–81.)
9. Zeng, Q., Wang, D. M., & Cai, Z. Y. (2011). Research and governance of coal fire. *Mining Safety & Environmental Protection*, *38*(01), 72–75. (曾强, 王德明, 蔡忠勇. 煤火研究与治理进展. 矿业安全与环保, 2011, 38(01): 72–75.)
10. Zhu, Y. Y., Gu, H. T., Zheng, Y., Tang, Y. Y., & Lang, F. K. (2025). Detecting coal fire areas by using Landsat 8 and Sentinel-2 images: A case study of Jiangjun Gebi mining area. *Remote Sensing Information*, *40*(02), 149–158. (朱艳茵, 顾皓天, 郑屹, 唐杨玉, 郎丰铠. 联合Landsat 8与Sentinel-2的煤火探测——以将军戈壁矿区为例. 遥感信息,2025,40(02):149–158.)
11. Zeng, Q., Tiyip, T., Wuttke, M. W., & Kurban, B. (2014). Characteristics of the heavy metal distribution in the soils of underground coal fire and the evaluation of its impact on environment. *Journal of China University of Mining and Technology*, *43*(04), 695–700. (曾强, 塔西甫拉提·特依拜, Manfred W. Wuttke, 布威萨热·库尔班. 地下煤火土壤重金属分布特征及其污染评价. 中国矿业大学学报, 2014, 43(04): 695–700.)
12. Du, X. M., Peng, S. P., Wang, H. Y., Bernardes, S., Yang, G., & Li, Z. P. (2015). Annual change detection by ASTER TIR data and an estimation of the annual coal loss and CO_2 emission from coal seams spontaneous combustion. *Remote Sensing, 7*(1), 319–341.
13. Zhang, X. Q., Zhang, Y., Yu, M. D., Li, X. F., Wang, T., & Hu, S. R. (2014). Damages and challenge for governing of coalfield fires in Xinjiang. *China Mining Magazine*, *23*(S2), 93–95. (张学庆, 张渝, 于明达, 李小飞, 王田, 胡社荣. 新疆煤田火区的危害与治理挑战. 中国矿业, 2014, 23(S2): 93–95.)
14. Li, H. M., Jiang, D. J., & Li, D. Y. (2014). Analysis of ground pressure and roof movement in fully-mechanized top coal caving with large mining height in ultra-thick seam. *Journal of China Coal Society*, *39*(10), 1956–1960. (李化敏, 蒋东杰, 李东印. 特厚煤层大采高综放工作面矿压及顶板破断特征. 煤炭学报, 2014, 39(10):1956–1960.)
15. Ma, L., Li, Z. B., Zhao, L., Zhang, Y. N., & Wang, Y. (2014). Coal spontaneous combustion prevention technology for great - inclined thick coal seam oblique–cut slice none coal pillar mechanized caving mining. *Safety in Coal Mines*, *45*(09), 141–143. (马砺, 李珍宝, 赵雷, 张嬿妮, 王雨. 大倾角厚煤层斜切分层无煤柱综放开采煤自燃防治技术. 煤矿安全, 2014, 45(09):141–143.)
16. Li, J. F., Ren, W. X., Guo, Q., Fang, S. Q., & Wang, Z. F. (2015). Application of super-high–water material in fire preventing and extinguishing of large dip angle fully–mechanized working face. *Safety in Coal Mines, 46*(10), 143–145. (李剑锋, 任万兴, 郭庆, 房树起, 王振峰. 超高水材料在大倾角综放工作面防灭火中的应用. 煤矿安全, 2015, 46(10):143–145.)
17. He, Y. (2014). *Research on fly ash slurry seriflux solidify filling technique in coal mine.* Master Thesis, Xi'an University of Science and Technology. (何毅. 矿井粉煤灰浆液固化充填技术研究. 西安: 西安科技大学, 2014.)
18. Zhang, D. (2014). *Study on the mining working face CO source and accumulation of close distance coal seams.* Master Thesis, Xi'an University of Science and Technology. (张铎. 近距离煤层群开采工作面CO来源及聚集规律研究. 西安: 西安科技大学, 2014.)
19. Xiao, J. F., Fan, S. X., Lu, P., & Yu, T. (2016). Research on interaction between gas extraction from high extraction roadway and spontaneous combustion of residual coal in the goaf for high gas and spontaneous combustion working face. *Journal of Safety Science and Technology*, *12*(02), 20–26. (肖峻峰, 樊世星, 卢平, 余陶. 高瓦斯易自燃工作面高抽巷瓦斯抽采与采空区遗煤自燃相互影响研究. 中国安全生产科学技术, 2016, 12(02): 20–26.)

20. Guo, H. X., Wang, W., & Mao, L. (2016). Fire prevention and control technology of coal mining face in shallow depth and close distance easy spontaneous combustion coal seam group. *Safety in Coal Mines, 47*(02), 148–150+154. (郭海相, 王伟, 毛龙. 浅埋深近距离易自燃煤层群开采工作面防灭火技术. 煤矿安全, 2016, 47(02):148–150+154.)
21. Li, Z. D., Wu Y. G., & Liu, S. (2014). Application of radon measurement by activated carbon in spontaneous combustion detection in mined-out areas in No.2 Jinggong mine. *Shanxi Coal, 34*(11), 68–70. (李正东, 吴玉国, 刘硕. 活性炭测氡法在平朔井工二矿小窑采空区自燃火源探测中的应用. 山西煤炭, 2014, 34(11): 68–70.)
22. Tang, X.F. (2014). *The study for rapid control technology of small coal mine fire district in Anjialing surface mine.* Master Thesis, China University of Mining and Technology. (汤笑飞. 安家岭露天矿小窑火区快速治理技术研究. 徐州: 中国矿业大学, 2014.)
23. Wen, H., Guo, J., Jin, Y. F. Zhang, D., Wang, T., & Liu, W. Y. (2016). Progress and trend of evaluation study on coal mine thermodynamic disasters in China. *Safety in Coal Mines, 47*(03), 172–174+178. (文虎, 郭军, 金永飞, 张泽, 王涛. 我国矿井热动力灾害评价研究进展及趋势. 煤矿安全, 2016, 47(03):172–174+178.)
24. Zeng, Q. (2012). *Study on the thermal dynamic characteristics of combustion system for coal fires in Xinjiang region.* Ph.D. dissertation, China University of Mining and Technology. (曾强. 新疆地区煤火燃烧系统热动力特性研究. 徐州: 中国矿业大学, 2012.)
25. Xiao, Y. (2013). *Study on the mechanical characteristics of coal-rock mass of coalfield fires with thermo-hydro-mechanical coupling for fissure seepage.* Ph.D. dissertation, Xi'an University of Science and Technology. (肖旸. 煤田火区煤岩体裂隙渗流的热-流-固多场耦合力学特性研究. 西安: 西安科技大学, 2013.)
26. Wang, W. F., Deng, J., Wang, C. P., & Fei, J. B. (2012). Design of wireless remote monitoring and early-warning system for coalfield fire. *Journal of Mine Automation, 38*(06), 1–3. (王伟峰, 邓军, 王彩萍, 费金彪. 煤田火灾无线远程监测预警系统设计. 工矿自动化, 2012, 38(06): 1–3.)
27. Deng, J., Xiao, Y., Lu, J. H., Wen, H., & Jin, Y. F. (2015). Application of composite fly ash gel to extinguish outcrop coal fires in China. *Natural Hazards, 2015*(79), 881–898.
28. Cao, K., Zhong, X. X., Wang, D. M., Shi, G. Q., Wang, Y. M., & Shao, Z. L. (2012). Prevention and control of coal eld re technology: A case study in the Antaibao open pit mine goaf burning area, China. *International Journal of Mining Science & Technology, 22*, 657–663.
29. Niu, H. (2009). *Study on comprehensive treatment technique for outcrop coal fire of 4# coal seam in Bozidun mine.* Master Thesis, Xi'an University of Science and Technology. (牛恒. 博孜敦4号煤层露头火灾的综合治理技术研究. 西安科技大学, 2009.)
30. Zhang, Y., Hu, S. R., Yue, T., Zhang, X. Q., Yu, M. D., & Li, X. F. (2014). Environmental effect analysis of coalfield fire in Xinjiang and its preventive. *Advanced Materials Research, 1010–1012*, 601–605.
31. He, Q. (2022). Research progress of underground coal fire detection technology in Western China under carbon peaking and carbon neutrality goals. *Coal Geology of China, 34*(04), 8–13. (贺强. 双碳目标下我国西部地区地下煤火探测技术研究进展. 中国煤炭地质, 2022, 34(04): 8–13.)
32. Stracher, G. B. (2007). Coal fires burning around the world: Opportunity for innovative and interdisciplinary research. *GSA Today, 17*, 36–37.
33. Song, Z. Y., & Kuenzer, C. (2014). Coal fires in China over the last decade: a comprehensive review. *International Journal of Coal Geology, 133*, 72–99.
34. Kurban, B. (2013). *Study on the characteristics of typical pollutants emission from coal fire—Illustrated with the Shui-xi-gou coal fire located in Jimsar County.* Master Thesis, Xinjiang University. (布威萨热·库尔班. 煤火典型污染物排放特征研究——以吉木萨尔县水西沟火区为例. 乌鲁木齐:新疆大学, 2013.)
35. Tian, J.J. (2020). Study on the harm and environmental impact of coal seam spontaneous combustion in Shanxi. *Energy Conservation & Environmental Protection*, (04), 77–78. (田晶晶. 山西省煤层自燃危害及环境影响研究. 节能与环保, 2020(04):77–78.)
36. Luo, S. Z. (2009). *Study on ecological restoration in coal fires area—A case of Wuda mining area in Inner Mongolia.* Master Thesis, Xinjiang University. (罗淑政. 煤火废弃地的生态恢复研究——以内蒙古乌达矿区为例. 乌鲁木齐: 新疆大学, 2009.)

37. Hower, J. C., Henke, K., O'Keefe, J. M. K., Engle, M. A., Blake, D. R., & Stracher, G. B. (2009). The Tiptop coal-mine fire, Kentucky: Preliminary investigation of the measurement of mercury and other hazardous gases from coal-fire gas vents. *International Journal of Coal Geology, 80*(1), 63–67.
38. Zhang, Y., Wang, F. S., Zhu, L. Q., Zhuo, J. K., & Hu, S. R. (2024). Analysis of the influence of coal field fire on environment. *Safety in Coal Mines, 55*(12), 132−142. (张渝, 王福生, 朱令起, 卓建坤, 胡社荣. 煤田火灾对环境的影响分析.煤矿安全,2024,55(12):132−142.)
39. Zhao, J. R. (2019). *Distribution of PAHs in coal gangue and its precipitation characteristics during spontaneous combustion.* Master Thesis, Taiyuan University of Technology. (赵京茹. 煤矸石中多环芳烃的分布及其在自燃过程中的析出特征. 太原: 太原理工大学, 2019.)
40. Liang, Y. C. (2018). *Emission characteristics of typical pollutants of coal fire in Wuda, Inner Mongolia.* Ph.D. dissertation, China University of Mining & Technology. (梁言慈. 内蒙古乌达煤火典型污染物排放特征研究. 北京: 中国矿业大学(北京), 2018.)
41. Dong, J. X. (2018). *Monitoring research of Daquanhu coal fire of Urumqi in Southern Junggar coalfield.* Master Thesis, Xinjiang University. (董敬宣. 准南煤田乌鲁木齐大泉湖火区监测研究. 乌鲁木齐: 新疆大学, 2018.)
42. Ni, S. Z., Shi, Q. D., & Wang, T. (2013). Application of local material in soil improvement of coalfield fire. *Journal of Anhui Agri, 41*(06), 2434−2435+2441. (倪绍忠, 师庆东, 王涛. 基于乡土材料在煤田灭火区土壤改良中的应用研究. 安徽农业科学, 2013, 41(06): 2434−2435+2441.)
43. Hu, W., Zhang, T., Liao, X. Y., Wu, Z. F., Mou, W. T., & Lou, K., et al. (2012). Functional diversity of soil microbial community in different combustion periods of underground coal mine fire. *Xinjiang Agricultural Sciences, 49*(03), 515−522. (胡雯, 张涛, 廖先燕, 吴尊凤, 牟文婷, 娄恺, 等. 地下煤火不同燃烧阶段上覆土壤微生物群落功能多样性. 新疆农业科学, 2012, 49(03): 515−522.)
44. Xu, W. D., & Chen, R. Y. (2006). Researching the environment hydrology function mode of the Coal−Gangue. *Clean Coal Technology*, (03), 108−112. (徐卫东, 陈润羊. 煤矸石的环境水文地质作用模式研究. 洁净煤技术, 2006, (03): 108−112.)
45. Zhang, Z. X., Ma, L., Liu, T. Y., Sun, G. H., Li, Y., & Bao, Y. F. (2023). Assessment of the ecological risk of heavy metal pollution in groundwater of typical lead-zinc mine and influenced area in Inner Mongolia. *Chinese Journal of Ecology, 42*(12), 2853−2863. (张紫翔, 马龙, 刘廷玺, 孙国华, 李耀, 包玉凤, 等. 内蒙古典型铅锌矿及其影响区地下水重金属污染生态环境风险评估. 生态学杂志, 2023, 42(12): 2853−2863.)
46. Zeng, Q., Nie, J., & Pu, Y. (2016). Characteristics of the distribution of typical heavy metals in the soils of underground coal fire. *Journal of China Coal Society, 41*(08), 1989−1996. (曾强, 聂静, 蒲燕. 地下煤火土壤典型重金属分布特征. 煤炭学报, 2016, 41(08): 1989−1996.)
47. Chen, J. (2005). *Occurrence law and environmental impact of polycyclic aromatic hydrocarbons in Huainan mining area.* Master Thesis, China University of Geosciences. (陈晶. 淮南矿区环境中多环芳烃分布赋存规律及环境影响. 北京: 中国地质大学(北京), 2005.)
48. Fan, J. S., Sun, Y. Z., Niu, H. Y., Hun, L. Y., Shang, Z. S., & LI, X. Y., et al. (2009). Organic pollution of coal gangue mountain in Kowloon Coal Mine. *Environmental Pollution and Control, 31*(01), 101−103. (樊景森, 孙玉壮, 牛红亚, 浑凌云, 尚忠森, 李新宇, 等. 九龙煤矿煤矸石山对环境的有机污染. 环境污染与防治, 2009, 31(01):101−103.)
49. Zhang, H. (2012). *Study on the characteristics of elements and PAHs of coal gangue of Duerping coal deposit.* Master Thesis, Taiyuan University of Technology. (张红. 西山杜儿坪矿煤矸石中元素与多环芳烃的地球化学特征研究. 太原: 太原理工大学, 2012.)
50. Wang, X.J., Wen, X.Q., & Yang, K. (2018). *Content and distribution characteristics of PHAs in spontaneously combustible gangue in Duerping mine.* In *Proceedings of the 2018 National Academic Annual Conference of Environmental Engineering* (Vol. 2, pp. 309−312+318). (王晓娟, 文雪琴, 杨凯. 杜儿坪矿自燃矸石中多环芳烃的含量及分布特征.《环境工程》2018年全国学术年会论文集(中册). 2018: 309−312+318.)
51. Jiang, N., Yang, F. J., Meng, Y. R., & Zhang, L. P. (2015). Study on polycyclic aromatic hydrocarbons (PAHs) content changing of coal gangue during spontaneous combustion. *Industrial Safety and Environmental Protection, 41*(04), 69−72. (姜楠, 杨锋杰, 孟彦如, 张丽萍. 煤矸石自燃过程中多环芳烃含量的变化研究. 工业安全与环保, 2015, 41(04): 69−72.)

52. Wang, L. C., & Chen, Y. (2013). The hazard evaluation and control scheme of small coal mine fire zones around Daliuta coal mine. *Safety in Coal Mines, 44*(08), 135−136+140. (王连聪, 陈洋. 大柳塔矿周边小窑火区危害评价及治理方案. 煤矿安全, 2013, 44(08): 135−136+140.)
53. Li, Y. (2014). *The research of the coal spontaneous combustion impact on land resources based on multispectral remote sensing*. Master Thesis, Taiyuan University of Technology, Taiyuan, China. (李燕. 基于多光谱遥感的煤层自燃对土地资源的影响研究. 太原: 太原理工大学, 2014.)
54. Zhang, Y., & Yang, W. (2017). Coal fire burning stage division and its characteristic analysis. *Coal Technology, 36*(06), 151−153. (张渝, 杨雯. 煤火燃烧阶段的划分及其特征分析. 煤炭技术, 2017, 36(06): 151−153.)
55. Mo, A., Yang, J. J., & Zhou, Y. Z. (2016). Relationship between soil seed bank and standing vegetation of abandoned coal fire area in Xigou coal mine. *Ecological Science, 35*(02), 66−74. (莫爱, 杨建军, 周耀治. 西沟煤矿煤火废弃地土壤种子库与地上植被关系研究. 生态科学, 2016, 35(02):66−74.)
56. Yang, J. J., Mo, A., Liu, W., Shi, Q. D., & An, W. M. (2015). Screening plant species for revegetation on the coal fire wasteland in Songshutou, Urumqi. *Chinese Journal of Ecology, 34*(06), 1499−1506. (杨建军, 莫爱, 刘巍, 师庆东, 安文明. 乌鲁木齐松树头煤田火区植被恢复的物种筛选. 生态学杂志, 2015, 34(06): 1499−1506.)
57. Du, C., Liang, H. D., & Li, Z. P. (2020). Characterization of sulfate salts in the coal fire sponge in Wuda coalfield, Inner Mongolia of China. *Environmental Chemistry, 39*(01), 264−270. (杜川, 梁汉东, 李展平. 内蒙古乌达煤火海绵体的硫酸盐表征. 环境化学, 2020, 39(01): 264−270.)
58. Gao, P. (2019). Evaluation of Polycyclic Aromatic hydrocarbon pollution in coal fire sponges of Wuda coalfield. *Inner Mongolia Coal Economy*, (11), 40−41. (高鹏. 乌达煤田煤火海绵体多环芳烃污染评价. 内蒙古煤炭经济, 2019, (11): 40−41.)
59. Cao, Y. Y., & Chen, F. (2017). Spontaneous combustion characteristics and environmental effects of Daquanhu coalfield in Urumqi. *Shandong Coal Science and Technology*, (05), 110−112. (曹远远, 陈飞. 乌鲁木齐大泉湖煤田自燃特征及环境影响. 山东煤炭科技, 2017, (05): 110−112.)
60. Li, Y. Y. (2000). Research on the effect of spontaneous combustion of gangue mountain on groundwater in coal washing water treatment. *Energy Environmental Protection*, (04), 29−30. (李艳英. 洗煤水治理矸石山自燃对地下水的影响研究. 煤矿环境保护, 2000, (04): 29−30.)
61. Zhao, Z. B., Liu, K. L., Xie, W., Pan, W. P., & Riley, J. T. (2000). Soluble polycyclic aromatic hydrocarbons in raw coals. *Journal of Hazardous Materials, 73*(1), 77–85.
62. Ao, W. H. (2005). *Study on the environmental geochemical effects of harmful substances in coal Gangue in Wuda mining area, inner Mongolia.* Master Thesis, China University of Geosciences. (敖卫华. 内蒙古乌达矿区煤矸石中有害物质环境地球化学效应研究. 北京: 中国地质大学(北京), 2005.)
63. Qi. D. X. (2005). Prevention and control of fire hazards in coalfields in Xinjiang. *China Coal, 31*(06), 34−36+51. (齐德香. 新疆煤田火灾的危害及其防治.中国煤炭, 2005, 31(06):34−36+51.)
64. Qi, J. D., & Yu, X. C. (2007). Present situation and comprehensive treatment of fire accident in Ningxia coal-field. *Shaanxi Coal*, (01), 36−38. (齐俊德, 禹学成. 浅谈宁夏煤田火灾现状及综合治理. 陕西煤炭, 2007, (01):36−38.)
65. Cao, D. Y., Fan, X. J., Wu, C. C., & Wang, G. L. (2009). Study on the fractures related with coalfield fire area in Wuda coalfield, Inner Mongolia. *Journal of China Coal Society, 34*(08), 1009−1014. (曹代勇, 樊新杰, 吴查查, 王国林. 内蒙古乌达煤田火区相关裂隙研究. 煤炭学报, 2009, 34(08):1009−1014.)
66. Liu, Z. J. (1959). On the characteristics, genesis and regularity of underground fire combustion of crested rocks. *Geological Review*, 5(05), 209−211+243. (刘志坚. 论烧变岩的特征、成因及地下火燃烧的规律性. 地质论评, 1959, (05): 209−211+243.)

67. Liu, C. L. (1988). On study of burnt deposits and burnt rocks and their significance. *Contributions to Geology and Mineral Resources Research*, (03), 54–61. (刘长龄. 论烧变矿床与烧变岩研究及其意义. 地质找矿论丛, 1988, (03): 54–61.)
68. Huang, L., & Liu, C. Y. (2008). Petrologic and REE geochemical characters of burnt rocks. *Journal of China University of Geosciences*, (04), 515–522. (黄雷, 刘池洋. 烧变岩岩石学及稀土元素地球化学特征. 中国地质大学学报, 2008, 33(04): 515–522.)
69. Huang, L., & Liu, C. Y. (2014). Products of combustion of the Yan'an formation coal seam and their characteristics in the Northeastern Ordos Basin. *Acta Geological Sinica*, *88*(09), 1753–1761. (黄雷, 刘池洋. 鄂尔多斯盆地北部地区延安组煤层自燃烧变产物及其特征. 地质学报, 2014, 88(09):1753–1761.)
70. Zhang, Y., Hu, S. R., Peng, J. C., Zhang, T. T., & Li, J. (2016). Metamorphic products of coal combustion and its macroscopic models in North China. *Journal of China Coal Society*, *41*(07), 1798–1805. (张渝, 胡社荣, 彭纪超, 张同同, 李瑾. 中国北方煤层自燃产物分类及宏观模型. 煤炭学报, 2016, 41(07): 1798–1805.)
71. Zhao, D. Q., Wang, Z. W., Zhao, X., & Li, Y. Z. (2013). Experimental study on burnt coal rock shear properties. *Opencast Mining Technology*, (09), 5–7. (赵德乾, 王振伟, 赵雪, 李营作. 烧变岩剪切特性试验研究. 露天采矿技术, 2013, (09):5–7.)
72. Tian, J., Li, H. Z., & Xia, D. (2013). Burnt rock characteristics and slope stability analysis in Xinjiang Bailitan open-pit mine. *Safety in Coal Mines*, 44(07), 223–225. (田军, 李海洲, 夏冬. 新疆白砾滩露天煤矿烧变岩特征及边坡稳定性分析. 煤矿安全, 2013, 44(07): 223–225.)
73. Wang, Z. W., Wang, H. Y., & Zhao, X. (2013). *Research on deformation evolution mechanism and stability control technology of slope of downpour calcined rock*. In *Proceedings of the 2013 Annual Conference of China Occupational Safety and Health Association* (pp. 442–447). (王振伟, 王海洋, 赵雪. 顺倾烧变岩边坡变形演化机理及稳定控制技术研究. 中国职业安全健康协会. 中国职业安全健康协会2013年学术年会论文集, 2013:442–447.)
74. Yang, W. F. (2007). Experimental research on permeability of burnt rock under servo–control condition. *Mining Research and Development*, *27*(01), 79–81. (杨伟峰. 烧变岩石伺服条件下渗透特性试验研究. 矿业研究与开发, 2007, 27(01):79–81.)
75. Liu, Z. W. (2005). Geological characteristic analysis and engineering property of baked metamorphic rock in North Shanxi region. *Electric Power Survey & Design*, (02), 27–30. (刘志伟. 陕北地区烧变岩的地质特性与工程性能分析. 电力勘测设计, 2005, (02): 27–30.)
76. Du, Z. N., Dang, X. Y., & Lu, N. (2008). Distribution characteristics of burnt metamorphic rocks in the Northern Shaanxi energy and chemical industry base, China and their hydrogeological significance. *Geological Bulletin of China*, *27*(08), 1168–1172. (杜中宁, 党学亚, 卢娜. 陕北能源化工基地烧变岩的分布特征及水文地质意义. 地质通报, 2008, 27(08): 1168–1172.)
77. He, W. Z. (2013). Water damage prevention and control technology in strong water-rich areas of burnt rock. Research on water damage prevention and control technology in coal mines. In *Proceedings of the Annual Conference of Shaanxi Coal Society*, (pp. 111–113). (贺卫中.烧变岩强富水区水害防治技术.煤矿水害防治技术研究——陕西省煤炭学会学术年会论文集, 2013:111–113.)
78. Wen, C. X., Liu, Y. J., & He, Z. H. (2014). Application of excavated rock drilling and curtain grouting technology in coal mine water control. *Energy Technology and Management*, *39*(05), 77–79. (温成新, 刘远军, 何泽宏. 烧变岩钻孔与帷幕注浆技术在煤矿治水中的应用. 能源技术与管理, 2014, 39(05): 77–79.)
79. Jiang, Z.Q. (2005). Characteristics of fritted rock and groundwater resources in Yushenfu mining area. *Shaanxi Coal*, *24*(04), 20–21+17. (蒋泽泉. 榆神府矿区烧变岩及地下水资源特征. 陕西煤炭, 2005, (04): 20–21+17.)
80. Hu, S. R., Jiang, D. C., Li, Z. G., & Jia, Y. P. (2001). The fire in coalfield and coal mine and its environment effect and prevention control. *Journal Geological Hazards and Environmental Preservation*, *12*(01), 21–23. (胡社荣, 蒋大成, 李泽光, 贾艳鹏. 煤田煤矿区火灾与环境效应及其防治对策. 地质灾害与环境保护, 2001, 12(01): 21–23.)

81. Dong, Z. W., Wu, X., Qi. Q. J., & Zheng, D. (2016). Research on the minimum unapplicable wind speed for coal pile compaction under different stacking conditions. *Journal of Engineering for Thermal Energy and Power*, *31*(10), 63−71. (董子文, 吴宪, 齐庆杰, 郑丹. 不同堆放条件下煤堆压实最低不适用风速研究. 热能动力工程, 2016, 31(10): 63−71.)
82. Hu, M. H., Wang, H. H., & Fan, X. S. (2001). Cause of spontaneous combustion of coal heap and measures for its prevention. *Industrial Safety and Environmental Protection*, *27*(01), 25−27. (胡明红, 王红汉, 范喜生. 煤堆自燃原因分析与防治措施. 工业安全与防尘, 2001, 27(01): 25−27.)
83. Zhang, Y. G. (1994). Petrographic composition and coal spontaneous combustion. *Journal of Jiaozuo Mining Institute*, *13*(02), 15−18. (张玉贵. 煤岩成分与煤的自燃. 焦作矿业学院学报, 1994, 13(02): 15−18.)
84. Wang, D. M. (2008). *Mine Fires*. China University of Mining and Technology Press. (王德明. 矿井火灾学. 徐州: 中国矿业大学出版社, 2008.)
85. Zhang, H. (2004). Spontaneous combustion mechanism of pyrite and its prevention. *Copper Engineering*, (03), 53−54+68. (张虹, 张春生. 黄铁矿自燃机理及其预防. 铜业工程, 2004, (03):53−54+68.)
86. Wang, X.S., & Zhang, G. S. (1990). *Prevention and control of mine fire*. China University of Mining and Technology Press. (王省身, 张国枢. 矿井火灾防治. 徐州: 中国矿业大学出版社, 1990.)
87. Küçük, A., Kadıoğlu, Y., & Gülaboğlu, M. Ş. (2003). A study of spontaneous combustion characteristics of a Turkish lignite: Particle size, moisture of coal, humidity of air. *Combustion and Flame, 133*(3), 255–261.
88. Wessling, S., Kessels, W., Schmidt, M., & Krause, U. (2008). Investigating dynamic underground coal fires by means of numerical simulation. *Geophysical Journal International, 172*(01), 439–454.
89. Cai, Q., Niu, C. C., Liu, Y. W., & Li, H. S. (2012). Theory of dual-porosity media in coal bed methane and its application. *Science & Technology Review*, *30*(24), 17−22. (蔡强, 牛丛丛, 刘曰武, 李海生. 煤层中双重孔隙介质渗流理论的应用. 科技导报, 2012, 30(24):17−22.)
90. Stracher, G. B., & Taylor, T. P. (2004). Coal fires burning out of control around the world: Thermodynamic recipe for environmental catastrophe. *International Journal of Coal Geology, 59*(1–2), 7–17.
91. Zhao, Y. S. (2010). *Multi-field coupling effect of porous media and its engineering response*. Science Press, Beijing, China. (赵阳升. 多孔介质多场耦合作用及其工程响应. 北京: 科学出版社, 2010.)
92. Ma, L., Liu, G., Xiao, Y., Lu, J. H., & He, Y. J. (2016) Research on multi-field coupling process of coalfield fire area development and evolution. *Science & Technology Review*, *34*(02), 190−194. (马砺, 刘庚, 肖旸,鲁军辉, 何勇军. 煤田火区发展演化的多场耦合作用过程. 科技导报, 2016, 34(02): 190−194.)
93. Wessling, S., Kuenzer, C., Kessels, W., & Wuttke, M. W. (2008). Numerical modeling for analyzing thermal surface anomalies induced by underground coal fires. *International Journal of Coal Geology, 74*(3–4), 175–184.
94. Drebenstedt, C., Gusat, D., & Wang, C. (2010). Numerical modelling of surface impact of coal fires on coal fire areas. Latest developments in coal fire research bridging the science, economics, and politics of a global dister. In *Proceeding of Second International Conference on Coal Fire Research* (pp 244−248). TU Bergakademie Freiberg Press.
95. Elick, J. M. (2013). The effect of abundant precipitation on coal fire subsidence and its implications in Centralia, PA. *International Journal of Coal Geology, 105*, 110–119.
96. Heffern, E. L., & Coates, D. A. (2004). Geologic history of natural coal-bed fires, Powder River basin, USA. *International Journal of Coal Geology, 59*(1–2), 25–27.
97. Rosem, A., Guan, H., & Veld, H. (2001). Simulation of spontaneous combustion to study the causes of coal result in the Rujigou Basin. *Fuel, 80*(1), 7–16.
98. Zeng, Q. (2009). Study on the regularity of gas/smoke flow and the characteristics of heat-transfer with spontaneous combustion in Xinjiang coalfield. *Journal of China Coal Society*, *34*(06), 792−796. (曾强. 新疆煤田火灾烟气流动与传热特性的研究. 煤炭学报, 2009, 34(06):792−796.)

99. Zeng, Q., & Chang, X. T. (2007). Study on the model of fire-heating airflow and its application to Xinjiang coal-field fires. *Journal of China Coal Society*, *32*(09), 955–958. (曾强, 常心坦. 新疆煤田火区火风压模式研究及其应用. 煤炭学报, 2007, 32(09): 955–958.)
100. Li, T. S., Zhou, X. Q., & Gu, H. J. (2005). Theoretical analysis of numerical value of the spontaneous combustion fire-heating air pressure for coalfield outcrop. *Journal of China Coal Society*, *30*(06), 737–740. (李唐山, 周心权, 谷红军. 煤田露头自燃火风压数值的理论分析. 煤炭学报, 2005, 30(06): 737–740.)
101. Wang, H. Y., Zhou, X. Q., Zhang, H. J., & Wang, S. F. (2010). Seepage–thermal dynamical coupling model for spontaneous combustion of coalfield outcrop and its application. *Journal of University of Science and Technology Beijing*, *32*(02), 152–157. (王海燕, 周心权, 张红军, 王素锋. 煤田露头自燃的渗流-热动力耦合模型及应用. 北京科技大学学报, 2010, 32(02): 152–157.)
102. Zhang, C. Y., Cao, P., Jin, J., Pu, C. Z., & Meng, J. J. (2013). Numerical simulation of fluid–structure interaction stability for Jinchuan deep roadway. *Science & Technology Review*, *31*(33), 31–36. (张春阳, 曹平, 靳瑾, 浦成志, 孟京京. 金川矿区深部巷道围岩流固耦合稳定性数值模拟. 科技导报, 2013, 31(33): 31–36.)
103. Nowak, T., Kunz, H., Dixon, D., Wang, W. Q., Görke, U. J., & Kolditz, O. (2011). Coupled 3-D thermo–hydro mechanical analysis of geo technological in situ tests. *International Journal of Rock Mechanics & Mining Sciences, 48*, 1–15.
104. Shaik, A. R., Rahman, S. S., Tran, N. H., & Tran, T. (2011). Numerical simulation of Fluid– Rock coupling heat transfer in naturally fractured geothermal system. *Applied Thermal Engineering, 31*, 1600–1606.
105. Zhu, W. C., Wei, C. H., Liu, J., Qu, H. Y., & Elsworth, D. (2011). A model of coal gas interaction under variable temperatures. *International Journal of Coal Geology, 86*(2–3), 213–221.
106. Zhang, H. B., Liu, J. S., & Elsworth, D. (2013). How sorption-induced matrix deformation affects gas flow in coal seams: A new FE model. *International Journal of Rock Mechanics & Mining Sciences, 45*(8), 1226–1236.
107. Xia, T. Q., Zhou, F. B., Liu, J. S., Kang, J. H., & Gao, F. (2014). A fully coupled hydro thermo-mechanical model for the spontaneous combustion of underground coal seams. *Fuel, 125*, 106–115.
108. Hu, X. C. (2018). *Micro physicochemical properties and macro thermal effect of coal influenced by methane-containing environment during oxidation.* Ph.D. dissertation, China University of Mining and Technology. (胡新成. 含瓦斯氧化气氛对自燃氧化过程中煤微观理化特性及宏观热效应影响研究.徐州: 中国矿业大学, 2017.)
109. Liu, H. W., Wang, F., & Ren, T. (2019). Research on the characteristics of the coal–oxygen reaction in a lean-oxygen environment caused by methane. *Energy & Fuels, 33*(9), 9215–9223.
110. Liu, J. (2016). Experiment study on spontaneous combustion of coal under different gas adsorption pressure. *Safety in Coal Mines*, *47*(05), 26–29. (刘敬. 不同瓦斯含量煤样自燃特性实验研究. 煤矿安全, 2016, 47(05): 26–29.)
111. Zhou, L. (2020). *Study on Gob coal spontaneous combustion early warning in high gassy and spontaneous combustion coal seam.* Ph.D. dissertation, Anhui University of Technology. (周亮. 高瓦斯易自燃煤层采空区遗煤自燃预警研究. 淮南: 安徽理工大学, 2018.)
112. Deng, J., Zhou, J. M., Bai, Z. J., Liu, L., & Wang, C. P. (2023). Effect of gas on microstructure and thermal reactivity of coal during low temperature oxidation. *Coal Science and Technology*, *51*(01), 304–312. (邓军, 周佳敏, 白祖锦, 刘乐, 王彩萍. 瓦斯对煤低温氧化过程微观结构及热反应性的影响研究. 煤炭科学技术, 2023, 51(01): 304–312.)
113. Lou, H. Z., & Jia, T. G. (2019). Study on thermal characteristics of coal spontaneous combustion in gas atmosphere by TG–DSC coupling techniques. *China Safety Sciences Journal*, *29*(11), 77–82. (娄和壮, 贾廷贵. TG-DSC联用研究瓦斯气氛对煤自燃热特性的影响. 中国安全科学学报, 2019, 29(11): 77–82.)
114. Zhang, Y. T., Zhang, Y. B., Li, Y. Q., Xue, J. H., Shi, X. Q., & Yang, C. P. (2022). Low-temperature oxidation characteristics of coals at the methane-containing atmosphere. *Journal of Thermal Analysis & Calorimetry, 147*(4), 3379–3389.

115. Zhang, Y. T., Zhang, Y. B., Li, Y. Q., Shi, X. Q., & Yang, J. (2021). Thermal effect of coal oxidation and evolution properties of key functional groups of coal oxidation at low gas atmospheres. *Journal of China University of Mining & Technology, 50*(04), 776−783. (张玉涛, 张园勃, 李亚清, 史学强, 杨杰. 低瓦斯气氛下煤氧化热效应和关键基团演变特性. 中国矿业大学学报, 2021, 50(04): 776−783.)
116. Lu, W., & Hu, Q. T. (2007). Relationship between the change rules of coal structure when being oxidation and spontaneous combustion process of coal. *Journal of China Coal Society, 32*(09), 939−944. (陆伟, 胡千庭. 煤低温氧化结构变化规律与煤自燃过程之间的关系. 煤炭学报, 2007, (09): 939−944.)
117. Xu, C. F., Yin, W. T., & Yao, H. F. (2018). On the optimal moisture content rate for the coal spontaneous combustion based on the activation energy index. *Journal of Safety and Environment, 18*(04), 1301−1306. (徐长富, 殷文韬, 姚海飞. 基于活化能指标的煤自燃最佳含水率研究. 安全与环境学报, 2018, 18(04): 1301−1306.)
118. Hao, Y., & Ye, Z. L. (2019). Research on index gas and activation energy of coal spontaneous combustion under different methane atmosphere. *Journal of Mine Automation, 45*(11), 65−69. (郝宇, 叶正亮. 不同甲烷气氛下煤自燃指标气体及活化能研究. 工矿自动化, 2019, 45(11):65−69.)
119. Ma, L., Guo, R. Z., Gao, Y., Ren, L. F., Wei, G. M., & Li, C. H. (2019). Study on coal spontaneous combustion characteristics under methane-containing atmosphere. *Combustion Science & Technology, 191*(8), 1456–1472.
120. Wang, D. M., Xin, H. H., Qi, X. Y., Dou, G. L., & Zhong, X. X. (2014). Mechanism and relationships of elementary reactions in spontaneous combustion of coal: The coal oxidation kinetics theory and application. *Journal of China Coal Society, 39*(08), 1667−1674. (王德明, 辛海会, 戚绪尧, 窦国兰, 仲晓星. 煤自燃中的各种基元反应及相互关系: 煤氧化动力学理论及应用. 煤炭学报, 2014, 39(08): 1667−1674.)
121. Lv, Z. G., Xu, Y. L., Liu, Z. J., Wu, J. D., & Li, M. J. (2022). Group evolution of long-flame coal oxidation process under various air ambience. *Safety in Coal Mines, 53*(02), 46−52. (吕志广, 徐永亮, 刘泽健, 吕晋东, 李敏杰. 不同气体环境长焰煤氧化进程基团演化特性. 煤矿安全, 2022, 53(02): 46−52.)
122. Liu, C., Zheng, W. C., Zhang, L., & Zhao, B. (2021). Influence of gas on evolution of functional groups during coal spontaneous combustion. *China Energy and Environmental Protection, 43*(07), 148−155. (刘超, 郑万成, 张镭, 赵波. 瓦斯对煤自燃过程中官能团演变的影响.能源与环保,2021,43(07):148−155.)
123. Xu, Q., Yang, S. Q., Tang, Z. Q., Hu, X. C., Song, W. X., Cai, J. W., et al. (2020). Optimum oxidation temperature of coal bed for methane desorption in the process of CBM ex-traction. *Fuel, 262*, Article 116625.
124. Xu, Q., Yang, S. Q., Hu, X. C., Song, W. X., Cai, J. W., & Zhou, B. Z. (2019). Low-temperature oxidation of free radicals and functional groups in coal during the extraction of coalbed methane. *Fuel, 239*, 429–436.
125. Hu, X. C., Yu, Z. Y., Cai, J. W., Jiang, X. Y., Li, P., & Yang, S. Q. (2022). The influence of methane on the development of free radical during low-temperature oxidation of coal in gob. *Fuel, 330*, Article 125369.
126. Wang, C. P., Deng, J., & Zhang, Y. N. (2016). Effect of micro structure characteristics on coal spontaneous combustion. *Journal of Xi'an University of Science and Technology, 36*(01), 8−12. (王彩萍, 邓军, 张嫣妮. 影响煤氧化自燃的微观结构特征分析. 西安科技大学学报, 2016,36(01): 8−12.)
127. Hu, X., Yu, Z., Cai, J., Gong, Y., Xie, X., & Yang, S. (2023). Crystalline structure development of coal during low-temperature oxidation under lean-oxygen environments caused by methane emission in gob. *Fuel, 335*, Article 127044.
128. Dong, X. W., Wang, F. S., & Meng, Y. M. (2014). Coal spontaneous combustion tendency affected by coal microscopic pore structure. *Coal Science and Technology, 42*(11), 41−45+49. (董宪伟, 王福生, 孟亚宁. 煤的微观孔隙结构对其自燃倾向性的影响. 煤炭科学技术, 2014, 42(11): 41−45+49.)

129. Tang, Y. B., Wang, J. F., Xue, S., Wu, J. M., & Yu, D. Y. (2016). Variation in spontaneous combustion properties of bituminous under inert environment of underground fire zone. *Journal of China Coal Society, 41*(02), 439−443. (唐一博, 王俊峰, 薛生, 邬剑明, 余大洋. 密闭火区内惰性环境下烟煤自燃特性变化.煤炭学报, 2016, 41(02): 439−443.)
130. Li, Z. W., Hao, Z. Y., Pang, Y., & Gao, Y. B. (2015). Fractal dimensions of coal and their influence on methane adsorption. *Journal of China Coal Society, 40*(04), 863−869. (李子文,郝志勇,庞源,高亚斌.煤的分形维数及其对瓦斯吸附的影响.煤炭学报,2015,40(04): 863−869.)
131. Cai, J. W., Yang, S. Q., Hu, X. C., Xu, Q., Zhou, B. Z., & Zhang, Z. C. (2019). The relationship between functional groups and gaseous productions and micropore structures development of coal oxidized at low temperature under methane-diluted atmospheres. *Combustion Science & Technology, 191*(8), 1337–1353.
132. Chen, M. Y., Yang, Y. P., Gao, C. H., Cheng, Y. P., Wang, J. C., & Wang, M. (2020). Investigation of the fractal characteristics of adsorption−pores and their impact on the methane adsorption capacity of various rank coals via N2 and H_2O adsorption methods. *Energy Science & Engineering, 8*(9), 3228–3243.

Chapter 2
Characteristic Parameters and Information Identification of Coal Fires

Based on the theory of complex systems, this chapter systematically expounds the nonlinear dynamic evolution characteristics of underground coal fires and their key information identification methods, and deeply analyzes the inherent complexity of its causal compounding, process dynamics, evolution nonlinearity and unpredictable results on this basis, focusing on the core process of coal spontaneous combustion, we will focus on key information carriers such as landmark gases, characteristic temperatures, oxidative thermal effects, thermal physical parameters and magnetic parameters of coal and rock. By clarifying the early warning effect of index gases, identifying the seven key characteristic temperature points of coal spontaneous combustion, revealing the oxidative heat release law and thermal physical change mechanism in different environments, and explaining the response characteristics of coal and rock magnetic parameters to thermal history, a set of multi-parameter and multi-index coal fire information identification system is constructed, aiming to accurately judge the development stage of coal fire, effectively implement early warning and precise prevention and control of disasters, and help reduce related carbon emissions. Provide a solid theoretical basis and systematic parameter support.

2.1 Complexity of Underground Coal Fire Development

2.1.1 Characteristics of Complex Systems

Uncovering the complexity of complex systems should first be understood by understanding them. Complex systems usually refer to nonlinear complex systems, which are essentially complex and adaptive systems, which have the common characteristics of the system such as integrity and emergence.

Y. Xiao et al., *Detection of Fire Source in Underground Coal Mine and its Applications*,
https://doi.org/10.1007/978-981-95-7064-5_2

(1) Integrity

Integrity means that any element change in a complex system will cause changes in other elements and the system as a whole, so that the system as a whole has new properties and new changes that a single element does not have. A single element within a system follows relatively simple and simple rules, and when multiple elements coexist and act on the system at the same time, complex and unpredictable effects can occur. The interconnection of elements in the overall system is the basis of the system, and it is also the internal cause of the wholeness and the formation of a complex system.

(2) Dynamic

Dynamics refers to the fact that complex systems are constantly in dynamic change. The complex interaction between the internal and external environments of complex systems makes complex systems continuously evolve between disordered and orderly, unbalanced and equilibrium, entropy increase and decrease, and other states that can be transformed into each other, forming a multi-factor and multi-level causal relationship network.

(3) Openness

Openness refers to the interaction between complex systems that are not in a closed state, but have a large number of complex and difficult linear relationships with the external environment. Usually, the external environment in which complex systems are located is also complex systems, and there is exchange and transformation of matter, information, energy, etc. between complex systems and the external environment. Openness also leads to the sudden emergence of system attributes in complex systems due to the sudden interaction between system elements and the external environment.

(4) Emergence

Emergence refers to the emergence of new properties that do not exist before the formation of the system after multiple elements form a complex system, which is an attribute at the system level. These emerging attributes are difficult to derive and predict, making complex systems highly uncertain and unpredictable, and increasing the difficulty of solving complex system problems.

(5) Uncertainty

Uncertainty refers to the type and likelihood of changes in complex systems that cannot be predicted and determined. Due to the nonlinear development process of complex systems and the phenomenon of emergence and self-organization, the evolution results are difficult to predict. The uncertainty of complex systems leads to the need to study them with a comprehensive method rather than a single method, and the law of probability statistics can be used to explore complex system problems.

(6) Self-organization

Self-organization refers to the fact that each element of a complex system forms structure and function according to a certain law under the condition of no intervention. The interaction between elements in a complex system will produce synergy, so that the stronger the self-organization function of the system from disorder to order, from unbalanced to balanced, from low to high, the stronger its ability to maintain and update functions, and the stronger the ability to cope with changes in the external environment.

2.1.2 Complexity Characteristics of Underground Coal-Fire Systems

Combined with the characteristics of complex systems, the typical complexity characteristics of underground coal fire systems are manifested in the complexity of underground coal fire disasters, the dynamics of disaster processes, the nonlinearity of disaster evolution, and the unpredictability of disaster results. The underground coal fire system is shown in Fig. 2.1.

(1) Complexity of disaster causes

Underground coal fire disasters are not the result of a single factor, but the result of physical factors such as coalbed methane and oxygen superimposed on climatic factors and management factors. The most important feature of underground coal fires is the "nesting" of disaster-causing factors, that is, the causal relationship is closely linked, which eventually leads to or aggravates the disaster, and the disaster-causing factors show compound overlap on multiple spatio-temporal scales. For example, in the process of underground coal fire, oxygen concentration affects the heat transfer of the combustion process and thus affects the temperature distribution, which in turn affects the rate of gas flow, forming a chain of "two-way causal relationship". The complex multi-field coupling relationship makes it difficult to control and solve underground coal fires by reductionist methods, and the causes of underground coal fire disasters are complex.

(2) Dynamics of the disaster process

Underground coal fires are essentially dynamic disasters that change over time. In underground coal, it is multi-field coupling of thermal-flow-solidification-chemical.

The kinetic system includes a variety of motion forms: heat transfer and conduction, fluid migration, deformation of solid rock formations, oxidative decomposition of coal, and these movements are transmitted through voids. In the process of underground coal fire control, fire extinguishing preparation, water injection, grouting and other links involve deformation, diffusion, seepage and other forms of movement. In

Fig. 2.1 Complex underground coal fire system [1]

the process of escalating disasters, it appears as a spiral, and various elements will overlap on multiple spatiotemporal scales. Therefore, the process of underground coal fire disasters is dynamic.

(3) Nonlinearity of disaster evolution

There are nonlinear changes in the deformation, fluid migration, heat transfer and other movements in the underground coal fire development space, especially the constitutive relationship of the rock mass involves the displacement, deformation, state transition, loading path, etc. after shear and tensile failure, which cannot be directly described by equations or inequalities. There is also non-linearity in the fire extinguishing project involving people, management and other elements. The nonlinearity of the multi-field coupling system and the treatment system determines that the occurrence and development of underground coal fire is prone to chaos, showing significant nonlinear characteristics.

(4) Unpredictability of disaster results

The interaction between the complex causes of underground coal fire disasters is a nonlinear model, which hinders the interpretation, control and prediction of the evolution of underground coal fire disasters. Moreover, with the continuous evolution of underground coal fire disasters, the expansion of the fracture area and the change of void field, the discontinuity, heterogeneity, anisotropy and unsteady changes of the rock strata in the coal fire generation space become more and more significant, making the results of underground coal fire disasters more difficult to predict. The links involving people, management and other elements in fire extinguishing projects are also difficult to predict. Therefore, the results of underground coal fire disasters are difficult to predict.

2.2 Characteristic Parameters and Stage Characteristics of Coal Fires

2.2.1 Index Gases

(1) Early warning method for the index gas of spontaneous combustion of coal

During the oxidation and heating process, coal releases a large amount of gaseous products such as CO, CO_2, and hydrocarbon compounds, and its concentration shows significant regular changes with the increase of coal temperature [2–4]. Based on this release characteristic, the index gas analysis method has been proposed and has become one of the main technical means for predicting coal spontaneous combustion in coal mines in our country [5–7]. By continuously monitoring the occurrence and concentration changes of specific gases in the downhole return air flow, this method can effectively determine the development stage of coal spontaneous combustion in the goaf, so as to achieve early prediction and forecasting.

In the application practice of index gases, CO has become the most basic and widely used index gas in the early forecast of coal mines at home and abroad because it can be detected at a low temperature of about 30 °C and its concentration increases steadily with the increase of temperature [8, 9]. With the in-depth study of the formation law of hydrocarbon gases, ethylene has been established as another key index gas. Its appearance indicates that coal spontaneous combustion has entered the stage of accelerated oxidation, which is an important signal that urgent prevention and control measures must be taken [10]. When acetylene is detected, it indicates that the coal temperature is approaching the ignition point, and spontaneous combustion disasters are about to occur [11]. In addition, to improve the accuracy of forecasting, many scholars have proposed to use gas ratio indicators as auxiliary criteria [12–14], and formulated corresponding index gas optimization principles. The most effective prediction index system is selected for coal types with different degrees of metamorphism and specific mine conditions.

Unlike external fires, coal spontaneous combustion disasters usually have a long incubation period, and they need to undergo a significant self-heating process before entering open flame combustion [15, 16]. This characteristic provides a critical time for underground personnel to identify the development stage of coal spontaneous combustion and take effective interventions. To accurately capture this period, various forecasting techniques have been developed in the scientific community, such as index gas analysis, distributed fiber optic temperature measurement, and odor detection [17–20]. Among them, the index gas analysis method has been widely used in practice due to its excellent practicability and reliability, and has become the core reliance on fire prevention in coalmines.

Under different temperature conditions during the oxidation and heating process, gaseous products such as CO, CO_2, H_2, and hydrocarbon compounds are released in a certain sequence, and the concentration of these gaseous products often shows a regular trend with the increase of coal temperature [21–23]. In view of the release characteristics of gaseous products during the oxidation of coal, some gaseous products are selected as index gases to indicate the development stage of coal spontaneous combustion, and this coal spontaneous combustion prediction method is the index gas analysis method [24–26]. By detecting the occurrence and concentration variations of the index gas in the underground airflow of the coalmine, the development stage of coal spontaneous combustion in the underground goaf is judged, so as to realize the prediction of coal spontaneous combustion. In China, a research team led by Academician Yingmin Qi has carried out systematic research on coal spontaneous combustion index gases since the 80 s of the twentieth centuries, and developed a special gas chromatograph for coalmines, which has been widely popularized and used in spontaneous combustion of coalmines [27, 28]. Subsequently, more scholars conducted in-depth research in this field. With the deepening of the understanding of the formation for gaseous products of coal spontaneous combustion, such as CO, C_2H_2 and C_2H_4 began to be selected as coal spontaneous combustion prediction index gases. In addition, by exploring the relationship between different index gases in the process of coal spontaneous combustion, many scholars have proposed various derivative indicators to predict coal spontaneous combustion, such as Morris ratio: $(\varphi(N_2) - 3.774^{*}\varphi(O_2))/(\varphi(CO) + \varphi(CO_2))$, Graham ratio: $\varphi(CO)/\Delta\varphi(O_2)$, Chain ratio: $\varphi(C_2H_6)/\varphi(CH_4)$, $\varphi(C_3H_8)/\varphi(CH_4)$, $\varphi(C_3H_8)/\varphi(C_2H_6)$, Alkane ratio: $\varphi(C_2H_6)/\varphi(C_2H_4)$ [29–31]. At present, the coal spontaneous combustion forecast indicator gases commonly used in global coalmines are shown in Table 2.1.

Although the commonly used coal spontaneous combustion forecast indicators in various countries are different, they are generally nothing more than CO, C_2H_4, C_2H_2 and their derivatives. Among them, CO is one of the most gaseous products produced

Table 2.1 Commonly used coal spontaneous combustion forecast index gas [32]

Country	Main index gases	Secondary indicators
China	CO, C_2H_4, C_2H_2	$CO/\Delta O_2$, C_2H_6/CH_4
Australia	CO, H_2	$CO/\Delta O_2$
United States	CO, H_2	$CO/\Delta O_2$
India	CO, $CO/\Delta O_2$	C/H
United Kingdom	CO, C_2H_4	$CO/\Delta O_2$
Russia	CO	C_2H_6/CH_4
Poland	CO	$CO/\Delta O_2$
Germany	CO	$CO/\Delta O_2$

during coal oxidation, and its concentration is extremely sensitive to changes in coal temperature, and the concentration of CO often shows a sharp increase trend with the increase of coal temperature [9, 36]. Therefore, CO is widely used in all countries in the world, and its role as a index gas has played an important role in the prediction of coal spontaneous combustion. However, some scholars have pointed out that the CO concentration monitored in the downhole return airflow is easily disturbed by factors, such as coalbed methane and diesel engine exhaust, which affects the accuracy of using CO to predict coal spontaneous combustion [15, 33]. In addition, CO gas can usually be detected at room temperature. If CO is used only as the index gas, it is difficult to judge the specific temperature of the coal body in the underground goaf, thus affecting the judgment of the development stage of coal spontaneous combustion [34]. The initial appearance temperature of C_2H_4 is usually 90–120 °C [35], at this time, the spontaneous combustion of coal has been in an accelerated stage, and prevention and control measures must be taken as soon as possible. The initial temperature of C_2H_2 is generally appeared 260–310 °C, which exceeds the ignition point of some low-metamorphic coals, that is, the application value of C_2H_2 in coal spontaneous combustion prediction is extraordinary limited. When only the commonly used index gas is used for coal spontaneous combustion forecasting, there are large blank forecast temperature ranges, that is, room temperature–90 and 120–260 °C, which greatly limits the accuracy of coal spontaneous combustion forecasting. Therefore, it is difficult to ensure the accuracy of the judgment of the development stage of coal spontaneous combustion using only one index gas. When using the index gas analysis method, a variety of index gases should be selected as much as possible to improve the accuracy of coal spontaneous combustion prediction.

(2) Coal temperature range of index gas ratio

In the process of natural heating, cooling with supported non-oxygen and re-ignition with air condition, at the outlet of the top of the apparatus, the concentration of the index gases and correspondence highest temperature of coal sample are exhibited in Figs. 2.2 and 2.3. As for the same coal sample, the productions of the index gases are almost identical, but not one to one correspondence and not monotone increase or decrease. As the temperature of coal sample goes up, oxygen consumption enhances and supplied airflow increases. Therefore, the undulatory property of index gases concentration varies greatly. The ratio of the index gases and correspondence highest temperature of coal sample in the apparatus is shown in Fig. 2.4. The inflexion point of variation curves of index gases ratio is correspondent with different temperatures, the different range of index gases ratio can identify different temperature ranges.

(3) Relation on index gases and temperatures of coal spontaneous combustion

① Relation on single index gas and temperatures of coal

With the increase of coal temperature, the index gas concentration has the corresponding trend. According to the results of coal samples as Nantun, Jier and Dongtan,

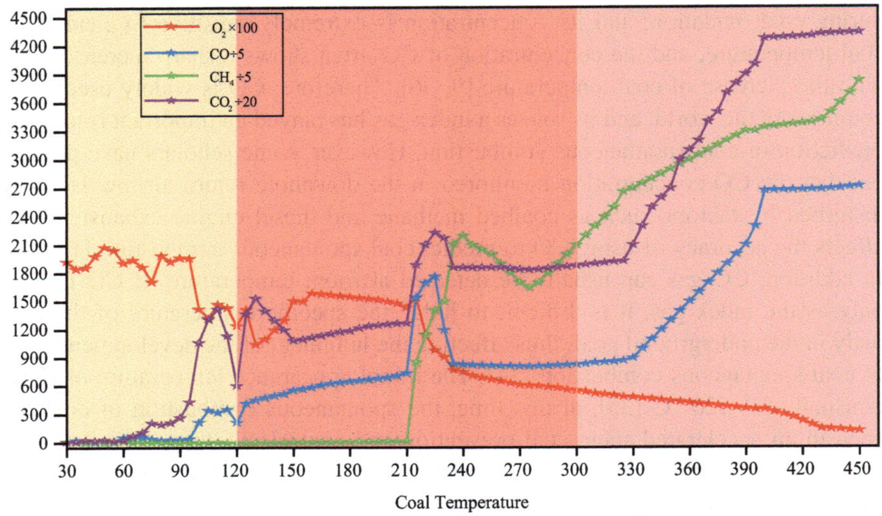

Fig. 2.2 Relation curve of index gases (O_2, CO, CH_4, CO_2) concentration and coal temperature [37]

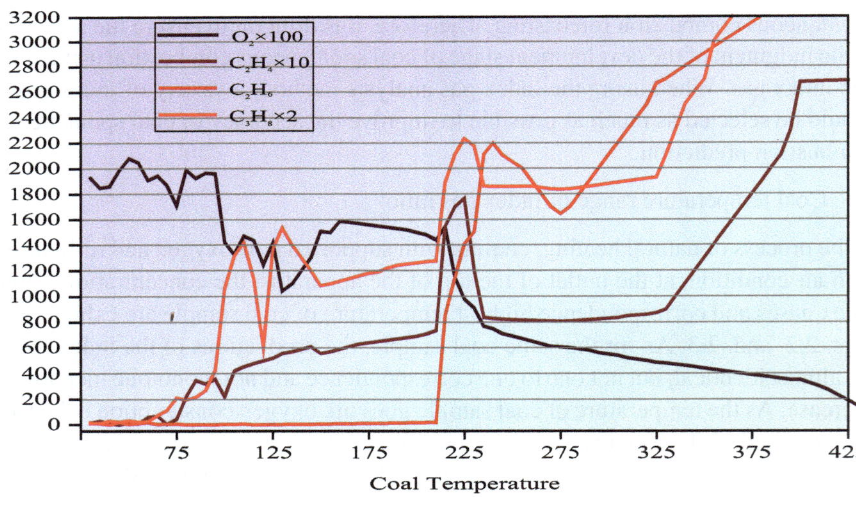

Fig. 2.3 Relation curve of index gases (O_2, C_2H_4, C_2H_6, C_3H_8) concentration and coal temperature [37]

the index gases, such as CO, C_2H_6, C_2H_4, and C_3H_8 are selected. The relations of index gases and coal temperature are described in Fig. 2.5.

Figure 2.5a shows that, in the different temperature ranges of the three coal samples, CO emerges at the same time so that the correspondence temperature at which CO emerges is ascertained, that is, 40–50 °C. And the concentration of CO

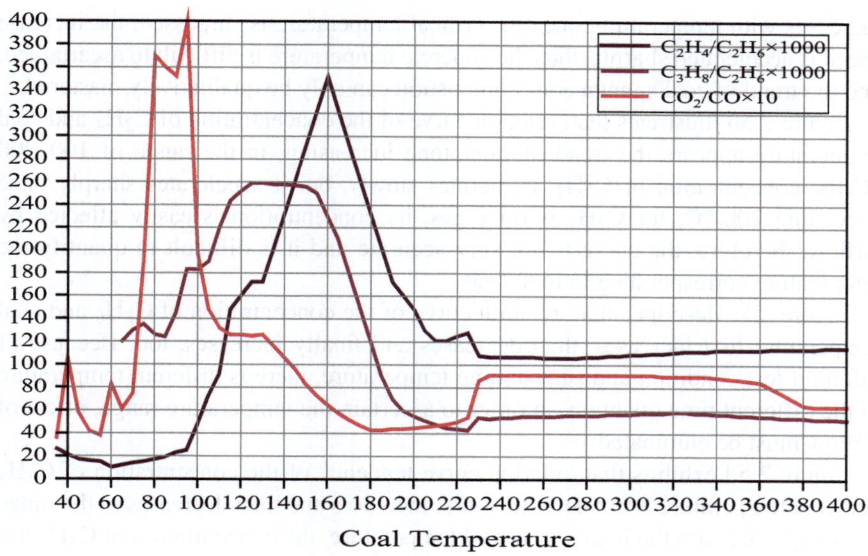

Fig. 2.4 Relation curve of the ratio of index gases and coal temperature [37]

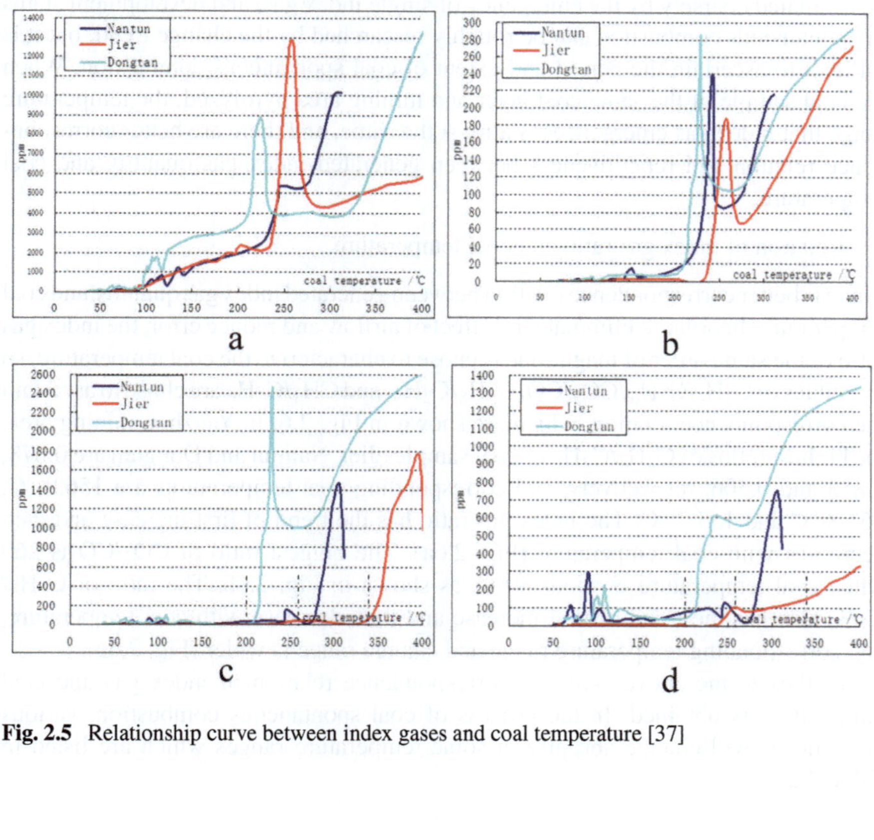

Fig. 2.5 Relationship curve between index gases and coal temperature [37]

increases with temperature, once the critical temperature is surpassed, the increase speed is accelerated sharply, thus the concrete temperature is difficult to ascertained, and the trend of coal spontaneous combustion can only be qualitatively mastered.

Figure 2.5b illustrates that, relation curve of the concentration of C_2H_4 and coal temperature appears the trend of monotone increasing, in the range of 100–200 °C the concentration of C_2H_4 accelerates slowly, but it accelerates sharply once exceeding 200 °C, for C_2H_4 is trace gas, its concentration is easily affected by airflow, therefore, the curve is not very accurate and it is difficult to quantify the temperature corresponding to index gas.

Figure 2.5c describes that, relation curve of the concentration of C_2H_6 and coal temperature first increases, then decreases and finally increases, too. Because of different test condition and surrounding temperature, there is different temperature in the different time quantum, in order to ascertain the temperature range, effect of airflow must be eliminated.

Figure 2.5d exhibits that, relation curve tendency of the concentration of C_3H_8 and coal temperature is not definite, it first increases and then decreases in the range of 60–120 °C, with the increase of coal temperature, the concentration of C_3H_8 has the trend of increase.

The above analysis indicates that, the extent of coal spontaneous combustion can be estimated coarsely by the emergency of single index gas, and development status of spontaneous combustion can be roughly researched by the change trend, but it is difficult to ascertain the period and extent of coal spontaneous combustion. When the coal sample of the same coal layer and mining area pyrolyzed, the temperature range that index gas emergences is almost the same, and there are better correspondence relation and repetitiveness between generated index gas quantity and coal temperature.

② Relation of index gas ratio and coal temperature

There is better correspondence relation between generated index gas quantity and coal temperature. In order to eliminate the effect of airflow and reduce error, the index gas ratio of the same order of magnitude is chose to characterize the coal temperature, so the values of C_2H_4/C_2H_6, CO_2/CO, C_3H_8/C_2H_6, and CH_4/C_2H_6 are chose to ascertain the correspondence relation, which are shown in Fig. 2.6. In Yanzhou mining area, the highest ratios of C_2H_4/C_2H_6 of coal samples Jier, Nantun and Dongtan are 0.878, 0.853 and 0.306, respectively, and corresponding coal temperatures are 156.9 °C, 161.8 °C and 155.7 °C. The index gas ratio has the trend of first increase and then decreases with coal temperature (Fig. 2.6a). The biggest ratio of CO_2/CO is 260 when coal temperature is about 80°C, as shown in Fig. 2.6b. The ratio of C_3H_8/C_2H_6 also has the trend of first increase and then decrease with coal temperature, and corresponding temperature range and change range is wider (Fig. 2.6c).

Synthesize the above analysis, correspondence relation of index gas and coal temperature is obtained. In the process of coal spontaneous combustion, various gas indexes will change abruptly in some temperature ranges which are listed in Table 2.2.

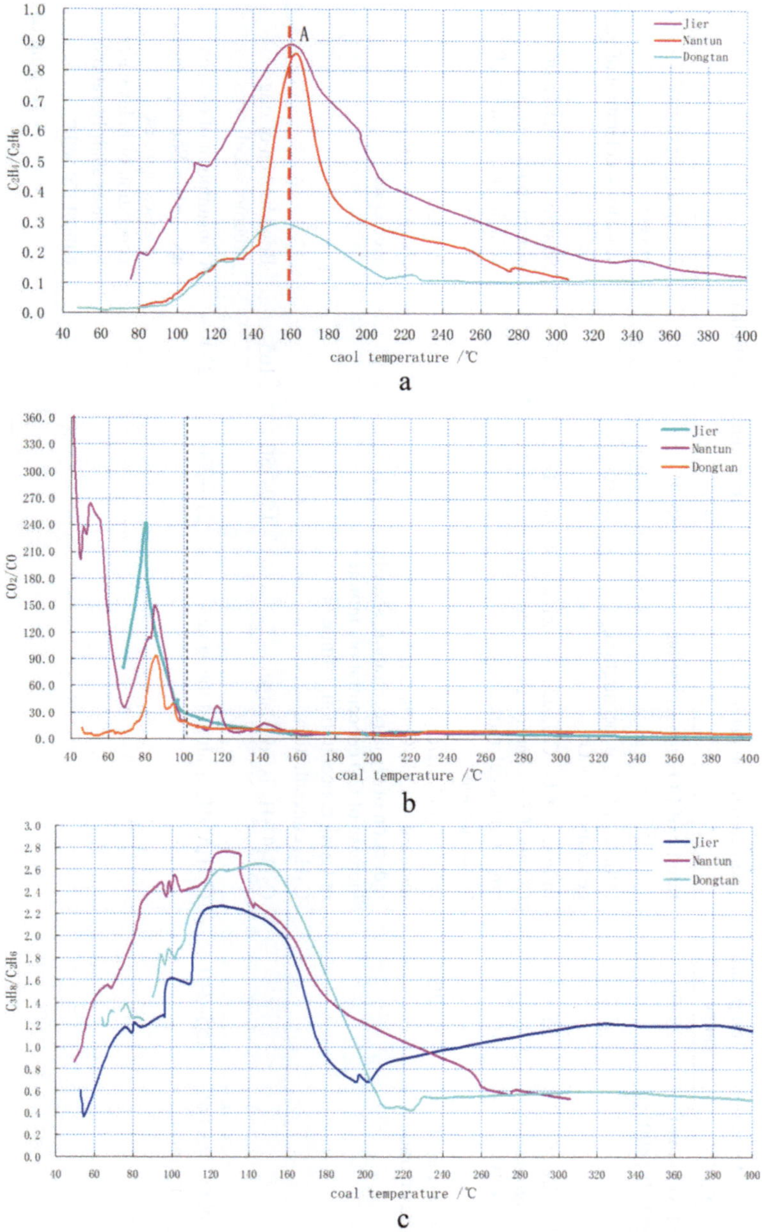

Fig. 2.6 Relationship between the ratio of index gases and coal temperature in Yanzhou mining area [37]

Table 2.2 Relation of characteristic temperatures and gas characterization in the process of coal spontaneous combustion [37] (the unit of temperature: °C)

Characteristic temperature	TG results		Experimental tests of coal spontaneous combustion			Remarks column
	Characterization parameter	Temperature range	Characterization parameter	Temperature range	Extrema	
High-adsorption temperature	Maximum of gravity ratio	25–30	Infinite of CO_2 concentration	35–45	40	Strong absorption of oxygen
Gas desorption temperature			Infinite of CH_4 concentration, generation of CO	50–65	55	Velocity acceleration of gas desorption
Critical temperature	Infinite of weightlessness velocity	55–90	Infinite of CO_2/CO, sharply increase of CO and CO_2 concentration, reduction amplitude acceleration of O_2 concentration	60–100	85	Sharply acceleration of consumption oxygen, velocity acceleration of chemical reaction
Reaction temperature	Inflexion of gravity ratio	115–200	Infinite of C_3H_8/C_2H_6, sharply increase of CH_4/C_2H_6, sharply reduction of O_2 concentration, sharply increase of CO concentration, infinite of C_3H_8 concentration, infinite of C_2H_6 concentration	100–150	130	Rupture and crack of side chain and bridge bond accelerate
Cracking temperature			C_2H_4/C_2H_6 is very large	150–180	165	Bond broken of ring structure accelerates
Active temperature	Infinitesimal of gravity ratio	155–240	C_2H_4/C_2H_6 is almost the same	210–250	230	Consumption of oxygen accelerates Sharply, release of gases increases, generation of index gases increases sharply, ratio is almost the same

(continued)

Table 2.2 (continued)

Characteristic temperature	TG results		Experimental tests of coal spontaneous combustion			Remarks column
	Characterization parameter	Temperature range	Characterization parameter	Temperature range	Extrema	
Temperature for maximum velocity of weight gain	Infinitesimal of weightlessness velocity	250–280	C_3H_8/C_2H_6, CO_2/CO is almost the same; concentrations of CO, CH_4, C_2H_4, C_2H_6, C_3H_8 increase sharply, O_2 concentration decreases sharply	250–340	280	Active structure increases, generation of index gas is almost the same, consumption of oxygen accelerates sharply
Ignition temperature	Infinite of gravity ratio	285–330	Consumption of oxygen accelerates sharply, rate of temperature change decrease sharply	>340	340	Beginning of coal combustion

As the characterization parameters of coal spontaneous combustion, index gases, such as C_3H_8, C_2H_6, C_2H_4, CO and ratios such as C_3H_8/C_2H_6, C_2H_4/C_2H_6, and CO_2/CO can suggest the characteristic temperature and the change situation of coal molecules in the process of natural heating up, so gas index can be used to predict the extent of coal spontaneous combustion and to make further research of mechanism of coal spontaneous combustion. As is known in the analysis of experimental data, there is great difference of various sampling gases concentration value in the furnace body, but the characterization parameters, ratios of gases concentration are almost the same, that is, the ratios of gases are not affected by sampling position and air volume/flow. Therefore, it is of much more practical use to estimate the extent of coal spontaneous combustion by ratios of gases.

2.2.2 Characteristic Temperature of Coal Spontaneous Combustion

(1) TG-DSC

At present, thermal analysis methods are divided into nine categories and 17 types, among which the main and commonly used thermal analysis methods are thermogravimetric analysis, differential thermal analysis and differential scanning calorimetry. Thermogravimetric analysis is usually divided into two types: non-isothermal thermogravimetric method and isothermal thermogravimetric method. The non-isothermal thermogravimetric method is to determine the relationship between the change of mass of substances and temperature under the program of temperature rise. Isothermal thermogravimetric method measures the relationship between the change of material mass and the temperature change at constant temperature. The isothermal method often needs to be experimented in a wide temperature range, which is more accurate but time-consuming, and is rarely used at present. The non-isothermal method is simple, practical and widely used. Xu et al. [38] selected Dongtan coal samples for TG analysis, and determined the temperature values of the maximum mass ratio, the inflection point of the mass ratio, the minimum point of the mass ratio, and the maximum point of the mass ratio, but did not determine the specific temperature range. Shu [39] selected coal samples from Jiaoping, Lingwu, and Shenfu for thermal analysis of coal spontaneous combustion process, and obtained several characteristic temperature parameters and weight change parameters of coal spontaneous combustion, but did not consider the influence of experimental conditions and determined the specific characteristic temperature range. Kök [40] conducted thermogravimetric experiments on 17 coal samples with a heating rate of 10 °C /min, and determined the reaction interval temperature, peak temperature, and burn-out temperature, but the division of the entire spontaneous combustion process of coal was unclear, and the specific temperature range was not determined. Biagini et al. [41] used biofuels such as coal, wood, pulp, and their components to perform thermogravimetric analysis and compared the onset temperature,

temperature for maximum rate of devolatilization, and final temperature of primary devolatilization, but did not make a detailed classification and determine the specific temperature range.

Using the TG209C thermogravimetric analyzer of Netzsch in Germany, 75 curves were obtained. From the conditions of different particle sizes, different oxygen concentrations and different heating rates of coal samples, it can be concluded that the characteristic temperature points are high adsorption temperature T_1, critical temperature T_2, cracking temperature T_3, active temperature T_4, growth rate T_5, ignition temperature T_6, and maximum weight loss rate point temperature T_7. The range of these characteristic temperatures can be determined through different experimental conditions, which provides a specific basis for the study of coal spontaneous combustion mechanism.

The TG and DTG curves of characteristic temperatures for coal samples of different particle sizes under heating rates of 10 °C/min and oxygen concentrations of 10% (Fig. 2.7). The characteristic temperatures of coal samples under different particle sizes, different heating rates and different oxygen concentrations are listed in Tables 2.3, 2.4 and 2.5.

① High adsorption temperature T_1

That is, the temperature at the initial weight gain on the TG curve, when the mass ratio of the coal sample reaches the maximum value, and the physical adsorption of the coal reaches equilibrium. In the process of oxygen consumption of coal, physical adsorption, chemical adsorption and chemical reactions are carried out according to their respective laws, and are not numerous, among which the physical adsorption of coal is a reversible process, the adsorption speed is relatively fast, easy to achieve equilibrium, and because it is an exothermic process, the equilibrium adsorption capacity will decrease with the rise of temperature, and chemical adsorption is in a secondary position when the temperature is low. At this time, the coal-oxygen compound is mainly based on physical adsorption, chemical adsorption and chemical reaction are slow, and there is basically no chemical reaction gas produced, and the physical adsorption capacity of the coal sample is greater than the amount of desorption gas, resulting in the weight gain of the coal sample. From Tables 2.3, 2.4 and 2.5, it can be seen that the high adsorption temperature of the coal sample is 25–30 °C.

② Critical temperature T_2

That is, the temperature of the first maximum point of weight loss rate on the DTG curve, and it is also the first temperature point that causes the automatic acceleration of coal-oxygen compound in the process of rising coal temperature from low to high. At this time, the coal desorption the primary gases (CO_2, CH_4, N_2, etc.) adsorbed and attached in the Microvoids. At the same time, the chemical reaction speed between coal and oxygen accelerates, consuming the oxygen adsorbed in the coal body, and the oxygen consumption rate increases, so that part of the active structure of coal molecules undergoes a multi-step reaction of coal-oxygen compound, and releases CO, CO_2 and other gases, so that the desorption and escape of the gas are greater

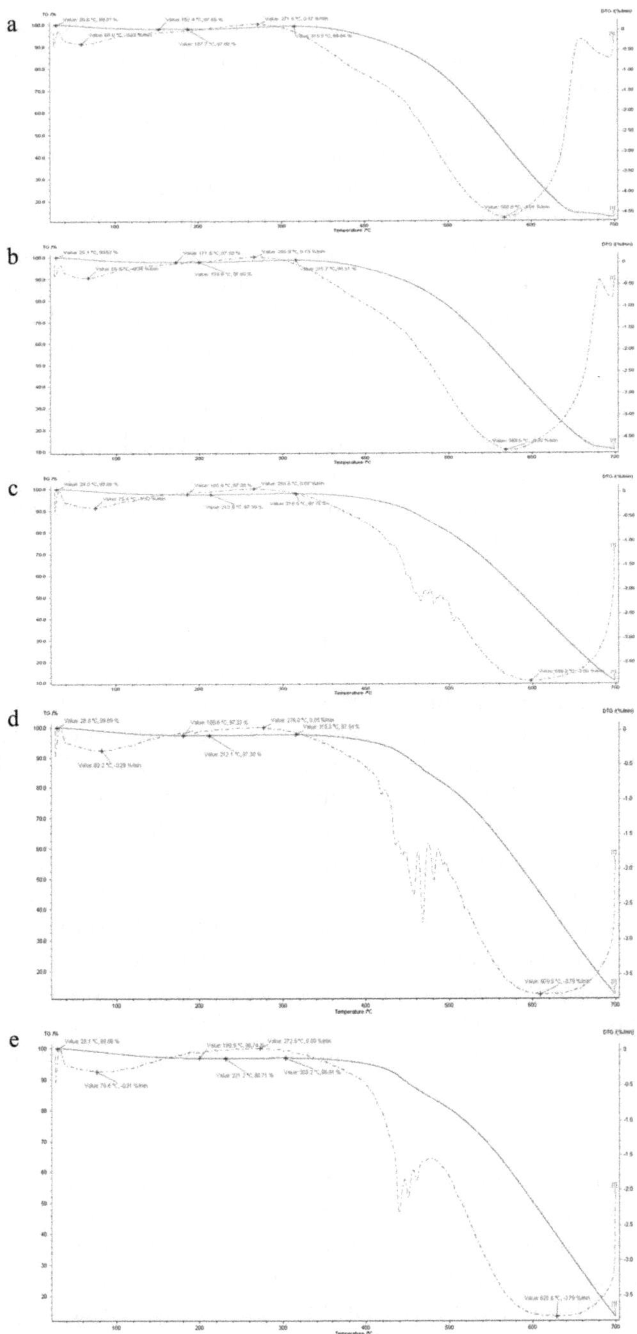

Fig. 2.7 TG and DTG curves of coal samples with different particle sizes [42]

Table 2.3 Characteristic temperatures under different experimental conditions (heating rate 5 °C/min) [42]

Serial number	Oxygen concentration/%	Coal sample size/mm	T_1/°C	T_2/°C	T_3/°C	T_4/°C	T_5/°C	T_6/°C	T_7/°C
1	5	0.088–0.098	27.9	56.9	133.7	178.9	265.5	308.0	550.2
2		0.098–0.105		57.3	143.2	174.7	258.1	306.9	558.7
3		0.105–0.150	28.3	64.0	154.1	192.5	264.9	305.8	583.6
4		0.150–0.200	26.5	64.7	158.2	191.6	275.3	308.8	578.7
5		0.200–0.300	28.6	65.2	187.1	220.1	265.7	305.5	603.0
6	9	0.088–0.098		58.0	132.6	164.3	260.9	308.7	532.1
7		0.098–0.105	26.0	59.1	119.9	160.0	260.7	308.4	533.4
8		0.105–0.150		63.0	132.7	164.5	260.8	309.9	553.4
9		0.150–0.200		57.0	143.3	174.2	260.4	312.2	565.6
10		0.200–0.300	26.9	73.2	156.5	180.9	260.5	311.7	575.0
11	13	0.088–0.098		56.0	120.0	158.4	259.4	300.4	508.9
12		0.098–0.105	25.8	61.2	119.8	162.1	252.2	298.6	505.0
13		0.105–0.150	25.0	60.6	148.0	197.0	263.3	290.2	507.0
14		0.150–0.200		64.2	155.0	195.9	260.4	306.3	539.0
15		0.200–0.300		64.9	159.6	206.4	259.1	306.0	546.3
16	17	0.088–0.098	26.0	55.0	115.3	160.0	253.0	300.1	500.7
17		0.098–0.105		57.9	125.5	160.0	253.0	299.8	506.2
18		0.105–0.150	29.3	60.0	126.2	166.6	252.7	306.9	534.8
19		0.150–0.200	28.6	60.4	136.6	175.6	252.9	301.0	525.4
20		0.200–0.300		60.2	160.1	173.2	252.9	301.0	528.0
21	21	0.088–0.098		55.8	117.8	155.5	258.2	301.3	494.1
22		0.098–0.105	25.9	56.0	133.5	166.1	251.5	286.0	477.7
23		0.105–0.150	26.4	61.8	124.3	164.3	264.8	300.0	515.3
24		0.150–0.200	26.0	61.9	145.3	165.9	251.9	291.9	501.9
25		0.200–0.300	27.1	64.9	153.6	189.4	252.2	299.1	519.5

than the amount adsorbed, the coal weight decreases rapidly, and the weight loss rate reaches the maximum. From a microscopic point of view, it is the starting temperature at which the total speed of the three-step reaction between bridge bonds and oxygen compounding in coal accelerates, and the desorption speed accelerates after the chemical adsorption of coal reaches the maximum equilibrium point. Generally speaking, the lower this temperature, the stronger the spontaneous combustion of coal. From Tables 2.3, 2.4 and 2.5, it can be seen that the critical temperature of the coal sample is about 55–88 °C.

③ Cracking temperature T_3

Table 2.4 Characteristic temperatures under different experimental conditions (heating rate 10 °C/min) [42]

Serial number	Oxygen concentration/%	Coal sample size/mm	T_1/°C	T_2/°C	T_3/°C	T_4/°C	T_5/°C	T_6/°C	T_7/°C
1	5	0.088–0.098	26.2	61.2	135.7	164.1	276.0	318.1	586.9
2		0.098–0.105		62.2	144.1	176.4	275.6	324.5	594.9
3		0.105–0.150		68.2	168.7	202.4	270.4	317.4	612.3
4		0.150–0.200	26.0	72.5	165.2	189.7	273.8	305.7	625.4
5		0.200–0.300	25.7	82.0	184.5	205.3	264.7	317.0	626.0
6	9	0.088–0.098	29.6	60.0	152.4	187.8	271.5	315.5	568.0
7		0.098–0.105	28.1	66.6	171.6	199.8	265.9	315.7	568.5
8		0.105–0.150	29.0	75.4	185.8	213.9	265.8	315.5	598.2
9		0.150–0.200	28.8	82.2	180.6	212.1	276.8	315.3	609.0
10		0.200–0.300	29.1	76.6	199.9	231.2	272.6	303.2	628.6
11	13	0.088–0.098	26.5	64.1	128.4	163.3	261.6	308.3	531.0
12		0.098–0.105		64.4	123.5	169.0	262.6	313.9	543.9
13		0.105–0.150	27.6	76.4	137.3	173.4	252.5	313.8	570.1
14		0.150–0.200		76.5	152.1	192.9	253.1	321.2	574.1
15		0.200–0.300		80.5	163.3	200.0	268.5	314.4	584.1
16	17	0.088–0.098		57.6	130.1	179.0	262.9	313.9	558.4
17		0.098–0.105		69.5	134.0	172.8	258.3	315.0	559.8
18		0.105–0.150	28.6	69.3	147.3	195.2	263.3	317.9	582.4
19		0.150–0.200		70.0	160.2	203.5	263.2	321.2	591.5
20		0.200–0.300	26.9	79.6	172.1	212.0	257.3	319.7	603.4
21	21	0.088–0.098	26.4	60.0	122.5	159.7	263.5	310.4	510.0
22		0.098–0.105		59.2	122.5	161.8	263.2	312.9	521.9
23		0.105–0.150		65.7	133.2	169.3	263.5	315.8	545.8
24		0.150–0.200	28.1	78.5	143.5	177.8	263.4	316.9	546.3
25		0.200–0.300	29.9	76.1	155.1	193.4	263.4	313.6	552.4

That is, the temperature at which the weight loss (TG curve) of the coal sample reaches the minimum value before the ignition temperature. It is the initial temperature at which the bridge bonds, alkyl side chains, oxygen-containing functional groups and some small molecules begin to crack or depolymerize and release small molecules as volatile substances in the coal-like molecular structure. At this temperature, the oxidation rate of the bridge and side chains in the molecular structure of coal is accelerated, which accelerates the growth rate of the active structure and the chemical reaction rate, and gaseous products, such as C_2H_6 and C_2H_4. At the same time, the oxygen absorption of coal is enhanced, the chemical adsorption capacity increases sharply, the weight loss rate slows down, and the desorption and escape rates of

Table 2.5 Characteristic temperatures under different experimental conditions (heating rate 20 °C/min) [42]

Serial number	Oxygen concentration/%	Coal sample size/mm	T_1/°C	T_2/°C	T_3/°C	T_4/°C	T_5/°C	T_6/°C	T_7/°C
1	5	0.088–0.098	26.5	75.9	165.2	188.6	285.7	324.9	622.9
2		0.098–0.105	25.8	77.3	157.6	170.4	264.6	318.3	609.2
3		0.105–0.150	27.2	78.7	184.4	208.8	268.8	321.9	621.5
4		0.150–0.200	27.1	88.2	198.8	236.0	263.2	322.5	624.0
5		0.200–0.300	26.3	88.1	199.8	224.6	263.4	316.7	497.0
6	9	0.088–0.098	27.1	70.3	138.1	188.3	262.1	329.5	594.5
7		0.098–0.105	26.0	80.0	143.6	194.9	261.8	326.3	594.2
8		0.105–0.150		87.4	168.0	210.8	260.7	328.2	607.7
9		0.150–0.200		87.1	186.7	227.5	260.0	321.1	622.1
10		0.200–0.300	27.1	88.0	199.9	242.7	259.7	324.6	637.7
11	13	0.088–0.098		81.3	143.0	190.3	269.0	329.4	581.1
12		0.098–0.105	29.7	80.4	150.9	193.9	269.1	330.7	590.6
13		0.105–0.150	29.7	82.6	171.9	216.3	268.5	330.9	596.6
14		0.150–0.200	29.5	82.6	178.1	222.7	268.7	330.3	609.7
15		0.200–0.300		83.2	196.4	239.0	269.8	328.9	627.6
16	17	0.088–0.098		75.9	142.8	188.8	280.0	321.9	579.1
17		0.098–0.105	26.7	75.1	146.9	186.4	280.4	324.5	580.7
18		0.105–0.150		87.5	167.2	210.7	280.3	320.4	595.8
19		0.150–0.200	29.7	87.0	175.8	221.9	281.2	323.4	609.9
20		0.200–0.300	29.5	87.7	189.0	226.9	279.5	326.2	610.7
21	21	0.088–0.098		73.5	140.9	176.4	274.4	321.1	551.9
22		0.098–0.105		72.0	142.3	176.6	274.4	324.1	562.4
23		0.105–0.150	28.1	83.9	159.6	200.2	273.8	326.6	579.1
24		0.150–0.200	29.5	85.2	180.5	217.2	272.7	322.7	591.0
25		0.200–0.300		85.7	192.0	226.3	273.0	324.1	593.7

gaseous products produced by oxidation reaction and cracking are basically equal to the binding speed of coal oxygen, forming a dynamic equilibrium, and the coal sample no longer loses weight. The cracking temperature of coal reflects the side chain of coal molecules breaking from the main structure and escaping in the form of gas products, and some of the reaction processes are as follows:

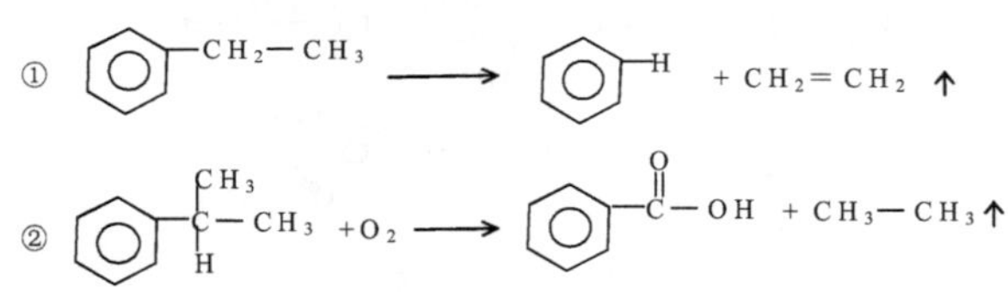

From Tables 2.3, 2.4 and 2.5, it can be seen that the cracking temperature of the coal sample is about 115–200 °C.

④ Active temperature T_4

That is, the temperature at which the coal sample keeps its weight unchanged from the cracking temperature to the starting point of weight gain. At this temperature, the bond breaking of macromolecules with ring structure in coal begins to accelerate, the growth rate of the active structure with strong oxygen absorption in coal molecules accelerates, the chemical adsorption capacity of coal to oxygen increases sharply, and the dynamic balance of coal is broken. It can be seen from Tables 2.3, 2.4 and 2.5 that the active temperature of the coal sample is about 155–239 °C, and the active temperature point decreases with the decrease of particle size.

⑤ Growth rate temperature T_5

That is, the temperature of the maximum weight gain rate of the coal sample. At this temperature, the breaking rate of cyclic macromolecules in coal molecules increases sharply, the number of active structures exposed increases sharply, the chemical reaction speed accelerates, and the adsorption capacity of oxygen by coal samples increases sharply, which is greater than the amount of gas produced by coal desorption and reaction. From Tables 2.3, 2.4 and 2.5, it can be seen that the growth rate temperature of coal samples is about 251–281 °C.

⑥ Ignition temperature T_6

That is, the temperature of the maximum value point of the mass ratio of the coal sample. Due to the increase of temperature, the number of active structures on the surface of the coal body increases sharply, and the aromatic ring structure that does not participate in the coal oxidation process at lower temperatures also begins to participate in the oxidation reaction, and the number of active structures in the coal and the adsorption capacity of oxygen in the coal reach the maximum value, so that the coal mass changes from a sharp decrease to an upward trend, and the weight increase of coal reaches the maximum. After that, the aromatic ring structure of the coal body rapidly oxidizes and decomposes, generating a large amount of CO, CO_2 and small molecule organic gases, releasing a lot of heat, and the mass of the coal body begins to decline sharply, indicating the comprehensive cracking of the aromatic nucleus of the thick ring and the discharge of a large amount of liquid volatiles, and also suggesting that the volatiles begin to burn and reach the initial combustion temperature of the coal sample. The reaction process when the ignition temperature is reached is as follows:

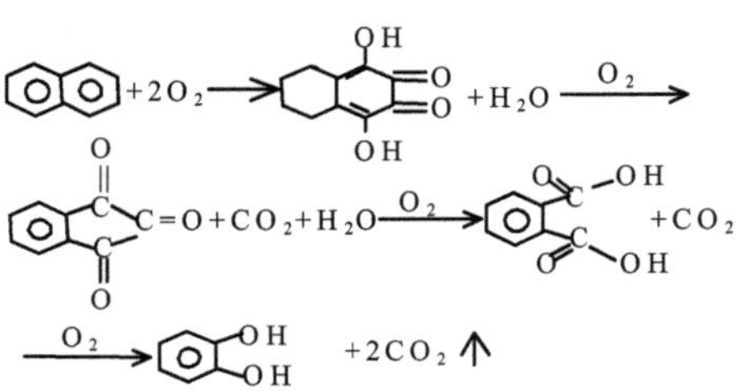

From Tables 2.3, 2.4 and 2.5, it can be seen that the ignition point temperature of the coal sample is about 286–331 °C.

⑦ Maximum weight loss rate point temperature T_7

That is, the maximum weight loss rate point of the coal sample during the whole process of spontaneous combustion ignition. At this time, the oxidation and ignition rate of coal is accelerated, and the weight loss of coal increases sharply, indicating that a violent chemical reaction has occurred inside the coal molecule, the CO production rate and oxygen consumption rate have increased sharply, the heating rate has accelerated sharply, a large amount of gas is generated, and the weight loss of coal samples is obvious. From Tables 2.3, 2.4 and 2.5, it can be seen that the maximum temperature point of weight loss of coal samples is about 494–638 °C.

Through TG analysis of coal samples, for different heating rates (5 °C/min, 10 °C/min, 20 °C/min), different oxygen concentrations (5%, 9%, 13%, 17%, 21%) and different particle sizes (0.088–0.098 mm, 0.098–0.105 mm, 0.105–0.150 mm, 0.150–0.200 mm, 0.200–0.300 mm) and other experimental conditions, 75 corresponding TG and DTG curves were obtained, and seven characteristic temperatures of the coal spontaneous combustion process were obtained through analysis, namely, high adsorption temperature T_1 (25–30 °C), critical temperature T_2 (55–88 °C), cracking temperature T_3 (115–200 °C), active temperature T_4 (155–239 °C), growth rate temperature T_5 (251–281 °C), ignition temperature T_6 (286–331 °C) and maximum weight loss rate point temperature T_7 (494–638 °C) reflect the influence of external conditions on the reaction process during the microscopic process of coal oxidation, which also provides a specific basis for the study of coal spontaneous combustion oxidation.

(2) C80 microcalorimetry

The oxidative exothermic characteristics of coal are one of the key factors affecting coal spontaneous combustion. Zhu et al. [43] analyzed and estimated the maximum and minimum exothermic intensity of coal oxidation at low temperature. Chen et al. [44] established the relationship between the maximum and minimum exothermic intensity and the actual exothermic intensity, and proposed a characteristic exothermic intensity to characterize the low-temperature oxidative exothermic intensity of coal. Deng et al. [45] calculated the actual exothermic intensity of coal oxidation and established an exponential relationship between exothermic intensity and temperature. Xiao et al. [46] proposed a calculation model for the heat production rate of coal adiabatic oxidation and combined with the heating rate to determine the tendency of coal to spontaneously combustion. Yang et al. [47] proposed a prediction model for the shortest ignition period of coal spontaneous combustion based on oxidative exothermic intensity. Zhou et al. [48, 49] found that the exothermic intensity of coal oxidation decreased with the increase of metamorphism, and analyzed the effect of air supply on exothermic intensity. Cheng et al. [50] found that the warming rate can shorten the endothermic stage, but has little effect on the overall heat release. Jia et al. [51] measured the spontaneous heat flow curves of coal with different water contents, indicating that the higher the water content, the more obvious the lag in oxidative heat release behavior of coal, and the lower the heat release. Wang et al. [52] divided the stages of low-temperature oxidation and exothermy of coal samples with different water contents, and believed that the rapid exothermic stage was the main cause of coal heat accumulation. Deng et al. [53] analyzed the oxidative exothermic characteristics of pre-oxidized coal and found that the exothermic intensity of treated coal was greater than that of raw coal in the early stage of oxidation and less than that of raw coal in the later stage. The oxygen concentration of coal spontaneous combustion in goafs is generally less than 18%, that is, the coal body is in a state of oxygen-poor oxidation. As a participant in coal oxidation, oxygen has a significant effect on coal spontaneous combustion gas products [54, 55], characteristic temperature [56, 57], and heat release [58, 59], mass changes [57, 60], kinetic parameters [57, 60], and combustion characteristics [61] etc. have an impact. Ren et al. [58] and Qi et al. [59] analyzed the trend of coal oxidation heat release in different oxygen concentrations and divided it into different stages of low-temperature oxidation of coal.

① Exothermic intensity

The heat flux curve during coal oxidation was measured using a C80 microcalorimeter. The magnitude of the heat flux value characterizes the oxidation heat release per unit time. Therefore, the heat flux value was used to characterize the oxidation heat release intensity of coal. The heat flux curves during the oxidation process of the two coal types under different oxygen concentrations are shown in Fig. 2.8.

The heat flow curve initially decreases during the initial stage. Upon reaching a certain temperature, the heat flow value begins to increase. The reasons for the initial decrease in the heat flow curve are complex, including factors such as water evaporation, gas desorption, and differences in thermal conductivity between the reference cell and sample cell filling media. According to the gradual self-activation theory of coal spontaneous combustion, coal contains multiple functional groups. Highly

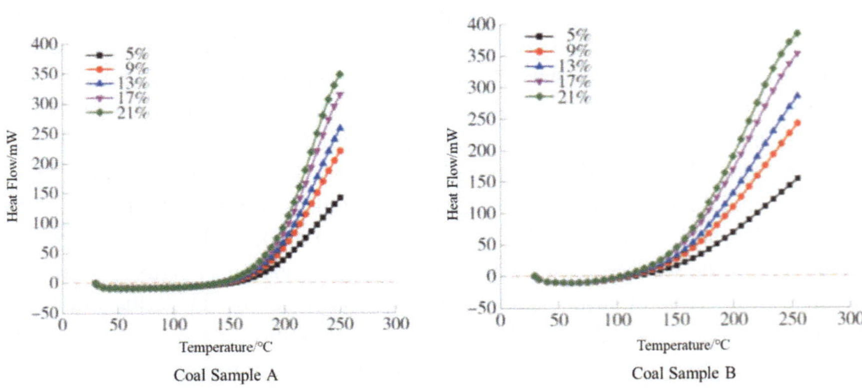

Fig. 2.8 Heat flow curves during oxidation of two coal types under different oxygen concentrations [62]

active functional groups react with oxygen first, while less active groups are gradually activated during oxidation and subsequently participate in the reaction. Therefore, as temperature rises, the number of active functional groups participating in the reaction increases, leading to greater exothermic intensity, manifested as a rise in heat flux values. Higher temperatures result in larger heat flux values. Conversely, lower oxygen concentrations yield lower heat flux values, indicating reduced exothermic intensity. This occurs because insufficient oxygen supply limits the number of active functional groups available for reaction.

② Initial exothermic temperature

Under oxygen-free conditions, the heat flow value of coal oxidation stabilizes near its minimum. Therefore, the temperature corresponding to the minimum heat flow point is defined as the initial exothermic temperature. The initial exothermic temperatures of the two coal types under different oxygen concentrations are shown in Fig. 2.9.

The initial exothermic temperature of the two coals increases with the decrease of oxygen concentration, and when the oxygen concentration decreases from 21 to 5%, the initial exothermic temperature of coal sample A and coal sample B increases by 18.80% and 3.39%, respectively. However, from the perspective of the change process of the initial exothermic temperature, it shows the nonlinear characteristics of fluctuating rise. For coal sample A, when the oxygen concentration was 17% and 9%, the initial exothermic temperature increased significantly, while the change was not significant in the range of 5%–9% and 13%–17%. For coal sample B, when the oxygen concentration is 17%, the initial exothermic temperature shows a decreasing trend, and then increases almost linearly with the decrease of oxygen concentration. During the oxidation process of coal, gas products such as CO and CO_2 are produced, and the release of these gas products will form resistance in the process of oxygen diffusion into the pores of the coal body. However, the rate and amount of gas products decreased, resulting in a decrease in the resistance of oxygen diffusion into the pores of the coal body, which was conducive to the diffusion of oxygen into the coal body.

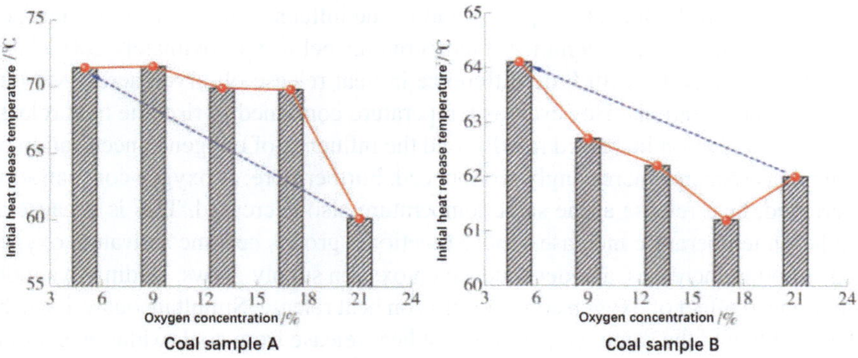

Fig. 2.9 Initial exothermic temperature of two coals with different oxygen concentrations [62]

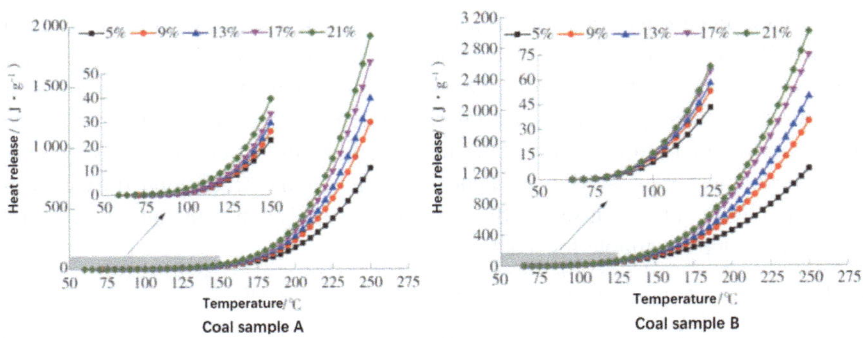

Fig. 2.10 Heat release in different oxygen concentrations of two types of coal [62]

Under the comprehensive effect, the initial exothermic temperature of coal oxidation fluctuated with the decrease of oxygen concentration.

③ Release heat

At the same time, the heat release value corresponding to the initial heat release temperature is used as the baseline to calculate the heat release, and the relationship between heat release after the initial heat release temperature and temperature and oxygen concentration is analyzed. The heat release in the environment of different oxygen concentrations of the two coals is shown in Fig. 2.10.

The exothermic heat release from coal oxidation exhibits similar patterns across different oxygen concentrations as temperature increases. Specifically, heat release gradually increases with rising temperature, and the rate of increase accelerates at higher temperatures. This occurs because higher temperatures activate more functional groups within coal molecules, intensifying the reaction. Simultaneously, the heat release decreases as the oxygen concentration decreases. When the temperature rises to 250 °C and the oxygen concentration is 5%, the cumulative heat release of coal sample A and coal sample B decreases by 56.84% and 58.57%, respectively. Furthermore, for both coal samples A and B, the influence of oxygen concentration on heat release is relatively minor at temperatures below approximately 200 °C and 150 °C, respectively, with little difference in heat release observed across varying oxygen concentrations. However, as temperature continued to rise, the heat release from coal oxidation increased rapidly, and the influence of oxygen concentration on heat release became increasingly pronounced. Furthermore, as oxygen concentration decreased, heat release at the same temperature also decreased. This is because as oxidation temperature increases, more functional groups become activated, oxygen consumption increases, and dependence on oxygen supply grows, leading to a more pronounced effect of oxygen concentration on heat release. Simultaneously, it can be observed that at the same temperature, the heat release from coal oxidation exhibits a nonlinear decrease trend as oxygen concentration decreases. For the experimental coal samples, when oxygen concentration is around 17% and 9%, the heat release shows a pronounced decreasing trend as oxygen concentration decreases.

2.2.3 *Thermal and Physical Property Parameters*

The study of the thermal and physical properties of coal is of great significance for the prevention of coal fire disasters. Thermal physical properties refer to the thermal properties of coal materials during the heating process. The thermal properties of coal have an important impact on coal combustion, thermal energy conversion and energy utilization efficiency. By studying and understanding the thermal and physical properties of coal, we can better ensure the production safety of coal mines, improve the efficiency of energy conversion, and reduce energy waste, so as to achieve sustainable energy development and provide cleaner energy supply. Deng et al. [63] studied the changes in thermal properties of coal at different oxygen concentrations and during low temperature oxidation. Xiao et al. [64] studied the thermal physical parameters of coal under different pre-oxidation conditions and found that the thermal diffusion coefficient and thermal conductivity increased with the increase of pre-oxidation temperature, and the specific heat capacity was the highest at 90 °C. Wang et al. [65] studied the heat conduction characteristics of bituminous coal under the condition of oxygen volume fraction. Zhang et al. [66] studied the changes in the thermal properties of different pre-oxidized coals. Xiao et al. [67] used the gray correlation analysis method to analyze the correlation between the thermal properties and volatilization of coal with different degrees of metamorphism. Xu [68] used the unsteady thermal conductivity product method to determine the uncertainty of the thermal properties of loose media. Hui et al. [69] used the C80 calorimetry experiment to determine the dynamic heat flow of coal samples during ignition, Zhang et al. [70] studied the pyrolysis and combustion characteristics of coal samples with different particle sizes, and Lu et al. [71] compared the thermal and physical properties of coal, granite, and sandstone.

(1) Experimental equipment and testing principle

The thermal and physical parameter test mainly includes three parameters: thermal conductivity, thermal conductivity (thermal diffusion coefficient), and specific heat capacity. The coefficient of measuring coal is generally used by the unsteady state method. Among them, laser flash is the most common and representative method of unsteady state method, and is generally recognized by international physics. The LFA457 laser thermal conductivity meter produced by NETZSCH in Germany can directly measure the temperature conductivity coefficient, and the specific heat capacity and thermal conductivity can be calculated by calculation.

The principle of the equipment is to irradiate the lower surface of the sample with a laser, the lower surface of the sample absorbs heat, the temperature rises, the heat is transferred from the lower surface to the upper surface, the upper infrared detection device of the sample measures the temperature change of the upper surface of the sample, and draws the temperature change curve corresponding to the time, and then obtains the temperature conductivity coefficient.

The thermal diffusion coefficient can be obtained from the thermal diffusivity model equation

$$\alpha = 0.13878L^2/t_{50} \tag{2.1}$$

In the formula:

t_{50}—half heating time, that is, the time it takes for the upper surface to reach the maximum temperature of 1/2, s;

L—thickness of the sample, mm.

The specific heat capacity coefficient of the sample is then calculated

$$c = \frac{\Delta U_{\text{std}} m_{\text{std}} c_{\text{std}}}{\Delta U m} \tag{2.2}$$

In the formula:

m—sample quality, g;

ΔU—the change in the electrical signal of the sample to which it is radiated;

c_{std}—specific heat capacity of standard sample, J/(g·K);

m_{std}—standard sample quality, g;

ΔU_{std}—the change in the electrical signal of the standard sample to be radiated.

The thermal conductivity is calculated based on the thermal diffusion coefficient and specific heat capacity

$$\lambda = \alpha c \rho \tag{2.3}$$

In the formula:

ρ—sample density, g/cm^3.

(2) Thermal physical parameters under different atmospheres

① Thermal diffusion coefficient

The thermal diffusion coefficients of the three coal samples during heating under different atmospheres are obtained by testing, as shown in Fig. 2.11. It can be obtained that under nitrogen and air conditions, the temperature is 30–210 °C, and the thermal diffusion coefficient decreases gradually, mainly due to the decrease of the average free path of phonons. Phonons are quantized elastic waves connected to molecular motion and interaction forces, which propagate in the form of waves in crystals. However, due to the increase of temperature, the vibration of the crystal structure in the coal increases, which reduces the average free path of phonons, resulting in a decrease in the thermal diffusion coefficient of the coal sample. When the coal temperature is 210–300 °C, the thermal diffusion coefficient continues to decrease under nitrogen conditions, but gradually increases under air conditions. This is mainly under air conditions, as the temperature rises, the oxygen-containing functional groups in the coal molecules react with oxygen, triggering the decomposition of carboxyl groups and producing small molecule compounds. These small molecules further react with oxygen, resulting in increased disorder inside the coal sample. This reaction process leads to a gradual increase in the thermal diffusion coefficient of the coal sample.

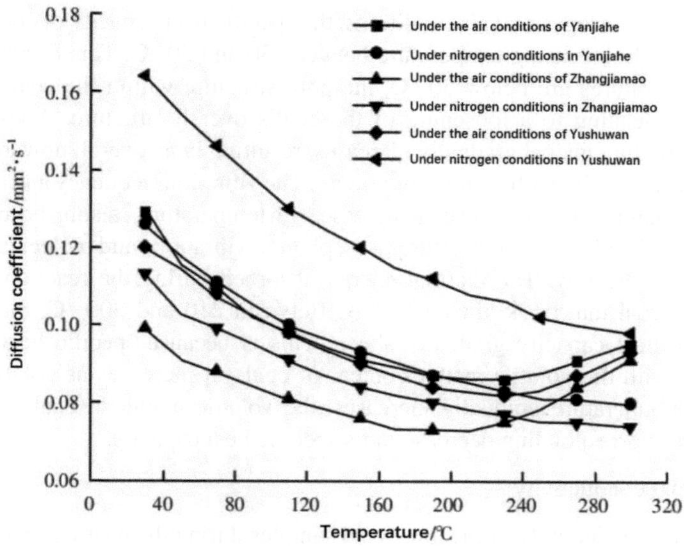

Fig. 2.11 Thermal diffusion coefficients of coal samples at different temperatures [72]

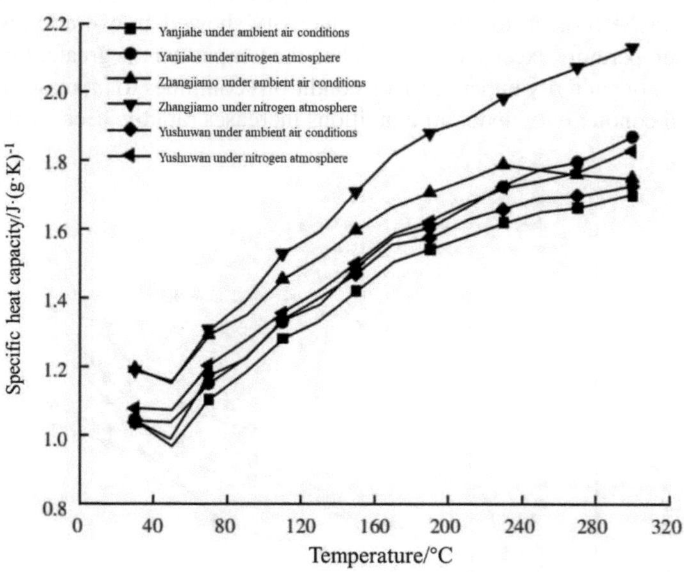

Fig. 2.12 Specific heat capacity at different temperatures [72]

② Specific heat capacity

The specific heat capacities of three coal samples during heating under different atmospheres were obtained through testing, as shown in Fig. 2.12.

Under both nitrogen and air conditions, the specific heat capacity of coal samples decreases with increasing temperature between 30 and 50 °C. This occurs because when temperatures are below 50 °C, the pore structure within the coal gradually diminishes, leading to a loosening of the coal's overall structure. Above 50 °C, the pore structure in coal gradually increases, resulting in an overall upward trend in specific heat capacity with rising temperature. The vibration frequency and amplitude of atoms around lattice nodes in coal increase with temperature, raising the total lattice energy of molecules. As temperature rises, phonon vibrations and molecular thermal motion also intensify. The thermal energy absorbed during the reaction is stored within the coal matrix as kinetic energy. Between 210 and 300 °C, the increase in specific heat capacity gradually slows. This is because specific heat capacity correlates with the volatile matter content of coal—it increases as volatile matter rises. As temperature gradually increases, the volatile matter in coal diminishes, leading to a corresponding decrease in its specific heat capacity.

③ Thermal conductivity

The thermal conductivity of the three coal samples during the heating process under air and nitrogen was obtained by the test, as shown in Fig. 2.13.

The thermal conductivity of coal samples at 30–50 °C under nitrogen and air conditions decreases with the increase of temperature due to the shrinkage of the pore structure. However, the thermal conductivity showed an overall upward trend when the temperature exceeded 50 °C. When the temperature is greater than 210 °C, the thermal conductivity under nitrogen conditions continues to increase slowly, and the thermal conductivity under air conditions increases rapidly, because the thermal

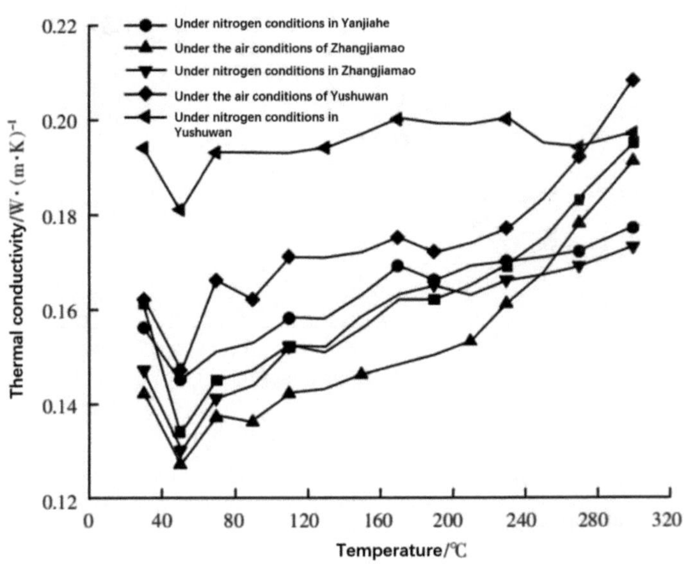

Fig. 2.13 Thermal conductivity at different temperatures [72]

diffusion coefficient gradually decreases, resulting in a gradual increase in specific heat capacity, because the change of thermal conductivity parameters is greater than the change of thermal diffusion parameters, so the downward trend caused by thermal diffusion parameters is offset.

In the nitrogen environment, the thermal diffusion coefficient gradually decreases with the increase of temperature, and the specific heat capacity and thermal conductivity gradually increase with the increase of temperature. Under air conditions, the thermal diffusion coefficient of 30–210 °C coal samples gradually decrease with the increase of temperature, and the specific heat capacity and thermal conductivity also increase with the increase of temperature. When the temperature is at 210–300 °C, the thermal diffusion coefficient increases gradually with the increase of temperature, the thermal conductivity increases rapidly, and the change of specific heat capacity tends to be stable. When the temperature exceeds 210 °C, the coal under air conditions will undergo oxidation reaction, which affects the change of thermal and physical parameters of coal.

2.2.4 *Magnetic Parameters of Coal and Rock*

Coal magnetic parameters refer to the magnetic properties of coal and rock materials under the action of magnetic fields, reflecting their internal structure and distribution of mineral components, and the magnetic parameters of coal and rock have an important impact on coal mining, processing and disaster prevention. By studying and understanding the magnetic parameters of coal and rock, we can better ensure the production safety of coalmines, improve resource utilization efficiency, reduce the occurrence of disasters, and achieve sustainable development of coalmines.

Qin [73] revealed the key laws of coal and rock magnetism with temperature: coal is weakly inversely magnetic at room temperature, turns paramagnetic at 200 °C, and increases the magnetic susceptibility at 300–400 °C due to the phase transition of magnetic minerals, and the coal can maintain high magnetism for a long time after combustion. Zhang et al. [74] studied the magnetic response between coal and rock through thermal experiments and confirmed that the magnetic properties and the magnetic susceptibility of coal are the most sensitive to temperature: it is inversely magnetic at 25–200 °C, the magnetic susceptibility increases sharply at 200–400 °C, and the increase slows down after 500 °C. Li [75] systematically analyzed the effects of temperature, particle size, and magnetic field on the magnetic susceptibility of coal, sandstone, and gangue, and showed that the magnetic susceptibility of the samples was significantly enhanced after high-temperature baking and cooling, but the starting temperature of the mutation was different (200 °C for coal, 300 °C for sandstone, and 400 °C for gangue), and this process was accompanied by the generation of strong magnetic ferromagnetic materials, and the magnetism obtained after cooling was stronger. It effectively supports the accuracy of magnetic method delineating the fire area.

(1) Magnetic parameters of coal and rock

The magnetic parameters (such as susceptibility, remanence, coercivity) and thermal behavior characteristics of samples (bituminous coal, mudstone and sandstone) under different heat treatment temperatures (0 °C, 300 °C, 500 °C, 700 °C) are analyzed through experimental tests, and the magnetic evolution and stage characteristics are revealed, which provides a basis for the boundary division of the fire zone.

① Hysteresis loop characteristics

As shown in Figs. 2.14 and 2.15, the coal rock is heat-treated at 0, 300, 500, and 700 °C, the hysteresis loop characteristics show obvious temperature dependence: the hysteresis loop of the original C is slender, the magnetization strength is extremely low, and the hysteresis phenomenon is not obvious. The loop of C_{300} is still thin, but the magnetization strength has been improved, and the hysteresis characteristics have been slightly enhanced, the loop of C_{500} becomes a full S-shape, the magnetization strength jumps significantly, the hysteresis phenomenon is obvious, and the magnetism changes from weak magnetism to strong magnetism. The C_{700} has a steeper and fuller loop, a higher level of magnetization strength, and a strong hysteresis characteristic, showing typical strong ferromagnetism. On the whole, with the increase of heat treatment temperature, the hysteresis loop of coal rock gradually changes from a slender and weak magnetic type to a full and strong magnetic type, and the magnetization strength first increases significantly and then remains high, and the coercivity first decreases and then increases, reflecting that the types, contents and magnetic states of magnetic minerals in coal and rock have evolved significantly with temperature.

As shown in Fig. 2.16, the hysteresis loop shape is smoother than that of coal samples because the rock sample itself contains more magnetic minerals and has strong stability. The hysteresis loop of the original mudstone and sandstone samples shows pinprick-like fluctuations, which is closely related to the incomplete crystal structure of the magnetic minerals in the rock, the different particle size and the uneven distribution of minerals, resulting in the nonlinear characteristics of local magnetic response under the action of magnetic field.

With the increase of heat treatment temperature, the hysteresis loop morphology of mudstone and sandstone samples changed significantly, and the hysteresis loop of rock samples was the smoothest after treatment at 500 °C, indicating that the magnetic response stability of the two rocks was enhanced under this temperature condition. This is due to the fact that the treatment temperature promotes the crystal structure of magnetic minerals to be perfected, and the lattice defects are reduced, which makes the behavior of magnetic minerals in the process of magnetization and demagnetization more consistent and the order is improved.

As shown in Fig. 2.17, the magnetic properties reflected by the hysteresis loop of the mudstone and sandstone samples changed significantly after different temperature treatments. From the perspective of the change of hysteresis loop parameters of rocks, with the increase of heat treatment temperature, the saturated magnetization strength and the residual magnetization strength both increase first and then decrease,

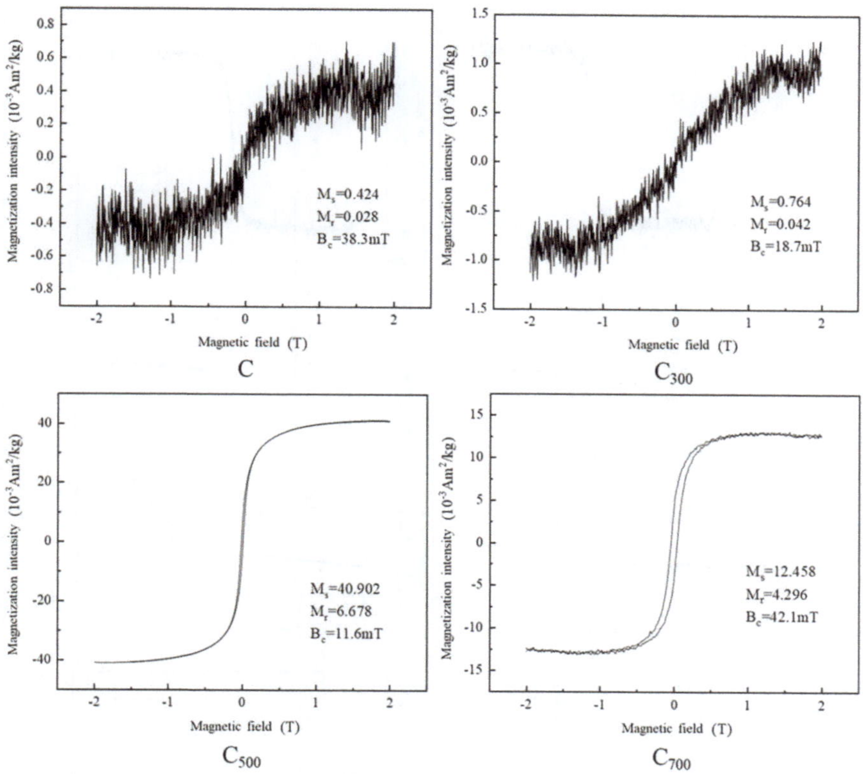

Fig. 2.14 Hysteresis loop of coal and rock under different heat treatment conditions

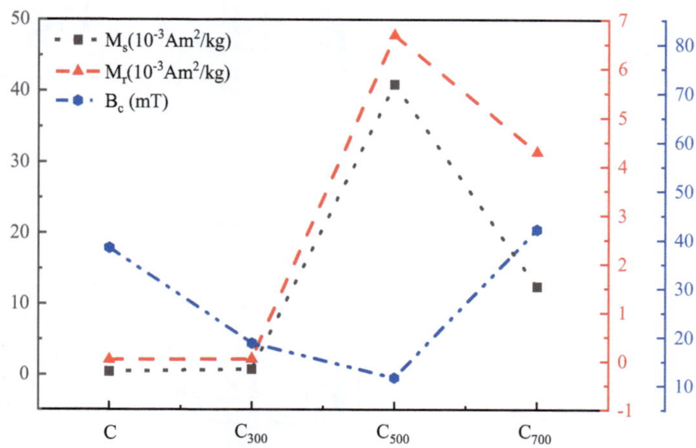

Fig. 2.15 Trends of M_s, M_r, and B_c of coal samples under different heat treatment conditions

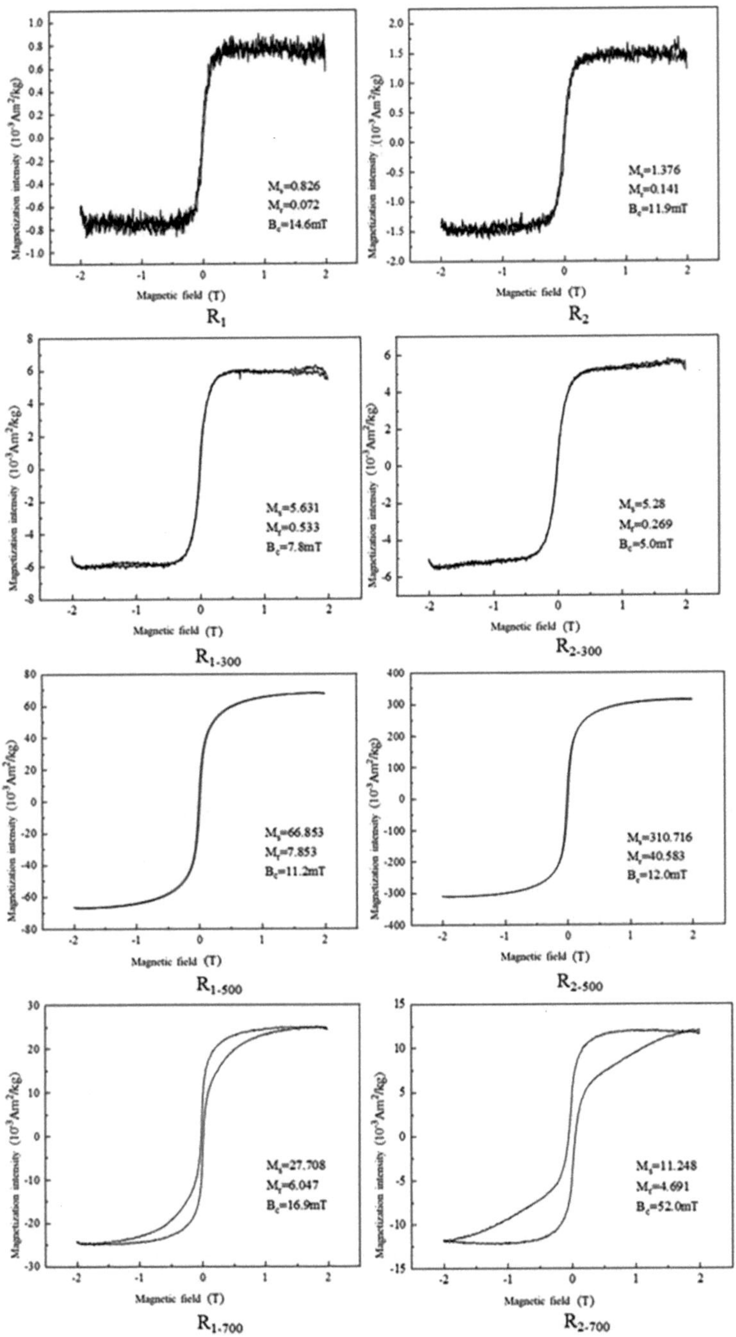

Fig. 2.16 Hysteresis loops of mudstone and sandstone under different heat treatment conditions

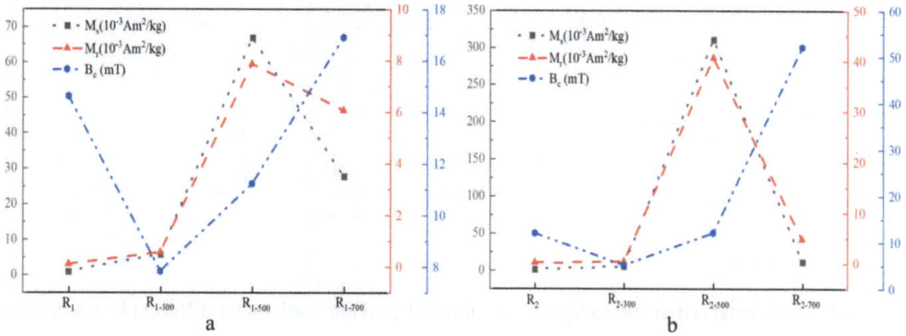

Fig. 2.17 Variation trends of M_s, M_r And B_c in different rocks under different heat treatment conditions

indicating that the magnetic minerals in the rock have undergone a transformation conducive to magnetic enhancement at lower heat treatment temperatures. The saturated magnetization strength and residual magnetization strength of mudstone and sandstone reached the maximum value after 500 °C treatment, indicating that the magnetic minerals in the rock were enriched or formed under the action of temperature, which significantly enhanced the magnetism. The closed area of hysteresis loops also tends to increase. These changes reflect the physicochemical changes of magnetic minerals in mudstone sandstone at different temperatures, including mineral phase transition, particle growth, aggregation, and structural changes, which together affect the magnetic characteristics of mudstone sandstone.

② Isothermal remanence and reverse demagnetization

The isothermal residual magnetism of bituminous coal raw coal samples and samples treated at 300 °C does not show a saturated state with the enhancement of the external field, indicating that the magnetic mineral content in the coal sample is less or the magnetic minerals are in a paramagnetic state, and a stable magnetic domain structure cannot be formed, so the common isothermal residual magnetic characteristics are not shown (Fig. 2.18). After 500 and 700 °C treatment, the isothermal remanence of coal samples showed a saturation state with the increase of the applied magnetic field, and the magnetization strength of coal samples increased rapidly before 0.3 T, and the saturation degree of saturation could reach 90% before 0.3 T in both 500 °C and 700 °C heat treatment, and the saturated isothermal remandate of coal samples treated at 500 °C was the strongest, with a value of 8.34×10^{-9} Am^2/kg.

On the other hand, all treated samples of rock samples can show saturation with the enhancement of the external field, indicating that the magnetic minerals in the rock are abundant and relatively stable, showing a stronger magnetic response than the coal sample. The saturated isothermal remanence of the mudstone sandstone samples treated at 500 °C was the strongest, indicating that the transformation or recrystallization of ferromagnetic and ferromagnetic minerals was the most significant under this temperature condition.

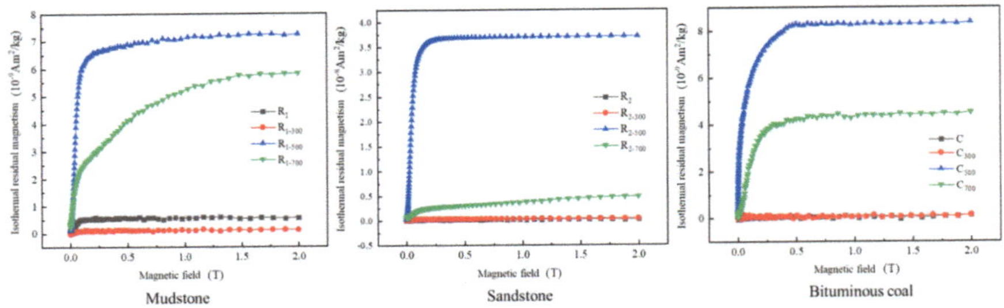

Fig. 2.18 Isothermal remanence acquisition curves of coal and rock under different heat treatment conditions

The raw coal sample and the coal sample treated at 300 °C are difficult to maintain the remanence stably due to the low content of internal magnetic minerals or the paramagnetic state, and the remanence fluctuates around 0 under the action of 0.2 T reverse magnetic field, showing the characteristics of demagnetization is not obvious, and it is difficult to judge the remanence coercivity of the sample (Fig. 2.19). The coal samples heat-treated at 500 c fand 700 °C showed significant demagnetization.

Both mudstone and sandstone samples exhibited pronounced demagnetization characteristics. The remanence of both the original mudstone and sandstone samples and those treated at 300 °C was reduced to zero below 50 mT, indicating poor remanence stability of magnetic minerals in the rocks under these conditions. The remanence coercivity of the 500 °C-treated mudstone and sandstone samples was highest, failing to reduce the remanence to zero before 0.2 T. The remanence coercivity was calculated using the slopes of the two data points closest to 0.2 T, yielding values of 214.2 and 326.1 mT, respectively. When the heat treatment temperature reached 700 °C, the remanence coercivity of the mudstone and sandstone samples decreased to 37.2 and 33.6 mT, respectively.

(2) Magnetic stage characteristics of coal and rock

MKF2-FA rotating multi-frequency Kappa bridge susceptibility meter and CS4 temperature control system to measure the magnetic susceptibility curve of coal

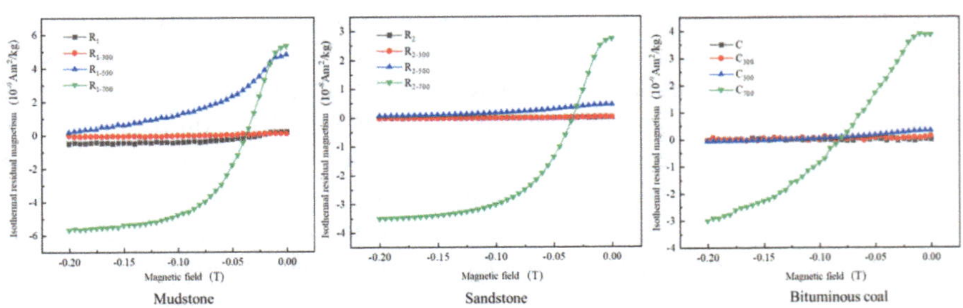

Fig. 2.19 Reverse demagnetization curves of coal and rock under different heat treatment conditions

samples and rock samples with temperature, the experimental system consists of three parts, namely the PC, the control unit and the measuring coil, and the maximum heating temperature is 700 °C. The coal and rock powder samples were conducted in a nitrogen atmosphere with a background magnetic field and frequency of 200 A/m and 976 Hz, respectively, and a gas flow rate of about 100 mL/min. Before the test, the prepared coal and rock samples were packed into quartz test tubes from sealed bags, and the sample loading height did not exceed 1 cm. Due to the lower ignition point of the coal sample than the rock sample, to ensure the safe completion of the experiment, the coal and rock samples were heated from room temperature to 350 °C and 700 °C in a nitrogen environment, respectively. During the experimental tests, the magnetic susceptibility data of the samples were continuously recorded from 25 °C to 350 °C and 700 °C, and then cooled to room temperature after heating to 350 °C and 700 °C, and the interval between the collected data was about 5 °C, and then the changes of the magnetic susceptibility of the samples during the heating and cooling processes were comparatively analyzed.

Since the sample will be affected by magnetic force, adjusting the current to change the magnetic induction strength, the gravitational change of the sample can be accurately measured by analytical balance, that is, the magnitude of the magnetic force. The formula for calculating the mass susceptibility is as follows:

$$\chi = \frac{2l\Delta mg}{MB^2} \tag{2.4}$$

In the formula, χ is the mass magnetic susceptibility of the sample, l is the height of the sample, Δm is the value of the change in the mass of the sample, g is the acceleration of gravity, M is the sample quality, B is the magnetic induction strength.

① Thermomagnetic curves of coal and rock at room temperature

The mass magnetic susceptibility of bituminous coal before heating is 4.17×10^{-8} m^3/kg. Bituminous coal contains a small amount of iron-containing oxides and a small amount of pyrite (FeS_2), and the iron-containing minerals exist in a relatively stable phase at room temperature, and their crystal structure and chemical composition determine that the coal sample has a certain initial magnetism (Fig. 2.20). This is mainly due to the fact that bituminous coal is susceptible to the thermal effect caused by temperature changes to change the magnetic moment arrangement inside the coal, resulting in different magnetic domain structure and magnetic domain wall behavior in bituminous coal. Since the substances in bituminous coal mainly include carbon, hydrogen, oxygen, nitrogen and sulfur and a small number of minerals, their magnetic characteristics mainly depend on the type and content of the minerals they contain. The amorphous carbon and microcrystalline graphite contained in coal are non-magnetic substances, and coal contains most of the organic substances, which are usually not magnetic.

The inverse relationship between the magnetic susceptibility and temperature of paramagnetic substances was determined by Curie law, indicating that the magnetic substances in coal were only reflected in a small part of the adsorbed paramagnetic

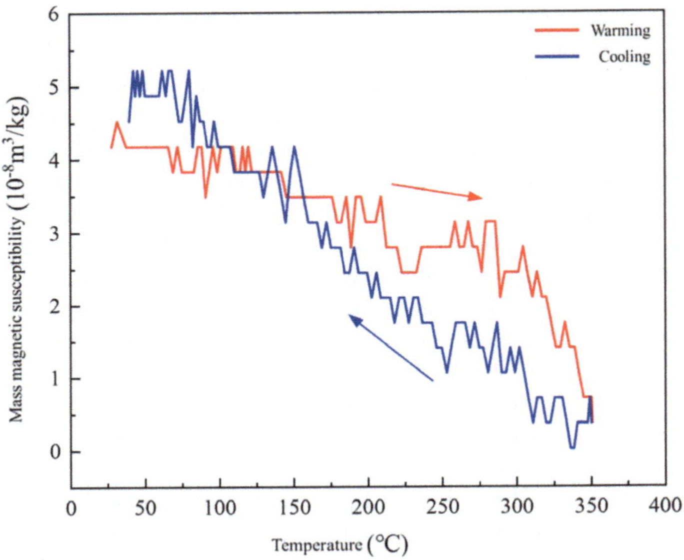

Fig. 2.20 Thermal magnetic curve of bituminous coal

metal ions, that is, the mass magnetic susceptibility of bituminous coal showed an overall downward trend during the heating process, and the mass magnetic susceptibility decreased to 0.35×10^{-8} m^3/kg, when the temperature rose to 350 °C and the mass magnetic susceptibility was reduced by 91.6%. During the cooling process, the thermal effect weakens, the disordered thermal motion of the magnetic domain decreases, and the magnetic domain states show a phenomenon of rearrangement and order, resulting in an increase in the mass magnetic susceptibility. The peak mass susceptibility during cooling was 5.21×10^{-8} m^3/kg, and the mass susceptibility was 4.52×10^{-8} m^3/kg when cooling to room temperature, and the mass susceptibility increased by 8.3% before and after heating, which was attributed to the change of the magnetic moment state in coal during the temperature rise and fall process. After cooling, the magnetic domains form a more orderly arrangement than the initial state, and the coal samples are more easily magnetized under the action of the external magnetic field, which is manifested as the increase of mass magnetic susceptibility.

Compared with the thermomagnetic curve of bituminous coal, the mass susceptibility of rocks shows unique characteristics with temperature during the heating process, among which the most significant features are the influence of blocking temperature, peak temperature and Curie point (Fig. 2.21).

Rock magnetism is mainly determined by the ferromagnetic mineral content in the rock. The mass susceptibility of common minerals such as quartz, calcite and potassium feldspar in rocks changed little during combustion, while ferromagnetic minerals had a greater influence on the mass susceptibility, and the valence state of iron had a significant effect on the mass susceptibility. The initial values of mass

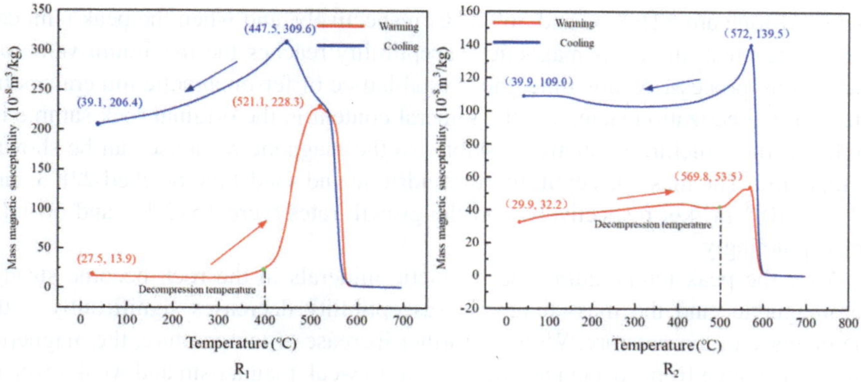

Fig. 2.21 Thermomagnetic curve of rock

magnetic susceptibility of mudstone and sandstone are 13.9 and 32.2 $\times$ 10^{-8} m^3/kg, respectively, due to the higher content of ferromagnetic minerals in sandstone compared with mudstone.

In the process of heating up of mudstone and sandstone, the mass magnetic susceptibility shows a specific critical temperature point with the change of temperature, that is, the blocking temperature, the blocking temperature of mudstone is about 400 °C, and the blocking temperature of sandstone is about 500 °C. Before the blocking temperature, the mass magnetic susceptibility of the rock sample remained basically stable, showing a low change rate, indicating that the mineral structure inside the rock was relatively intact and did not change significantly, and the minerals contained in the rock showed weak magnetism. The magnetic moment direction of the magnetic mineral particles in the rock is effectively blocked in a certain direction, and the magnetic minerals in the rock are in a stable magnetic state, which is constrained by various factors such as the crystal structure inside the rock, the boundary of particles and the interaction between particles, and the magnetic susceptibility of rock mass changes relatively steadily at this stage.

When the temperature gradually rises to the blocking temperature, the thermal effect overcomes the constraints, so that the magnetic domain wall can begin to move, the direction of the magnetic moment becomes easier to adjust, the weak magnetic minerals in the rock undergo phase change, the minerals begin to change from non-magnetic and weakly magnetic minerals to ferromagnetic and sub-ferromagnetic, forming more magnetic minerals make the mass magnetic susceptibility increase rapidly, and the magnetic susceptibility of the rock begins to increase with the increase of temperature, which is due to the lattice structure of ferromagnetic minerals begins to adjust. The magnetic moment arrangement tends to be ordered under the action of temperature and external magnetic field, and the magnetic moment direction inside the magnetic domain is highly consistent, showing a strong magnetic susceptibility, and the magnetic susceptibility increases significantly at this stage, reflecting the thermally driven magnetization effect in minerals. The peak temperatures of mudstone and sandstone are 521.1 °C and 569.8 °C, respectively, and when the peak temperature is reached, the mass magnetic susceptibility reaches the maximum value of the heating process. At this time, the crystal lattice of ferromagnetic minerals is in the best magnetization state, and the mineral content in the original rock sample is high and the structure is relatively stable, so the magnetic response can be shown at this time. The mass susceptibility of mudstone and sandstone reached 228.3 and 53.5 × 10^{-8} m^3/kg, respectively, and the growth rates were 1542.4% and 66.1%, correspondingly.

After the peak temperature, the magnetic minerals in the rock become strong paramagnetic, and the mass magnetic susceptibility decreases significantly with the increase of temperature. With the further increase of temperature, the magnetic minerals change from strong para-magnetism to weak magnetism and weak inverse magnetism, and the magnetic moment direction gradually becomes disordered due to thermal motion, the response to the external magnetic field weakens, and the mass susceptibility of rock gradually decreases. In the cooling process, the magnetic susceptibility of the rock did not completely return along the heating path, but showed a significant upward movement relative to the warming process. The mass susceptibility of the cooling process is higher than that of the heating process, because the thermal energy of the cooling process decreases, the ferromagnetism or ferromagnetism is re-established, and below the Curie temperature, the ferromagnetism and ferromagnetism will resume spontaneous magnetization, the mass susceptibility is enhanced, and the high magnetic response is retained. The magnetic moments that were originally out of order due to high temperature are rearranged in an orderly manner, forming a new magnetic domain structure, which enhances the remand magnetic effect of rock samples and shows higher mass magnetic susceptibility values. When cooled to room temperature, the mass susceptibility of mudstone and sandstone was 206.4 and 109.0 × 10^{-8} m^3/kg, correspondingly, and the mass susceptibility growth rates of mudstone and sandstone before and after heating were 1384.9% and 238.5%, respectively.

By comparing the thermomagnetic curves of coal samples and rock samples, the growth rate of coal samples is lower than that of rock samples regardless of initial mass susceptibility, peak mass susceptibility and mass susceptibility. The main difference between coal and rock is that the organic matter content in coal reaches 70 to 90%, while the organic matter content in rock is less, and organic matter is not a magnetic substance. The rock has a high mineral content, rich in transition metal elements such as iron and manganese, and the elements will undergo phase change during the heating process, producing strong magnetic substances. Compared with coal samples, rocks have a denser structure, and the magnetic interaction between particles is more likely to occur when heated, and it is easier to form new magnetic domains, which increases the mass magnetic susceptibility.

② Thermomagnetic curves of high-temperature heat-treated coal

As shown in Fig. 2.22, comparing the thermomagnetic curves of the coal samples before and after treatment, it is found that the overall trend of raw coal and coal samples after high temperature heat treatment is similar, and both show the decrease

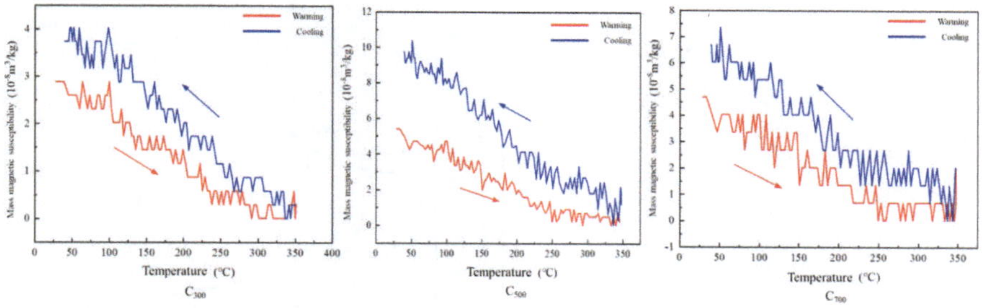

Fig. 2.22 Thermal magnetic curves of bituminous coal after different heat treatments

of mass susceptibility during heating and the increase of mass magnetic susceptibility during cooling. The magnetic domain structure in the coal sample is more active after heat treatment, and the thermal movement destroys the original magnetic domain wall and other structures of the coal sample, and the magnetic domain can be adjusted and optimized in a larger range during the cooling process. Therefore, the magnetic domain structure of the cooled coal sample is more orderly and stable than at the beginning of the experiment, the magnetic response is stronger, and the mass magnetic susceptibility is higher.

With the increase of heat treatment temperature, the blocking temperature of mudstone shows a trend of advance and then delay with the increase of heat treatment temperature, while the blocking temperature of sandstone advances with the increase of heat treatment temperature, which is due to the difference in the initial content of magnetic minerals in mudstone and sandstone, when heat treatment at 300 °C, some ferromagnetic and ferromagnetic minerals are generated in the rock samples, and when the heat treatment temperature reaches 500 °C, there are more paramagnetic minerals in mudstone than ferromagnetic and ferromagnetic, which delays the blocking temperature, as shown in Fig. 2.23.

After 700 °C heat treatment, the mass susceptibility of mudstone no longer peaks during the heating process, and continues to decrease with the increase of temperature, which is attributed to the fact that 700 °C exceeds the Curie temperature of the magnetic minerals in the mudstone, the original magnetic minerals are decomposed in large quantities, and the high temperature promotes the transformation of magnetic minerals into strong paramagnetic minerals or the formation of non-magnetic and weak inverse magnetic minerals. Therefore, the mass magnetic susceptibility decreases with the increase of temperature. According to the sandstone is stronger than the mudstone mineral thermal stability, 700 °C heat treatment is not enough to completely decompose the magnetic minerals, and still retains ferromagnetic and ferromagnetic, so there is still a heating peak in sandstone after 700 °C treatment.

When the heat treatment temperature is lower than the peak temperature, the crystalline water and organic matter in the rock sample are partially removed, and the non-magnetic and weakly magnetic minerals in the rock sample are transformed

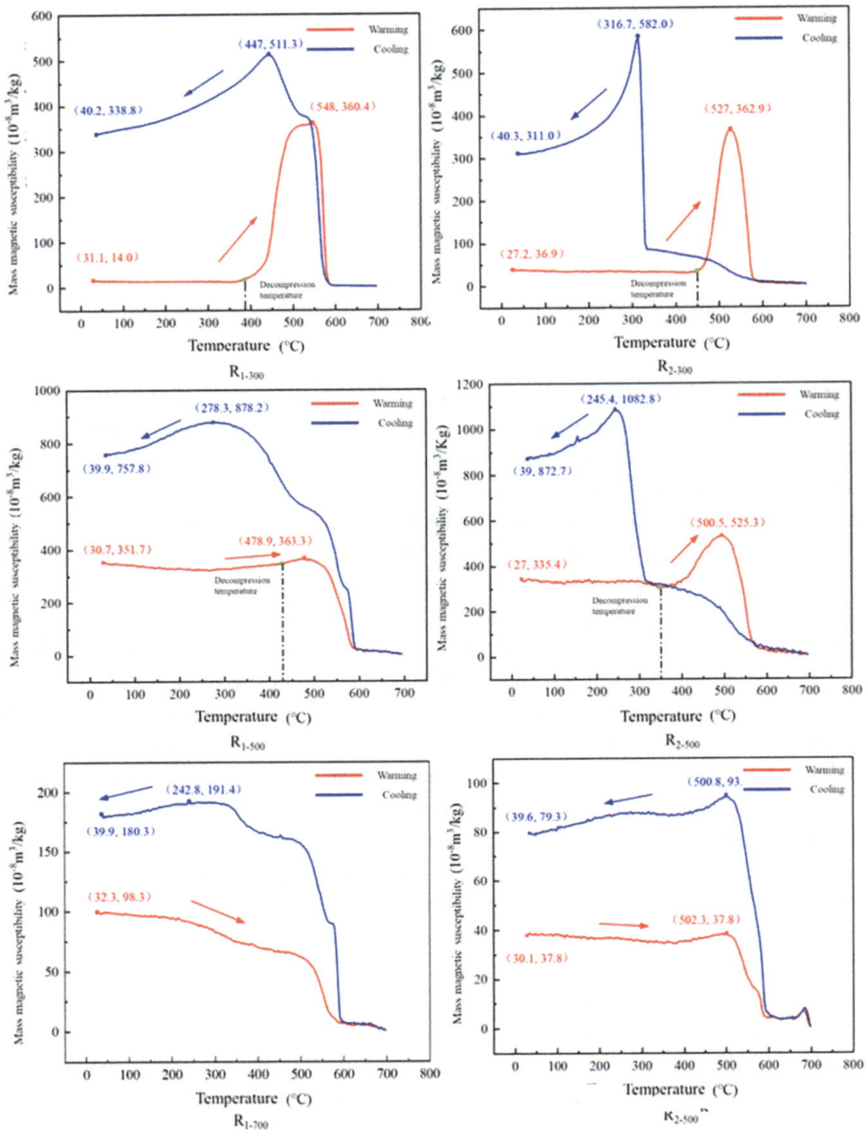

Fig. 2.23 Thermomagnetic curves of rock samples under different heat treatment conditions

into ferromagnetic and sub-ferromagnetic minerals with stronger magnetic properties. The magnetic susceptibility of the peak mass and the mass susceptibility after cooling both increased. When the heat treatment temperature is greater than the peak temperature, the magnetic minerals originally contained in the rock undergo phase change and decomposition, and transform into minerals with relatively weak magnetism, which shows a downward trend between the initial mass susceptibility

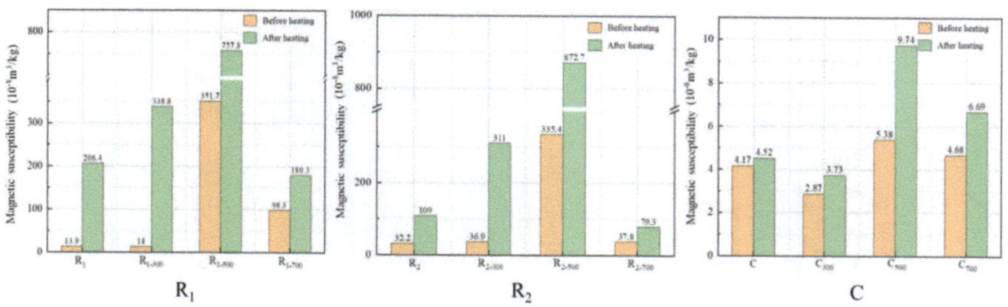

Fig. 2.24 Comparison of mass magnetic susceptibility of coal and rock before and after heating under different heat treatment conditions

and the mass susceptibility after cooling, and the peak mass susceptibility decreases or there is no peak. Under different heat treatment conditions, the initial mass susceptibility and peak mass susceptibility of mudstone and sandstone after 500 °C treatment were higher than those under other conditions, and the mudstone was 351.7, 363.3 and 757.8 × 10^{-8} m^3/kg, respectively, and 335.4, 525.3 and 872.7 × 10^{-8} m^3/kg, correspondingly. Considering the thermomagnetic curve and the change of magnetic parameters with heat treatment temperature, it is reflected that when the heat treatment temperature is less than the peak temperature, the residual magnetization strength, saturated magnetization strength and saturated isothermal remanence of rock samples show an increasing trend, while when the heat treatment temperature is greater than the Curie temperature, the residual magnetization strength, saturated magnetization strength and saturated isothermal remnant magnetism decrease.

From Figs. 2.24 and 2.25, it can be seen that the mass susceptibility of coal samples before heating under different heat treatment conditions is less than 6 × 10^{-8} m^3/kg, and the mass susceptibility values after heating are less than 10 × 10^{-8} m^3/kg, and the magnetic response is weak, while the maximum mass susceptibility of mudstone and sandstone before heating under different heat treatment conditions reaches 351.7 and 335.4 × 10^{-8} m^3/kg, respectively and the maximum reached 757.8 and 872.7 × 10^{-8} m^3/kg, correspondingly, which was significantly higher than that of coal samples. The mass susceptibility of coal samples before and after heating under different heat treatment conditions was lower than that of rock samples, which was mainly affected by the difference in magnetic mineral content of coal samples and the difference in magnetic mineral formation during heat treatment. The increase of mass susceptibility before and after heating is also less than that of rock samples, and the increase of mass susceptibility of coal samples before and after heating under different heat treatment conditions is less than 85%, while the increase of rock samples is above 80%, with a maximum increase of 2320.0%, which shows that rocks play a key role in the magnetic anomaly in the coalfield fire area, and their high magnetic response can be used as a magnetic detection signal.

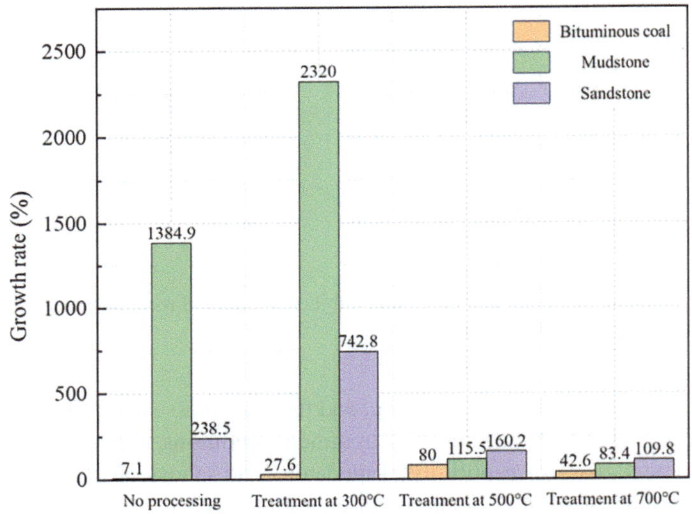

Fig. 2.25 Comparison of the increase of mass magnetic susceptibility before and after heating under different heat treatment conditions

References

1. Yang, Y. L. (2023). *Research on characteristics of multi-field coupled complex system and governance strategies for underground coal fires*. Master Thesis, Central South University, Changsha China (杨雅兰. 地下煤火多场耦合复杂系统特征及治理策略研究. 长沙:中南大学, 2023)
2. Ma, T., Zhai, X. W., Xiao, Y., Bai, Y. E., Shen, K., Song, B. B., et al. (2023). Study on the influence of key active groups on gas products in spontaneous combustion of coal. *Fuel, 344*, Article 128020.
3. Xiao, Y., Ren, S. J., Deng, J., & Shu, C. M. (2018). Comparative analysis of thermokinetic behavior and gaseous products between first and second coal spontaneous combustion. *Fuel, 227*, 325–333.
4. Su, H. T., Zhou, F. B., Li, J. S., & Qi, H. N. (2017). Effects of oxygen supply on low-temperature oxidation of coal: A case study of Jurassic coal in Yima, China. *Fuel, 202*, 446–454.
5. Ren, W.X., Guo, Q., Shi, J. T., Lu, W., & Sun, Y. (2021). Construction of early warning indicators for coal spontaneous combustion based on statistical characteristics of index gases. *Journal of China Coal Society, 46*(06), 1747–1758 (任万兴, 郭庆, 石晶泰, 陆伟, 孙勇. 基于标志气体统计学特征的煤自燃预警指标构建. 煤炭学报, 2021,46(06):1747–1758)
6. Zhang, D., Cen, X. X., Wang, W. F., Deng, J., Wen, H., & Xiao, Y. (2021). The graded warning method of coal spontaneous combustion in Tangjiahui Mine. *Fuel, 288*, Article 119635.
7. Lu, W., Wen, J. M., Wang, D., Zhang, W. R., Hu, X. M., Wu, H., et al. (2021). Study on gas transport characteristics of positive pressure beam tube system and its application to coal spontaneous combustion early warning. *Journal of Cleaner Production, 316*, Article 128342.
8. Guo, J., Wen, H., Zheng, X. Z., Liu, Y., & Cheng, X. J. (2019). A method for evaluating the spontaneous combustion of coal by monitoring various gases. *Process Safety and Environmental Protection, 126*, 223–231.
9. Wen, H., Yu, Z. J., Fan, S. X., Zhai, X. W., & Liu, W. Y. (2017). Prediction of spontaneous combustion potential of coal in the gob area using CO extreme concentration: a case study. *Combustion Science and Technology, 189*(10), 1713–1727.

10. Zhang, Y. T., Zhang, Y. B., Li, Y. Q., Li, Q. P., Zhang, J., & Yang, C. P. (2020). Study on the characteristics of coal spontaneous combustion during the development and decaying processes. *Process Safety and Environmental Protection, 138*, 9–17.
11. Zhou, X. X., Zhang, X. F., & Guo, Q. (2024). Classification of spontaneous combustion hazard levels of coal with different metamorphism degrees. *Combustion Science and Technology, 196*(10), 1459–1472.
12. Wang, C. P., Deng, Y., Zhang, Y. T., Xiao, Y., Deng, J., & Chi, M. S. (2021). Coal oxidation characteristics and index gases of spontaneous combustion during the heating and cooling processes. *Fuel, 307*, Article 121806.
13. Guo, X. Y., Deng, C. B., Zhang, X., & Wang, Y. S. (2019). Formation law of hydrocarbon index gases during coal spontaneous combustion in an oxygen-poor environment. *Energy Sources, Part A: Recovery, Utilization, and Environmental Effects, 41*(5), 626–635.
14. Zhao, J. Y., Zhang, Y. L., Song, J. J., Guo, T., Deng, J., & Shu, C. M. (2023). Oxygen distribution and gaseous products change of coal fire based upon the semi-enclosed experimental system. *Energy, 263*, Article 125721.
15. Zhang, G. S. (2020). *Ventilation safety*. China University of Mining and Technology Press (张国枢. 通风安全学. 中国矿业大学出版社, 2000)
16. Zhao, X. G., Dai, G. L., Qin, R. X., Zhou, L., Li, J. L., & Li, J. H. (2023). Study on oxidation kinetics of low-rank coal during the spontaneous combustion latency. *Fuel, 339*, Article 127441.
17. Chen, H., & Yang, Y. L. (2013). Research status of predicting coal spontaneous combustion. *Safety in Coal Mines, 44*(09), 194–197 (陈欢, 杨永亮. 煤自燃预测技术研究现状. 煤矿安全, 2013,44(09):194–197)
18. Yang, H. M. (2005). *Research of technology for mine fire prediction by using smell identification*. Master Thesis, Liaoning Technical University, Liaoning China (杨宏民. 煤矿火灾预测预报气味识别技术的研究. 辽宁工程技术大学, 2005)
19. Yang, H.M., & Luo, H. Z. (2007). Experimental research on identification of mine fire by using bionic smell sensors array. *Journal of China Coal Society*, (08), 822–826 (杨宏民, 罗海珠. 仿生气味传感器阵列判识煤矿火灾的实验研究. 煤炭学报, 2007(08): 822–826)
20. Wen, H., Wu, K., Ma, L., Wang, W. F., & Wang, T. (2014). Application of distributed optical fiber temperature measurement system in monitoring goaf coal spontaneous combustion. *Safety in Coal Mines, 45*(05), 100–102 (文虎, 吴慷, 马砺, 王伟峰, 王涛. 分布式光纤测温系统在采空区煤自燃监测中的应用. 煤矿安全, 2014,45(05):100–102)
21. Niu, B. S., Yan, H. W., Liu, P. J., Chen, Z. Y., Peng, C., Wang, X. T., et al. (2023). Experimental investigation on spontaneous combustion oxidation characteristics and stages of coal with different metamorphic degrees. *Environmental Science and Pollution Research, 30*(3), 8269–8279.
22. Li, J. H., Li, Z. H., Yang, Y. L., Duan, Y. J., Xu, J., & Guo, R. T. (2019). Examination of CO, CO_2 and active sites formation during isothermal pyrolysis of coal at low temperatures. *Energy, 185*, 28–38.
23. Ma, D., Qin, B. T., Zhong, X. X., Sheng, P., & Yin, C. G. (2024). Effect of flammable gases produced from spontaneous smoldering combustion of coal on methane explosion in coal mines. *Energy, 279*, Article 128125.
24. Zhu, H. Q., Sheng, K., Zhang, Y. L., Fang, S. H., & Wu, Y. L. (2018). The stage analysis and countermeasures of coal spontaneous combustion based on “five stages” division. *PLoS ONE, 13*(8), Article e202724.
25. Singh, A. K., Singh, R. V. K., Singh, M. P., Chandra, H., & Shukla, N. K. (2007). Mine fire gas indices and their application to Indian underground coal mine fires. *International Journal of Coal Geology, 69*(3), 192–204.
26. Wang, K., Li, Y., Zhang, Y. N., & Deng, J. (2023). An approach for evaluation of grading forecasting index of coal spontaneous combustion by temperature-programmed analysis. *Environmental Science and Pollution Research, 30*(2), 3970–3979.
27. Qi, Y. M. (1999). The modern development and application of fire prevention and control technologies in coal mines in China. *Coal*, (02), 4–6 (戚颖敏. 我国煤矿火灾防治技术的现代发展与应用. 煤, 1999(02):4–6)

28. Luo, H. Z., & Qian, G. Y. (2003). Research on the indicators of natural ignition indicator gases of various coal types. *Safety in Coal Mines*, (S1), 86–89 (罗海珠, 钱国胤. 各煤种自然发火标志气体指标研究. 煤矿安全,2003,(S1):86–89)
29. Morris, R. (1988). A new fire ratio for determining conditions in sealed areas. *Minerals Engineering (London), (United Kingdom), 147*(317)
30. Wang, F. S., Wang, J. T., Gu, L., Sun, P., Sun, C., & Wang, C. (2018). Study on multi-parameter index system for prediction and forecast of coal spontaneous combustion. *Journal of Safety Science and Technology, 14*(06), 45–51 (王福生, 王建涛, 顾亮, 孙鹏, 孙超, 王川. 煤自燃预测预报多参数指标体系研究. 中国安全生产科学技术, 2018,14(06):45–51)
31. Muller, S., Ryan, L., & Hollyer, J. (2017). *Review of oxygen deficiency requirements for graham's ratio*. Proceedings of the 17th Coal Operators' Conference
32. Liang, Y. T., Zhang, J., Wang, L. C., Luo, H. Z., & Ren, T. (2019). Forecasting spontaneous combustion of coal in underground coal mines by index gases: A review. *Journal of Loss Prevention in the Process Industries, 57*, 208–222.
33. Dong, X. W., Wen, Z. C., Wang, F. S., & Meng, Y. N. (2019). Law of gas production during coal heating oxidation. *International Journal of Mining Science and Technology, 29*(4), 617–620.
34. Zhang, Y.G., Qian, G. Y., & Tang, X. Y. (1995). Analysis of olefins indicators in coal spontaneous combustion and their petrological factors. *Journal of Henan Polytechnic University (Natural Science)*, (06):29–34 (张玉贵, 钱国胤, 唐修义. 煤自然发火烯烃指标及其煤岩学因素分析. 焦作矿业学院学报, 1995(06):29–34)
35. Liang, Y. T. (2008). Gas index method to predict coal spontaneous combustion. *Coal Science and Technology*, (06), 5–8 (梁运涛. 煤炭自然发火预测预报的气体指标法. 煤炭科学技术, 2008(06):5–8)
36. Wang, J. F., Zhang, Y. L., Xue, S., Wu, J. M., Tang, Y. B., & Chang, L. P. (2018). Assessment of spontaneous combustion status of coal based on relationships between oxygen consumption and gaseous product emissions. *Fuel Processing Technology, 179*, 60–71.
37. Xiao, Y., Ma, L., Deng, J., Wen, H., Wang, Z. P., & Mei, G. D. (2007). *Experimental study on correspondence relation of index gases of coal spontaneous combustion and characteristic temperatures*. International Academic Conference on Mining Science and Safety Technology
38. Xu, J. C. (2001). *Theory for determining hazardous areas of coal spontaneous combustion*. China Coal Industry Press (徐精彩. 煤自燃危险区域判定理论. 煤炭工业出版社, 2001)
39. Shu, X. Q. (1994). Thermal analysis research on spontaneous combustion of coal. *Coal Geology of China, 2*(6), 25–29 (舒新前. 煤炭自燃的热分析研究. 中国煤田地质, 1994, 2(6): 25–29)
40. Kök, M. V. (2005). Temperature-controlled combustion and kinetics of different coal samples. *Journal of Thermal Analysis and Calorimetry, 79*, 175–180.
41. Biagini, E., Barontini, F., & Tognotti, L. (2006). Devolatilization of biomass fuels and biomass components studied by TG/FTIR technique. *Industrial & Engineering Chemistry Research, 45*(13), 4486–4493.
42. Xiao, Y., Ma, L., Wang, Z. P., Deng, J., Wang, W., & Xiang, X. (2007). Research on characteristic temperature in coal spontaneous combustion with thermal gravity analysis method. *Coal Science and Technology*, (05), 73–76 (肖旸, 马砺, 王振平, 邓军, 王威, 向欣. 采用热重分析法研究煤自燃过程的特征温度. 煤炭科学技术, 2007,(05):73–76)
43. Zhu, H. Q., Zhao, H. R., Li, K., Wang, X. K., & Li, S. X. (2019). Study on characteristic parameters laws of gas coal spontaneous combustion. *Safety in Coal Mines, 50*(02), 19–24+29 (朱红青, 赵鸿儒, 李珂, 王晓宽, 李世祥. 气煤自燃特征参数规律研究. 煤矿安全, 2019,50(02):19–24+29)
44. Chen, X. K., Yi, X., & Deng, J. (2005). Experiment study of characteristic self-heating intensity of coal. *Journal of China Coal Society*, (05), 81–84 (陈晓坤, 易欣, 邓军. 煤特征放热强度的实验研究. 煤炭学报, 2005, (05):81–84)
45. Deng, J., Xiao, Y., Li, Q. W., Lu, J. H., & Wen, H. (2015). Experimental studies of spontaneous combustion and anaerobic cooling of coal. *Fuel, 157*, 261–269.
46. Xiao, B. Q., Wang, Q., Li, D. J., & Wu, S. Y. (2020). Spontaneous combustion tendency of coal using adiabatic spontaneous combustion test method. *Coal Quality Technology, 35*(02), 87–92 (肖兵球, 王芹, 李冬军, 吴抒轶. 采用绝热式自燃测试方法分析煤的自燃倾向.煤质技术, 2020, 35(02):87–92)

47. Yang, Y. L., Li, Z. H., Hou, S. S., Gu, F. J., Gao, S. Y., & Tang, Y. B. (2014). The shortest period of coal spontaneous combustion on the basis of oxidative heat release intensity. *International Journal of Mining Science and Technology, 24*(1), 99–103.
48. Zhou, X. H., Bai, G., Nie, R. S., Song, D. P., Sun, B. T., & Li, A. (2018). Experimental study on spontaneous combustion characteristic parameters of coal in different metamorphic grades. *Journal of Liaoning Technical University (Natural Science), 37*(01), 37–42 (周西华, 白刚, 聂荣山, 宋东平, 孙宝铁, 李昂. 不同变质程度煤自燃特征参数实验研究. 辽宁工程技术大学学报(自然科学版), 2018, 37(01):37–42)
49. Zhou, X. H., Niu, Y. P., & Bai, G. (2020). Experimental study on the influence of air supply on characteristics of lignite spontaneous combustion. *Mining Safety & Environmental Protection, 47*(01), 31–35 (周西华, 牛玉平, 白刚. 供风量对褐煤自燃特性参数影响的实验研究. 矿业安全与环保, 2020, 47(01): 31–35)
50. Cheng, G. Y., Zhou, Y. F., Cheng, Y., & Cao, J. (2016). Differential scanning calorimetry experiment and kinetics research for Jurassic coal in western Inner Mongolia. *China Coal, 42*(11), 91–95+121 (程根银, 周逸飞, 程宥, 曹健. 蒙西侏罗纪煤差示扫描量热试验及动力学研究. 中国煤炭, 2016, 42(11): 91–95+121)
51. Jia, T. G., Lou, H. Z., Liu, J., & Qu, G. N. (2020). Experimental study on thermal characteristics of spontaneous combustion process of coal with different moisture. *Journal of China Coal Society, 45*(S1), 346–352 (贾廷贵, 娄和壮, 刘剑, 曲国娜. 不同水分含量煤自燃过程热特性实验研究. 煤炭学报, 2020, 45(S1): 346–352)
52. Wang, Y. C., Yuan, Q., Xiao, Y., & Deng, J. (2018). Effect of moisture on exothermic characteristics of oxidation process of Baijiao anthracite. *Journal of Xi'an University of Science and Technology, 38*(05), 721–727 (王亚超, 袁泉, 肖旸, 邓军. 水分对白皎无烟煤氧化过程放热特性的影响. 西安科技大学学报, 2018, 38(05):721–727)
53. Deng, J., Zhao, J. Y., Zhang, Y. N., Wang, C. P., & Wang, K. (2016). Contrast experimental research on spontaneous combustion of oxidized-coals. *Journal of Xi'an University of Science and Technology, 6*(02), 157–162 (邓军, 赵婧昱, 张嬿妮, 王彩萍, 王凯. 煤样两次程序升温自燃特性对比实验研究. 西安科技大学学报, 2016, 36(02): 157–162)
54. Jin, Y. F., Li, H. T., Yue, T. F., & Yuan, L. (2014). Optimal selection for gas prediction index of coal spontaneous combustion under different conditions of oxygen concentration. *Safety in Coal Mines, 45*(11), 27–30 (金永飞, 李海涛, 岳宁芳, 袁林. 变氧浓度环境下煤自然发火气体预报指标优选. 煤矿安全, 2014, 45(11): 27–30)
55. Xu, Y. L., Song, Z. P., Wang, S. K., Jing, G. S., & Yu, M. G. (2016). Research on oxidation characteristics of coal at low temperature affected by oxygen-lean environment. *Journal of Safety Science and Technology, 12*(11), 82–87 (徐永亮, 宋志鹏, 王少坤, 荆国松, 余明高. 贫氧环境对煤低温氧化特性影响研究. 中国安全生产科学技术, 2016, 12(11):82–87)
56. Liu, F. F., & Li, H. T. (2017). Critical temperature of coal spontaneous combustion and influential factors under varying ambient oxygen. *Coal Technology, 36*(04), 165–168 (刘峰峰, 李海涛. 变氧浓度环境下煤自燃临界温度及影响因素分析. 煤炭技术, 2017, 36(04): 165–168)
57. Deng, J., Li, Q. W., Xiao, Y., & Wen, H. (2017). The effect of oxygen concentration on the non-isothermal combustion of coal. *Thermochimica Acta, 653*, 106–115.
58. Ren, L. F., Deng, J., Li, Q. W., Ma, L., Zou, L., Wang, B. L., et al. (2019). Low-temperature exothermic oxidation characteristics and spontaneous combustion risk of pulverised coal. *Fuel, 252*, 238–245.
59. Qi, G. S., Wang, D. M., Zheng, K. M., Xu, J., Qi, X. Y., & Zhong, X. X. (2015). Kinetics characteristics of coal low-temperature oxidation in oxygen-depleted air. *Journal of Loss Prevention in the Process Industries, 35*, 224–231.
60. Zhu, H. Q., Shen, J., Wang, H. Y., Chang, M. R., & Hu, R. L. (2014). Thermo gravimetry experiment study on relationship between oxygen concentration and characteristics of coal oxidation. *Safety in Coal Mines, 45*(05), 16–19 (朱红青, 沈静, 王海燕, 常明然, 胡瑞丽. 氧气浓度与煤氧化特征关系的热重实验研究. 煤矿安全, 2014, 45(05):16–19).
61. Li, H. J., Chi, H. Y., Hu, S., Song, G. X., Leong, H. N., Xu, K., et al. (2020). Effect of oxygen concentration on combustion behavior of single coal pellets from three different ranks in a concentrating photothermal reactor. *Fuel, 269*, Article 117372.

62. Li, Q. W., Ren, L. F., & Ren, S. J. (2020). Study on low-temperature oxidation characteristics of coal under oxygen deficient atmosphere. *Safety in Coal Mines*, *51*(11), 34–38 (李青蔚, 任立峰, 任帅京. 煤低温贫氧氧化放热特性研究. 煤矿安全, 2020, 51(11):34–38)
63. Deng, J., Ren, S. J., Xiao, Y., Li, Q. W., Shu, C. M., & Bai, G. Y. (2019). Thermophysical properties of coal during low temperature oxidation under different oxygen concentrations. *Thermochimica Acta, 676*, 186–197.
64. Xiao, Y., Yin, L., Deng, J., Wang, C. P., & Shu, C. M. (2019). Thermophysical parameters of coal with various levels of preoxidation. *Journal of Thermal Analysis and Calorimetry, 135*(5), 2819–2829.
65. Wang, X. T., Zhao, Y. T., Ma, D., Wang, P., & Zeng, J. F. (2023). Study on heat transfer characteristics of bituminous coal spontaneous combustion under variable oxygen volume fraction condition. *Safety in Coal Mines*, *54*(03), 123–130 (王新堂, 赵玉桃, 马德, 王鹏, 曾俊峰. 变氧体积分数条件下烟煤自燃传热特性研究. 煤矿安全, 2023, 54(03):123–130)
66. Zhang, X. H., Zhou, S. L., Tuo, L. L., & Hou, Q. Y. (2019). Transfer characteristics of coal under different preoxidation degree. *Journal of Xi'an University of Science and Technology*, *39*(05), 761–766 (张辛亥, 周山林, 拓龙龙, 侯钦元. 不同程度预氧化煤传热特性. 西安科技大学学报, 2019, 39(05): 761–766)
67. Xiao, Y., Chen, L. G., Li, Q. W., Zhang, X. Y., Bai, Z. J., & Wang, K. (2018). Identification and determination of the thermophysical properties of coal at lower temperature. *Journal of Safety and Environment*, *18*(06), 2190–2194 (肖旸, 陈龙刚, 李青蔚, 张馨悦, 白祖锦, 王凯. 低温条件下煤的热物性参数试验研究. 安全与环境学报, 2018,18(06):2190–2194)
68. Xu, S., Gao, W., & Chen, W. T. (2019). Uncertainty analysis of thermal properties of coal body by unsteady thermal conductivity product method. *Coal Quality Technology*, *34*(06), 30–34+38 (徐尚, 高伟, 陈文涛. 非稳态导热乘积法测量煤体热物性的不确定度分析. 煤质技术, 2019, 34(06):30–34+38)
69. Hui, S. (2019). *Adiabatic self-duration of coal studied by ISO-temperature-difference heating experience*. Master Thesis, Xi'an University of Science and Technology, Xi'an China (回硕. 基于等温差引领法的煤绝热自然发火期研究. 西安科技大学, 2019)
70. Zhang, J. Z., Li, H. B., Zhao, R. D., Wu, J. L., He, T., & Wang, Z. Q. (2019). Effect of coal particle size on its physical properties and combustion characteristics. *Thermal Power Generation*, *48*(11), 37–42 (张金芝, 李海宾, 赵瑞东, 武景丽, 何涛, 王志奇. 煤颗粒物性及热转化特性随粒径变化规律研究. 热力发电, 2019, 48(11):37–42)
71. Lu, J. H. (2016). *Research on Thermal physical parameters and destruction characteristics of coal and rock of coalfield fire area*. Master Thesis, Xi'an University of Science and Technology, Xi'an China (鲁军辉. 煤田火区煤岩体热物性参数及热破坏特性研究. 西安科技大学, 2016)
72. Ren, L. F., Wong, T. F., Tao, F., Li, Q. W., Ma, T., & Zhai, X. W. (2024). Study on changing law of thermophysical parameters in oxidation process of coal. *Coal Technology*, *43*(09), 123–127 (任立峰, 翁腾飞, 陶帆, 李青蔚, 马腾, 翟小伟. 煤氧化过程中热物性参数变化规律研究. 煤炭技术, 2024, 43(09): 123–127).
73. Qin, Z. (2018). *Study on the law and mechanism of high temperature magnetic anomalies in coal measures stratum*. Master Thesis, Xi'an University of Science and Technology, Xi'an China (秦政. 煤岩高温磁异常规律及机理研究. 西安科技大学, 2018)
74. Zhang, X. H., Qin, Z., Lu, M. M., & Bai, Z. J. (2018). Effect of temperature on magnetic properties of coal and rock. *Safety in Coal Mines*, *49*(12), 27–30 (张辛亥, 秦政, 卢苗苗, 白祖锦. 温度对煤岩磁性影响的实验研究. 煤矿安全, 2018, 49(12):27–30)
75. Li, J. W. (2020). *Research on Concealed fire detection of coal seam in NaLinMiao mine based on Magnetic property of coal and rock*. Master Thesis, Xi'an University of Science and Technology, Xi'an China (李经文. 基于煤岩磁性的纳林庙矿煤层隐蔽火灾探测研究. 西安科技大学, 2020)

Chapter 3
Overview of Coal Fire Detection Technologies

Coal fire, as a spontaneous combustion phenomenon in underground coal seams, not only causes huge waste of resources and economic losses, but also releases a large amount of toxic and harmful gases and greenhouse gases, seriously damaging the ecological environment and threatening the safety of mining areas. Effective detection, localization and monitoring of coal fire areas is the foundation and key to disaster prevention and implementation of fire extinguishing projects. Coal fire detection technology has developed from a single means to a comprehensive technical system of multidisciplinary interdisciplinary, multi-platform collaboration and multi-information integration, aiming to overcome the limitations of a single technology and improve the accuracy, timeliness and applicability of detection. This chapter will systematically expound the core content of the coal fire detection technology assembly, first outline the current mainstream and cutting-edge detection methods, and then deeply analyze the applicable conditions and selection basis of various technologies in complex practical application scenarios, so as to provide theoretical support and practical guidance for the scientific formulation of detection schemes.

3.1 Coal Fire Detection Methods

3.1.1 Surface Temperature Detection Method

In the process of coal fire combustion, the change of temperature field is the most intuitive, which can be divided into oxidative self-heating period, coal seam combustion period and extinguishing period. The core principle of coal fire temperature detection method is to use the heat released by coal fire combustion to cause significant temperature anomalies in the surrounding environment (surface, soil, rock formation or underground space), and capture this temperature difference through special equipment, and then locate the location, scope and development state of coal fire.

Y. Xiao et al., *Detection of Fire Source in Underground Coal Mine and its Applications*,
https://doi.org/10.1007/978-981-95-7064-5_3

At present, the temperature measurement technology used to detect the location of coal spontaneous ignition sources at home and abroad mainly includes temperature sensor method, infrared temperature measurement method and optical fiber method.

The temperature sensor method directly or indirectly senses the temperature anomaly generated by coal fire through various temperature sensors, and converts the physical quantity of temperature into quantifiable and transmittable electrical signals or optical signals, so as to achieve accurate detection of the location, range and temperature of coal fires. Its core logic is to use the sensitivity of sensors to temperature changes to capture the ambient temperature anomalies caused by the heat released by the oxidation of coal fire.

Infrared temperature measurement is a non-contact temperature measurement technique that works based on the principle that "any object with a temperature above absolute zero (−273.15 °C) continuously radiates infrared (infrared radiation) outward, generating a thermal image of the temperature distribution on the surface of the corresponding object [1]. By detecting the infrared energy radiated by an object, its surface temperature can be inverted. Because of its advantages of non-contact, fast and visualization, it is widely used in coal fire surface detection, large-scale census and dynamic monitoring.

Fiber optic temperature sensors (such as distributed fibers) use Raman scattering effect: when laser pulses propagate in optical fibers, they collide with fiber molecules to produce scattered light (including Rayleigh scattering, Raman scattering, etc.), in which the frequency of Raman scattered photons changes with temperature [2]. In coal fire detection, the fiber is buried on the surface, in a shallow hole or laid along a roadway, and the laser pulse is fired from one end of the fiber, and the detector at the other end receives the scattered light, which can be inverted at each point along the fiber by analyzing the intensity ratio of the scattered light from the Stokes to the Anti−Stokes.

3.1.2 Remote Sensing Detection Method

The application of remote sensing technology in the field of coal fire has a history of more than 50 years. After the coal fire slowly spontaneously ignites underground or burns under the action of external factors, a large amount of heat generated is transferred to the surface through heat conduction, heat convection and heat radiation propagation, forming thermal anomalies and thermal radiation anomalies. The remote sensing method is based on the thermal radiation and spectral characteristics of the fire area, and uses satellites, aircraft, and other remote sensing platforms to detect and extract the spectral anomaly information of the fire area on different remote sensing images, so as to delineate the scope of the fire area and study the dynamic development law of the fire area through regular observation [3]. According to different sensor platforms, remote sensing technology is divided into satellite remote sensing and UAV remote sensing Fig. 3.1.

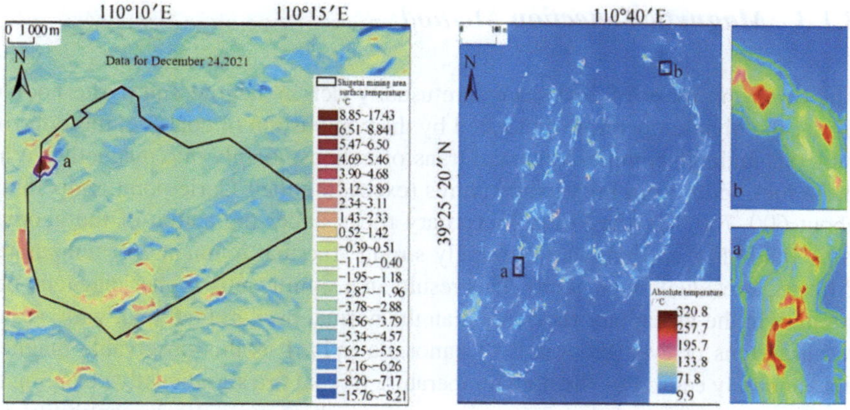

Fig. 3.1 Thermal infrared temperature inversion of satellite remote sensing and UAV remote sensing [4]

Satellite remote sensing detection is a comprehensive technology that uses artificial earth satellites to carry various sensors to receive and record electromagnetic radiation signals reflected or emitted by ground objects on the earth surface (including atmosphere, land, ocean, etc.) from space from a long distance, and extract ground object information after data processing and analysis. It is a core component of the modern earth observation system, with wide coverage, periodic observation, and non-contact detection. Dong et al. [5] used Landsat image data to invert the daytime surface temperature of the Daquanhu fire area of the Zhunnan coalfield, and analyzed the temporal and spatial variation characteristics of the temperature field in the fire area. Zhang et al. [6] used the Landsat series of thermal infrared images as the data source to extract the combustion range of underground coal fires in Wuda coalfield. Wang et al. [7] combined the surface deformation information obtained by the PSI method with the detected thermal anomalies for coal fire detection. Yu et al. [8] proposed a method for identifying coal fires that combines multi-temporal thermal infrared remote sensing surface temperature with multi-temporal interferometric synthetic aperture radar settlement information.

UAV remote sensing method is a technology that uses UAV as a flight platform, equipped with various remote sensing sensors, and extracts the features and spatial distribution information of features through data processing and analysis by obtaining electromagnetic wave information of features. It is mainly composed of three parts: UAV platform, remote sensing sensor, data processing and control system. The UAV is lightweight and convenient to carry around, making it convenient for staff to perform refined aerial photography only for specific areas and specific areas, and its multiple mode switching can obtain multi-source data, including optical images and thermal infrared images, which are widely used [9, 10].

3.1.3 Magnetic Detection Method

The overlying rocks of coal seams are usually rich in siderite and pyrite nodules. Under the high temperature generated by the combustion of coal fire, these ferrous minerals undergo complex chemical transformations, mainly into magnetite. When these rocks cool from high temperatures (exceeding their Curie point temperature, about 600–700 °C), they form incendiary rocks under the action of the geomagnetic field [11], and produce extremely strong "thermal remnant" (or temperature difference residual magnetism). This results in a significantly higher magnetic field strength in the incendiary rock area than the normal background field, with outliers reaching tens or even thousands of nanotesla (nT). It is this observable magnetic field anomaly caused by the high temperature of coal fire significantly changing the magnetic properties of the surrounding rock (such as magnetic susceptibility, and residual magnetization strength) that constitutes an extremely important geophysical feature of coalfield fires. By measuring this magnetic field change in the surface system and interpreting the collected data invertedly, the range of the fire zone from the combustion center to the peripheral extinguishing cooling zone can be effectively delineated, so as to reveal the spatial distribution characteristics of the underground fire area. Hopper [12] collected 99 samples of normal and burned sandstone, siltstone, and shale in a fire area in Wyoming, USA, and measured the magnetic susceptibility, and the results showed that there was a significant difference in the magnetic susceptibility between the burned rock and the normal rock, with a maximum difference of nearly three orders of magnitude. The magnetic susceptibility of most rocks is positively correlated with the magnetite content. Yu [13] used numerical methods to conduct forward simulation studies on magnetic anomalies based on gradient and multi-layer burned rock anomalies based on secondary and tertiary bodies. Sternberg and Lippincott [14] detected coalfield fire areas by magnetic method in three fire areas in North Dakota, USA, and observed a maximum magnetic anomaly of 2200 nT. At the same time, the magnetic properties of the rock samples were measured and found that the burnt rocks in the overfired area had high magnetic susceptibility and high residual magnetization strength, while the magnetic properties of the rock samples in the active fire area were smaller. Bandelow and Gielisch [15] pointed out that magnetic method is an effective method for detecting exerted rocks in coalfield fire areas. Hu et al. [16] proposed the ground-based "radon – magnetism" joint technology detection principle to accurately detect the location and range of small kiln fire sources, and used the ground-based "radon – magnetism" joint technology to process and analyze the collected data, and comprehensively delineated the location and range of hidden fire sources in small kilns. Based on decades of research on the magnetic anomalies of high-temperature rocks, magnetic detection technology has been widely and maturely used in the field of coalfield fire detection due to its effective capture of this critical physical response.

3.1.4 Electrical Detection Method

The high temperature generated by coal fire can not only change the temperature field and magnetic field of the fire area, but also the electric field of the fire area [16]. Based on the core principle that high temperature of coal fire significantly changes the electrical characteristics of coal and rock (resistivity reduction, polarization enhancement, and dielectric constant change), the electric method detection realizes the accurate identification and dynamic monitoring of underground fire areas by systematically observing the surface or underground electric field/electromagnetic field response. In 1964, Zhang [17] observed the natural potential anomaly in the Rujigou fire area in Ningxia, China and after analysis, he believed that the natural potential anomaly came from two aspects: one was the redox electric field formed by coal combustion. The second is the carbon cyanine matter formed after the high-temperature decomposition of coal. Duba [18] studied the conductivity of coal samples and their cracking products with temperature, and found that the resistivity of coal samples increased at low temperature and decreased sharply by about 5 orders of magnitude in the high temperature stage. Schaumann et al. [19] used aeromagnetic and ground transient electromagnetic methods to detect the Wuda fire area in Inner Mongolia, respectively, and revealed areas with low resistivity. Revil et al. [20] conducted a natural potential survey in the Lewis coalfield fire area in the United States and observed a natural potential anomaly with a maximum amplitude of − 50 mV. The high-density electrical method was also used to detect the coalfield fire area in Colorado, USA, and significant low-resistance anomalies were found in the geoelectric profile.

In coal fire detection, electrical methods are widely used, mainly including transient electromagnetic method and high-density electric method. The transient electromagnetic method (TEM) is a method of emitting a pulsed magnetic field underground using an ungrounded circuit or a grounding wire source, and using a coil or grounding electrode to observe the secondary eddy current field during the interval of a pulsed magnetic field [21]. The basic principle is that a transmitting coil with a certain current is set up on the ground or in the air, and an electromagnetic field is generated in the surrounding space, and an induced current is generated in the underground conductive rock ore body. By measuring the change of the secondary field of the medium with time after power failure, the apparent resistivity is obtained, and the water-rich areas, cracks, and fire areas of the underground coal seam are analyzed and located according to the apparent resistivity [23].

Based on the difference in resistivity of underground coal, the change of apparent resistivity of underground coal and rock section is calculated through the data collected by the measurement electrodes on each measurement line, and the apparent resistivity section map is obtained, and the burial depth, influence range, and development degree of underground coalfield fire area are determined after inversion [24–26]. Fig. 3.2

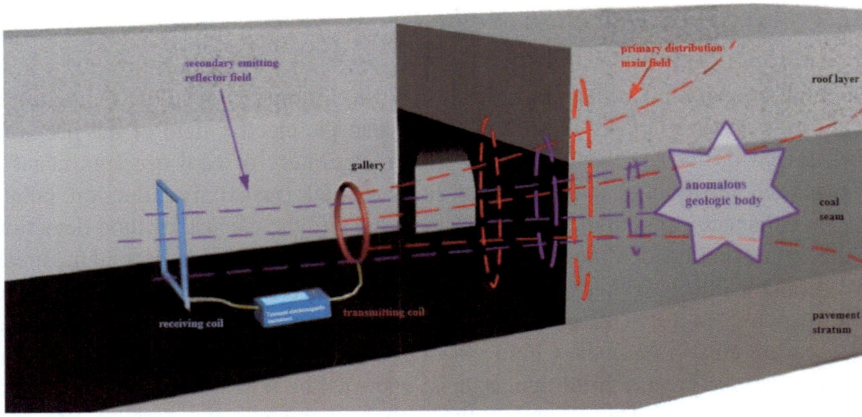

Fig. 3.2 Transient electromagnetic method working schematic diagram [27]

3.1.5 Characteristic of Gas Detection Method

In the process of combustion/smoldering, coal fire will release specific gas components, and its concentration change and component ratio directly reflect the state of the fire area.

During oxidative combustion of coal heating, a variety of complex reactions will occur in the internal structure, resulting in the formation of a series of index gases such as CO, C_2H_4, C_2H_6, and C_3H_8 [28, 29], which migrate along the cracks in the fire area to the surface under the impetus of fire wind pressure, forming a gas concentration field that diffuses from the combustion center of the fire area to the surrounding gradient. Coal will have different gas output at different stages of combustion, and the amount of gas produced has a certain relationship with the coal temperature. Xiao et al. [30] used adiabatic oxidation experiments to determine the correspondence between the Index gas and the characteristic temperature, and gave the variation characteristics of the gas at each stage. Guo et al. [31] used adiabatic oxidation to construct a prediction model of CO/CO_2, Graham coefficient, etc. and coal temperature. Ren et al. [32] obtained the gas volume fraction with coal temperature through the program heating experiment, obtained the constitutive equation of "gas– temperature", constructed a coal spontaneous combustion warning system based on gas statistical characteristics, and divided the coal spontaneous combustion risk stages.

The radon measurement method uses radon as the only gaseous daughter body produced by radioactive elements during the decay process, and the rapid migration rate of radon and its isotopes is fast due to the high temperature generated by underground fire areas. When oxidation heating or spontaneous combustion occurs in the underground coal seam, the precipitation rate of the natural radioactive radon in the surrounding and overlying rock layers increases, and the radioactive anomaly is formed due to the ion exchange action during radon decay to the surface and forms a radioactive anomaly, which can be detected as information reflecting temperature. Taiyuan University of Technology proposed a method based on surface radon detection to locate underground high-temperature heat sources based on the characteristic that radon precipitation increases with the increase of coal temperature [34]. Wen et al. [35] studied the precipitation law of radon during the spontaneous combustion of loose coal and found that the radon concentration increased linearly at low temperature and increased sharply in the range of 50–70 °C. As the temperature continued to rise, the radon concentration continued to increase and stabilized in the range of 100–120 °C.

3.1.6 Acoustic Detection Method

The principle of temperature measurement based on acoustics was proposed by Newton at the end of the seventeenth century, but due to the limitations of technology and productivity at that time, acoustic temperature measurement was not paid attention to and developed. It was not until the seventies and eighties of the last century that acoustic temperature measurement was widely studied and developed [36]. Among the acoustic temperature measurement techniques, the sound velocity method and the frequency method are the two most commonly used, and their cores are based on the correlation between acoustic wave characteristics and medium temperature. The principle of temperature measurement is to measure the sound velocity on the path by reasonably arranging the sound wave propagation path in the area to be measured, and finally reversing the temperature distribution of the area to be measured. Under the premise of obtaining the acoustic propagation distance and the time of acoustic wave flight, the mine acoustic temperature measurement technology obtains the temperature of the measured medium by arranging multiple pairs of acoustic wave transceivers to construct multiple acoustic wave propagation paths, and calculating the acoustic propagation velocity under different media through various noise suppression algorithms [38]. Li et al. [39] studied the propagation characteristics of acoustic waves in coal and rock masses, analyzed the effect of temperature on wave velocity, and concluded that temperature was negatively correlated with the wave velocity of longitudinal and transverse waves in coal bodies, laying the foundation for the popularization and application of acoustic temperature measurement technology in mines. Guo et al. [40] combined acoustic temperature measurement technology and coal spontaneous combustion theory to prospect the prospect of acoustic signals in the field of coal temperature detection, summarized and discussed the signal processing method and temperature field reconstruction feasibility of acoustic emission signals in coal and rock temperature sensing, and provided new technical ideas for the field of coal temperature inversion detection in coalmines. The main advantages of acoustic temperature measurement by acoustic detection method include wide measurement range, sensitive response, large measurement space, and continuous real-time temperature monitoring.

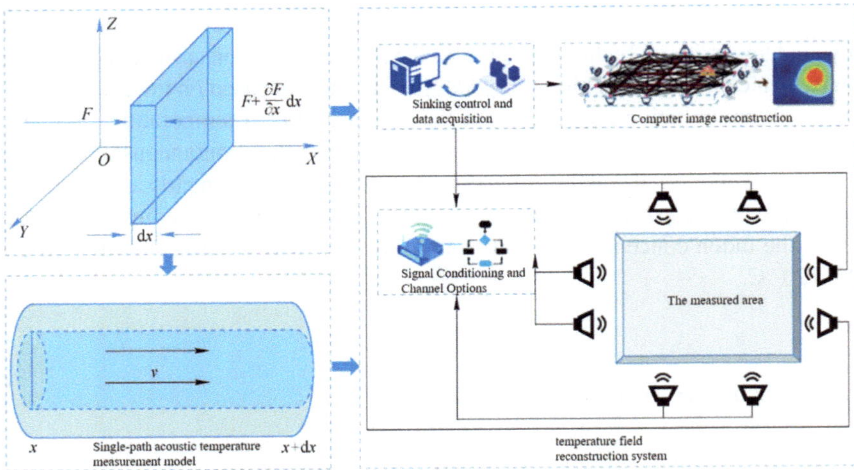

Fig. 3.3 Diagram of temperature measurement system with acoustic [41]

The principle of frequency method is to use the correlation between sound wave frequency (especially resonant frequency) and temperature. For example, the resonant frequency of sound waves in a resonant cavity depends on the speed of sound and the length of the cavity (fixed), and the speed of sound changes with temperature, so the resonant frequency indirectly reflects the temperature, or the temperature change of the medium causes its vibration frequency to shift, and the temperature is pushed back by the frequency offset Fig. 3.3.

3.1.7 Geological Radar Detection Method

Ground penetrating radar (GPR) is a modern non-destructive detection instrument with extremely wide practical applications. Geological radar is generally composed of radar host, radar antenna, transmission line and display, of which the radar antenna includes the transmitting antenna and the receiving antenna.

The working principle of geological radar non-destructive detection technology is Maxwell electromagnetic field theory. The geological radar emits electromagnetic waves through the transmitting antenna (T) to the ground, which is received by the receiving antenna (R). The greater the difference in the dielectric properties of the two substances, the stronger the electromagnetic wave reflection signal, and the combustion of coal fire will form two typical geological structures: cavity and fracture area. Shao et al. [43] used a combination of geological exploration and borehole temperature measurement to determine the scope and development degree of the Antaibao 9# coal fire area, and finally determined the distribution range and development degree of the fire area through the inversion of the temperature field. Yang [44] used geological radar to detect and study the anomalous geological structure in the Fenxi mining

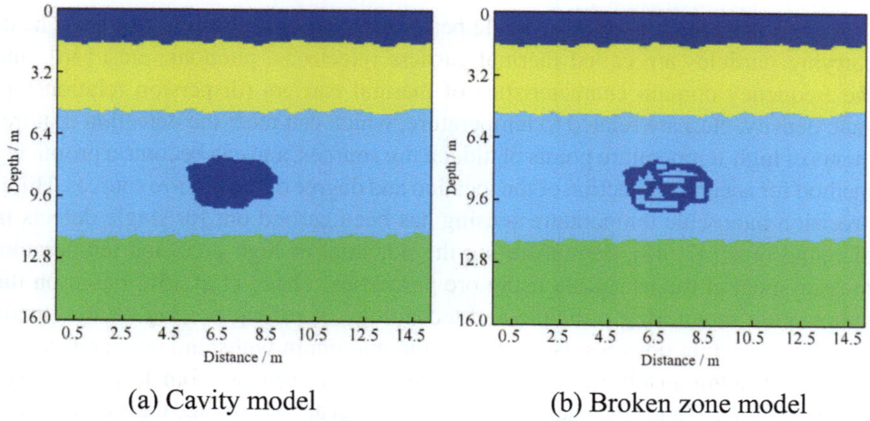

Fig. 3.4 Schematic diagram of typical geological structure formed by coal fire combustion[4]

area, Shanxi, China. Wang [45] applied geological radar to the goaf and the filling of the goaf to detect it.

Geological radar detection technology has the following advantages: (1) Non-destructive detection: electromagnetic wave carrier realizes non-intrusive detection without damaging the target medium. (2) Ultra-high resolution: high-frequency antennas (up to GHz level) support centimeter-level shallow recognition (such as cracks, voids). (3) Strong anti-interference: signal modulation technology and filtering algorithm effectively suppress environmental noise. (4) Efficient and lightweight: lightweight equipment (single person can be operated) + real-time acquisition (>50 lines/second), covering $\geq 1\ km^2$ in a single day. (5) Real-time imaging: the supporting software automatically completes gain adjustment, filtering and three-dimensional visualization, and outputs intuitive profiles on site. (6) Precise positioning: integrated high-precision positioning (RTK /differential GPS), the horizontal error is < 0.1 m, and the depth error is < 5%. (7) Cost-effective: the cost per point is lower than that of drilling/geophysical exploration in the well, and the maintenance is simple and the reuse rate is high Fig. 3.4.

3.1.8 Quantum Technology Detection Method

Quantum detection coal fire is a new generation of high-precision detection technology developed based on quantum physics principles (such as quantum sensing, quantum entanglement, and atomic magnetometer), which realizes early identification and localization by detecting changes in weak physical fields (such as magnetic field, electric field, and temperature field) caused by coal fire.

Heat conduction in all substances is the result of the interaction of microscopic particles carrying thermal energy inside the substance. Microscopic particles carry heat from one region to another while performing irregular motion, and these heat-carrying particles are called thermal carriers (electrons, phonons, etc.) [46], and the frequency domain characteristics of thermal carriers (dispersion relationship, state density, etc.) are related to temperature, which can meet the detection requirements of high-temperature points of hidden fire sources, and can become a promising method for accurate detection of the location and degree of hidden fire sources. High-precision nanoscale temperature sensing has been carried out for single defects in diamond ores [47, 48], demonstrating the potential of high-precision temperature measurement at the nanoscale using ore properties. Zhang et al. [50] based on the temperature monitoring principle and technology of rolling bearing rotating parts based on quantum dot sensors, improved the maximum withstand temperature and stability of cadmium telluride quantum dot temperature sensors from the perspectives of control time, ligand, and packaging process. Lin et al. [51] studied the measurement of the energy spectra and electron spin quantum states of diamond nitrogen-vacancy color centers to obtain information on physical quantities such as temperature and electromagnetic field in their vicinity.

In the field of mine ignition detection, the superposition and entanglement characteristics of quantum states can theoretically accurately capture small changes in abnormal temperature regions, providing a new means for detecting deeply hidden ignition sources Fig. 3.5.

3.2 Applicable Conditions of Coal Fire Detection Technology

3.2.1 Applicable Conditions of Temperature Detection Technology

The surface temperature measurement method has the best effect on the exploration of the coalfield fire area of shallow buried coal seams, and the heat generated by coal combustion will cause temperature abnormalities to the surface. When the coal fire enters the stage of open flame combustion or high-temperature oxidation (the surface temperature is usually $\geq$ 50 °C, or even hundreds of degrees Celsius), the temperature difference between the surrounding normal area (such as soil temperature 10–30 °C) is significant (usually $\geq$ 10 °C), and the temperature measurement method (such as infrared thermal imager, surface sensor) can clearly identify the high temperature anomaly area. Table 3.1 below shows the application of temperature detection methods in different combustion processes.

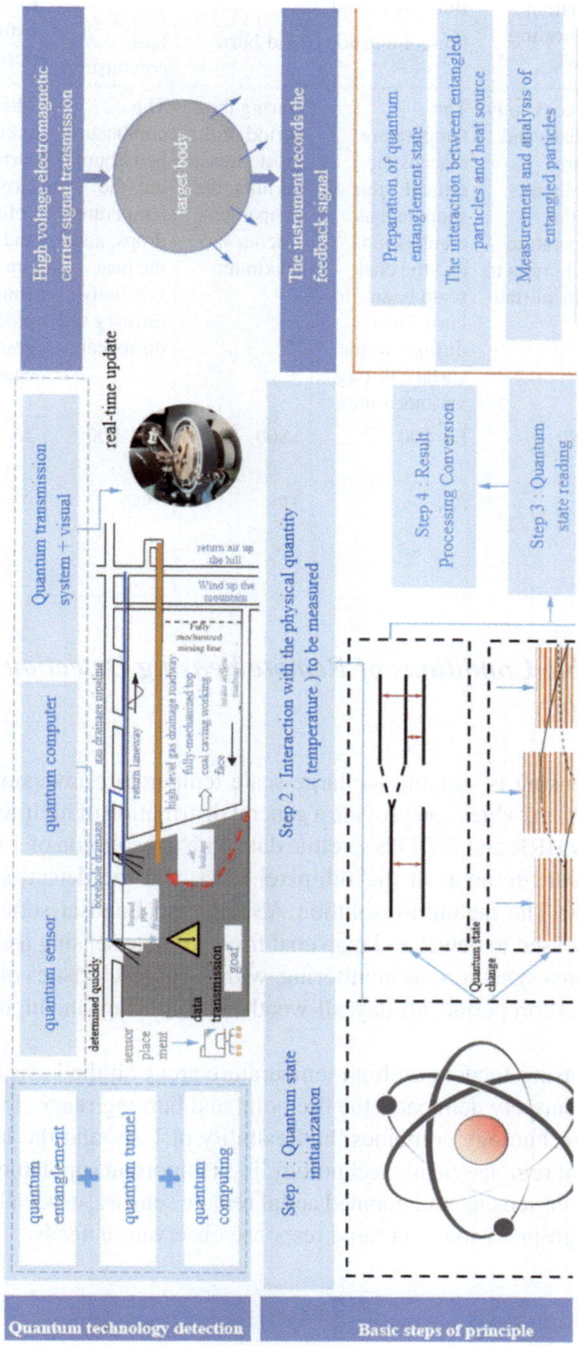

Fig. 3.5 Conceptual diagram of the ignition detection process of quantum technology[41]

Table 3.1 Applicability of different combustion processes

Combustion process	Oxidation self-heating period	Burning period			Combustion extinguishing period
		Pre-combustion	Mid-burn	Late combustion	
Peculiarity	The coal body oxidizes and absorbs endotherms, and the temperature slowly rises to a critical state	The temperature reaches the critical value of spontaneous combustion, and the coal seam begins to burn. Heat diffuses to the surface through various routes	During the period of the most intense burning, the temperature reaches its maximum	The combustion body burns out, the temperature drops, and the heat gradually diffuses and disappears	The coal seam is emptied, forming a large area of collapse area, and the coal fire combustion process is gradually ended
Temperature/°C	35–80	80–800	>800	100–800	35–100
Whether it can be detected by temperature measurement	No	Yes	Yes	Yes	No

3.2.2 *Applicable Conditions of Remote Sensing Detection Technology*

Satellite remote sensing is suitable for large-scale temperature inversion, with an accuracy of about 30 m, which can provide a general distribution of high-temperature areas. NOAA – AVHRR and MODIS satellite data with a resolution of 1 km require the detection of coal fire areas at the sub-pixel level, and the detection accuracy of coal fires is low. The medium-resolution ASTER and Landsat series of satellites are commonly used to detect and map coal fires. Remote sensing technology is suitable for large-area synchronous monitoring, with the characteristics of fast information acquisition, short period, all-day, all-weather monitoring, and high economic benefits.

UAV remote sensing focuses on high-temperature areas on the basis of satellite remote sensing, accurately delineates the fire point and burning range, UAV remote sensing detection technology combines the flexibility of UAV and the information acquisition ability of remote sensing technology, is an important supplement to traditional satellite remote sensing and manned aerial remote sensing, especially suitable for small-range, high-precision, and rapid response observation needs.

3.2.3 Applicable Conditions of Magnetic Detection Technology

The magnetic method [51] has the advantages of high positioning accuracy, fast measurement speed, high resolution, flexible and convenient measurement, and is not affected by rock cracks and tectonic changes. It is suitable for areas with a small detection range and high detection accuracy. The magnetic method is sensitive to magnetic interference and needs to be applied in an environment with few interference sources and stable magnetic field background: ① Less surface interference: avoid the presence of a large number of artificial magnetic objects (such as iron, cables, magnetic construction waste) or strong magnetic natural substances (such as magnetite debris, magnetic soil) on the surface, otherwise it will mask the real magnetic anomalies of the target body. ② Stable background of regional magnetic field: The strong background magnetic field fluctuations caused by regional tectonic activity and magmatic activity should be weak (or can be eliminated by data correction), otherwise it will interfere with the identification of target anomalies (for example, the background field needs to be stripped off first in strong regional magnetic anomalies).

3.2.4 Applicable Conditions of Electrical Detection Technology

The applicability of transient electromagnetic method (TEM) in coal fire detection depends on the geological background, fire area characteristics and technical parameter configuration, and its core advantage is the high sensitivity response to electrical anomalies, which is sensitive to low resistance bodies, and the effective detection depth can reach more than 300 m. The dehydration of coal-fired rocks and mineral transformation (pyrite → magnetite) form a low resistance body (resistivity can be reduced to 1–10 Ω·m), while the water-filled fracture or water-rich area further aggravates the low resistance anomaly, and the attenuation of the TEM response signal slows down, and the anomaly characteristics are obvious.

The high-density electrical method (ERT) is the preferred technology for detecting shallow-to-medium-deep coal fires (<100 m) due to the need for clear electrical differences in coal fire detection.

3.2.5 Applicable Conditions of Characteristic Gas Detection Technology

Coal releases recognizable signature gas during combustion, and the gas can be effectively transported to the detection point through the permeable channel, and the index gas detection technology is applicable to most mines. Similarly, when oxidation and heating or spontaneous combustion occur in underground coal seams, the precipitation rate of natural radioactive radon in the surrounding and overlying rock layers increases, and radioactive anomalies are formed due to the ion exchange action during radon decay, which is most suitable for large-scale detection and detection of ignition sources at a depth of 500–800 m [52].

3.2.6 Applicable Conditions of Acoustic Detection Technology

Based on the principle of acoustic temperature measurement, coal fire combustion needs to cause detectable changes in the physical properties of the surrounding medium, which is the premise of changing the sound velocity, the sound wave needs to be able to propagate from the transmitter to the coal fire area and return to the receiving end, and the signal strength meets the detection requirements, the propagation medium needs to have a certain continuity, the coal fire needs to have a certain spatial scale (horizontal range > several meters), if it is a small local combustion (such as $< 1\mathrm{m}^2$ point ignition source), the sound wave abnormality caused by it is easy to be masked by background noise and difficult to identify. In summary, acoustic temperature measurement is mainly suitable for medium and large coal fire areas, and the propagation medium in the goaf has a certain continuity.

3.2.7 Applicable Conditions of Geological Radar Detection Technology

Geological radar is mainly suitable for shallow (<20 m) dry and low-conductivity media environment (such as sandstone and limestone), and can identify combustion fracture zones (chaotic reflection), burned caverns (strong reflection flat layers) and grouting body distribution (high amplitude continuous reflection) in coal fire detection. Its centimeter-level resolution can accurately depict the shallow structure of the fire zone, but the deep fire zone (>30 m), water-rich clay layer or high-temperature ash area is not suitable due to signal attenuation or shielding effect, so it is necessary to combine magnetic method and transient electromagnetic method (TEM) to break through the depth limitation, and combine RTK positioning in complex terrain to ensure accuracy.

3.2.8 Applicable Conditions of Quantum Detection Technology

Quantum detection coal fire is suitable for three core scenarios: early weak signal recognition, complex environment monitoring, and high-precision positioning, especially in deep hidden fire areas, strong electromagnetic interference or high temperature environments. Its technical feasibility depends on equipment miniaturization, cost reduction and data processing capacity improvement, while practical application requires comprehensive decision-making based on geological conditions, economic costs and monitoring objectives.

References

1. 文虎, 马民, 费金彪. 基于红外成像技术的煤矿火灾治理. 煤炭科学技术, 2010, 38(01): 28−30. Wen, Hu., Ma, M., & Fei, J.B. (2010). Mine fire disaster control based on infrared imaging technology. *Coal Science and Technology*, 38(01): 28−30.
2. 冯玉祥, 汪雨辰, 童家欢, 吕立冬. 基于动态标定的拉曼分布式光纤测温系统研究. 物理学报, 2025, 74(07): 118−126. Feng, Y.X., Wang, Y.C., Tong, J.H., & Lv, L.D. (2025). Research on Raman distributed fiber temperature measurement system based on dynamic calibration. *Acta Physica Sinica*, 74(07): 118−126.
3. 邵振鲁, 王德明, 王雁鸣. 煤田火灾探测方法研究进展. 煤矿安全, 2012, 43(08): 189−192. Shao, Z.L., Wang, D.M., & Wang, Y.M. (2012). Research progress of coalfield fire detection method. *Safety in Coal Mines*, 43(08): 189−192.
4. 孙长斌. 无人机热红外遥感技术对矿区煤火探测识别研究. 中国煤炭, 2023, 49(S2): 85−94. Sun, C.B. (2023). Research on coal fire detection and identification in mining areas based on UAV thermal infrared remote sensing technology. *China Coal*,2023,49(S2): 85−94.
5. 董敬宜, 曾强, 李根生, 赵龙辉. 准南煤田大泉湖火区地表温度反演及时空变化特征.中国矿业, 2018, 27(01): 160−164. Dong. J.X., Zeng, Q., Li, G.S., & Zhao, L.H. (2018). Analysis of the chang characteristics of time and space and inversing of surface temperature of Daquanhu coal fire in south Junggar coalfield. *China Mining Magazine*, 27(01): 160−164.
6. 张志敏, 江利明, 柳林, 赵龙辉. 利用Landsat热红外影像探测地下煤火区范围——以乌达煤田为例. 测绘通报, 2018, (03): 93−97. Zhang, Z.M., Jiang, L.M., Liu, L., & Wang, H.S. (2018). Detecting the underground coal fire by using Landsat thermal infrared imagery-taking Wuda coalfield as an example. *Bulletin of Surveying and Mapping*, (03): 93−97.
7. Wang, T., Wang, Y., Zhao, F., Feng, H., Liu, J. L., & Zhang, L. X. (2022). Aspatio – temporal temperature–based thresholding algorithm for underground coal fire detection with satellite thermal infrared and radar remote sensing. *International Journal of Applied Earth Observation and Geoinformation, 110*, Article 102805.
8. Yu, B., She, J., Liu, G., Ma, D., Zhang, R., Zhou, Z., et al. (2022). Coal fire identification and state assessment by integrating multitemporal thermal infrared and InSAR remote sensing data: A case study of Midong District, Urumqi, China. *ISPRS Journal of Photogrammetry and Remote Sensing, 190*, 144–164.
9. Wang, Y., Tian, F., Huang, Y., Wang, J., & Wei, C.(2015).Monitoring coal fires in Datong coalfield using multi-source remote sensing data. *Transactions of Nonferrous Metals Society of China*, 25 (10): 3421−3428.
10. Li, F., Yang, W., Liu, X., Sun, G., & Liu, J. (2018). Using high-resolution UAV-borne thermal infrared imagery to detectcoal firesin Majiliang mine,Datong coalfield, Northern China. *Remote Sensing Letters*, 9(01): 71−80

11. Pal, S. K., Vaish, J., Kumar, S., & Bharti, A. (2016). Coal fire mapping of EastBasuria Colliery, Jharia coalfield using vertical derivative technique of magnetic data. *Journal of Earth System Science, 125*(01), 165–178.
12. Hooper, R. L. (1987). Factors affecting the magnetic susceptibility of baked rocks above a burned coal seam. *International Journal of Coal Geology, 9*(02), 157–169.
13. 于景村. 多层烧变岩磁异常正演模拟研究. 中国矿业大学学报, 1999, 28(04): 86−88. Yu, J.C. (1999). Forward modeling of magnetic anomaly of the multilayer burned coal tanks. *Journal of China University of Mining & Technology*, 28(04): 386−388.
14. Sternberg, R., & Lippincott, C. (2004). *Magnetic surveys over clinkers and coal seam fires in Western North Dakota.* Proceedings of the 2004 Denver Annual Meeting.
15. Bandelow, F.K., & Gielisch, H. (2004). *Modern exploration methods as key to fighting of uncontrolled coal fires in China.* Proceedings of the 2004 Denver Annual Meeting.
16. 胡相明, 洪海博, 董浩, 吴佰谦, 孙自超, 王伟, 等. "测氡-磁"法探测小窑自然发火技术的应用研究. 煤炭科学技术, 2025, 53(02): 200−210. Hu, X.M., Hong, H.B., Dong, H., Wu, B.Q., Sun, Z.C., & Wang ,W., et al (2025). Research on the application of "radon measurement-magnetic" method to detect spontaneous combustion technology in small coal mines. *Coal Science and Technology*, 53(02): 200−210.
17. 张秀山. 新疆煤田火烧区特征及灭火问题探讨. 中国煤田地质, 2004, 16(01): 18−21. Zhang, X.S. (2004). Probe into Xinjiang coalfield underground combustion area features and fire-fighting problems. *Coal Geology of China*, 16(01): 18−21.
18. Duba, A. G. (1977). Electrical conductivity of coal and coal char. *Fuel, 56*(04), 441–443.
19. Schaumann, G., Siemon, B., & Yu, C. (2008). Geophysical investigation of Wuda coal mining area, inner mongolia: Electromagnetics and magnetics for coal fire detection. *Spontaneous Coal Seam Fires: Mitigating a Global Disaster International Research for Sustainable Control and Management, ERSEC Ecological Book Series,· 04*, 336–350.
20. Revil, A., Karaoulis, M., Srivastava, S., & Byrdina, S. (2013). Thermoelectric self-potential and resistivity data localize the burning front of underground coal fires. *Geophysics, 78*(05), B259–B273.
21. 于景邨, 刘志新, 刘树才, 汤金云. 深部采场突水构造矿井瞬变电磁法探查理论及应用. 煤炭学报, 2007, (08):818−821. Yu, J.C., Liu, Z.X., Liu, S.X., & Tang, J.Y. (2007). Theoretical analysis of mine transient electrom agnetic method and its application in detecting water burst structures in deep coal stope. *Journal of China Coal Society*, (08): 818−821.
22. Pantzer, A., Vakahy, A., Eliyahou, Z., Levi, G., Horvitz, D., & Kohn, A. (2014). Dopant mapping in thin FIB prepared silicon samples by off-axis electron holography. *Ultramicroscopy, 138*(01), 36–45.
23. 李富, 廖国忠, 刘鑫明. 矿井瞬变电磁法在煤矿陷落区探测中的应用. 煤矿安全, 2013, 44(04):155–158. Li, F., Liao, G.Z., & Liu, X.M. (2013). Application of mine transient electromagnetic method in coal mine collapse area detection. *Safety in Coal Mines*, 44(04): 155−158.
24. 董浩斌, 王传雷. 高密度电法的发展与应用. 地学前缘, 2003, (01):171–176. Dong, H.B., & Wang, C.L. (2003). Development and application of ^{2}D resistivity imaging surveys. *Earth Science Frontiers*.(01): 171−176.
25. 邵振鲁, 王德明, 王雁鸣. 高密度电法探测煤火的模拟及应用研究. 采矿与安全工程学报, 2013, 30(03):468–474. Shao, Z.L., Wang, D.M., & Wang, Y.M. (2013). Simulation of high-density electrical method indetecting coal fires and its application. *Journal of Mining & Safety Engineering*, 30(03):468−474.
26. 宋吾军, 王雁鸣, 邵振鲁. 高密度电法与磁法探测煤田火区的数值模拟. 煤炭学报, 2016, 41(04):899–908. Song, W.J., Wang, Y.M., & Shao, Z.L. (2016). Numerical simulation of electrical resistance tomography method andmagnetic method in detecting coal fires. *Journal of China Coal Society*, 41(04): 899−908.
27. 辛邈. 煤自燃分段电学特性及瞬变电磁探测技术研究. 中国矿业大学(北京), 2017. Xin, M.(2017). *Segmented electrical characteristic of coal spontaneous combustion and transient electromagnetic detection technology on localization of combustion area in coal Mine.* Ph.D. dissertation, China University of Mining and Technology, Beijing China.

28. 陆伟. 煤自燃过程气态产物产生机理. 湖南科技大学学报(自然科学版), 2009, 24(02):10–14. Lu, W. (2009). Theory of gases generation during the process of spontaneous combustion of coal. *Journal of Hunan University of Science and Technology(Natural Science Edition)*, (02): 10−14.
29. 邓军, 李贝, 李珍宝, 张莹, 关欣杰. 预报煤自燃的气体指标优选试验研究. 煤炭科学技术, 2014, 42(01):55–59,79. Deng, J., Li, B., Li, Z.B., Zhang, Y., & Guan, X.J. (2014). Experiment study on gas indexes optimization for coal spontaneous combustion prediction. *Coal Science and Technology*, 42(01): 55−59, 79.
30. 肖旸, 王振平, 马砺, 翟小伟. 煤自燃指标气体与特征温度的对应关系研究. 煤炭科学技术, 2008, (06):47–51. Xiao, Y., Wang, Z.P., Ma, L., & Zhai, X.W. (2008). Research on correspondence relatinship between coal spontaneous cambustion index gas and feature tamperature. *Coal Science and Technology*, (06):47−51.
31. Guo, J., Wen, H., Zheng, X.Z., Liu, Y., & Cheng, X.J. (2019).A method for evaluating the spontaneous combustion of coal by monitoring various gases. *Process Safety and Environmental Protection*, (126): 223−231.
32. 任万兴, 郭庆, 石晶泰, 陆伟, 孙勇. 基于标志气体统计学特征的煤自燃预警指标构建.煤炭学报, 2021, 46(06):1747–1758. Ren, W.X., Guo, Q., Shi, J.T., Lu, W., & Sun, Y. (2021). Construction of early warning indicators for coal spontaneous combustion based on statistical characteristics of index gases. *Journal of China Coal Society*, 46(06):1747−1758.
33. 费金彪, 文虎, 金永飞. 测氡法在浅埋煤层火区探测中的应用. 西安科技大学学报, 2018, 38(01):26–30. Fei, J.B., Wen, H., & Jin, Y.F.(2018). Application of radon method in detection of fire area in shallow coal seam. *Journal of Xi'an University of Science and Technology*, 38(01): 26−30.
34. Hao, Y., & Wu, J.(2003). Study on mechanism of detecting underground fire by Radon measurement technique. *China Coal Society*. 28 (03), 260–263.
35. Wen, H., Cheng, X., Xu, Y., Ma, L., Guo, J., & Jiang, X. (2019). Law of radon precipitation and migration in loose coal during spontaneous combustion process. *China Coal Society., 44*(09), 2816–2823.
36. David, C. (1987). Acoustic pyrometry−a sound way to measure gas temperature. *Research Gate, 68*(07), 67.
37. 邓军, 屈高阳, 任帅京, 王彩萍, 王津睿, 赵小勇, 等. 松散煤体中声波传播特性及主要路径实验研究. 煤炭学报, 2023, 48(03):1238–1245. Deng, J., Qu, G.Y., Ren, S.J., Wang, C.P., Wang, J.R., & Zhao, X.Y., et al. (2023). Experimental study on acoustic wave propagation characteristics and main paths in loose coal. *Journal of China Coal Society*, 48 (03): 1238−1245.
38. 苗锦, 刘志强, 张跟鹏. 基于互相关的时延估计方法及其精度分析. 舰船电子工程, 2008, (06):98–100. Miao, J., Liu, Z.Q., & Zhang, G.P. (2008). Cross correlation method and precision analysis of time-delay estimation. *Ship Electronic Engineering*, (06) : 98−100.
39. 李祥春, 聂百胜, 杨春丽, 毛燕军, 王龙康, 孙琦, 等. 煤岩体声波波速随温度变化规律试验研究. 煤炭科学技术, 2016, 44(05):140–144. Li, X.C., Nie, B.S., Yang, C.L, Mao, Y.J., Wang, L.K., & Sun, Q., et al (2016). Experiment study on acoustic wave velocity law withtemperature variation in coal and rock mass. *Coal Science and Technology*, 44 (05) : 140−144.
40. 郭军, 王凯旋, 蔡国斌, 金彦, 刘荫. 声发射信号研究进展及其在煤温感知领域应用前景. 煤炭科学技术, 2022, 50(11):84–92. Guo, J., Wang, K.X., Cai, G.B., Jin, Y., & Liu, Y. (2022). Research progress of acoustie emission signal and its application prospect in coal temperature sensing field. *Coal Science and Technology*, 50 (11): 84−92.
41. 邓军, 王津睿, 任帅京, 何骞, 王彩萍, 屈高阳. 声波探测技术在矿井领域中的应用及展望. 煤田地质与勘探, 2023, 51(06):149–162. Deng, J., Wang, J.R., Ren, S.J., He, Q., Wang, C.P., & Qu, G.Y. (2023). Application and prospect of acoustic detection in the mining sector. *Coal Geology & Exploration*, 51(06):149−162.
42. 胡明顺, 潘冬明, 董守华, 李娟娟. 煤火区浅部松散煤体及采空区的探地雷达响应特征. 物探与化探, 2012, 36(04):598–601+606. Hu, M.S., Pan, D.M., Dong, S.H., & Li, J.J. (2012). Gpr response characteristics of shallow loose coal seam and goaf in the coalfire combustion area. *Geophysical & Geochemical Exploration*, 36(04):598−601+606.

43. 邵光强, 赵猛超, 曹凯. 火区自然发火范围及发展程度的判定方法. 煤矿安全, 2012, 43(10):181–184. Shao, G.Q., Zhao, M.C., & Cao, K. (2012). Judgment method of spontaneous combustion range and development degree of fire area. *Safety in Coal Mines*, 43(10):181−184.
44. 杨立彪. 运用矿井探地雷达探测汾西矿区的异常地质构造. 矿业安全与环保, 2012, 39(05):83–84+96. Yang, L.B.(2012). Detection of abnormal geological structures in Fenxi Mining Area by mine ground-scanning radar. *Mining Safety & Environmental Protection*, 39 (05):83−84+96.
45. 王雪涛. 地质雷达探测技术在乌东煤矿的应用实践. 煤炭科技, 2016, (04): 92−95. Wang, X.T. (2016). Practical application of ground penetrating radar technology in Wudong coal mine. *Coal Science & Technology Magazine*, (04): 92−95.
46. Lin, E.K., Lee, H.J., Lynn, G.W., Wu, W.L., & O'Neill, M.L. (2002). Structural characterization of a porous low-dielectric-constant thin film with a non-uniform depth profile. *Applied physics letters*, 81(04): 607−609.
47. Toyli, D. M., Delas, C. C. F., Christle, D. J., Dobrovitski, V. V., & Awschalom, D. D. (2013). Fluorescence thermometry enhanced by the quantum coherence of single spins in diamond. *Proceedings of the National Academy of Sciences, 110*(21), 8417–8421.
48. Neumann, P., Jakobi, I., Dolde, F., Burk, C., Reuter, R., & Waldherr, G. (2013). High-precision nanoscale temperature sensing using single defects in diamond. *Nano letters, 13*(06), 2738–2742.
49. 张盼,闫柯,朱永生,洪军,梁潘婷.面向旋转部件监测的耐高温量子点传感器研究.机械工程学报, 2020, 56(10):12–18+26. Zhang, P., Yan, K., Zhu, Y.S., Hong, J., & Liang, P.T. (2020). Research on the heat resistant quantum dots sensor for rotating components monitoring. *Journal of Mechanical Engineering*, 56(10): 12−18+26.
50. 林豪彬, 张少春, 董杨, 郑瑜, 陈向东, 孙方稳. 基于金刚石氮-空位色心的温度传感.物理学报, 2022, 71(06):59–68. Lin, H.B., Zhang, S.C., Dong, Y., Zheng, Y., Chen, X.D., & Sun, F.W. (2022). Temperature sensing with nitrogen vacancy center in diamond. *Acta Physica Sinica*, 71(06):59−68.
51. Zhang, B.F., Xiao, F., & Jin, W.B.(2023). Burnt coal field detection viamagnetic exploration. *Environmental Earth Sciences*, 82(07):160.
52. 张辛亥, 牛恒, 费金彪. 测氡法探测煤层自燃火源在巴依里矿的应用. 煤矿安全, 2009, 40(07): 29−30. Zhang, X.H., Niu, H., & Fei, J.B. (2009). Application of radon detection method for identifying spontaneous combustion sources in the Bayili mine. *Safety in Coal Mines*, 40(07): 29−30.

Chapter 4
Infrared Remote Sensing Detection Method

Based on the infrared radiation characteristics of objects, infrared remote sensing detection technology realizes target identification, monitoring and analysis by receiving and analyzing infrared band electromagnetic waves emitted or reflected by objects. Infrared remote sensing detection method realizes non-contact detection by receiving thermal radiation information of the target object. Its principle is based on the law of blackbody radiation, which uses infrared sensors to convert radiation energy into electrical signals, which can be divided into two detection methods: active and passive. This technology has the advantages of working around the clock, penetrating smoke, etc., but is disturbed by atmospheric absorption and ambient temperature. Typical equipment includes satellite payloads (such as MODIS), airborne thermal imagers, etc., and the core performance depends on temperature resolution vs. spatial resolution. Infrared remote sensing is widely used in coalfield fire area monitoring, military (night target recognition), environmental monitoring (forest fire warning) and fields, and the case shows that it is irreplaceable in high-precision thermal field inversion. In the future, with the improvement of detector sensitivity, infrared remote sensing will play a greater role in refined monitoring.

4.1 Principle of Infrared Remote Sensing Detection

4.1.1 Theoretical Basis of Infrared Rays

Infrared (IR) is an electromagnetic wave with a wavelength between visible light and microwaves, with a wavelength range of 0.78–1000 μm, and is named infrared because it is located outside the visible red light band. Its essence is invisible light radiated by objects due to thermal motion, usually produced by molecular vibrations, rotational energy level transitions, or accelerated movement of charged particles [1]. According to the classification of the International Illumination Commission (CIE)

Y. Xiao et al., *Detection of Fire Source in Underground Coal Mine and its Applications*,
https://doi.org/10.1007/978-981-95-7064-5_4

Table 4.1 Infrared wavelength range [1]

Sub-band	Wavelength range (μm)	Peculiarity
Near-infrared (NIR)	0.78–1.4	It is close to visible light and has significant reflection characteristics, and is often used in vegetation monitoring and night vision equipment
Shortwave infrared (SWIR)	1.4–3	It has both reflective and thermal radiation properties, and is used in mineral exploration and military reconnaissance
Medium wave infrared (MWIR)	3–8	The main radiation band for high-temperature targets (e.g., engine tailfrogs)
Long wave infrared (LWIR)	8–15	The main band of thermal radiation of objects at room temperature (human body, ground) is suitable for security and medical care
Far infrared (FIR)	15–1000	Low temperature radiation requires special detectors (e.g. terahertz technology)

and the International Organization for Standardization (ISO), the specific breakdown of infrared wavelengths is shown in Table 4.1.

4.1.2 Microscopic Generation Mechanism of Infrared Radiation

The essence of infrared radiation is the energy transition caused by the thermal movement of molecules. Molecules have two forms of motion: vibration (reciprocating motion of atoms in equilibrium positions) and rotation (rotation of molecules around the center of mass), and their energy is quantized (there are discrete energy levels). When molecules absorb external energy (such as thermal energy), they will transition from low energy levels to high energy levels. When transitioning from high energy level to low energy level, energy will be released, some of which will be radiated in the form of electromagnetic waves, and if the radiation wavelength is in the infrared band, it is infrared radiation.

Vibration transition: mainly corresponding to the mid-infrared band (2.5–25 μm), the chemical bonds of different molecules (such as C–H, O–H) have different vibration frequencies, and the infrared wavelength of radiation has "fingerprinting" (for example, the O–H bond of water has strong radiation peaks at 6.3 μm and 18 μm).

Rotational transition: corresponding to the far infrared band (>25 μm), the molecular rotational energy level difference is small, and the radiation energy is weak, mainly seen in gas molecules (such as CO_2, H_2O).

The infrared radiation of objects at room temperature is mainly vibration–rotational transitions driven by molecular thermal motion, so infrared radiation is also called "thermal radiation" [2].

4.1.3 Physical Basis of Thermal Infrared Temperature Inversion Algorithm

In nature, all objects exchange energy with other objects or environments through thermal radiation. When other heat transfer mechanisms such as conduction and convection are ignored, the evolution of the thermodynamic state of an object mainly depends on the net difference between its own radiation emission and external radiation absorption. If the radiation energy emitted by an object in a specific time is equal to the external radiation energy it absorbs, the object is in thermal equilibrium, and its thermal state can be uniquely characterized by temperature T. This law is strictly described by the law of thermodynamics, which provides a theoretical basis for infrared remote sensing detection.

In infrared remote sensing, the thermal radiation characteristics of the detection target directly reflect its temperature and material composition. The infrared radiation energy emitted by the target is received through the sensor and its temperature or thermodynamic parameters are inverted based on the laws of blackbody radiation (Planck law, Stephen–Boltzmann law, Wien law of displacement). When the target is in local thermal equilibrium, its radiation spectral distribution is consistent with the blackbody model, while the difference in emissivity (ε) is used to distinguish between different materials [3]. This method of analyzing the target state through radiation energy difference is the core principle of quantitative monitoring of infrared remote sensing.

(1) Planck law

Planck law, also known as the law of black body radiation, is used to describe the relationship between the radiation intensity and wavelength of a black body at any temperature. The energy (spectral radiant brightness) radiated from the unit surface area of the black body per unit of surface area (spectral radiant brightness) can be expressed as a function of temperature and wavelength. For a vacuum or a medium with a consistent refractive index, Planck law can be simplified to the following form:

$$M_\lambda(T) = 2\pi hc^2\lambda^{-5}\left[e^{(hc/\lambda kT)} - 1\right]^{-1} \tag{4.1}$$

where, h is Planck constant, the value is 6.626×10^{-34} J·s, k is the Boltzmann constant, the value is 6.626×10^{-23} J· K^{-1}, c is the speed of light, the value is 2.998×10^8 m·s^{-1}, λ is the wavelength (m), and T is the thermodynamic temperature K.

After the remote sensing image is calibrated by radiation, the radiation intensity of the object is obtained. Radiant emission is the electromagnetic wave power radiated in all directions per unit area of the surface of the blackbody:

$$M_\lambda(T) = \int_\Omega B_\lambda(T)n \cdot n_s d\Omega = \int_\Omega B_\lambda(T)\cos\theta d\Omega \tag{4.2}$$

For Lambert forms, $B_\lambda(T)$ has nothing to do with Ω, i.e.:

$$M_\lambda(T) = \int B_\lambda(T)\cos\theta d\Omega \tag{4.3}$$

Thereupon:

$$B_\lambda(T) = 2hc^2\lambda^{-5}\left[e^{(hc/\lambda kT)} - 1\right]^{-1} \tag{4.4}$$

From Planck law, the following three basic conclusions can be drawn: ① At each wavelength width, the radiation intensity increases with temperature. ② Correspondingly, the shorter the wavelength, the more energy is radiated. ③ The shorter the wavelength, the higher the radiation intensity until it reaches its peak. However, surface objects are not ideal black bodies, in fact, ideal black bodies do not exist, and the surface emissivity of physical objects is ε ratio of the radiation of the object to the radiation of the corresponding black body under the same conditions. Namely:

$$\varepsilon_j = \frac{L_j}{L_j^{BB}} \tag{4.5}$$

(2) Stefan–Boltzmann law

Simply known as Stephen Law, it is a famous law in the field of modern thermodynamics. In 1879, physicist Joseph Stephen summarized the law after years of sorting out and summarizing various experimental data. In 1884, physicist Ludwig Boltzmann started from the theory of thermodynamics and experimentally hypothesized that light energy (electromagnetic wave radiation) would be used instead of gas as the working medium of the electric heating engine, and finally successfully derived the same conclusion as Stephen induction.

The content of Stephen law is: the radiation degree j^* of a black body per unit time (proportional to the fourth power of its own thermodynamic temperature T, i.e.:

$$j^* = \varepsilon\sigma T^4 \tag{4.6}$$

Thereinto:

j^* is the radiation of the body, in W/m^2;

T is the thermodynamic temperature of the black body, and the unit is Kelvin (K);

ε is the emissivity of the object, and the emissivity of the black body is 1.

The coefficient σ is the Stefan constant, which is measured in $W/(m^2 \cdot K^4)$, which can be calculated by formula (4.6). From Eq. (4.5), if the temperature of a black body doubles, its radiation emission will increase by 16 times.

$$\sigma = \frac{2\pi^5 k^4}{15c^2h^3} = 5.67032 \times 10^{-8} \tag{4.7}$$

This law shows that the higher the thermodynamic temperature of an object, the stronger the electromagnetic wave energy it radiates. Affected by the emissivity, the radiation capacity is strongest when an object is a black body. Objects with the same temperature will have completely different radiation emission degrees due to different emissivity.

(3) Wien law of displacement

To find the partial differentiation of the wavelength of the Planck spectral distribution function and make it zero, we can obtain that the peak wavelength of the spectral radiation flux density of the black body is satisfied between the peak wavelength Ge and the absolute temperature r of the black body:

$$\lambda_m T = 2897.8\mu m \tag{4.8}$$

The peak wavelengths of spectral radiation are all in the infrared region within the temperature range that can actually be reached. For example, at 300K room temperature, the peak wavelength is 9–66 μm, so the 8–14 μm infrared band is sometimes called the thermal infrared band.

The spectral radiation flux density of the peak wavelength is proportional to the fifth power of the absolute temperature, i.e.:

$$W_{\lambda m} = bT^5 \tag{4.9}$$

where, b is $1.2862 \times 10^{-15}(W \cdot cm^{-2} \cdot Sr^{-1} \cdot \mu m^{-1} \cdot K^{-5})$

(4) Kirchhoff law

At a certain temperature, the ratio of the radiant emissivity F_X, r to its absorption rate A_X, r is a universal function *E* (λ, T) related to temperature and wavelength, independent of the properties of the object. i.e.

$$F_{\lambda,T}/A_{\lambda,T} = E(\lambda,T) \tag{4.10}$$

It shows that the ratio of the radiant emission F_λ, T and its absorption rate A_λ, T of any object is equal to the radiant emission of the black body *E* (λ, T) at the same temperature. For the actual object, the ratio of its radiation emissivity to the radiation emission of the black body at the same temperature is called the emissivity or specific emissivity of the object

$$\varepsilon_{\lambda,T} = F_{\lambda,T}/E(\lambda,T) \tag{4.11}$$

It can be seen that, $\varepsilon_{\lambda,T} = A_{\lambda,T}$ means that the specific emissivity of the object is equal to the absorption rate of the object.

4.1.4 Technical Classification

(1) Active and passive infrared remote sensing

Infrared remote sensing detection technology is divided into active infrared remote sensing and passive infrared remote sensing according to the detection principle, and the core difference between the two is whether they rely on artificial radiation sources for detection. Active IR remote sensing systems actively emit infrared radiation (such as lasers or modulated infrared light) to irradiate targets and receive their reflected or scattered signals, which is not limited by natural light conditions and can achieve all-weather detection, especially at night or in low-light environments [4]. Due to the need to carry radiation sources, active systems have higher power consumption and more complex structures, but their advantage lies in the ability to obtain multi-dimensional information such as distance and speed of targets, and typical applications include laser ranging (LiDAR), military target reconnaissance, and active infrared night vision.

In contrast, passive infrared remote sensing does not rely on artificial radiation sources but detects by receiving natural radiation emitted or reflected by the target itself. This method has a simple system structure and low power consumption, but the detection ability is greatly affected by the ambient temperature and radiation characteristics, especially at night, if the target temperature is close to the ambient temperature, the detection sensitivity will be significantly reduced. Passive infrared remote sensing is widely used in meteorological satellites (such as MODIS), surface temperature monitoring, forest fire early warning, and other fields, and its data is mainly used to invert the thermodynamic properties of targets rather than geometric information. The key comparison between the two is shown in Table 4.2.

(2) Typical differences in different infrared bands

The infrared band can be divided into shortwave infrared (SWIR, 1.4–3 μm), medium-wave infrared (MWIR, 3–8 μm) and long-wave infrared (LWIR, 8–15 μm) according to the wavelength, and the detection capabilities and application scenarios of each band are different due to their interaction with the atmosphere and target radiation characteristics [5]. The shortwave infrared band has both reflection and thermal radiation characteristics, can penetrate mist, and is suitable for high-resolution imaging and reflectance spectroscopy, but its signal is susceptible to atmospheric water vapor absorption. This band is particularly important in mineral exploration, such as identifying mineral characteristic spectra through HyMap satellite data, while healthy plants have high reflectivity in the 1.55 μm and 2.1 μm bands in vegetation health monitoring, which can be used to assess vegetation growth status. In addition, shortwave infrared is also used in the field of military night vision enhancement, usually combined with laser illumination technology to improve imaging quality.

Table 4.2 Key comparison on the two methods of detection [12]

Characteristic	Proactive	Passive
Radiation source	Artificial launch	Natural radiation (sun/features)
Diurnal ability	Valid around the clock	Nighttime dependent on target temperature
Data dimensions	Distance and speed information can be obtained	Radiation intensity/temperature information only

The medium-wave infrared band is extremely sensitive to the radiation of high-temperature targets (such as combustion flames or engine tail flames), and its atmospheric window transmittance of 3–5 μm is high, which is suitable for high-temperature detection. In the missile warning system, DSP satellites use the medium-wave infrared band to capture the tail flame radiation of rocket engines to achieve early warning [6, 7]. In the industrial field, medium-wave infrared thermal imaging is used to detect equipment overheating faults, such as abnormal hot spots in power equipment. Forest fire monitoring also relies on this band, because the radiation intensity of the fire point in the 3–5 μm band is much higher than that of normal temperature features, which can effectively identify fire hot spots.

The long-wave infrared band mainly receives thermal radiation from objects at room temperature (such as the human body, surface, and buildings), and its atmospheric window transmittance of 8–14 μm is excellent, and it is less interfered with by the atmosphere. This band is widely used in surface temperature inversion, such as the TIRS band of the Landsat satellite, which calculates land and ocean temperatures from long-wave infrared data. In the field of security surveillance, passive long-wave infrared thermal imaging can detect human or vehicle in a completely lightless environment, and is often used for border patrols and night surveillance. Medical diagnostics also use long-wave infrared imaging to detect abnormal body temperature, such as the use of thermal imaging cameras in fever screening. The choice of different bands depends on the target temperature and detection needs, with medium-wave infrared preferential use for high-temperature targets, long-wave infrared for room-temperature features, and short-wave infrared data for tasks requiring reflection characteristic analysis. Multi-band collaborative detection (such as MODIS satellites equipped with shortwave, medium wave and longwave infrared bands at the same time) can further improve the comprehensive application capability of remote sensing data.

(3) Imaging and non-imaging techniques

According to the sensor type, it is divided into imaging infrared remote sensing technology and non-imaging infrared remote sensing technology. Imaging infrared remote sensing technology outputs a two-dimensional infrared image of the target, reflecting the radiation intensity or temperature distribution of the target through grayscale differences [8]. Its equipment has infrared cameras, such as uncooled microbolometer cameras, for security and night driving. Scanning imagers, satellite-mounted infrared scanning radiometers, and Landsat TIRS sensors to obtain images of a wide range of surface temperatures. Hyperspectral imagers, such as airborne mid-infrared hyperspectrometers, simultaneously acquire spatial images, and spectral information for mineral identification.

Non-imaging infrared remote sensing technology. It does not generate images, but only quantitatively measures the infrared radiation parameters of the target (such as radiation intensity, and spectral characteristics). Its equipment has an infrared radiometer, which measures the radiation brightness of a specific wavelength band and is used for atmospheric composition analysis. Infrared spectrometer to obtain the infrared absorption/emission spectra of the target, such as detecting CO_2 and SO_2

concentrations in industrial exhaust gases. Infrared thermometer, a single point to measure the target temperature, such as a portable infrared thermometer.

4.2 Characteristics of Infrared Remote Sensing Detection

4.2.1 Advantages of Infrared Remote Sensing Detection

One of the most significant advantages of infrared remote sensing technology is its ability to work around the clock. Unlike visible light remote sensing, infrared systems do not rely on external light conditions but achieve observation by detecting the thermal radiation of the target itself. This feature allows it to work continuously at night, during dusk and even polar night environments. The detection process is shown in Fig. 4.1.

The infrared sensors carried by meteorological satellites can monitor global cloud maps and sea surface temperature changes 24 h a day, while traditional optical satellites cannot obtain effective data in the absence of light. The infrared band (especially the long-wave infrared of 8–14 μm) is less affected by atmospheric scattering and can still maintain observation ability under weather conditions such as thin clouds and light fog, while visible light imaging will cause severe image quality degradation due to strong Rayleigh scattering in these scenarios.

When it comes to penetrating smoke and dust, infrared remote sensing shows unique physical advantages. Because the wavelength (micron range) of infrared radiation is much larger than the diameter of smoke particles (0.1–10 μm), it is significantly less affected by Mie scattering than visible light. Actual observations show that at the scene of forest fires, infrared sensors can penetrate a 10-km-thick smoke layer to accurately locate the location of the fire point, while visible light cameras can only identify a range of less than 1 km under the same conditions [9–11]. This

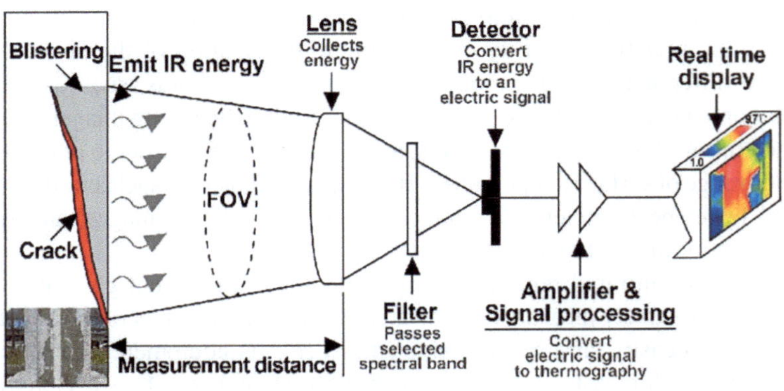

Fig. 4.1 Ignition source detection diagram of infrared remote sensing technology [16]

feature makes infrared technology the preferred solution for special scenarios such as volcanic ash monitoring and battlefield environmental reconnaissance. The U.S. Geological Survey used the thermal infrared band of the ASTER sensor to successfully map the temperature field of lava flows during volcanic eruptions, while the optical images during the same period were completely invalid due to volcanic ash occlusion.

Passive infrared remote sensing does not require external light sources to bring multiple technical advantages. By receiving the infrared energy radiated by the target, the system not only avoids the energy consumption problem of active lighting systems (such as lidar) but also ensures the concealment of observation—a feature that is particularly important in military reconnaissance and wildlife monitoring. Any object with a temperature above absolute zero spontaneously radiates infrared light, which allows the infrared system to directly reflect the true thermal state of the target.

4.2.2 Technical Limitations of Infrared Remote Sensing

(1) Atmospheric absorption effect (water vapor/CO_2)

The most important limitation faced by infrared remote sensing is the problem of selective absorption by the atmosphere. Gas molecules such as water vapor (H_2O), carbon dioxide (CO_2), and methane (CH_4) in the earth atmosphere have a strong absorption effect on specific infrared bands, forming several "atmospheric absorption zones" [12]. Among them: water vapor produced strong absorption in the 2.5–3 μm, 5–7 μm and > 18 μm bands. CO_2 has significant absorption peaks around 4.3 μm and 15 μm, and ozone forms an absorption band at 9.6 μm. These absorption effects cause severe attenuation of the surface radiation signal received by the satellite. Taking medium-wave infrared (3–5 μm) as an example, when the atmospheric water vapor content reaches 20 g/cm2, the surface radiation energy received by the sensor is only 30%–40% of the actual emission [13–15].

(2) Trade-off between resolution and temperature sensitivity

Infrared sensors face a physical trade-off between spatial resolution (IFOV) and temperature resolution (NETD): the improvement of spatial resolution requires the reduction of the instantaneous field of view, resulting in a decrease in the radiation energy received by a single pixel. The improvement of temperature sensitivity requires increasing the photosensitive area or integration time of the detector. Infrared wavelengths (3–12 μm) are 6–24 times higher than visible light (0–5 μm), with significantly reduced resolution for the same aperture size [15].

(3) Ambient temperature interference problem

Infrared remote sensing measurements are highly susceptible to interference from changes in ambient temperature, which are mainly reflected in three levels [16]:

(1) Background radiation interference

All objects with temperatures above absolute zero continuously radiate infrared energy, creating a complex background noise field. In a typical scenario, the difference in emissivity between surface vegetation (emissivity $\varepsilon \approx 0.92$–0.97) and artificial buildings ($\varepsilon \approx 0.6$–0.9) will lead to measurement bias. More seriously, the sky radiance can contribute 15–40% of the total signal in the 8–14 μm band, and its intensity fluctuates with changes in atmospheric temperature and humidity. For example, when the near-surface temperature rises from 20 °C to 30 °C, the downward radiation of the atmosphere increases the radiation brightness received by the sensor by about 3.2 W/(m2·Sr·μm), which is equivalent to the overestimation of the surface target apparent temperature by 4–6 K [17–20].

(2) Instrument thermal disturbance

The thermal stability of the infrared sensor itself directly affects the measurement accuracy. Joule heat when the detector is working can cause a temperature rise of 3–8 °C, increasing the dark current by an order of magnitude. Optical systems can produce non-uniform thermal deformation under sunlight conditions. The US LandSat-9 mission has recorded that the reference temperature of the TIRS-2 sensor fluctuates by 0–8 K within 30 min when the satellite enters the sunshine area, and it needs to be stabilized within ± 0.05 K through real-time thermoelectric cooling (TEC) [21].

(3) Cross-radiation coupling

When observing a high-temperature target, its intense radiation produces stray light within the optical system. The test data showed that the lens surface scattering caused the apparent temperature of adjacent pixels to increase by 2–8 °C when imaging the target at 500 °C [22]. The military infrared seeker found in laboratory tests that when there is an 800 K hotspot in the field of view, the temperature measurement error in the surrounding 10pixel areas reaches ± 7 K.

4.2.3 Applicability of Infrared Remote Sensing with Other Remote Sensing Methods

(1) Infrared and visible light remote sensing

As the two main technical means of optical remote sensing, infrared and visible light remote sensing have significant differences in detection mechanism and application scenarios. Infrared remote sensing detects by receiving the thermal radiation of the target itself, and its operating band (0.78–1000 μm) determines its unique observation advantages. Visible light remote sensing (0.38–0.78 μm) relies on the reflection characteristics of sunlight, which is in stark contrast in the observation method [23]. These two technologies present obvious complementary characteristics in practical applications. Infrared breaks through the light limit to achieve all-day observation, while visible light provides morphological information that is more in line with human vision. Comparison of typical application scenarios is shown in Table 4.3.

Table 4.3 Comparison of typical application scenarios [18]

Application requirements	Infrared remote sensing applicability	Visible light remote sensing limitations
Surface temperature inversion	Direct measurements (e.g., geothermal anomaly monitoring)	Completely unattainable
Night target identification	By thermal radiation imaging (military, wildlife)	Additional lighting is required
Combustion source detection	Can penetrate smoke to identify fire points (3–5 μm band)	Fails under smoke cover
Vegetation health assessment	Combined heat stress index (CWSI)	Only the apparent form can be obtained

(2) Infrared and microwave remote sensing

Infrared and microwave remote sensing represent two types of detection technologies with significant differences in the electromagnetic spectrum, which form a distinct complement in terms of physical mechanism and application scenarios. Infrared remote sensing (3–14 μm) is based on the thermal radiation characteristics of the target and has sub-Kelvin temperature sensitivity, Microwave remote sensing (1mm–1m) detects by measuring the dielectric characteristics and roughness of the target, showing a unique all-weather observation ability, microwave remote sensing uses microwaves with a longer wavelength in the electromagnetic spectrum (0.3–300 GHz) for observation, and its detection mechanism is based on the theory of interaction between objects and microwave radiation [24–27]. According to Rayleigh Gins law, the microwave radiation intensity of an object is positively correlated with physical temperature and emissivity, and this radiation characteristic provides theoretical support for passive microwave remote sensing [28]. The detection process follows the radiation transport theory and involves the absorption, scattering, emission, and transmission of microwaves in the atmospheric–surface–sensor path. Thanks to the penetrating characteristics of microwaves, this technology has all-weather observation capabilities and can still provide reliable observation data even in adverse weather conditions. Taking passive microwave remote sensing as an example, it detects the bright temperature information of the natural radiation of the target and combines the radiation transport equation to invert the parameters, such as atmospheric temperature and humidity, soil moisture, and precipitation intensity. Active microwave remote sensing uses pulse emission and echo analysis, combined with scattering matrix theory to obtain surface dielectric constants and structural characteristics. These properties make microwave remote sensing particularly suitable for observation in dynamic environments, such as cloud and rain processes and hydrological changes. The new generation of fusion sensors (such as NASA TEMPO program) are breaking through the boundaries of traditional technologies to achieve collaborative observation of infrared and microwave. Table 4.4 compares the performance of the two types of techniques in typical feature observation in detail.

Table 4.4 Comparison of complementarity application scenarios [23]

Observation target	Infrared solutions	Microwave solutions
Sea surface temperature	High accuracy (0–0.5K) but affected by clouds	All-weather but low accuracy (1–2K)
Soil moisture	Indirect inversion (requires thermal inertia model)	Direct measurement (strong L-band penetration)
Polar ice sheet monitoring	Surface temperature measurement (but limited by polar night)	Round-the-clock thickness monitoring (SAR interference)
Urban heat island	10m resolution thermal field distribution	Only kilometer-level average temperatures can be obtained

4.3 Infrared Remote Sensing Detection Equipment

4.3.1 System Composition

(1) Optical system

The design of infrared optical systems is a complex and precise project that requires comprehensive consideration of material properties, optical properties, and practical application needs. As the core component of the optical system, the design of the lens group must strictly follow the special requirements of infrared optics. Germanium lenses are the preferred choice for medium-to long-wave infrared systems due to their excellent light transmission in the 2–16 μm band, but their temperature sensitivity requires the system to be equipped with a temperature compensation mechanism [29]. Zinc selenide lenses are often used in multispectral systems due to their wide spectral transmission range (0.5–20 μm) and low absorption loss [30]. Modern infrared optical systems are increasingly using aspherical lens designs, which can effectively reduce system size and weight while improving aberration correction. In addition, the introduction of diffractive optical elements (DOE) has opened up new possibilities for infrared optical systems, allowing for special optical functions that are difficult to achieve with traditional refractive optics.

Filter technology is also evolving rapidly. In addition to traditional bandpass filters, tunable filters show great potential in the field of infrared remote sensing. Electrically tuned filters based on the Fabry–Perot interference principle can achieve fast band switching, providing a new solution for hyperspectral imaging. In recent years, the emergence of meta surface filters has revolutionized filtering technology, and this filter based on sub-wavelength structure can achieve performance indicators that are difficult to achieve with traditional thin-film filters. In terms of system integration, advancements in optical coating technology enable more precise control of multi-layer anti-reflective and filter films, significantly improving the overall transmittance of optical systems. It is worth noting that optical systems for different application scenarios also need to consider environmental adaptability, including dustproof, anti-fog, impact resistance, etc., which need to be fully considered in the optical design stage.

(2) Detector type (cooled/uncooled)

According to different working principles, infrared detectors can be divided into two categories: cooled type and uncooled type. Although the refrigerated detector has excellent performance, it usually needs to work at a temperature below 200 K to suppress the increase of dark current and noise, due to the need to configure a chiller, etc., the refrigerated detector is often expensive to manufacture, the whole machine life is short, the device volume, weight, and power consumption are large, which seriously limits the development of infrared detectors [31–33]. Uncooled detectors solve these shortcomings to some extent.

The development of refrigerated infrared detectors has entered the fourth generation of technology, and new class II superlattice detectors are gradually replacing traditional tellurium mercury detectors [34]. Smith and Mailhiot applied InAs/GaSb class II superlattice materials to infrared detectors in 1987, the combination of class II superlattice materials is the energy band of one material that does not contain another material, divided into "staggered type" and "staggered type", the emergence of class II superlattice infrared materials overcomes the problems of cadmium mercury telluride infrared detectors, and at the same time has its advantages: by adjusting the energy band structure and band gap of the material, it can operate in any wavelength range of 3–30 μm, It is conducive to realizing the needs of two-color and multi-color detection, Through material design, Auger composite and dark current can be suppressed, the working temperature of the focal plane can be increased, the satellite power consumption and resources can be saved, and the cooling pressure can be effectively reduced [35–38]. It has high quantum efficiency, can absorb forward incidence, shorten detector integration time, and fast response time. Therefore, the penetration current is small, and the detection rate can be obtained, especially beneficial for long waves and very long waves. The use of MBE for material growth has good uniformity and low cost, which is conducive to the development of uniform large-scale devices, so class II superlattices will become the preferred material for third-generation infrared detectors [39]. This detector based on the InAs/GaSb superlattice structure offers better uniformity and manufacturability while maintaining extremely high detection performance. In terms of refrigeration technology, the emergence of micro pulse control chillers has greatly improved the reliability and service life of refrigerated detectors, and their design without moving parts has made the mean time between failures reach more than 30,000 h [40–42]. It is worth noting that the new J–T throttle chiller can achieve rapid refrigeration start-up, which has unique advantages in military emergency applications.

An uncooled detector is a detector that works at room temperature and converts the infrared thermal radiation signal incident from the outside into an electrical signal. According to the different working principles, it can be divided into pyroelectric uncooled infrared detectors, thermopile uncooled infrared detectors, microbolometer uncooled infrared detectors (Fig. 4.2) and thermal diode uncooled infrared detectors [43]. Among them, the microbolometer uncooled infrared detector is the mainstream uncooled detector with the most reliable performance and the highest share in the market. Uncooled detectors have been widely used in military and civilian fields due to their small size, low cost, no need for refrigeration, and reliable performance.

With the rapid development of large-scale integrated circuits and microelectromechanical systems, its market share is growing, among which vanadium oxide uncooled detectors are the mainstream uncooled detectors with the most mature technology and the highest market share. Vanadium oxide-based microbolometer arrays can already achieve a pixel spacing of 17 μm, and NETD (noise equivalent temperature difference) is better than 30 mK. In addition, the new silicon-based infrared detector is fully compatible with traditional IC processes through the CMOS–MEMS process, significantly reducing manufacturing costs. In terms of material innovation, two-dimensional materials (such as graphene, black phosphorus) infrared detectors have shown unprecedented response speed and detection rate, pointing out a new direction for the development of uncooled detectors in the future [44]. In addition, the concept of intelligent detector directly integrates some signal processing functions on the detector chip, which significantly improves the response speed and energy efficiency ratio of the system.

(3) Signal processing module

The signal processing module of the infrared remote sensing system is the core data processing center of the whole system, and its technical principle involves the cross-integration of multiple disciplines. The module mainly undertakes functions such as signal conditioning, data conversion, image processing and intelligent analysis, and its technical implementation principles can be elaborated in depth from the following levels:

(1) Analog signal conditioning technology

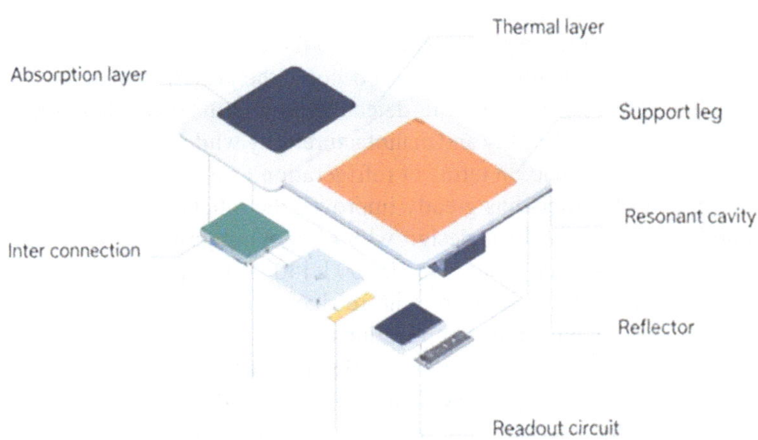

Fig. 4.2 Schematic diagram of a microbolometer

In the original signal processing stage of the detector output, low-noise amplification technology and adaptive filtering technology are mainly applied. The low-noise preamplifier adopts an instrumentation amplifier architecture that effectively suppresses common-mode interference through a differential amplification circuit with a common-mode rejection ratio (CMRR) of more than 100 dB. At the same time, the reset noise is eliminated by using correlated double sampling (CDS) technology, which effectively eliminates the reset noise of the detector node capacitor by sampling twice before and after the signal reading [45]. Adaptive filtering enables dynamic range adjustment through a programmable gain amplifier (PGA), which automatically adjusts the magnification based on signal strength, ensuring that the analog-to-digital converter (ADC) is always operating in a linear range.

(2) Digital signal processing technology

The analog-to-digital conversion process adopts oversampling and noise shaping technology, pushes the quantized noise to the high frequency band through the σ–δ modulator, and then uses the digital extraction filter to obtain a high-resolution digital signal. The non-uniformity correction (NUC) algorithm establishes a gain-offset correction model for each pixel based on the calibration data collected by the reference radiation source, and its mathematical expression is:

$$V_{corrected}(i,j) = G(i,j) \times V_{raw}(i,j) + O(i,j) \tag{4.12}$$

where, G (i, j) and O (i, j) represent the gain and offset coefficients of each cell, respectively. The three-dimensional block matching (BM3D) algorithm is used for time-domain noise reduction, which can effectively maintain image details while suppressing random noise by searching for similar image blocks in the spatio–temporal domain for collaborative filtering [46].

(3) Real-time image processing technology

FPGA-based hardware acceleration architecture adopts pipeline parallel processing technology to decompose image processing tasks into multiple stages that can be executed in parallel. Among them, the blind element compensation algorithm replaces the abnormal pixel values through neighborhood pixel interpolation, and its core is to establish a dead pixel map and update it dynamically. Image enhancement uses the technical route of adaptive histogram equalization (AHE) combined with guided filtering to enhance contrast while avoiding over-amplifying noise. Motion object detection uses the Gaussian mixed model (GMM) background modeling method to describe the brightness change characteristics of each pixel by establishing multiple Gaussian distributions [47–49].

(4) Intelligent analytics technology

The deep learning algorithm adopts a hybrid architecture of convolutional neural networks (CNNs) and Transformers. Among them, the improved network based on U–Net realizes super-resolution reconstruction of infrared images, and retains multi-scale features through jump connections [50]. The target recognition module adopts a lightweight network improved by YOLOv5 architecture, which reduces the computational amount through deep separable convolution. Anomaly detection uses an algorithm combining autoencoder (AE) and generative adversarial network (GAN) to identify abnormal heat sources by reconstructing errors. These algorithms are deployed through quantitative perception training (QAT), converting 32-bit floating-point models into 8-bit integer models to adapt to the processing power of edge computing devices.

(5) System and processing technology

Multi-sensor data fusion adopts the method of combining Kalman filtering and deep learning to realize the feature-level fusion of visible light and infrared images. The system-on-chip (SoC) integrates ARM cores and neural network accelerators, reducing data handling overhead through in-memory computing (PIM) technology. Power management uses dynamic voltage-frequency adjustment (DVFS) technology to adjust the operating voltage and clock frequency of the operating unit in real time based on the processing load [51]. Security protection is built on a trusted execution environment (TEE) that protects the confidentiality of critical algorithms and data through hardware isolation.

(6) Emerging technology applications

Quantum signal processing technology has begun to be applied to high-end infrared systems to improve signal integrity through quantum decoherence suppression algorithms. The photonic computing architecture uses an optical matrix multiplier to accelerate convolutional operations, which improves the energy efficiency ratio of traditional electronic computing by two orders of magnitude [51]. Neuromorphic computing mimics the event-driven nature of biological nervous systems, triggering operations only when significant signal changes are detected, significantly reducing power consumption. The integration and application of these cutting-edge technologies is driving the development of infrared signal processing in the direction of higher performance and lower power consumption.

From the perspective of implementation mechanism, modern infrared signal processing modules have developed into a complex system integrating analog circuit design, digital signal processing, computer vision and artificial intelligence. Its technological evolution presents the following obvious characteristics: the processing link is migrated from the back-end to the front-end, the algorithm complexity develops from linear to non-linear, the computing architecture changes from centralized to distributed, and the implementation mode is optimized from general to specialized. These technological advancements have collectively promoted the continuous improvement of the performance of infrared remote sensing systems, making them achieve significant breakthroughs in target recognition accuracy, environmental adaptability, and real-time processing speed.

4.3.2 Key Technical Parameters

(1) Temperature resolution (NETD)

Noise Equivalent Temperature Difference (NETD) is a key index to characterize the temperature resolution ability of an infrared system, which is defined as the target temperature difference corresponding to the system output signal-to-noise ratio (SNR) of 1 under specific test conditions [52]. Its physical significance reflects the smallest temperature difference that the system can resolve, which directly affects the temperature measurement accuracy of the system. Its mathematical expression is:

$$\mathrm{NETD} = \frac{V_n \cdot \Delta T}{\Delta V} \tag{4.13}$$

Thereinto:
Vn is the total noise voltage of the system (μV);
ΔT is the temperature difference of the target blackbody (K);
ΔV is the signal voltage change (μV) corresponding to the temperature difference.
The typical value range is shown in Table 4.5.
Alternative routes:

$$\mathrm{NETD} \propto \frac{1}{\sqrt{A_d \cdot \tau_0 \cdot D^* \cdot \sqrt{t_i}}} \tag{4.14}$$

(2) Spatial resolution

The Instantaneous Field of View (IFOV) represents the spatial angle corresponding to a single detector element of the system, which determines the system's ability to resolve spatial details. Its size directly affects the spatial resolution of the image. The calculation formula is:

$$\mathrm{IFOV} = 2\arctan\left(\frac{d}{2f}\right) \approx \frac{d}{f}\text{(Small angle approximation.)} \tag{4.15}$$

Thereinto:
d is the detector pixel size (μm);
f is the focal length of the optical system (mm).
Table 4.6 is a comparison table of typical parameters.
Modulation Transfer Function (MTF) Impact:

Table 4.5 Typical value range of detectors

Classify	Numeric value (mK)
Refrigerated detector	< 20
Uncooled detector	30–50
Commercial thermal imager	50–100

Table 4.6 Comparison of typical parameters [21]

System type	Pixel size (μm)	Focal length (mm)	IFOV (mrad)	Ground resolution at 300m (m)
Telephoto monitoring	17	150	0–113	0.034
Airborne reconnaissance	12	500	0–024	0.007
Satellite remote sensing	10	2000	0–005	0.0015

$$MTF_{system} = MTF_{optics} \times MTF_{detector} \times MTF_{motion} \tag{4.16}$$

The MTF of the optical system is limited by the diffraction limit:

$$MTF_{diff} = \frac{2}{\pi}\left[\arccos(f/f_c) - (f/f_c)\sqrt{1-(f/f_c)^2}\right] \tag{4.17}$$

where, f_c is the cut–off frequency.

(3) Frame rate and response time

Frame Rate: The number of frames per second (Hz) that the system outputs a complete image.

Response Time: How quickly the detector responds to changes in radiation, including:

Rise time τr: 10–90% response;

Decline time τf: 90–10% recovery;

Time constant τ: 63–2% response.

Theoretical model:

The detector response follows the RC circuit model:

$$V(t) = V_0\left(1 - e^{-t/\tau}\right) \tag{4.18}$$

where $\tau = R_{th}C_{th}$, R_{th} is the thermal resistance, and C_{th} is the heat capacity.

Analysis of frame rate limiting factors:

$$f_{max} = \frac{1}{t_{int} + t_{read} + t_{proc}} \tag{4.19}$$

where, t_{int}: Crediting time.

t_{read}: Read out the time.

t_{proc}: processing time.

Typical system parameter comparison is shown in Table 4.7.

$$t_{int} = \min\left(t_{max}, \frac{Q_{sat} - Q_{dark}}{I_{photon}}\right) \tag{4.20}$$

Table 4.7 Comparison of typical system parameters [26]

Detector type	Response tim (ms)	Maximum frame rate (Hz)	Application scenarios
Refrigeration type (MCT)	0.1–1	100–1000	High-speed target tracking
Non-refrigeration (VOx)	5–15	30–60	Security monitoring
Amorphous silicon	10–20	15–30	Industrial Inspection

where Q_{sat} is the saturated charge and I_{photon} is the photon flux.

(4) Coupling relationship for parameters

Comprehensive performance balance equation:

$$\text{Performance index} = \frac{1}{\text{NETD}} \times \frac{1}{\text{IFOV}} \times f \times \exp\left(-\frac{\tau}{t_f}\right) \quad (4.21)$$

where t_f is frame period.

4.4 Application Cases of Infrared Remote Sensing Detection

Coal is one of the important energy sources in many countries in the world, and coal resources are very abundant in northern China. Due to the distribution of coalfields in arid and semi-arid zones, coupled with the long history of mining, underground coal seam fires are very serious. It not only causes a large loss of resources, but also pollutes the environment for a long time [53]. Therefore, the investigation and prevention of coal seam fires have become a major issue to be solved urgently. In order to meet the needs of the state for coal seam fire prevention and control planning and engineering construction, the method of underground coal seam fire detection has been studied.

The application of infrared remote sensing technology in the field of coalmines has formed an all-round and multi-level monitoring system, showing significant technical advantages and practical value. In terms of monitoring coalmine fire areas, thermal infrared remote sensing can accurately identify surface temperature anomalies through the 8–14 μm band, and the Zhundong coalfield in Xinjiang, China successfully tracked the expansion trend of coalfield spontaneous combustion zone using Landsat-8 TIRS data, with a temperature resolution of 0.5 K [54]. In view of the major safety hazard of goaf subsidence, the fusion analysis of thermal infrared and visible light data combined with InSAR technology enables monthly subsidence monitoring with centimeter-level accuracy in the Datong mining area of Shanxi, China. In the field of gas monitoring, a breakthrough has been made in hyperspectral detection technology based on the methane 7.7 μm characteristic absorption spectral line, and the "Gaofen-5" satellite has the ability to invert the concentration of methane columns in mining areas, providing a new technical means for preventing gas accidents [55].

As two important technical means for monitoring coalmine fire areas, UAV infrared remote sensing and satellite infrared remote sensing have their own advantages in detection capabilities, application scenarios and technical characteristics, and together constitute a multi-scale and three-dimensional fire monitoring system.

(1) DJ1 M210 V2 drone and Zenmuse XT 2 camera

Remote sensing detection of coalfield fires at home and abroad has a history of more than 30 years, and the theoretical experience is relatively complete and the practical experience is relatively rich. A large number of scholars have conducted research on the delineation of coalfield fire areas, and have also produced a lot of results. In terms of technical means, UAV infrared remote sensing is mainly equipped with high-resolution thermal imagers (such as FLIR Tau2) to obtain centimeter-level precision thermal infrared images through low-altitude flight (50–500 m). Its typical working mode is grid inspection, which adopts dual light (visible light + infrared) simultaneous acquisition, and the temperature sensitivity can reach 0.05 °C. The application of the Integrator UAV system developed by the American company Insitu in Australian coalmines has shown that it can identify hot spots in the fire area of 0.5 m^2 [56]. The process mainly includes surface temperature inversion of thermal infrared remote sensing images, ground interpretation of high-resolution data + DEM data, ground temperature/gas survey, interpretation and delineation of fire area boundaries, etc.

The DJI M210 V2 drone has a battery life of about 34 min (no load) when equipped with two TB55 batteries, can carry a load such as Zenmuse XT 2 (total weight ≤ 6.14 kg), and supports 24 min of continuous flight to complete monitoring of an area of about 10km^2. With a maximum flight altitude of 5000 m (altitude) and a wind resistance of 10 m/s, it is suitable for high-altitude stable operations in complex terrain in coalfields. It supports GPS/GLONASS dual-mode positioning, and with the waypoint planning function, it can realize S-shaped route scanning or hovering observation in key areas. Equipped with OcuSync 2.0 image transmission system, the transmission distance is up to 5 km, and it supports real-time back transmission of thermal infrared images and visible light images, which is convenient for real-time decision-making by the ground command center [57]. The M210 V2 can rise to altitudes of more than 500 m and overlook large coalfields, avoiding obstruction from monitoring by ground obstacles (e.g., buildings, vegetation).

With a high resolution of 640 × 512, the Zenmuse XT2 camera can identify temperature differences of 0.05 °C and clearly capture hot and low oxidation zones in coalfield fire areas. Low-gain mode coverage–40 °C to 550 °C for simultaneous monitoring of open and hidden fire zones. Available in four focal lengths: 9 mm, 13 mm, 19 mm, and 25 mm, adapting to different monitoring distances [58]. Automatically locks the highest temperature point in the screen and continuously tracks it, suitable for dynamic fire zone movement monitoring. Thermal infrared raw data and

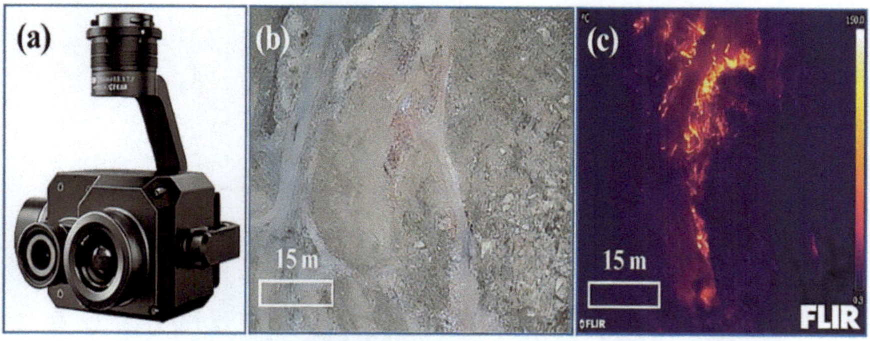

Fig. 4.3 Schematic diagram of Zenmuse XT2 detection coalfield fire area

visible light images are stored independently, and radiation calibration and temperature inversion can be performed later with FLIR Tools or DJI Pilot software. It can penetrate the thick smoke generated by coalfield fires, directly detect the high-temperature radiation of the flame and the heat conduction of underground hidden fire sources. By presetting the temperature threshold, the fire area is quickly delineated, and the temperature gradient distribution map is generated. The detection diagram is shown in Fig. 4.3.

In the process of detecting coalfield fire areas, the M210 V2 UAV is often used to collect thermal infrared and visible light data synchronously with the Zenmuse XT2 camera. At the edge of the fire zone or near the hot spot, you can switch to hover mode for fine scanning. The ground station uses DJI Pilot software to view the temperature distribution map in real time, trigger high temperature alarms, and mark the location of suspected fire points. With the core advantages of high-precision thermal imaging, long endurance, strong wind resistance and real-time data processing, it has become the best choice for coalfield fire area monitoring, and its detection diagram is shown in Fig. 4.4.

(2) Satellite infrared payload system

Modern satellite infrared remote sensing systems have formed a multi-platform and multi-spectral comprehensive observation network, and their technological development shows a trend of paying equal attention to high spatiotemporal resolution and high spectral resolution. The high-resolution (between 0.5 and 2.5m) satellite remote sensing images that are more practical for the investigation and monitoring of coalfield fire areas mainly include domestic Gaofen 1, Gaofen 2, Resource 3, Gaojing 1, Beijing 2, etc., foreign satellites have worldview, pleiades, geoeye, quickbird and other data, and the multispectral and thermal infrared bands of the medium and low-resolution satellite Landsat8. Satellite remote sensing images have the advantages of relatively large coverage and easy access to archived data in major fire areas and surrounding areas, which are convenient for analyzing the evolution history of fire areas [59]. Satellite remote sensing can also preliminarily interpret the distribution

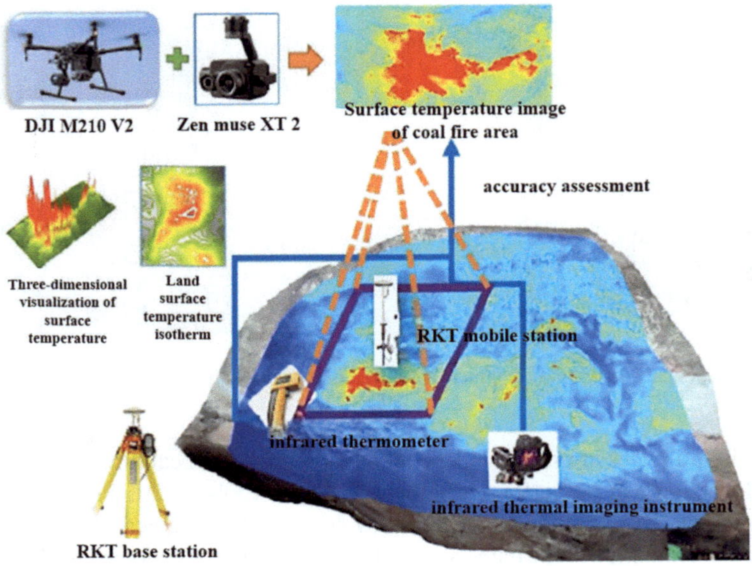

Fig. 4.4 Schematic diagram of M210 V2 equipped with Zenmuse XT2 synchronous detection of coalfield fire area

range, geographical location, topography, land use type, vegetation and ecological environment background information of the fire area.

MODIS (Medium Resolution Imaging Spectrometer) is the core payload of Terra/Aqua satellites, and the infrared channel design of MODIS represents the latest level of current spaceborne remote sensing. Its technical features include 20 infrared bands (3.7–14.4 μm) in 36 spectral channels, a double-sided scanning mirror design, a scanning width of up to 2330 km, and a temperature sensitivity of 0.05 K (at 300 K). The newly released Collection 7 data product improves the accuracy of surface temperature inversion to ± 1 K, which is widely used in global climate change research. The VIIRS (Visible Infrared Imaging Radiometer Kit) released in 2022 is a successor model that improves spatial resolution to 375 m (infrared band).

ASTER (Advanced Spaceborne Thermal Emission and Reflection Radiometer) The ASTER system developed by METI in Japan is known for its unique multi-angle observation capabilities. Equipped with a stand-alone thermal infrared imager (TIR, 8–12 μm), spatial resolution up to 90 m (thermal infrared band), temperature resolution is better than 0.3K. In the field of mineral resource exploration, ASTER thermal infrared data successfully identified the alteration zone distribution at the Escondida copper deposit in Chile through emissivity inversion [60]. The ASTER Global Emissivity Database (GEDv4), updated in 2023, has become a standard reference for geological research. Satellite and UAV detection of the coalfield fire area is shown in Fig. 4.5.

(3) Gaofen-5 satellite

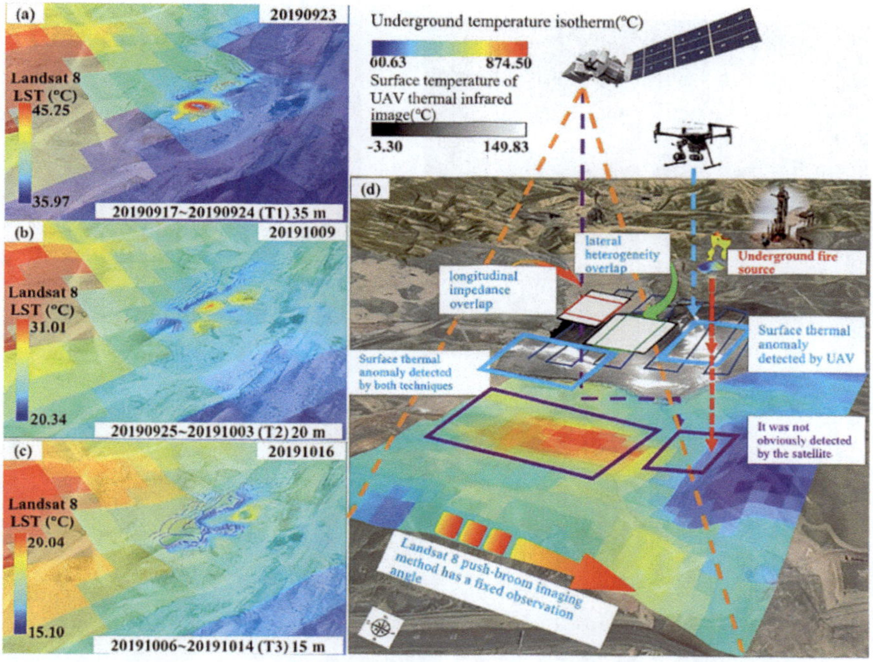

Fig. 4.5 Schematic diagram of satellite and UAV detection of coalfield fire area

The visible shortwave infrared hyperspectral camera of the Gaofen-5 satellite covers 400–2500 nm, has 330 spectral channels, and can recognize the characteristic wavelength band of 3.7–4.8 μm. Atmospheric trace gas differential absorption spectrometer monitors SO_2, NO_2 and other pollutants, and greenhouse gas detector inverts CO_2 concentration to assist in the analysis of combustion intensity in coalfield fire area. The multi-angle polarization detector analyzes the distribution of haze particles from 9–11 observation angles, distinguishing between smoke and heat radiation in the fire area. In the Shenfu coalfield in Shaanxi, China, the open flame point was identified through the 3.7 μm band, combined with the InSAR deposition data and NDVI vegetation index, and the accuracy rate reached 91% [61]. Using the spacecraft as the carrier, thermal infrared and high-resolution remote sensing are used to detect the combustion traces and phenomena of the fire area, combined with surface temperature, wind and other environmental characteristics, to determine the distribution of the fire area, the detection time is short, and it can effectively detect large-scale coalfield fires, and the thermal infrared band has low resolution and limited accuracy. The detection of the coalfield fire area is shown in Fig. 4.6.

(4) Portable thermal imager and spectrometer

The technological evolution of portable devices focuses on AI integration and Internet of Things integration, and the application scenarios continue to expand, FLIR T1020 scientific research-grade thermal imager, this handheld portable infrared device has

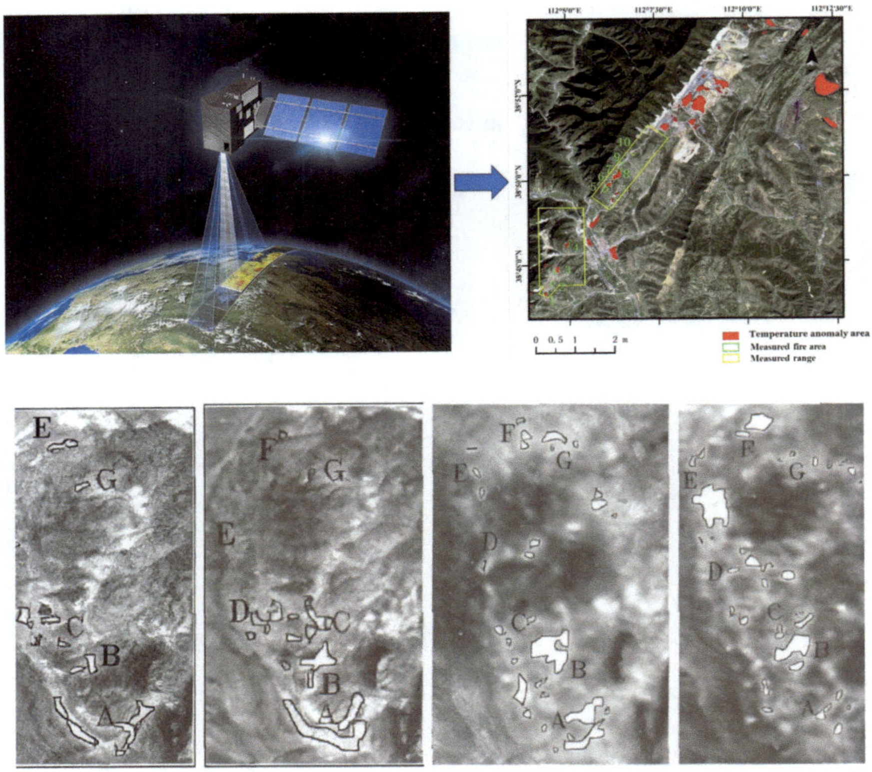

Fig. 4.6 Schematic diagram of satellite detection of coalfield fire area [31]

a resolution of 1024 × 768 infrared, and the Dequan Coalmine in northern Shaanxi, China quickly determines the type and spatial distribution of air leakage channels through handheld thermal imaging cameras and gas chromatographs, and combines fire prevention and extinguishing projects to achieve overall temperature control in the fire area [62]. The detection of the coalfield fire area is shown in Fig. 4.7.

(5) Emerging application directions

The FLIR ONE Pro series of mobile phone thermal cameras (160 × 120 resolution) enables consumer-grade applications through the USB-C interface, with shipments exceeding 2,000,000 units in 2022. Nippon Avionics Inf ReC R500EX series (5.1–5.4 μm band) has been successfully applied to semiconductor production line inspection, and the defect recognition rate has increased to 99.9%. The comparative analysis of technological development is shown in Table 4.8.

The current technological development trend shows the following characteristics: satellite systems are developing in the direction of hyperspectral, airborne platforms are accelerating drones, and portable devices are deeply integrated with AI. Together, these equipment systems constitute an infrared remote sensing observation network

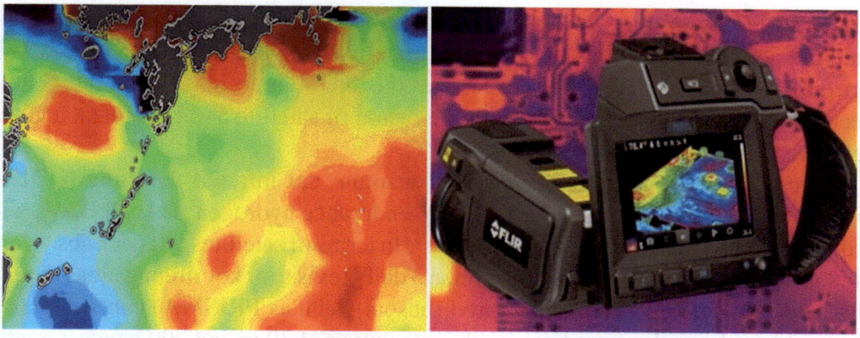

Fig. 4.7 Schematic diagram of portable thermal imager detecting coalfield fire area

Table 4.8 Comparative analysis of technological development [35]

Technical indicators	Satellite payload	Airborne systems	Portable equipment
Typical spatial resolution	100m–1km	0.1m–5 m	0.5mm–5 cm
Temperature sensitivity	0.05–0.5K	0.02–0.1 K	0.01–0.05 K
Number of spectral channels	5–20	1–256	1–8
Single coverage	100–3000km	1–50 km	0.01–0.1 km
Typical price range	$1–500 million	$0.5–5 million	$10,000–500,000

from macro to micro, and play an irreplaceable role in climate change monitoring, national defense and security, smart cities and other fields [63]. The core difference between drones and satellites is reflected in four aspects: first, the spatiotemporal resolution, drones can achieve sub-meter resolution (up to 0.01 m) and on-demand observation, while satellite data is limited by orbit, with resolutions mostly in the range of 15–100 m, and the revisit period is several days to weeks. A comparative study by the University of West Virginia in the United States shows that drones have a 98% success rate in detecting fire areas of less than 10 m^2, while satellites can only identify fire areas larger than 400 m^2. The second is the detection depth, drones can detect shallow fire areas 0.5–2 m below the surface due to low-altitude operations, while satellites mainly reflect surface heat radiation [64]. The third is the operating cost, the cost of a single UAV operation is about RMB 5,000–20,000 yuan, while the cost of satellite data acquisition is almost zero (public data). Finally, there are applicable scenarios, UAVs are suitable for fine investigation in key areas of mining areas, and the single coverage area usually does not exceed 10 km^2. Satellites are suitable for rapid screening in tens of thousands of square kilometers.

In terms of technology integration, the latest development trend is to build a "satellite–aircraft collaboration" monitoring network. The "space–air–ground" integrated system developed by China Coal Science and Industry Group first uses satellites to screen large-scale fire areas, and then dispatches drones to finely verify suspicious areas. In the application of the Jungger coalfield in Inner Mongolia, the system improves the positioning accuracy of the fire area from ± 50 m to ± 0.5 m by the satellite alone, and shortens the response time by 80%. The Fire Sat system developed by NASA in the United States realizes the three-dimensional temperature field reconstruction of the fire area through the fusion of satellite and UAV data, and the temperature inversion error is less than 1 °C.

In terms of UAVs, the application of quantum dot infrared detectors (QDIPs) will make NETD exceed 0.01 °C, and long-endurance hydrogen fuel cell UAVs will expand the single operation range to 50 km^2. In terms of satellites, Planet Labs "Constellation" program will achieve more than 3 daily revisits of key mining areas around the world, while hyperspectral infrared sensors (such as PRISMA) can simultaneously obtain thermal infrared data with a resolution of 30 m and 200 bands. Greatly improve the accuracy of fire area identification [65]. The deep integration of these two technical means is promoting the fundamental transformation of coalmine fire monitoring from "post-event disposal" to "pre-warning".

References

1. Dong, B. (2023). *Application of thermal infrared remote sensing data in geothermal resource exploration——An example of Kangding area in the western Sichuan plateau.* Master Thesis, Chengdu University of Technology. (董奔. 热红外遥感数据在地热资源探测中的应用. 成都理工大学, 2023)
2. Du, X. M. (2015). *Thermal infrared remote sensing based coal spontaneous combustion detection method and change detection for coal fire areas–A case study of the Wuda coal field*. Ph.D. dissertation, China University of Mining and Technology. (杜晓敏. 热红外遥感煤火探测方法及煤火变化检测. 中国矿业大学(北京), 2015).
3. Wang, K. (2020). *Thermal infrared remote sensing technology for geothermal resources detection based on multi-source & multi-temporal datain Dandong, Liaoning*. Ph.D. dissertation, Jilin University. (王康. 基于多源多时相热红外遥感技术的丹东地热资源探测方法研究. 吉林大学, 2020).
4. Ji, M., Xu, Y. M., Mo, Y. P., Zhang, Y., Zhou, R. Y., & Zhu, S. Y. (2023). Land surface temperature retrieval method based on UAV thermal infrared remote sensing and synchronized atmospheric profiles. *Journal of Geo-information Science*, *25*(12), 2456−2467. (吉蒙, 徐永明, 莫亚萍, 张杨, 周睿禹, 祝善友. 基于无人机热红外遥感和大气同步廓线的地表温度反演方法. 地球信息科学学报, 2023, 25(12):2456–2467).
5. Zhu, J. S., Ren, H. Z., Ye, X., Zeng, H., Nie, J., & Jiang, C. C., et al. (2021). Ground validation of land surface temperature and surface emissivity from thermal infrared remote sensing data:A review. *National Remote Sensing Bulletin*, *25*(08), 1538−1566. (朱金顺, 任华忠, 叶昕, 曾晖, 聂婧, 蒋晨琛, 等. 热红外遥感地表温度与发射率地面验证进展. 遥感学报, 2021, 25(08):1538–1566).
6. Yao, Y. Y. (2022). *Retrieval of surface temperature form thermal infrared dataand its application*. Master Thesis, Dalian Maritime University, Dalian China. (姚雨莹. 热红外数据陆地表面温度反演及应用. 大连海事大学, 2022).
7. Teng, S. L., Kong, J. A., & Shi, N. R. (1985). *Theory of microwave remote sensing*. Wiley−Interscience.
8. Pampaloni, O., & Paloscia,O. (2023). *Microwave radiometry and remote sensing of the earth's surface and atmosphere*.CRC Press.
9. Gui, R. D., Howar, D. J., & Hong, D. C. (1979). Adual-channel micro wave radio meter form easure mento fpreci pitable water vapor and liquid. *IEEE Transactions on Geoscience Electronics, 17*(04), 129–136.

10. Lin, M. S., & Jin, Y. J. (2022). Present and future marine micro wave satel lite missions in China. *Remote Sensing, 14*(06), 1330.
11. Smith, D. L., & Mailhiot, C. (1987). Proposal for strained type II super lattice infrared detectors. *Journal of Applied Physics, 62*(06), 2545–2548.
12. Zeng, G. H., Shi, Y. L., & Zhuang, J. S. (2011). Principles, status and prospect of type II superlattice infrared detectors. *Infrared Technology*, *33*(06), 311−314. (曾戈虹, 史衍丽, 庄继胜. II类超晶格红外探测器的机理、现状与前景. 红外技术, 2011, 33(06):311–314).
13. Grein, C. H., Cruz, H., & Flatte, M. E. (1994). Theoretical performance of very long wavelength InAs/InGa Sbsuper lattice based infrared detectors. *Journal of Applied Physics, 65*(20), 2530–2532.
14. Li, J. Y., & He, J. Y. (2020). Progress and prospects of the combined application of microwave remote sensing and infrared hyperspectral remote sensing. *China Journal of Space Science*, *45*(02), 413−423. (李旌阳, 何杰颖. 微波遥感与红外高光谱遥感联合应用的进展与展望. 空间科学学报, 2025, 45(02):413–423).
15. Liu, D. H., Qiu, J. T., Yan, B. K., & Tian, F. (2018). Application of thermal inferred hyperspectral remote sensingin ore exploration. *Geological Review*, *64*(05), 1190−1200. (刘德长, 邱骏挺, 闫柏琨, 田丰. 高光谱热红外遥感技术在地质找矿中的应用. 地质论评, 2018, 64(05):1190–1200).
16. Wan, Y.Q., Jiang, X. G., & Xu, K. L. (2020). Application of drone remote sensing technology in coalfield fire area investigation and monitoring. *Coal Geology of China*, *32*(06), 78−85. (万余庆, 江晓光, 徐凯磊. 无人机遥感技术在煤田火区调查监测中的应用. 中国煤炭地质, 2020, 32(06):78–85). (张兵, 柳钦火, 李晓明, 刘良云, 杨必胜, 高连如, 等. 遥感科学的内涵与基础性问题.遥感学报, 2025, 29(01):1–48).
17. Zhang, B., Liu, Q. H., Li, X. M., Liu, L. Y., Yang, B. S., & Gao, L. R., et al. (2020). The core concepts and fundamental issues of remote sensing science. *National Remote Sensing Bulletin*, *29*(01), 1−48.
18. Sun, Q. Y., Chai, R. Q., Xie, S. W., Xie, L. L., Fan, B., & Chen, R. M. (2023). Demands and development of spaceborne infrared detectors. *Laser & Infrared*, *53*(10), 1469−1475. (孙启扬, 柴瑞青, 谢圣文, 谢莉莉, 樊奔, 陈瑞明. 航天遥感红外探测器需求与发展.激光与红外, 2023, 53(10):1469–1475).
19. Xu, K. L., Jiang, X. G., Wan, Y. Q., Yang, C., & Qiang, J. H. (2022). Thermal infrared remote sensing monitoring technology of UAV in coalfield fire area. *Coal Technology*, *41*(07), 120−123. (徐凯磊, 江晓光, 万余庆, 杨创, 强建华. 煤田火区无人机热红外遥感监测技术. 煤炭技术, 2022, 41(07):120–123).
20. Chen, Y. H., Chen, H. D., & Fu, Y. T. (2015). Study on background of space infrared optical system caused by self-emission. *Shanghai Infrared and Remote Sensing Society*. In *Proceedings of the 2015 Infrared and Remote Sensing Technology and Application Symposium and Interdisciplinary Forum. Shanghai Institute of Technical Physics, Chinese Academy of Sciences University of Chinese Academy of Sciences* (pp. 177−182). (陈永和, 陈洪达, 傅雨田. 红外遥感器自身引起的背景辐射研究. 上海市红外与遥感学会. 2015年红外、遥感技术与应用研讨会暨交叉学科论坛论文集. 中国科学院上海技术物理研究所, 中国科学院大学, 2015:177–182).
21. Yang, X., Mao, N. Q., & Xu, S. Y. (2014). Research of cryocooler application in space infrared sensor system. *Vacuum and Cryogenics*, *20*(02), 113−115+124. (杨雪, 茅年清, 徐圣亚. 制冷机在空间红外遥感领域的应用研究. 真空与低温, 2014, 20(02):113–115+124).
22. Zhu, M. G., Qian, Q. Y., & Wang, J. (2014). Reliability assurance technology of infrared focal Plane assembly. *Aerospace Shanghai(Chinese & English)*, *31*(04), 50−53. (朱美光, 钱秋瑛, 王佳. 红外探测器组件可靠性保证技术. 上海航天, 2014, 31(04):50–53).
23. Xu, L. N., Deng, X. G., Jin, Z. L., & Wang, Z. M. (2016). Application prospect of type II super–lattice infrareddetector in remote sensing. *Laser & Infrared*, *46*(09), 1035−1039. (徐丽娜, 邓旭光, 金占雷, 王智谋. II类超晶格红外探测器在遥感中的应用前景. 激光与红外, 2016, 46(09):1035–1039).
24. Xu, L. N., Dong, J., Dai, L. Q., & Jia, H. L. (2018). Study and analysis of the detectivity of infrared detector. *Laser & Infrared*, *48*(04), 497−502. (徐丽娜, 董杰, 戴立群, 贾慧丽. 红外探测器探测率的研究与分析. 激光与红外, 2018, 48(04):497–502).

25. Su, Y. D., He, H., Tan, C. J., Hui, M. Y., Su, R. J., & Wang, H. W., et al. (2025). Multi-source remote sensing detection method for hidden fire sources in coal mines. *Surveying and Mapping of Geology and Mineral Resources, 41*(02), 6–12. (苏亚东, 何浩, 谭崇玖, 惠梦圆, 苏润杰, 王宏伟, 等. 煤矿隐蔽火源多源遥感探测方法. 地矿测绘, 2025, 41(02):6–12).
26. Ji, H. L. (2012). *Application of multi-scale thermal infrared remote sensing data in coalfield fire area information extraction.* Master Thesis, Xinjiang University, Xinjiang China. (姬洪亮. 多尺度热红外遥感数据在煤田火区信息提取中的应用. 新疆大学, 2012).
27. Sun, C. B, (2023). Research on coal fire detection and identification in mining areas basedon UAV thermal infrared remote sensing technology. *China Coal, 49*(S2), 85–94. (孙长斌. 无人机热红外遥感技术对矿区煤火探测识别研究. 中国煤炭,2023,49(S2):85–94).
28. Zhang, Z. K. (2023). *Design and implementation of temperature monitoring system for spontaneous combustion coal gangue pile.* Master Thesis, China University of Mining and Technology. (张振科. 自燃煤矸石山温度监测系统的设计与实现. 中国矿业大学, 2023).
29. Zhang, D. Y., & Pan, Z. X. (2015). Application of remote sensing technique to monitor mining activities. *Bulletin of Surveying and Mapping*, (06), 72–74. (张端阳, 潘振祥. 遥感技术在矿业活动动态监测中的应用. 测绘通报, 2015, (06):72–74).
30. Qin, K. (2013). *GEOSS-based thermal anomalies spatio−temperal analysis for seismicity: method and verification.* Ph.D. dissertation, China University of Mining and Technology. (秦凯. 基于GEOSS的地震遥感热异常时空分析方法与检验.中国矿业大学(北京), 2013).
31. Wang, K., Yang, P., Lu, W. S., Zhu, L. Y., & Yu, G. M. (2020). Current status and development trend of UAV remote sensing applications in themining industry. *Chinese Journal of Engineering, 42*(09), 1085–1095. (王昆, 杨鹏, 吕文生, 诸利一, 于广明. 无人机遥感在矿业领域应用现状及发展态势.工程科学学报, 2020, 42(09):1085–1095).
32. Xie, H. B., Yang, X., Tan, D. J., Luo, Z. F., & Z, L. (2014). Research on ecological environment change in mining concentrated area based on GIS and RS. *Journal of Henan Polytechnic University (Natural Science), 33*(03), 388–394. (谢洪斌, 杨雪, 谭德军, 罗真富, 曾琳. 基于GIS和遥感技术的矿业集中区生态环境变化研究. 河南理工大学学报(自然科学版), 2014, 33(03):388–394).
33. Hou, C. H. (2021). *Heterogeneous characteristics of surface thermal environment in mining-intensive areas and driving mechanism.* Ph.D. dissertation, North China University of Science and Technology. (侯春华. 矿业开发密集区地表热环境异质性特征与产生机制研究. 华北理工大学, 2021).
34. Zhang, C. W., & Gao, C. J. (2023). Analysis of practical application of remote sensing technology inatmospheric environment monitoring. *Heilongjiang Environmental Journal, 36*(02), 156–158. (张川伟, 高朝继. 遥感技术在大气环境监测中的实践应用分析. 黑龙江环境通报, 2023, 36(02):156–158).
35. Yuan, X. (2019). *Monitoring and analysis of urban and rural local ecological environment based on thermal infrared remote sensing of very high resolution.* Ph.D. dissertation, South China University of Technology. (袁旭. 基于超高分辨率热红外遥感的城乡局部生态环境监测与分析研究. 华南理工大学, 2019).
36. Chen, B. M. (2020). Progress in the application of remote sensing technology to ecological environment monitoring and law enforcement supervision. *Mining and Metallurgical Engineering, 40*(04), 165–168+173. (陈博明. 遥感技术在生态环境监测及执法中的应用进展. 矿冶工程, 2020, 40(04):165–168+173).
37. Yi, K. P., & Bao, Y. L. (2016). A review on applications and contributionsof remote sensing techniques in fire monitoring. *World Forestry Research, 29*(04), 36–41. (伊坤朋, 包玉龙. 遥感技术在火监测中的应用研究综述. 世界林业研究, 2016, 29(04):36–41).
38. Lei, L. Q., Yue, Y. Z., Sun J. L., & Li, Y. Q. (2002). Research on the application of remote sensing techniques in monitoring environmental pollution in coalmine. *Environmental Protection*, (02), 33–36. (雷利卿, 岳燕珍, 孙九林, 李永庆. 遥感技术在矿区环境污染监测中的应用研究. 环境保护, 2002, (02):33–36).
39. Zhang, J. L., & Cao, D. Y. (2005). Application of hyperspectral remote sensing technology to environment monitoring in coalmine area. *Journal of Natural Disasters*, (04), 158–162. (张杰林, 曹代勇. 高光谱遥感技术在煤矿区环境监测中的应用. 自然灾害学报, 2005, (04):158–162).

40. Zhao, S. H., Wang, Q., You, D., Yao, Y. J., Zhu, L., & Sui, X. X., et al. (2015). Preliminary analysis on the application of satellite infrared remote sensing techniques in environmental protection of china. *Journal of Geo-Information Science, 17* (07), 855–861. (赵少华, 王桥, 游代安, 姚云军, 朱利, 随欣欣, 等. 卫星红外遥感技术在我国环保领域中的应用与发展分析. 地球信息科学学报, 2015, 17(07):855–861).
41. Ding, F., Tang, G., Zhang, T., Wu, W. L., Zhao, B. J., Miao, J. W., et al. (2025). Spaceborne lightweight and compact high-sensitivity uncooled infrared remote sensing camera for wildfire detection. *Remote Sensing, 17*(08), 1387–1387.
42. Sadovnikov, S., Yakovlev, S., Kravtsova, N., Romanovskii, O. A., & Tuzhilkin, D. A. (2025). Dual-channel infrared OPO lidar optical system for remote sensing of greenhouse gases in the atmosphere: Design and characteristics. *Sensors International*, 6100307–100307.
43. Ma, H., Liu, Z., Ran, Y., Wu, X. J., & Feng, H. H. (2024).Satellite detection of sea fog and low clouds in the Arctic in the daytime during the summer through multiple sources of infrared remote sensing. *Global and Planetary Change*, 243104615–104615.
44. Appavou, M., Ribeiro, L., Nicaise, P., Jie, Hu., Jean, M. M., & Josiane, F., et al. (2024). Design and simulation of TiN-based suspended meander kinetic inductance detectors (KIDs) for visible and near-infrared astronomy applications. *Journal of Low Temperature Physics, 216*(1–2), 320–327.
45. Teng, W., Wang, Y. J., Feng, Z., Han, F., Liu, J. L., & Zhang, L. X., et al. (2022). A spatio–temporal temperature-based thresholding algorithm for underground coal fire detection with satellite thermal infrared and radar remote sensing. *International Journal of Applied Earth Observation and Geoinformation, 110.*
46. Hanel, R. A., Conrath, B. J., Jennings, D. E., & Samuelson, R. E.(1993). Exploration of the solar system by infrared remote sensing. *Choice Reviews Online, 30*(09), 30–49+72.
47. Nobileau, D., & Antoine, D. (2004). Detection of blue-absorbing aerosols using near infrared and visible (ocean color) remote sensing observations. *Remote Sensing of Environment, 95*(03), 368–387.
48. Perido, J., Day, K. P., Beyer, D. A., Cothard, N. F., Hailey, D. S., Leduc, H. G., et al. (2024). Parallel-plate capacitor titanium nitride kinetic inductance detectors for infrared astronomy. *Journal of Low Temperature Physics, 216*(1–2), 39–49.
49. Ulbricht, G., Lucia, D. M., Piercy, J., Oisín, C., Colm, B., Cáthal, M., et al. (2024). Optimizing Ti/TiN multilayers for UV, optical and near-IR microwave kinetic inductance detectors. *Journal of Low Temperature Physics, 216*(1–2), 175–184.
50. Foote, L., Albert, C., Baselmans, J., Beyer, A. D., Cothard, N. F., Day, P. K., et al. (2024). High-sensitivity kinetic inductance detector arrays for the PRobe far-infrared mission for astrophysics. *Journal of Low Temperature Physics, 214*(3–4), 219–229.
51. Chao, J. Q., Zhao, Z. F., Lai, Z. B., Xu, S. G., Liu, J. Y., Li, Z. Y., et al. (2023). Detecting geothermal anomalies using Landsat 8 thermal infrared remote sensing data in the Ruili Basin, Southwest China. *Environmental Science and Pollution Research International, 30*(11), 32065–32082.
52. Hang, Z., Li, C.L., Hao, X., Lin, C., & Zheng, Y. Q., et al. (2023). Dark background correction of the infrared detector for hyperspectral remote sensing application. *Sensors and Actuators: A. Physical, 349.*
53. Tomoki, T., Kenji, K., Shunsuke, H., Oguri, S., Nishinomiya, Y., Kusaka, A., et al. (2022). Development of Al–Nb hybrid lumped-element kinetic inductance detectors for infrared photon detection. *Journal of Low Temperature Physics, 209*(3–4), 441–448.
54. Anonymous. (2021). Far-infrared microwave kinetic inductance detector (FIR MKID) Array. *NASA Tech Briefs, 45*(12), 36–36.
55. Asgari, M., Viti, L., Zannier, V., Sorba, L., Vitiello, M. S., et al. (2021). Semiconductor nanowire field-effect transistors as sensitive detectors in the far-infrared. *Nanomaterials, 11*(12), 3378–3378.
56. Mezzena, R., Faverzani, M., Ferri, E., Giachero, A., Margesin, B., Nucciotti, A., et al. (2020). Development of microwave kinetic inductance detectors for IR single-photon counting. *Journal of Low Temperature Physics, 199*(1–2), 73–79.

57. Zobrist, N., Coiffard, G., Bumble, B., Swimmer, N., Steiger, S., Daal, M., et al. (2019). Design and performance of hafnium optical and near-IR kinetic inductance detectors. *Applied Physics Letters, 115*(21), 213503–213503.
58. Liu, K. Y., Wang, W. H., Yu, Y. F., Hou, X. Y., Liu, Y. P., Chen, W., et al. (2019). Graphene-based infrared position-sensitive detector for precise measurements and high-speed trajectory tracking. *Nano letters, 19*(11), 8132–8137.
59. Wan, B., & Small, W. G. (2019). Passive airborne fourier transform infrared remote detection of methanol by use of wavelet analysis as a feature extraction method. *Analytical Letters, 52*(14), 2251–2265.
60. Beldi, S., Boussaha, F., Hu, J., Monfardini, A., Traini, A., Levy-Bertrand, F., et al. (2019). High Q-factor near infrared and visible Al_2O_3-based parallel-plate capacitor kinetic inductance detectors. *Optics express, 27*(09), 13319–13328.
61. Wang, K., Jiang, Q. G., Yu, D. H., Yang, Q. L., Wang, L., Han, T. C., et al. (2019). Detecting daytime and nighttime land surface temperature anomalies using thermal infrared remote sensing in Dandong geothermal prospect. *International Journal of Applied Earth Observations and Geoinformation, 80*, 196–205.
62. Beldi, S., Boussaha, F., Chaumont, C., Mignot, S., Reix, F., Tartari, A., et al. (2018). Design of near infrared and visible kinetic inductance detectors using MIM capacitors. *Journal of Low Temperature Physics, 193*(3–4), 184–188.
63. Szypryt, P., Meeker, S. R., Coiffard, G., Fruitwala, N., Bumble, B., Ulbricht, G., et al. (2017). Large-format platinum silicide microwave kinetic inductance detectors for optical to near-IR astronomy. *Optics express, 25*(21), 25894–25909.
64. Guo, W., Liu, X., Wang, Y., Wei, Q., Wei, L. F., Hubmayr, J., et al. (2017). Counting near infrared photons with microwave kinetic inductance detectors. *Applied Physics Letters, 110*(21), 212601–212601.
65. Qiao, S., Feng, K. Y., Li, Z. Q., Fu, G. S., & Wang, S. F. (2017). Ultrahigh, ultrafast and large response size visible-near-infrared optical position sensitive detectors based on CIGS structures. *Journal of Materials Chemistry, 5*(20), 4915–4922.

Chapter 5
Temperature Detection Method

Accurate detection of coal fire temperature is the core technical link in the prevention and control of spontaneous combustion disasters in coalmines. The current main methods include contact and non-contact: the contact temperature sensor method (such as thermocouple and thermal resistance) relies on the characteristics of the physical parameters of the component to change with temperature, which is direct and reliable, but has the limitations of dense deployment, high cost and susceptibility to environmental interference. According to Stephen – Boltzmann law, the non-contact infrared temperature measurement method can quickly obtain large-area thermal images by capturing the infrared radiation on the surface of objects to invert the temperature distribution, which is convenient to operate, but the ability to detect internal or hidden fire sources is insufficient. Based on the principle of Raman scattering effect and optical time-domain reflection (OTDR), distributed optical fiber temperature measurement technology realizes long-distance, continuous and real-time temperature monitoring, with high spatial resolution, anti-electromagnetic interference, corrosion resistance, etc., especially suitable for complex hazardous areas such as underground goafs and roadways, but its optical fiber deployment and protection requirements are high. The supporting equipment has also developed from handheld infrared thermometers and thermal imagers to distributed fiber optic temperature measurement systems (DTS) integrating pulsed lasers, wavelength division multiplexers and high-speed acquisition cards. In the future, temperature detection technology is developing in the direction of multi-source integration, high-precision positioning and intelligent early warning to improve the reliability and efficiency of coal fire prevention and control.

Y. Xiao et al., *Detection of Fire Source in Underground Coal Mine and its Applications*,
https://doi.org/10.1007/978-981-95-7064-5_5

5.1 Temperature Detection Methods and Technologies

In the process of coal fire combustion, the change of temperature field is the most intuitive, and its development can be divided into three stages: oxidative self-heating stage, coal seam combustion period and extinguishing period. At this stage, the temperature measurement technology used at home and abroad to detect the location of coal spontaneous ignition mainly covers temperature sensor method, infrared temperature measurement method and optical fiber method. The temperature sensor method uses the characteristics of component performance parameters to change with temperature to detect the temperature abnormal area, and then locate the hidden fire source. This method is suitable for the detection and research of fire areas, such as shallow coal-bearing outcrops and open-pit coal yards with limited coal storage. For example, in some shallow coalfield outcrop areas, by arranging temperature measurement sensors, the abnormal temperature increase caused by coal spontaneous combustion can be effectively captured, so as to preliminarily determine the scope of the fire area. However, this method has certain limitations in practical application, its workload is large, it needs to be intensively arranged in a large area, the investment cost is high, and it is greatly affected by environmental factors, and the stability is not good.

Infrared temperature measurement method detects the radiation power signal in the fire area through a thermal imager, and then generates a thermal image map corresponding to the temperature distribution of the surface of the object. Compared with the temperature sensor method that requires burying the probe, the infrared temperature measurement method is easier to operate, and the temperature information of a large area can be quickly obtained without direct contact with the detected object. At present, it is one of the most commonly used temperature detection methods. According to Stephen − Boltzmann law, the overall radiation intensity $E_{(T)}$ (unit: W/m^2) and the absolute temperature T (unit: K) of the object at temperature T satisfy the relationship $E_{(T)} = \sigma T^4$, where the σ is the Stephen Boltzmann constant, and the value is 5.67×10^{-8} $W{\cdot}m^{-2}{\cdot}K^{-4}$. This principle provides a solid theoretical basis for infrared temperature measurement. In practical application scenarios, such as monitoring the spontaneous combustion temperature field of coal piles, outcrops, coal pillars and close coal seams, the infrared temperature measurement method shows good results. However, it also has certain shortcomings, and is not suitable for temperature detection inside loose coal bodies or goafs, because the thermal radiation signals in these areas are easily blocked or interfered with, resulting in the inability to accurately obtain temperature information.

Distributed fiber temperature measurement technology uses the response characteristics of specific optical signals transmitted by optical fibers to temperature, and realizes distributed temperature measurement through signal demodulation at one end of the optical fiber. When the optical signal is transmitted in the optical fiber, it can be divided into scattered light and transmitted light after temperature modulation, among which scattered light is more widely used in temperature measurement. This technology enables continuous and real-time temperature monitoring of areas distributed along optical fibers, with the advantages of accurate positioning and high sensitivity. In some coalmines, underground roadways or goafs can be laid to monitor the temperature changes of coal in real time and detect potential coal spontaneous combustion hazards in time. However, the arrangement of optical fibers is difficult, and in complex underground environments, it is easily affected by environmental factors such as ground pressure, rockfalls, and closures or mining processes, resulting in optical fiber damage or inaccurate measurement data. The temperature detection method is the most intuitive and physically significant method to infer the state, scope and development trend of underground coal fire directly by measuring the temperature in the surface, shallow soil or hole. Its theoretical basis is the theory of heat conduction (Fourier law) and the law of thermal radiation (Stephen – Boltzmann law, Wien law of displacement). As a high-temperature heat source, the heat generated by underground coal fire is transferred upward through conduction, convection and radiation, causing abnormalities in surface and shallow ground temperature, and the detection of these anomalies is the fundamental purpose of this method.

5.1.1 Contact Temperature Measurement Method

According to incomplete statistics, 90% of underground fires in coalmines originate from spontaneous combustion of coal left in goafs. Therefore, it is extremely important to strengthen the prediction and prediction of spontaneous combustion of coal in goafs, and scholars at home and abroad have carried out a lot of research from the perspective of temperature and characteristic gases. Zhang [1] et al. established a spontaneous ignition model of coal left in the goaf according to the ventilation characteristics of the roadway and the working face, and combined with numerical simulation technology to develop the spontaneous combustion characteristics of the coal left in the goaf under the conditions of not leaving the roadway, not taking anti-leakage measures and taking anti-leakage measures in the roadway, and proposing a hierarchical prevention and control technology system for fire extinguishing along the empty roadway from the perspective of temperature control. Yang [2] et al. obtained the oxygen concentration distribution of the goaf on both sides of the isolated coal pillar under the influence of mining by on-site temperature measurement and sampling analysis of the new and old goafs of the 2603 and 2601 faces of Zhangcun Coalmine, and revealed the characteristics of secondary oxidation of residual coal in the goaf. Liu [3] constructed an experimental system for simulating the spontaneous combustion of residual coal in the goaf, and judged the development status of spontaneous combustion of residual coal in the goaf by the change of the concentration of the characteristic element radon. Zhang [4] used CO as the index gas to study the effect of high-efficiency composite foam on the prevention and control of spontaneous combustion of coal left in goafs by combining numerical simulation and physical experiments. Zhu [5] et al. used CO, C_2H_6, and C_2H_4 to judge the combustion status and development direction of the fire area by using the iconic gases produced in the low-temperature oxidation stage of coal. In addition, Zheng et al.

[6] reviewed the development status and application effects of common coalfield fire monitoring technologies, including magnetic detection method, resistivity method, radio wave method, geological radar method, gas measurement method, isotope radon measurement method, temperature method, remote sensing method, etc., and proposed that low-altitude remote sensing technology with UAV mounted infrared thermal imager is the development direction of coalfield fire detection. Yang et al. [7, 8] and Qin [9] et al. used infrared thermal imaging to carry out experimental research on the imaging effect and detection accuracy of high-temperature coal bodies in the lane gang under different parameters and industrial and mining conditions.

The above-mentioned scholars have made rich research results in the detection of spontaneous combustion of residual coal from multiple angles, but the development degree of spontaneous combustion of residual coal in the empty area is in the stage of indirect prediction, and there is no direct observation data to explain. The reality that operators cannot enter the goaf for direct detection has attracted much attention to the emergence of non-contact temperature measurement methods. This method allows operators to directly scan the remaining coal in the goaf behind the two smooth grooves with an infrared thermal imager at the end of the working face, so as to quickly capture the high temperature point and its location within the scanning range, and then take more targeted prevention and control measures. At present, some scholars have studied the effects of infrared detection under different particle sizes, moisture contents, and heat source depths [10, 11], but the prediction effect and accuracy of spontaneous combustion of residual coal based on infrared detection distance have been rarely reported.

Infrared rays are the most common electromagnetic waves in nature, with a wavelength range of 0.76–100 m, which can be divided into four types according to wavelength: near-infrared, mid-infrared, far-infrared and extreme far-infrared. Infrared detection involves capturing the heat radiated by the object and converting electrical signals, and displaying infrared thermal images through the display screen through amplification, conversion, or standard video signals [12–14]. As shown in Fig. 5.1, under a fixed scanning angle, when the detection distance is different, the scanning range between the infrared thermal imager and the object is different, and the infrared light entering the thermal imaging lens increases with the increase of the detection distance, and the more factors affect the detection accuracy of the target object [15–17].

Theoretically, after excluding the influence of environmental factors and human operation reasons on the accuracy of the detection results, the calculation formula of the detection distance is shown in Eq. 5.1.

$$\begin{cases} \mathrm{I} = \frac{\mathrm{Y \cdot F}}{\mathrm{N \cdot a}} \\ \mathrm{F = b \cdot m} \end{cases} \tag{5.1}$$

where, I is the detection distance of the infrared thermal imager, m; Y is the length of the short side of the target object, m; F is the product of b and m; N is the number of pixels occupied by the short-side imaging of the optical axis section of the detectable

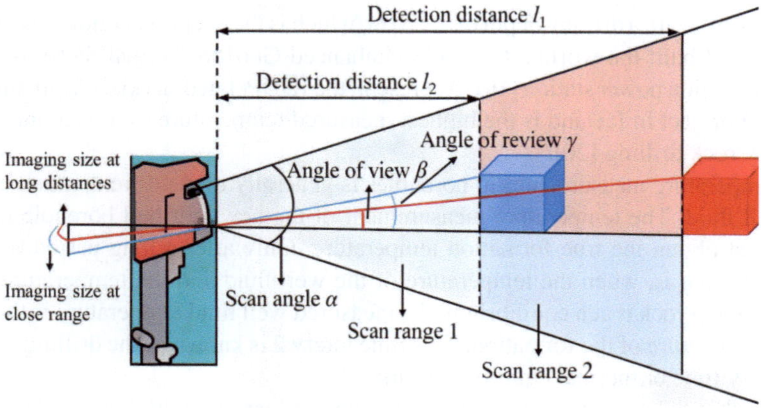

Fig. 5.1 Schematic diagram of infrared temperature measurement

target; a is the cell spacing, μm; b is the distance between the lens of the infrared thermal imager and the short side of the target object, m; m is the focal length of the lens, m.

5.1.2 Shallow Ground Temperature Detection Method

(1) Drilling temperature measurement method:

The direct underground temperature measurement method in coalmines is a commonly used temperature measurement method, which mainly evaluates the temperature of key areas such as mining faces and main ventilation roadways to understand the underground thermal environment conditions. Use a temperature measurement instrument with high accuracy, such as a digital thermometer, and use a temperature probe or detector to measure the temperature directly at a specific location in the coalmine. The probe is generally connected to the temperature display instrument by a connecting wire. Generally, regular measurements are adopted, such as daily, weekly or monthly measurements to grasp the changes in downhole temperature. It is necessary to choose the appropriate measurement position to avoid the influence of local temperature anomalies. At the same time, the measurement data should be recorded and analyzed to provide a basis for subsequent improvement of the thermal environment.

The research history of underground direct temperature measurement in coalmines can be traced back to the period of human development of geothermal resources, and some geological tectonic movements have caused the earth to enrich a large amount of internal heat in a certain area, and reach the level that human beings can develop and utilize, which constitutes geothermal resources or geothermal energy resources [18]. On April 11, 2004, the Sutz project in France completed the construction of the

GPK4 well, with a drilling depth of 5,270 m, which is the deepest dry hot rock drilling so far, and built the world's first EGS (Enhanced Geo158 thermal Systems, EGS) demonstration power station [19]. At present, the IDDP-1 geothermal well in the deep drilling project in Iceland is the highest measured temperature ever encountered by hot dry rock drilling [20].

Temperature measurement in boreholes is generally to measure the temperature of well fluid. The temperature measurement of the newly drilled borehole usually does not obtain the true formation temperature. Only after a long period of static stability, that is, when the temperature of the well fluid and the temperature of its surrounding rock reach equilibrium, the measured well fluid temperature is the original temperature of the formation. This time interval is known as the drilling thermal recovery time or thermal equilibrium time.

The thermal recovery time of the borehole is generally quite long, and it is very complicated to give a quantitative evaluation. BULLSRD [21] theoretically estimated the thermal recovery time of the borehole, and he believed that the disturbance of the temperature field of the borehole and the surrounding rock was caused by a constant heat source along the borehole axis. At a given point in the borehole, the action of the heat source begins from the moment the drill bit touches this point until the end of the drilling and well fluid circulation, and it is assumed that: (1) the heat diffusivity of the well fluid in the borehole is equal to the surrounding rock of the borehole. (2) Drilling is continuous and uniform. (3) Consider heat exchange only by conduction. The change rate of rock temperature at the axis distance r of the borehole is ΔT in the time t after the start of drilling

$$\Delta T = -\frac{Q_1}{4\pi k} E_i\left(-\frac{r^2}{4\chi \iota}\right) \tag{5.2}$$

where, Q1 is the heat source intensity along the drilling axis, J/(s· cm); k is the thermal conductivity of the deer, J/(cm· s· K); χ is the thermal diffusivity of the rock, m^2s; E_i is the exponential integral function, that is,

$$E_i(-Z) = \int_z^{\infty} \frac{e^{-Z}}{Z} dZ \tag{5.3}$$

Among them, Z is a dimensionless index indicator, and its value varies from 1 to 10^{-15}. If Z is less than 0.03 and the allowable error is 1%, then $E_i(-Z) \approx \ln Z + 0.577$, then the following relationship is

$$\frac{\Delta T}{\Delta T_o} = \frac{\ln\left(1 + \frac{t_1}{t_2}\right)}{\ln\left(\frac{4X\iota_1}{a^2}\right) - 0.577} \tag{5.4}$$

where: $\Delta T0$. is the initial coal body temperature change rate, t_1 is the drilling time, d; a is the drilling radius, cm; t_2 is the thermal equilibrium time, d.

Usually when $\Delta T/\Delta T_0$ is known, the thermal equilibrium time can be deduced. This method requires the help and retention of boreholes, except for special circumstances, generally need to be washed, and it is unrealistic to measure steady-state temperature at the bottom of the well. Therefore, there are two parts of the data that are more reliable for borehole temperature measurement in geological exploration: ① the bottom of the hole, and ② the temperature of the neutral point or neutral section. The ground temperature at these two sites is minimally disturbed by drilling and recovers the fastest, and the temperature at these two points may be very close to the original ground temperature within a few days after the drilling is stopped.

Jaeger [21] argued that the heat exchange between the borehole fluid and the surrounding rock occurs radially in the borehole at the point close to the bottom of the borehole, and as long as there is no heat exchange along the borehole, there is the following relationship:

$$\frac{T - T_v}{T_0 - T_v} = \frac{4\alpha}{\pi^2} \int_0^\infty \frac{\exp(-\tau u^2)du}{u\left\{\left[uJ_0(u) - (\alpha - hu^2)J_i(u)\right]^2 + \left[uY_0(u) - (\alpha - hu^2)Y_i(u)\right]^2\right\}} \tag{5.5}$$

where, $J_i(u)$ and $Y_i(u)$ ($i = 0, 1$) are the first and second types of Bézier functions, respectively; Tv is the original ground temperature; T_0 is the temperature of the pore liquid when $t = 0$; T is the temperature of the pore liquid at a certain time; $h = k/(aH)$ represents the contact between the well liquid and the surrounding rock, and H is the heat transfer washing of the well wall. $\alpha = 2\rho c/(\rho_1 c_1)$, which represents the comparative relationship between the surrounding rock and the thermal properties of the well fluid, ρ, c and ρ_1, c_1 represent the density and specific heat of the surrounding rock and well fluid, respectively; $\tau = X\,t/a^2$, t is the time from the start of temperature measurement to the drilling point, and h, α and τ are all factor less quantities.

Order, $\frac{T-T_v}{T_0-T_v} = n$, then $n = 1 - \frac{T-T_v}{T_0-T_v}$. It is the degree to which the well liquid temperature recovers at a certain time at the point near the bottom of the well, that is, $n = 1 - F(h, \alpha, \tau)$. For drilling, the well fluid is mainly water, which has good thermal contact with the well wall, so $h = 0$, then the $F(h, \alpha, \tau)$ function becomes $F(0, \alpha, \tau)$. After calculation in Fig. 5.2, the $F(0, \alpha, \tau)$ function to τ relation function is obtained [21].

For aquifer boreholes with consistent diameter, well fluid temperature T can be measured over a series of time t in rock formations with known or hypothesized thermal properties and higher temperatures than well fluid. In this way, the value of $F(0, \alpha, \tau)$ can be calculated with the help of Fig. 5.2, and the n value of each time can be obtained from this. Although in the actual drilling clock, the value of T_0 cannot be measured, but as long as a series of T_t, α, n_t are measured, it can be measured through $T = T_0 + (T_v - T_0)n$ parses T_v and T_0 comes.

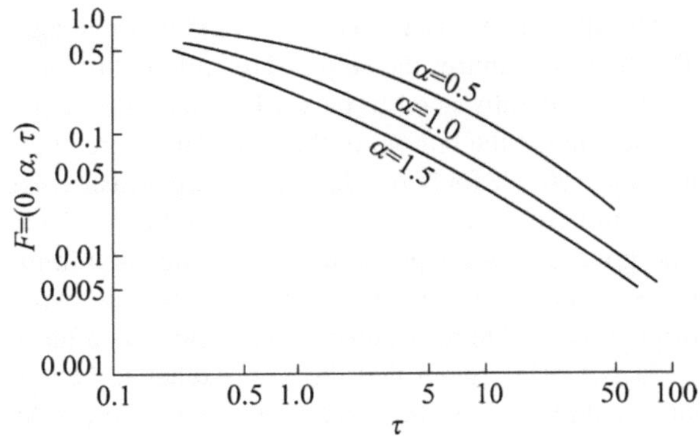

Fig. 5.2 The relationship function diagram of F (0, α, τ) function to τ is obtained

(2) Distributed optical fiber temperature measurement technology

With the increasing development of science and technology and the continuous deepening of research, a large number of research results have been obtained in fiber optic temperature measurement technology [22].

The initial research on optical fiber temperature measurement was based on the temperature sensor developed by Th. Bosselmann, mainly on the fluorescence lifetime, which is a device with a core diameter of 0.6 mm and a temperature probe diameter of 1.5 mm, with a temperature measurement range of 273–343 K [23]. At the beginning of the twentieth century, Indian physicist Raman discovered that the frequency of light waves changes when they are scattered, which later generations called the Raman effect (Raman scattering). British physicist Rayleigh discovered another type of scattered light, Rayleigh scattering, in 1900 while studying light waves. The discovery of two types of scattered light laid the foundation for the research and development of optical fiber temperature measurement technology. In 1987, Chongqing University successfully developed a laser pulser with a spatial resolution of 4 m and a temperature measurement length of 1 km [24]. A year later, the university developed a fiber optic temperature measurement system, with a temperature distance of an astonishing 30 km, but the spatial resolution and temperature resolution were only 4 m and 2 °C.

In recent years, optical fiber temperature measurement equipment has begun to be widely promoted as an emerging equipment in our country's coal industry for temperature monitoring such as underground goafs, conveyor belts, and underground cables. Optical fibers have the advantages of corrosion resistance, electromagnetic interference resistance, high sensitivity, and continuous monitoring. The United Kingdom is the first country to use optical fiber temperature measurement technology to be applied to the coal industry, and Chinese coal industry has also begun to vigorously develop optical fiber temperature measurement technology in recent years, and has achieved good results. Li [25] used fiber optic temperature measurement technology to monitor the temperature in real time in the goaf of the 15,104 fully mechanized mining face, Mapu Coalmine, divided the scope of spontaneous combustion into "three zones" and the safe propulsion speed, and provided a feasible solution for the prevention and control of spontaneous combustion of residual coal, which achieved the expected goal. Wang [26] et al. of Anhui University of Science and Technology used the layout scheme of laying "W" shaped optical fibers along the strike to monitor and monitor the temperature of the goaf in a large area, which more comprehensively reflected the temperature distribution law of the goaf. Wei [27] et al. used industrial control technology to integrate and integrate optical fiber temperature measurement equipment with nitrogen production equipment in Guotun Coalmine, Shanxi, China, and developed a complete set of systems for real-time temperature monitoring and early warning in goafs and intelligent nitrogen injection through automatic control technology, which achieved the purpose of suppressing the spontaneous combustion of coal left in the goaf.

(1) Selection of mining fiber optic temperature measurement technology

The distributed optical fiber system uses optical fiber to realize temperature measurement and positioning, and the main technical means are to use the scattering principle of light and optical time-domain reflectometry technology (OTDR) to obtain the specific information of the temperature of coal residual in the goaf. When a beam of light enters the fiber, scattering occurs during propagation, i.e., forward scattered light and backscattered light. Backscattered light includes Rayleigh scattering, Raman scattering, and Brillouin scattering.

For Rayleigh scattering and Brillouin scattering temperature measurement techniques, the occurrence of external stress and the change of temperature have a great influence on the frequency of scattered light. Due to the huge pressure generated by the overlying rock in the goaf of the working face, if the optical fiber is squeezed and deformed, the reflection ability disappears, which is equivalent to losing the ability to measure temperature. However, in the process of laying, installing and using optical fibers in coalmines, it is inevitable to be subjected to external forces, and if the above two reflection technologies are used to monitor the temperature change of the goaf, it will inevitably cause large errors and cannot meet the requirements of underground temperature measurement. However, Raman scattering technology is not affected by external stress interference and is only sensitive to temperature, because temperature measurement is measured by the energy change generated by the collision of photons inside the optical fiber. Therefore, Raman scattering fiber temperature measurement technology is the best choice when measuring goaf temperature in coalmines.

(2) Raman scattering temperature measurement technology

Raman scattering in optical fibers is the scattering of frequency changes caused by the collision of light beams with optical fiber molecules. When the incident photon absorbs energy, it is converted into high-frequency anti-Stokes scattering photons, When the incident photon loses energy, it is converted into a low-frequency Stokes – Raman scattering photon [28]. The Raman scattering level diagram of the photon is shown in Fig. 5.3.

From the Raman scattering quantum theory, it can be seen that when the incident light collides with the medium molecules, the photon loses part of the energy, and the molecule transitions from the low-energy state to the excited state. The photon gains a

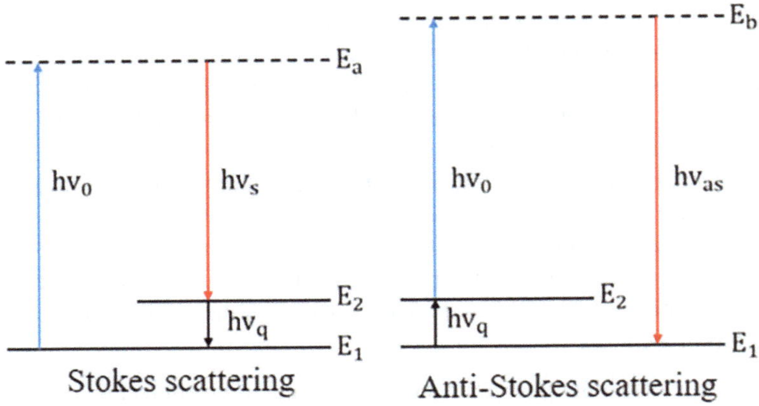

Fig. 5.3 Schematic diagram of Raman scattering energy level

part of the energy, and the molecule transitions from the excited state to the low energy state, at which point the scattered light is Anti – Stokes. Anti – Stokes scattered light is more sensitive to temperature than Stokes scattered light. The temperature of the optical fiber system at any time can be obtained by the following equation:

$$T = \frac{h\Delta\gamma/kT_0}{h\Delta\gamma - kT_0 \ln\left[\frac{P_{as}(T)/P_s(T)}{P_{as}(T_0)/P_s(T_0)}\right]} \tag{5.6}$$

where, h is Planck common practice, J·s; k is the Moorsmann constant, J/K; $\Delta\gamma$ is the frequency of the offset wave, kHz; Pas is the frequency of Anti – Stokes scattered light, dB; T_0 is the absolute temperature at a certain time, K; T is the absolute temperature at any moment, K.

The temperature T at any time after the light intensity of the Anti – Stokes and Stokes scattered light can be measured by the above equation, and then the temperature values of all measurement points on the entire fiber can be obtained [29].

(3) Optical time domain (OTDR) reflection technology

The basic principle of optical time domain (OTDR) reflection is the theory of coordinate positioning of the temperature of the measured area by the distributed optical fiber temperature measurement system, which is called Optical Time Domain Reflectometer, which was first proposed by Barnoski in 1977. It was initially used in the field of optical communication, and later gradually applied to mine fire prevention.

As shown in Fig. 5.4, the pulsed laser emits a very narrow high-intensity light pulse inside the optical fiber, and a part of it is scattered during the propagation process, and this part of the scattered light will return to the emitter of the light source along the path of the source and enter the detector through the coupler, and finally the measured position is located through the signal captured by the detector [30]. Assuming that scattering occurs at point M in the temperature measurement

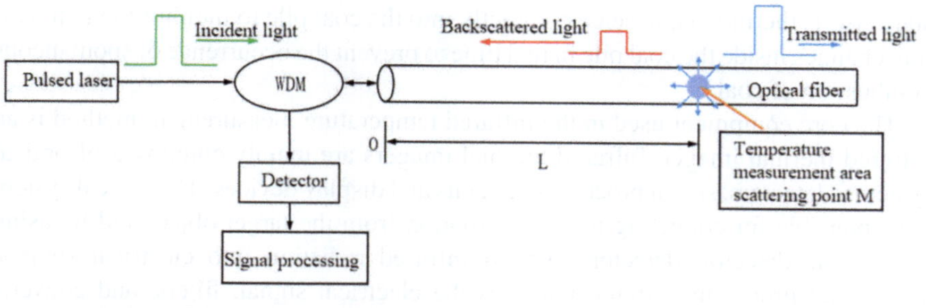

Fig. 5.4 Optical time-domain reflection flow diagram

area, the distance between points OM is set to L, and the light pulse returns to point O after backscattering when passing through point M, and the time taken is set to t when the light pulse passes through a distance of 2L, then the following relationship exists:

$$2L = v \times t \tag{5.7}$$

$$v = C/n \tag{5.8}$$

where, L is the distance from the temperature measurement to the starting point, m; v is the speed at which light travels through the optical fiber, m/s; C is the speed of light propagation in a vacuum, m/s; n is the refractive index of the optical fiber, 1.4–1.6.

5.2 Temperature Detection Equipment

While temperature detection methods continue to develop, the corresponding detection equipment is also constantly updated and improved. For different temperature detection technologies, various types of detection equipment have been developed.

For the temperature measurement sensor method, common equipment includes thermocouple sensors, thermal resistance sensors, etc. Thermocouple sensors work based on the thermoelectric effect, when two conductors of different materials form a closed loop and the two contacts have different temperatures, a thermal potential is generated in the loop, and the temperature is estimated by measuring the thermal potential. Thermal resistance sensors measure temperature by using the characteristics of the resistance value of conductors or semiconductors with temperature, which has the advantages of high measurement accuracy and good stability. These sensors can be made into different shapes and specifications according to actual needs to adapt to various complex detection environments. In some small open-pit coal yards, the staff inserts thermocouple sensors directly into the coal pile to monitor the temperature changes inside the coal pile in real time to prevent the occurrence of spontaneous combustion of coal.

The core equipment used in the infrared temperature measurement method is an infrared thermal imager. Infrared thermal imagers are mainly composed of optical systems, detectors, signal processing circuits and display devices. The optical system is responsible for collecting infrared radiation from the target object and focusing it onto the detector, detectors convert infrared radiation into electrical signals, The signal processing circuit amplifies the electrical signal, filters, and converts analog-to-digital conversion, and finally generates an intuitive thermal image on the display device. With the continuous advancement of technology, the performance of infrared thermal imaging cameras has been significantly improved, and their resolution is getting higher and higher, and it can more clearly distinguish areas with small temperature differences, The temperature measurement range is also constantly expanding, which can meet the temperature measurement needs in different scenarios. For example, some high-end infrared thermal imaging cameras have temperature measurement accuracy of up to ± 0.1 °C, which can accurately detect subtle temperature changes in coal fire areas. In practical applications, operators only need to hold an infrared thermal imager and scan the target area to quickly obtain the temperature distribution of the area, providing an efficient and convenient means for the detection of underground coal fires.

The distributed optical fiber temperature measurement system is mainly composed of optical fiber, light source, signal demodulator and data processing software. The optical signal emitted by the light source is transmitted to the monitored area through the optical fiber, and during the transmission process, the optical signal is moderated by temperature to produce scattered light, which carries the temperature information and then returns to the signal demodulator through the optical fiber. The signal demodulator analyzes and processes the scattered light, demodulates the temperature information, and transmits it to the data processing software. The data processing software stores, analyzes, and displays the temperature data in an intuitive way for the user. The system can achieve long-distance and wide-range temperature monitoring, and can flexibly adjust the laying path and monitoring range of optical fibers according to actual needs. In the monitoring of the underground goaf of the coalmine, the optical fiber is laid along the roadway around the goaf, which can monitor the temperature change of the coal body in the goaf in real time.

5.2.1 Ground Temperature Measurement Equipment

(1) Infrared detection principle

Infrared rays are a type of electromagnetic radiation emitted from the inside of matter, The intensity and spectral composition of thermal radiation depend on the temperature of the radiator. After the spontaneous combustion of coal, the infrared radiation energy increases [31], but this trend is strong in some bands and not obvious in others. Areas with large variations in infrared radiation energy are usually referred to as detectable bands. The detectable wavelength of coal spontaneous combustion is mainly 8.0–15.1 μm.

According to the Stephen – Boltzmann law of radiation, as long as any object is above absolute zero (0 K), molecular vibrations and lattice vibrations will be generated, and infrared electromagnetic waves will be emitted outward, forming an infrared radiation field. The relationship between the radiation field and the temperature is shown in the following equation:

$$E = \varepsilon\alpha T^4 \tag{5.9}$$

where, E is the radiation flux density, W/m^2; ε is the radiation coefficient; α is the Stephen – Boltzmann constant, 5.67×10^{12} $W/(m^2 \cdot K^4)$; T is the temperature.

From Eq. (5.9), it can be seen that the relationship between infrared radiation flux density and temperature is to the fourth power, and temperature change can cause a significant change in radiation flux density. Therefore, the infrared radiation flux of the test coal body can accurately invert the temperature value of the coal body. The infrared detectors for coalmines mainly include infrared thermometers and infrared imagers.

(2) Handheld infrared thermometer: When choosing, you need to pay attention to the optical resolution (D:S), temperature measurement range (recommended above – 20–600 °C), spectral response (8–14 μm), emissivity adjustment function and laser aiming method. It is an essential tool for rapid screening in the field.

How it works: All objects with temperatures above absolute zero are constantly emitting infrared radiation energy into the surrounding space. The magnitude of the infrared radiation energy of an object and its distribution by wavelength are closely related to its surface temperature. Therefore, by measuring the infrared energy radiated by the object itself, its surface temperature can be accurately determined.

Handheld infrared thermometers have a key measurement attribute, the D:S attribute, which is the ratio of distance to spot diameter [32]. The thermometer optical system collects energy (90% of the total energy) from the circular measurement spot and focuses it on the detector. The larger the D:S ratio, the better the resolution of the infrared thermometer and the smaller the diameter of the measured spot. To ensure the accuracy of the measured temperature, it is necessary to control that the distance between the thermometer and the measured object is less than the product of the width of the object and the D:S ratio. If this distance is exceeded, the measurement spot will overflow the surface of the measurement object, which will be disturbed by the background temperature and cause a large deviation (Fig. 5.5).

(3) Infrared thermal imager

   (1) Types: It is divided into refrigerated type (using indium antimonide InSb or mercury cadmium telluride MCT detector, high sensitivity, used for precision scientific research) and uncooled type (using vanadium oxide

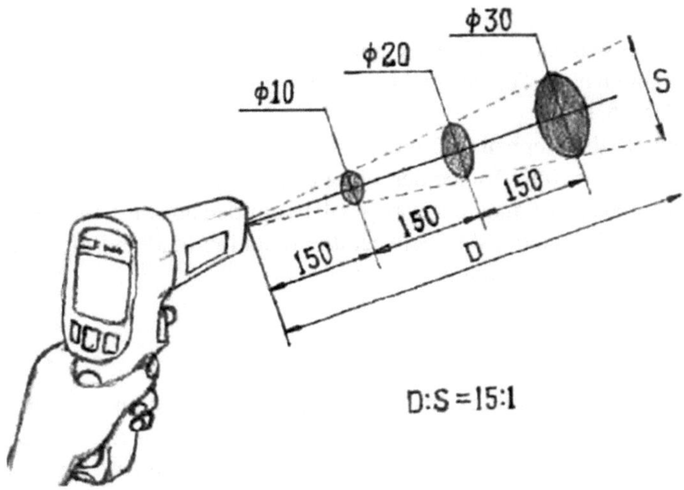

Fig. 5.5 Schematic diagram of handheld infrared thermometer

microbolometer detector, small size, low power consumption, low cost, suitable for engineering census) [33].

(2) Key parameters: thermal sensitivity (NETD), the smaller the value, the better, high-quality equipment can reach <0.03 °C, spatial resolution (IFOV), and field of view (FOV). Thermal radiation video recording for dynamic analysis.

(3) Composition: It is composed of a multi-channel data collector, a thermocouple/thermal resistance sensor array, a solar power supply system and a wireless data transmission module (4G/5G or LoRa).

(4) Principle: The infrared imager can test the surface temperature field of the mine roadway to judge the location of the high temperature point in the coal body near the roadway. Objects usually radiate energy in the form of electromagnetic waves, and the thermodynamic temperature of the object is directly proportional to the radiation energy. As a commonly used temperature measurement technology, infrared thermal imaging thermometer mainly uses infrared detection technology to obtain thermal information on the infrared radiation state of the equipment, and forms a temperature value that can be displayed after conversion and processing. During the actual measurement period, the thermometer can measure the average temperature of the measured area and determine whether there are abnormal problems in the operation of the equipment based on the analysis of temperature parameters [34]. The infrared radiation of the equipment is received by using the infrared thermometer as the medium of air, and the radiation energy can be converged with the help of the optical system, and the energy

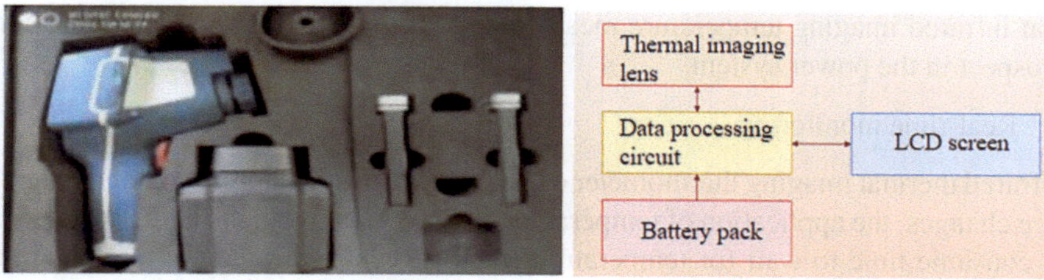

Fig. 5.6 Infrared thermal imaging working principle

radiation is converted into electrical signals by the detector, and the temperature data of the measured area is intuitively presented after processing. The working principle of infrared thermal imaging is shown in Fig. 5.6.

(5) Advantages

① Non-contact temperature measurement

The traditional contact temperature measurement method requires direct contact with the test object through the temperature measurement instrument to obtain temperature data, which is easy to cause interference to the measured object, especially when the temperature is high or the object is in motion, the impact is more significant. The application of infrared imaging temperature measurement technology realizes non-contact temperature measurement by receiving infrared radiation emitted by the object, without direct contact with the measured object, which can effectively avoid interference with the measured object and significantly improve the safety of the measurement process [34]. In addition, infrared thermal imaging technology can also achieve accurate temperature measurement of equipment in motion, irregular shape, or difficult to contact directly (such as high-voltage equipment in power systems, transmission lines and transformers), and can complete temperature monitoring without downtime. This technology not only improves the efficiency of temperature measurement, but also greatly reduces the risks caused by manual operation.

② Large-scale temperature measurement

Conventional temperature measurement equipment is limited by its effective range, making it difficult to achieve accurate measurements of extremely hot or cold objects. Infrared imaging thermometers exhibit a wider temperature measurement range, covering a temperature range from very low to very high, enabling accurate temperature measurement of equipment in operation. In the power system, many critical equipment (such as transformers, cables, and insulators) require continuous temperature monitoring. With the help of infrared imaging temperature measurement technology, by receiving the infrared signal radiated by the object, high-temperature equipment can be monitored quickly and accurately, regardless of the size and geometry of the target object. It is precisely because of its wide temperature adaptability that infrared imaging temperature measurement technology has a wide application prospect in the power system.

③ Real-time monitoring

Infrared thermal imaging thermometer can achieve real-time monitoring of temperature changes, the application of temperature measurement methods in the past needs to consume time to wait for temperature stability, while infrared thermal imaging thermometer can obtain the temperature distribution image of the target object in a short time. During the operation of the power system, temperature abnormalities can be detected in time and potential faults or fire risks can be effectively prevented [35].

5.2.2 *Drilling Temperature Measurement Equipment*

(1) Portable digital drilling thermometer

To meet the needs of field work, we have designed a portable digital temperature measurement instrument powered by dry batteries based on the principles of simple structure, easy to carry and use, power saving, and accurate and reliable [36].

The 87 drilling digital thermometer takes AD590, LH0044 and ICL7106CPL as the main body, plus some CMOS digital circuits, and uses coding methods to expand the scope of use of 7106, thereby simplifying the structure of the instrument and reducing the cost, and its measurement results are typed out by liquid crystal display and micro printer.

System composition: It is composed of temperature sensor, temperature measurement circuit, A/D conversion circuit, decimal point drive circuit, display circuit, coding circuit, printing interface and printer (Fig. 5.7) [37].

1) Main performance

Temperature measurement range: −55–150 °C. The measurement error of 0 °C– + 50 °C is ±0.15 °C, +50 °C– +100 °C is ±0.3 °C. Display LCD display, maximum display number is 199.9, resolution is 0.1 °C, automatic indication of

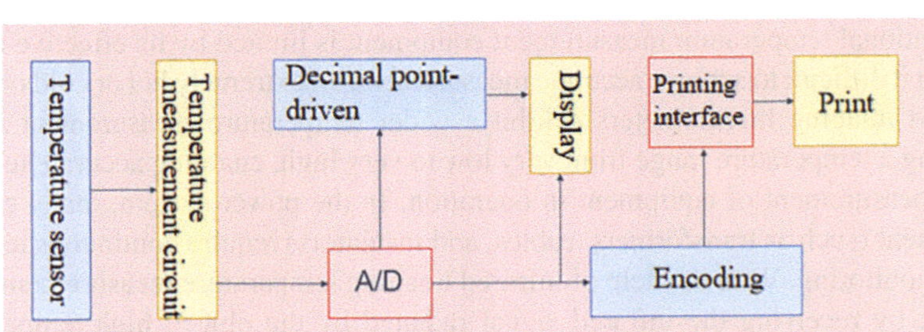

Fig. 5.7 Working principle diagram of the 87 drilling digital thermometer

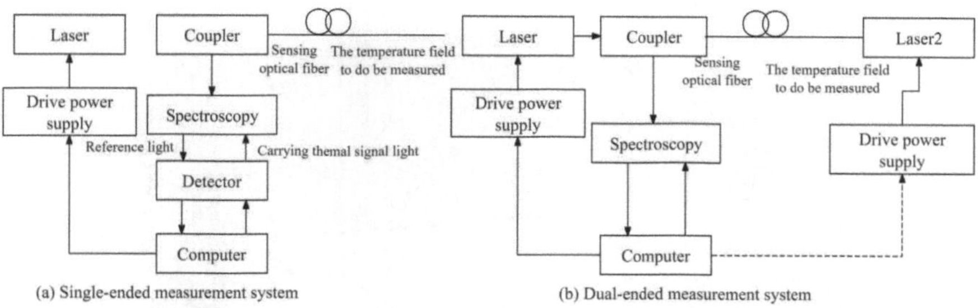

Fig. 5.8 Principle of distributed fiber optic temperature measurement technology based on scattered light [39]

polarity and overload, manual control printing, power supply ± 9 V. Weight: 2 kg, volume: 280 mm × 240 mm × 110 mm.

(2) Distributed optical fiber temperature measurement system (DTS)

Distributed fiber temperature measurement technology uses the response of specific optical signals transmitted by optical fibers to temperature, and realizes distributed temperature measurement through signal demodulation at one end of the optical fiber.

After the signal is modulated by temperature, it can be divided into scattered light and transmitted light, of which scattered light is more commonly used. Figure 5.8 shows the principle of distributed fiber optic temperature measurement technology based on scattered light [38].

1) Basic components of distributed optical fiber temperature measurement system

The main components of the optical fiber temperature monitoring device realized by Raman scattering technology include pulsed laser generator wavelength division multiplexer, dual-channel photoelectric detection element, fast information acquisition interface and computer system. According to the technical specifications of the fiber optic temperature monitoring system, parameter planning and hardware selection are implemented around different components. Through this process, the effective monitoring of the temperature of the coalmine goaf not only effectively reduces the system development cost, but also uses a bidirectional high-speed data acquisition card design with a USB standard interface, which greatly improves the portability of the system. The main components of the new Raman distributed temperature sensor for monitoring the temperature of the goaf are [40].

① Pulsed lasers

In the design of fiber optic temperature measurement systems, selecting the appropriate pulsed laser is crucial as it directly determines the system performance and monitoring effectiveness. The selection of pulsed laser equipment requires a comprehensive evaluation of four core elements, namely: the main wavelength of the pulsed light, the pulse width, the repetition frequency, and the power value of the pulsed light. When selecting a pulsed laser, the core wavelength setting is crucial. Pulsed light

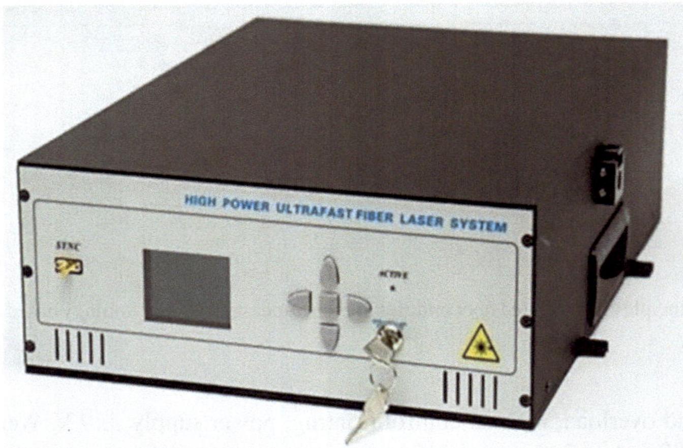

Fig. 5.9 Pulsed laser

transmission properties of different lengths vary within temperature-sensitive fibers, which is directly related to the boundaries of system detection and its sensitivity. Normally, when the wavelength of the pulsed laser decreases, the phenomenon of spontaneous Raman scattering becomes more intense, but its propagation range also decreases. Therefore, in the process of selecting a model, it is necessary to balance the brightness of autonomous Raman scattering and the transmission distance. Currently, there are three common pulsed lasers with three output center wavelengths, namely 850, 1330, and 1550 nm. In order to meet the conditions of monitoring at least 10 km, it was decided to use a pulsed laser emitter with a center wavelength of 1550 nm to ensure more accuracy and stability during long-distance detection. The selected pulsed laser is shown in Fig. 5.9.

The output power of the laser has a significant impact on the intensity of Raman scattered light. When the output power is too high, it can cause stimulated Raman scattering, which cannot be used to effectively demodulate temperature information. Therefore, when designing this system, it is necessary to avoid excessive laser power to avoid affecting the temperature demodulation effect of the system.

Furthermore, the pulse duration emitted by the pulsed laser equipment is one of the key factors affecting the spatial resolution ability of the system. The pulse width determines the temporal resolution of the laser, i.e. how quickly the system responds to temperature changes. The reduction of pulse width can improve the spatial resolution of the system. This is because the pulse width is smaller, and the emitted laser pulse can more accurately locate the temperature change point in the temperature-sensitive fiber, improving spatial resolution. The relationship between the spatial resolution of the system and the pulse width can be expressed as:

$$\delta L_\tau = \frac{1}{2} \cdot \frac{c}{n} \cdot \Delta t \tag{5.10}$$

In Eq. (5.10), the speed of light propagation in a medium with a refractive index of n is expressed in terms of c/n, and the duration of the pulse is expressed in Δt; If the system needs to achieve a spatial resolution of 1m, then the value of t needs to be set to 10 ns.

② WDM multiplexer

WDM multiplexers play a key role in fiber optic temperature measurement systems, and their functions include not only fusing optical signals of different wavelengths, but also separating optical signals of different wavelengths according to their wavelengths. Pulsed light waves transmitted through optical fibers trigger Rayleigh scattering, Brillouin scattering, and Raman scattering. The scattering effect will lead to signal attenuation and interference in optical communication, which in turn will affect the performance of system operation and monitoring efficiency. Among the many scattered light, the system needs Raman scattered light signals, so it is necessary to use wavelength division multiplexing equipment to exclude other optical signals of different wavelengths, leaving only Raman scattered light.

Given that the luminous flux generated by Rayleigh scattering significantly exceeds that generated by Raman scattering, and that the distributed fiber temperature detection technology relies on the decoding of the optical flux, the selected WDM equipment needs to have excellent isolation performance to efficiently filter out non-target wavelength optical signals, allowing only 1450 nm Stokes light and 1650 nm Anti − Stokes light to penetrate. Only in this way can the system have an excellent signal-to-noise ratio, and then ensure that the temperature measurement accuracy and stability of the optical fiber temperature detection system are maintained. The principle of the WDM multiplexer is shown in Fig. 5.10.

③ Photodetector

Stokes and Anti − Stokes radiation are separated by wavelength division multiplexing equipment and then conducted to the photoelectric detection device. The

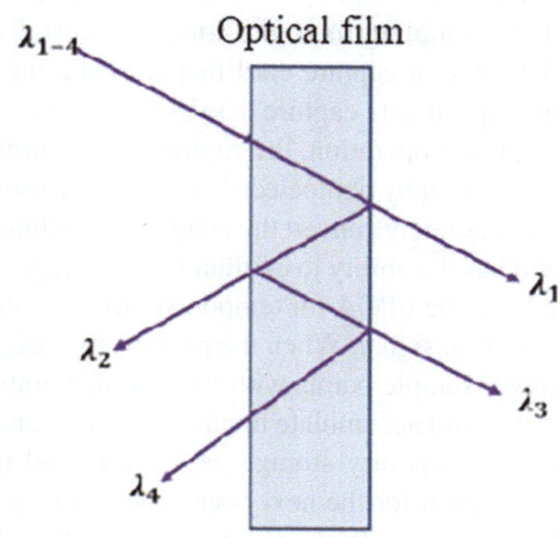

Fig. 5.10 Schematic diagram of a WDM multiplexer

function of the photoelectric sensor is to convert the light pulse into an electrical pulse, but the Stokes and Anti − Stokes light intensity generated during the scattering process is not high, so the signal may be too weak in the process of direct conversion into electrical pulses, which will affect the ratio of signal to noise. Therefore, it is extremely important to choose photoelectric sensors with high magnification capabilities when designing system architecture. The temperature monitoring scheme used in this paper implements a dual-channel distributed fiber optic temperature measurement device, which is equipped with an avalanche photodiode (APD) module designed for distributed temperature measurement for optical signal detection. This module uses an avalanche photodiode as a key component, which converts the input light signal into a current signal and amplifies the avalanche photodiode is a special type of diode whose internal structure enables the electronic avalanche effect that occurs under the action of a high electric field to quickly amplify the input signal. Its mechanism of action relies on the amplification of the avalanche effect, and once photons collide with atoms inside the semiconductor, electrons and holes will appear in pairs. When a strong electric field is applied, the number of electron–hole pairs increases rapidly, resulting in a significant current response.

The selected dual-channel APD module for distributed temperature measurement has a spectral response range of 900–1700 nm, a minimum reverse breakdown voltage of 40 V, and a response of 8.5 A/W. This module has the advantages of high gain, high sensitivity, and fast response, and is suitable for distributed fiber optic temperature measurement systems.

④ High-speed capture card

The function of the high-speed acquisition card is to receive the analog signals generated by the photoelectric detector, convert these signals into digital formats, and then transmit them to the main control computer. Since the system needs to record two optical signals at the same time, it is necessary to use a high-speed acquisition card that can process two analog signals in parallel. To ensure that the system has good spatial recognition capabilities, the collection equipment needs to be equipped with a high sampling frequency. In addition, in order to ensure the synchronization of the sampling time, the collector must also have dual external trigger functions. Therefore, a capture card that can take into account all requirements and support high-speed data capture is selected in this study.

During operation, the capture card simultaneously acquires dual-channel analog signals sent by photoelectric sensors and uses analog-to-digital converters to quickly and accurately convert these signals into digital form. In addition, the data acquisition card has the ability to cumulatively average out. During initial activation, data is first fed into the FPGA for temporary storage, and then paused in anticipation of the next activation signal. When the next event is triggered, the FPGA will match the current trigger sample points with the sample points that originally triggered the cache one by one and accumulate them. After the accumulation is completed, the data is stored in the temporary storage area again, and the system returns to standby mode in preparation for the next operation signal to be activated. This cycle continues until the cumulative times reach a preset value. When the cumulative frequency reaches

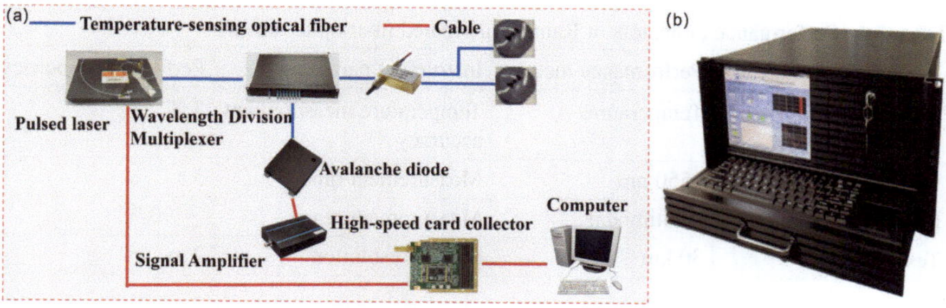

Fig. 5.11 New distributed fiber optic temperature measurement system for mining: **a** Integrated design scheme of distributed optical fiber temperature measurement system; **b** New Raman distributed fiber optic sensor

the predetermined requirements, the accumulated data will be recorded in the FIFO cache space. Finally, the data is transmitted to the master computer via the USB20 port for further processing and analysis.

2) Distributed optical fiber temperature measurement system

The basic principle of operation of the device is as follows: First, a high-power pulsed laser emitter (center wavelength: 1550 nm) produces an elongated pulse of light with adjustable pulse width and frequency, which is then transmitted into the sensitive optical fiber, causing a uniformly distributed Raman scattering phenomenon throughout the optical fiber's path. The Raman Anti – Stokes signal with a wavelength of 1450 nm is screened from backward Raman scattered light by wavelength division multiplexing equipment. The signal produced by Raman Stokes scattering is 40–50 dB lower than that of Rayleigh scattering. Therefore, the system uses a signal booster to realize the interconversion and amplification of photoelectric signals. Finally, the converted voltage data is collected through a high-speed capture card and a computer to complete the decoding of the temperature. The time-domain multiplexing of optical fiber sensors can be completed through the photoelectric modulator, allowing the Raman backscattered light signals in several sensing optical cables to be collected in parallel, forming a distribution characteristic diagram of the signal and a high-resolution Raman distributed optical fiber sensing device diagram, as shown in Fig. 5.11. The performance parameters of the instrument are shown in Table 5.1.

3) Distributed fiber temperature measurement advantages

The system has continuous distributed measurement capabilities, which can realize real-time monitoring of the temperature of each point on the entire optical fiber deployment path, breaking through the limitations of traditional point temperature measurement in space coverage, and greatly improving the monitoring range and data integrity. Secondly, this technology has high spatial resolution and long-distance monitoring capabilities, and can achieve meter-level or sub-meter-level temperature

Table 5.1 Performance indicators of Raman distributed fiber optic sensors

Instrument parameters	Performance metrics	Instrument parameters	Performance metrics
Detection parameters	Temperature	Temperature measurement accuracy	3 °C
Detection wavelength	1550 nm	Measurement time	5 s
Fiber type	Multimodal	Measurement range	≤ 90 °C
Test distance	30 km	Spatial resolution	2 m

localization within a range of several kilometers or even tens of kilometers, which is especially suitable for temperature safety monitoring in large-scale and long-distance facilities (such as oil pipelines, tunnels, and cable corridors). Third, the system is inherently resistant to electromagnetic interference, and uses optical fiber as the sensing medium, which has the characteristics of high temperature resistance, corrosion resistance, flame and explosion resistance, and can operate stably and reliably in harsh environments such as strong electromagnetic fields, flammable, explosive or corrosive. In addition, distributed fiber optic temperature measurement also has the characteristics of strong real-time, convenient installation, and low maintenance cost, and can work continuously for a long time after a single deployment, greatly reducing the operation and maintenance burden of the whole life cycle [41].

References

1. Zhang, G. J., Lu, X. W., & Wang, W. (2023). Study on law and prevention and control technology of goaf residual coal spontaneous combustion of gob-side entry retaining in combustible coal seam. *Coal Technology*, *42*(04), 108−113 (张广杰, 芦晓伟, 王文. 自燃煤层沿空留巷采空区遗煤自燃规律及防控技术研究. 煤炭技术, 2023, 42(04):108–113)
2. Zhang, T. M., Yang, S. Q., & Luo, R. J. (2023). Spontaneous combustion and oxidation characteristics of coal relics in old and new mining areas of "Y" Type comprehensive mining face and gas concentration distribution law. *Coal Technology*, *42*(03), 178−181 (张天明, 杨胜强, 罗仁俊. "Y"型综采面新老采空区遗煤自燃氧化特点及瓦斯浓度分布规律. 煤炭技术, 2023, 42(03):178–181)
3. Liu, B. C. (2022). *Experimental simulation on radon precipitation law during spontaneous combustion of coal in goaf*. Master Thesis, Inner Mongolia University of Science and Technology Baotou China (柳博聪. 采空区遗煤自燃过程氡析出规律实验模拟研究. 包头:内蒙古科技大学, 2022)
4. Zhang, X. (2021). Study on prevention and control of coal spontaneous combustion in goaf of fully mechanized caving face. *Coal Technology*, *40*(04), 104−108 (张欣. 综放工作面采空区遗煤自燃防控研究. 煤炭技术, 2021, 40(04):104–108)
5. Tan, B., Zhu, H. Q., Wang, H. Y., X, J. Y., Zou, J., & Niu, H. Y., et al. (2014). Combustion state and surface temperature field evolution of closed firing zone in top-coal caving region of coal drift. *Journal of Central South University (Science and Technology)*, *45*(03), 946−951 (谭波, 朱红青, 王海燕, 徐纪元, 邹俊, 牛会永, 等.巷道高冒封闭火区燃烧状态及表面温度场演变规律. 中南大学学报:自然科学版, 2014,45(03):946–951)
6. Zheng, X. Z., Jia, Y. X., Guo, J., Wen, H., & Wang, B. Y. (2019). Research status and prospect of coalfield fire monitoring technologics. *Journal of Mine Automation*, *45*(05), 6−10,61 (郑学

召, 贾勇骁, 郭军, 文虎, 王宝元. 煤田火灾监测技术研究现状及展望. 工矿自动化, 2019, 45(05):6–10,61)
7. Yang, Y. C., Mi, W. S., Li, Q. X., Sun, Y., & Li, W. L. (2017). Application effect simulation of infrared thermal imager in Gudging spontaneous combustion in goaf. *Coal Technology*, *36*(01), 147–150 (杨永辰, 米万升, 李奇贤, 孙越, 李丽伟. 红外热成像仪在判断采空区自燃中的应用效果模拟研究. 煤炭技术, 2017, 36(01):147–150)
8. Yang, Y. C., Mi, W. S., Cao, S. F., & Mao, H. Y. (2018). Simulation research on prediction of roadway fire source based on infrared detection technology. *Coal Technology*, *37*(12), 124–126 (杨永辰, 米万升, 曹少方, 毛宏远. 基于红外探测技术的巷道火源预测预报模拟研究.煤炭技术, 2018, 37(12):124–126)
9. Qin, R. X., Tao, Y., He, Z. L., & Chen, K. (2014). Detection and analysis of high temperature area of coal near roadway by infrared thermal imaging. *Coal Geology & Exploration*, *42*(04), 90–92 (秦汝祥, 陶远, 何宗礼, 陈凯. 近巷煤体高温区域红外成像探测与分析. 煤田地质与勘探, 2014, 42(04):90–92)
10. Wen, H., Mi, W. S., Cheng, X. J., Fan, S. X., & Wang, H. (2022). Experimental study on influencing factors of residual coal heat transferrin goaf based on infrared imaging. *Journal of Thermal Analysis and Calorimetry*, *147*(23)
11. Mi, W.S. (2018). *The study of forecast of tunnel fire source based on infrared detection technology*. Master Thesis, Hebei University of Engineering Handan China (米万升. 基于红外探测技术的巷道火源预测预报研究. 邯郸:河北工程大学, 2018)
12. Feng, Z. Y., Sun, H., Zhang, Y. J., Ma, W., Zhai, X. W., & Hu, M., et al. (2023). Research and judgment of coal spontaneous combustion dangerous area in goaf of fully mechanized mining face with large mining height. *Coal Technology*, *42*(04), 144–147 (冯自宇, 孙昊, 张彦吉, 马巍, 翟小伟, 胡冕, 等. 大采高综采工作面采空区煤自燃危险区域研究与判定. 煤炭技术, 2023, 42(04):144–147)
13. Zhou, X., Zhu, Y., Zhang, J. L., Qin, J. S., & Wang, Y. B. (2022). Study on prediction model of coal spontaneous combustion based on PSO–XGBoost. *Mining Safety & Environmental Protection*, *49*(06), 79–84 (周旭, 朱毅, 张九零, 秦思佳, 王艺博. 基于PSO-XG Boost的煤自燃程度预测研究. 矿业安全与环保, 2022, 49(06):79–84)
14. Yan, J. H., Wang, F. T., Zhang, S. H., Zhang, Y., Hu, J. N., & Diao, W. J. (2022). Study on geothermal extraction method and control factors of extraction effect in high temperature goaf. *Mining Safety & Environmental Protection*, *49*(06), 85–90 (闫璟泓, 王方田, 张少华, 张源, 胡剑宁, 刁望杰. 高温采空区地热抽采方法及抽采效果控制因素研究. 矿业安全与环保, 2022, 49(06):85–90)
15. Liu, D. M. (2022). Study on the distribution of "three zones" of spontaneous combustion in goaf of ultra-long downward working face with igneous rock intruding into coal seam prone to spontaneous combustion. *Mining Safety & Environmental Protection*, *49*(04), 135–139 (柳东明. 火成岩侵入易自燃煤层超长俯采工作面采空区自燃"三带"分布范围研究. 矿业安全与环保, 2022, 49(04):135–139)
16. Yang, F. Q., Fan, J. F., Wang, Z. F., Hao, H. W., & Wang, G. T. (2022). Study on distribution law of "three zones"of spontaneous combustion in goaf of 40101 fully mechanized working face in Gaojialiang Coal Mine. *Mining Safety & Environmental Protection*, *49*(02), 96–101 (杨富强, 范军富, 王兆峰, 郝海伟, 王国涛. 高家梁煤矿40101综采工作面采空区自燃"三带"分布规律研究. 矿业安全与环保, 2022, 49(02):96–101)
17. Yang, H. Y., Liu, Y. B., Li, Y., Pan, K. R., Wang, H., & Cao, S. G., et al. (2022). Study on strata stress and deformation response of contiguous seams undercutting roof for entry retaining. *Mining Safety & Environmental Protection*, *49*(01), 8–13,19 (杨红运, 刘延保, 李勇, 潘瑞凯, 王辉, 曹树刚. 近距离煤层群切顶留巷覆岩应力及变形响应研究. 矿业安全与环保, 2022, 49(01):8–13,19)
18. Li, C.H. (2014). Types and development prospects of geothermal resources. *Management & Strategy of Qinghai Land & Resources*, (04), 48–53 (李长辉. 地热资源类型及发展前景. 青海国土经略, 2014, (04):48–53)
19. Lu, C., &Wang, G. L. (2015). Current status and prospect of hot dry rock research. *Science & Technology Review*, *33*(19), 13–21 (陆川, 王贵玲. 干热岩研究现状与展望. 科技导报, 2015, 33(19):13–21)

20. Elders, W. A., Friðleifsson, G. Ó., & Albertsson, A. (2014). Drilling into magma and the implications of the Iceland deep drilling project (IDDP) for high temperature geothermal system s worldwide. *Geothermics, 49*, 111–118.
21. Geothermal Research Papers Collection, Institute of Geology. (1978), *Chinese Academy of Sciences*. Science Press, Beijing Chian. 47−50, 52, 116−124 (中国科学院地质研究所地热组著, 地热研究论文集. 北京:科学出版社, 1978:47-50，52,116–124)
22. Zhang, Y. (2005). *Fluorescent fiber temperature sensor*. Master Thesis, Harbin Engineering University Harbin China (张杨. 荧光光纤温度传感器. 哈尔滨工程大学, 2005)
23. Bosselmann, T., Reule, A., Schroder, J., Kersten, R. T., & Kist, R. (1984). Fiber-optic temperature sensor using fluorescence decay time. *International Society for Optics and Photonics, 0514*, 151–154.
24. Lu, P., Lalam, N., Badar, M., Liu, B., & Chorpening, B. T. (2019). Distributed optical fiber sensing: Review and perspective. *Applied Physics Reviews, 6*(04), 41302.
25. Li, J. P. (2016). Application of distributed optical fiber temperature measurement system in goaf fire early warning. *Inner Mongolia Coal Economy*, (24), 140−143 (李江平. 分布式光纤测温在采空区中的应用. 内蒙古煤炭经济, 2016(24):140–143)
26. Wang, M. Z., Lu, C., & Xia, T. F. (2014). Application of distributed fiber temperature measurement technology in the coal mine gob fire pre-warning. *China Energy and Environmental Protection*, (03), 62−64 (王明重, 吕辰, 夏腾飞. 分布式光纤测温技术在采空区火灾预警中的应用. 中州煤炭, 2014(03):62–64)
27. Wei, X. K., Wang, B., & Xu, Z. D. (2021). Research on the combined application of distributed optical fiber temperature measurement system and nitrogen generation device in coal mines. *Inner Mongolia Coal Economy,* (15), 38−39 (魏勋阔, 王兵, 徐召栋. 分布式光纤测温系统与制氮装置在煤矿中的联合应用研究. 内蒙古煤炭经济, 2021(15):38–39)
28. Cheng, G. X. (2008). *Raman Brillouin scattering, 2nd edition*. Science Press, Beijing China (程光煦. 拉曼布里渊散射第2版. 北京:科学出版社, 2008)
29. Zhai, Y. Z., Zhao Y. M., Xu S. R., & Wang, D. W. (2007). Distribute optical fiber temperature measurement system. *Instrument Technique and Sensor*, *295*(09), 36−38 (翟延忠, 赵玉明, 许舒荣, 王大文. 分布式光纤测温系统. 仪表技术与传感器, 2007, 295(09):36–38)
30. Jiang, F. T. (2018). *Temperature alarm method and temperature precision improvement for distributed fiber Raman temperature sensor system*. Master Thesis, Shandong University Qingdao China (江凤婷. 分布式拉曼测温系统报警方法的改进及提高温度精度的研究. 山东大学, 2018)
31. Song, Z. K., Gao, B., Wang, J. S., Yin, Z., & Kong, T. Y. (2020). Discussion on the use of hand-held infrared thermometer. *Chinese LABAT Man, 62*(01), 7−12+43 (宋准科, 高宾, 王金生, 尹政, 孔天仪. 手持式红外测温仪使用方式的探讨. 蓄池, 2025, 62(01):7–12+43)
32. Yang, Y. C. (1988). Analysis and countermeasures of spontaneous fire locations in gently inclined fully mechanized caving faces. *Safety in Coal Mines*, (04), 26−28 (杨永辰. 缓斜综放工作面自然发火地点的分析及对策. 煤矿安全, 1998, (04):26–28)
33. Kuenzer, C., Zhang, J., Tetzlaff, A., Dijk, P. V., Voigt, S., Mehl, H., et al. (2007). Uncontrolled coal fires and their environmental impacts: Comparing two underground coal mine fires in China. *International Journal of Coal Geology, 72*(02), 87–100.
34. Li, B., Chen, J. W., Liu, Z. Y., Bai, J., Fan, L., & Zhang, C., et al. (2023). Quality detection of protective coating for power equipment based on lock-infrared thermography. *Nondestructive Testing*, *45*(08), 73−78 (李波, 陈俊卫, 刘卓毅, 白洁, 樊磊, 张聪, 等. 基于锁相红外热成像技术的电力设备防护涂层质量检测. 无损检测, 2023, 45(08):73–78)
35. Shao, J., Hu, W. Y., Jia, F. M., Hu, Y. R., & Ma, X. W. (2013). Application of infrared thermal imaging technology to condition-based maintenance of power equipment. *High Voltage Apparatus*, *49*(01), 126−129+133 (邵进, 胡武炎, 贾风鸣, 胡育蓉, 马晓薇. 红外热成像技术在电力设备状态检修中的应用. 高压电器, 2013, 49(01):126–129+133)
36. Vocker, M., & Hutter S. (2011). Instruments and methods for the investigation of subsurface coal fires. *In Coal and peat fires: A global perspective*, Volume 1, Elsevier.
37. Zhao, Y. X., & Hu, Z. Y. (1991). 87-type borehole digital thermometer. *Chinese Journal of Geophysics*, (04), 495−500 (赵永信, 胡忠义. 87型钻孔数字测温仪. 地球物理学报, 1991, (04):495–500)

38. Xie, J. W., Lu, X., Shangguan, K. F., Yang, S. J., Fan, Z. D., & Xing, Z. (2014). Application of distributed optical fiber temperature detection technology in spontaneous combustion monitoring of large angle flammable coal seam goaf. *Safety in Coal Mines*, *45*(11), 118–121 (谢俊文, 卢熹, 上官科峰, 杨世杰, 范振东, 邢震. 分布式光纤测温技术在大倾角易燃煤层采空区自燃监测中的应用. 煤矿安全, 2014, 45(11):118–121)
39. Zhao, Y. S., Xu, X. L., Cao, S. L., & Zhu, P. Q. (2020). A review of underground coal fire detection technology and engineering application. *Journal of Mining Science and Technology*, *10*(01), 24–39 (赵禹深, 许献磊, 曹仕龙, 朱鹏桥. 地下煤火探测技术及工程应用综述. 矿业科学学报, 2025, 10(01):24–39)
40. Zhong, X. X., Wang, D. M., & Zhu, H. J. (2010). Application of distributed fiber temperature measurement technology in coal mine. *Journal of Mining & Safety Engineering*, *27*(01), 119–123 (仲晓星, 王德明, 朱红菊. (2010).分布式光纤测温技术在煤层自然发火监测中的应用.采矿与安全工程学报,27(01),119–123)
41. Tyler, S. W., Selker, J. S., Hausner, M. B., Hatch, C. E., Torgersen, T., & Thodal, C.E., et al. (2009). Environmental temperature sensing using Raman spectrum DTS fiber-optic methods. *Water Resources Research*, *45*(04).
42. Huang, S. L., Liang, D., & Liu, G. (1991). A study on the distributed fiber optic temperature sensor. *Chinese Journal of Scientific Instrument*, (04), 25–30 (黄尚廉, 梁大巍, 刘龚. 分布式光纤温度传感器系统的研究. 仪器仪表学报, 1991(04):25–30)
43. Zhang, Z. X., Liu, T. F., Zhang, B. X., Chen, Y., & Chen, X. Z. (1995). Laser Raman distributed fiber temperature sensor system. *Acta Optica Sinica,* (11), 1585–1589 (张在宣, 刘天夫, 张步新, 陈阳, 陈晓竹. 激光拉曼型分布光纤温度传感器系统. 光学学报, 1995(11):1585–1589)
44. . Zhang, Z.X., Liu, H. L., Xu, H. F., Wang, J. F., & Dai, B. Z. (2005). *Development of long distance distributed optical fiber Raman temperature sensor technology.* 20th Anniversary of the Establishment of the Optoelectronic Technology Professional Committee of the Chinese Optical Society and the 11th National Academic Conference on Optoelectronic Technology and Systems, Beijing, China (张在宣, 刘红林, 徐海峰, 王剑锋, 戴碧智. 远程分布式光纤拉曼温度传感器技术的进展:中国光学学会光电技术专业委员会成立二十周年暨第十一届全国光电技术与系统学术会议, 中国北京, 2005)
45. Zhang, Z. X., Jin, S. Z., Wang, J. F., Liu, H. L., Sun, Z. Z., &Gong, H. P., et al. (2010). Distributed optical fiber Raman photon sensor research review. *Chinese Journal of Lasers*, *37*(11), 2749–2761
46. Wang, J. F., Liu, H. L., Zhang, S. Q., Yu, X. D., Sun, Z. Z.., & Jin S. Z., et al. (2013). New type distributed optical fiber temperature sensor (DTS) based on Raman scattering and its application. *Spectroscopy and Spectral Analysis*, *33*(04), 865–871

Chapter 6
Indicator Gas Detection Method

Indicator gas monitoring technology is a commonly used technique for monitoring coal spontaneous combustion fires. The gases released during coal spontaneous combustion include CO_2, CO, CH_4, H_2, C_2H_4, C_2H_2, and other higher hydrocarbon gases [1–3]. These indicator gases originate from the pyrolysis and decomposition of oxygen-containing hydrocarbons [4]. Common indicator gases for monitoring coal spontaneous combustion are CO, C_2H_4, and Graham's ratio. Graham's ratio, typically ranging between 0 and 0.4, represents the ratio of carbon monoxide produced to oxygen consumed by volume [5]. Additionally, other gases such as C_2H_2, C_2H_6, C_3H_8, and C_4H_{10} are often selected as indicator gases. Another proposed indicator is the CO/CO_2 ratio, which can indicate the development of underground coal fires [6]. Research has identified over ten indicator gases suitable for monitoring the early stages of coal spontaneous combustion. The selection of these gases varies depending on specific conditions, including the scale and circumstances of the coal seam fire [7].

6.1 Index Gas Detection Methods and Technologies

6.1.1 Principle of Index Gas Selection

After coal seam mining, coal oxidizes and exotherm malizes in contact with oxygen, and the temperature rises caused by heat accumulation, which greatly increases the risk of spontaneous combustion of coal. According to the temperature rise and gas release characteristics in the natural process of coal at this stage, it is the basis of the natural prediction method to identify the natural state in advance, and to identify and warn the spontaneous ignition of coal in advance. There are mainly indicator gas analysis method and temperature measurement method [8–10].

Y. Xiao et al., *Detection of Fire Source in Underground Coal Mine and its Applications*,
https://doi.org/10.1007/978-981-95-7064-5_6

The temperature measurement method [11] is direct and clear, but it is difficult to bury the temperature sensor and buried pipe wiring, and the range that can be controlled by the distribution point is narrow, and it is difficult to obtain the temperature change of the stope and goaf quickly, comprehensively and accurately, and the one-time consumption of materials increases the production cost of tons of coal.

The gas analysis method predicts the early fire according to the change law of a series of gases that reflect the natural combustion characteristics of coal produced during the oxidation process of coal [12–15]. Coal sample heating experiments and practices have proved that the gas composition is closely related to the temperature of coal. Analyze the composition of gases produced in coal samples under different temperature conditions, find out the correspondence between coal temperature and gas composition, and use easy-to-detect, representative and regular gases, i.e., index gases, as the basis for predicting coal spontaneous combustion. By analyzing the gas composition at a certain point in the mine, the temperature range of the wind flow passing through the regional coal can be obtained. This method, also known as the indirect temperature method, is the most applicable and effective method.

The optimization and application of coal spontaneous combustion index gas have always been a problem studied and discussed. Based on the analysis of a large number of experimental data, Luo and Qian focused on the composition and change of gas products in the process of oxidation and spontaneous combustion of coal of different coal types in our country, and proposed the quantitative relationship between index gas and coal temperature and the early prediction of spontaneous ignition index.

However, coal pyrolysis produces a variety of gases, and the relationship between the amount of gas generation and the temperature of the coal varies depending on the quality of the coal [16]. The generation of each gas is affected by a variety of factors, especially in actual production, the index gas is restricted by geological conditions, mining conditions, ventilation conditions and other factors. At the same time, the location and time of gas sample collection will also affect the determination of the development degree of coal spontaneous combustion.

6.1.2 Index Gas Monitoring Method

(1) Bundle tube monitoring system. The bundle tube monitoring is buried in the goaf through the mining of the working face [17], and then the air from the monitoring site is pumped to the analytical instrument for analysis with the help of a pump, and its working principle is shown in Fig. 6.1.

Beam tube monitoring can realize continuous monitoring of the gas at the measurement point with high reliability. However, because the monitoring of the bundle pipe relies on the pipeline to pump the gas at the measurement point, there are some defects: ① The bundle pipe is easy to be broken by the collapse of the overlying rock layer in the goaf, resulting in air leakage, and the sampling gas is easily polluted by the gas in the goaf [18]. ② The pipeline is too long, resulting in a lag in the time

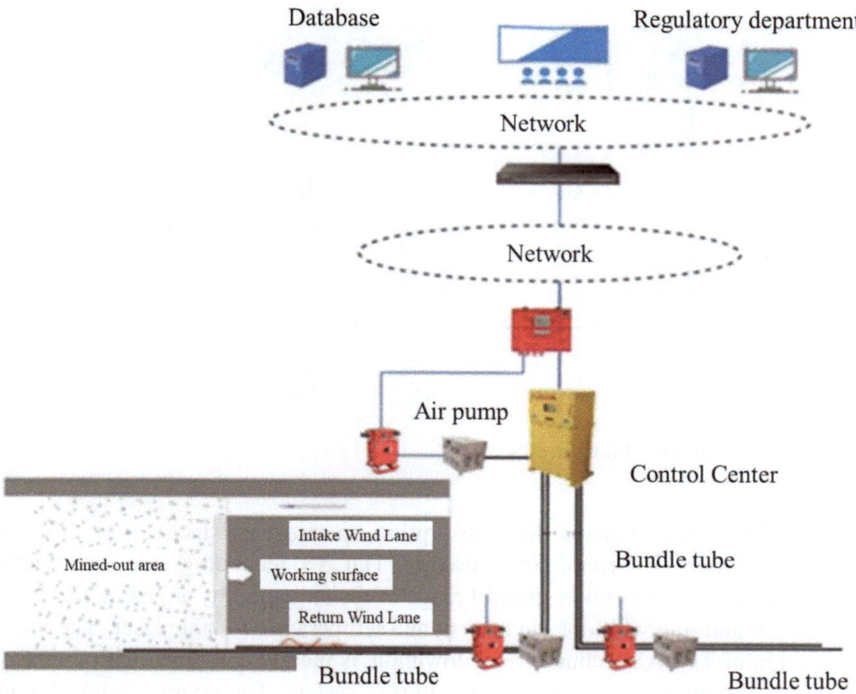

Fig. 6.1 Working principle of the bundle tube monitoring system

of monitoring data. The current commonly used negative pressure beam pipe monitoring system will enter the pipeline after leakage, resulting in the distortion of the detection gas, and the negative pressure system has little transportation power, which cannot achieve the purpose of rapid gas transportation. Based on this, Lu et al. developed a positive pressure beam tube monitoring system, in which the beam tube leaks gas from the inside out when a leak occurs, which will not affect the concentration of the gas components to be detected. In addition, the conveying power of the positive pressure bundle tube is much greater than that of the negative pressure pumping, with fast gas delivery speed and short lag time [26]. Wang et al. developed a high-positive pressure beam pipe monitoring system for gas transmission and control common pipelines, and studied the optimal pipe diameter and optimal output pressure of gas pipelines [19–21].

With the increase of mining depth, the monitoring of the bundle pipe gradually has problems such as long pumping distance, high resistance and difficult maintenance of the bundle pipe. In addition, the gas distribution in the goaf is uneven, and the gas collected by the measuring point cannot fully reflect the real gas composition and concentration in the goaf, and the gas information in the goaf cannot be accurately grasped. Therefore, it is urgent to systematically optimize the existing defects to meet the needs of monitoring the risk of coal spontaneous combustion in goafs [22–25].

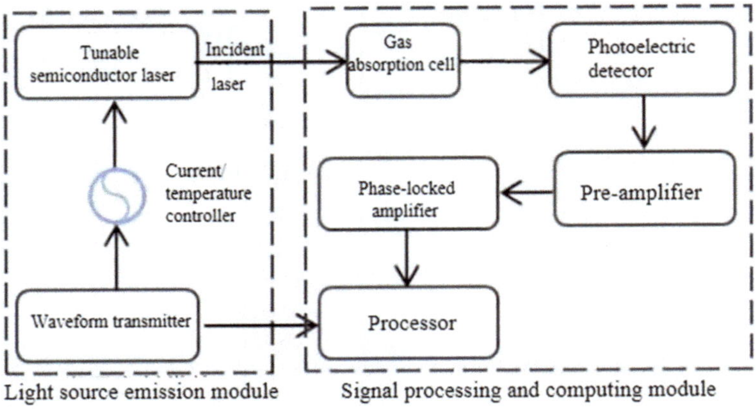

Fig. 6.2 TDLAS technical fundamentals

(2) Tunable semiconductor laser absorption spectroscopy technology. Tunable Diode Laser AbsOrption Spectroscopy (TDLAS) technology scans within the characteristic absorption spectral range of the target gas by adjusting the wavelength and line width of the current and temperature tunable laser. The application of TDLAS technology in downhole is mainly combined with the beam tube to form a laser beam tube monitoring system. The system pumps the gas at the measurement point to the gas absorption tank for scanning with the help of a beam tube, and finally obtains the concentration of gas components at the measurement point. The basic principle is shown in Fig. 6.2.

For the goaf environment, the gas monitoring system developed by Shen and others based on TDLAS technology, combined with the Internet of Things and optical fiber transmission technology, can realize real-time monitoring of downhole gas at kilometers away. Aiming at the phenomenon of mutual interference between absorption lines in multi-gas detection, Wang et al. solved the problem of overlapping interference between CO and CH4 absorption lines based on the support vector regression model. Yao et al. developed a TDLAS system based on frequency division multiplexing and normalized harmonic detection, which realizes the simultaneous detection of multi-component gases by applying modulated signals of different frequencies to lasers of different wavelengths [29, 30].

With the breakthrough of TDLAS technology in key technologies such as anti-interference algorithm and multi-component synchronous detection, it has effectively overcome the bottleneck of gas monitoring in the complex environment of goafs. This technology has the advantages of high sensitivity and high accuracy, and is not interfered with by background gases, which can realize real-time dynamic monitoring of multi-component gases and meet the needs of coal spontaneous combustion risk monitoring in goafs.

6.1.3 Indicator Gas Sampling Method

(1) Direct sampling method.

The core goal of direct sampling is to maintain the original gas components as much as possible, and directly obtain "in-situ" gas samples from coal fire areas (such as boreholes, crannies, cracks, ventilation pipe outlets, and closed wall observation holes), which can best directly reflect the gas composition at a specific point and a specific time. At the heart of the operation is the rapid and efficient capture and isolation of the target gas from the outside air, and any delay or poor sealing can lead to air mixing, severely diluting key trace gases (e.g., CO, H_2, olefins) and potentially introducing oxidation reactions that distort the analytical results. Therefore, the selection of sampling points is crucial for the representativeness of the results, considering factors such as airflow paths and temperature gradients.

There are two main types of commonly used sampling tools. Bag sampling (e.g., Teflon PTFE bags or aluminum foil composite air bags) takes advantage of the material very low surface adsorption and gas permeability to effectively prevent gas components (especially organic volatiles VOCs, acid gases such as H_2S) from being adsorbed or lost/infiltrate. It has a flexible capacity (a few hundred milliliters to several liters) and is easy to operate and can be filled with negative pressure at a pump or on-site, but care must be taken to avoid breakage, trace penetration of light gases such as H_2, and electrostatic effects. Glass syringe sampling (with three-way valve and rubber cap, commonly 50 mL–1L) can seal the gas in the inert glass cavity at the moment of sampling, with excellent air isolation effect, extremely low adsorption for most gases and completely impermeable. It is especially suitable for sampling small flow and low-pressure points (such as small cracks, negative pressure zones), and the samples extracted are almost non-diluted, especially suitable for short-term storage and transportation of trace active gases, but require high operating skills.

This method is mainly used for underground close-range fire source detection, which is its most typical and important scenario. For example, the internal gas of the goaf is directly collected through embedded pipes, observation holes or cracks, and the concentration and ratio of O_2 consumption and $CO/CO_2/H_2/CH_4$/olefins (such as ethylene C_2H_4, propylene C_3H_6) and other landmark gases (such as Graham ratio) are the key basis for judging the location, combustion stage and fire trend of the hidden ignition source. Sampling in the detection hole of the confined wall can monitor the changes in the gas composition in the confined area, and evaluate the sealing effect, fire control status or recurrence risk. For the high temperature anomalies on the roadway wall and coal pillar surface, deep gas can be collected through drilling holes or cracks to find evidence of spontaneous combustion. In addition, because of its relatively simple sampling process and direct sample use, it is also the preferred pre-sampling method for high-precision analysis of high-purity samples in laboratories that require high-purity samples, such as gas chromatography (GC), gas chromatography—mass spectrometry (GC—MS), and Fourier transform infrared spectroscopy (FTIR).

The advantages of direct sampling are that the operation is relatively simple and intuitive, the technical threshold is low, the initial investment cost is low (the tool itself is cheap), and the collected samples can usually be directly used in most laboratory gas analysis instruments without complex pretreatment, which can provide the laboratory with high-precision raw samples and provide accurate gas composition data at specific sampling points under the premise of avoiding contamination and timely analysis. However, this method also has significant limitations and challenges. The main disadvantage is that it is easily disturbed by ambient airflow, and the wind flow caused by underground ventilation, air leakage, and thermal power in the fire area will dilute the sampling gas, so that the concentration of key index gases (especially low-concentration warning gases) is low or even undetectable, which seriously reduces the detection sensitivity and accuracy, especially when the sampling point is improperly selected or slow to operate. The aging requirements of gas components are high, and some active gases (especially olefins such as ethylene and propylene) and acid gases (such as H_2S) will undergo chemical reactions (oxidation, polymerization) or adsorption loss after sampling, usually requiring laboratory analysis within 24 h (cold chain transportation can be slightly extended). This method is essentially point sampling, which only represents the information of a single time point and spatial location, and it is difficult to capture the spatial heterogeneity and temporal dynamic changes of gas distribution in the fire area. There are also safety risks of high-temperature burns, toxic and harmful gas poisoning (high concentration of CO, H_2S, etc.), and even gas explosions near the fire area (especially in areas with suspected open flames or high temperatures), which need to strictly abide by safety regulations and have reliable protection. In addition, the method is highly dependent on the sampling point, requiring the existence of accessible channels with gas outflow (boreholes, cracks, pipelines, etc.), which cannot be applied to areas with good closure and no gas escape, and it is difficult to achieve real-time continuous monitoring [31].

In summary, the direct sampling method is still an indispensable basic method in coal fire gas detection because of its simple operation, economic practicality, and good sample compatibility, especially for gas samples that can provide near-original state for high-precision analysis in the laboratory, especially suitable for downhole close-range fixed-point sampling (such as goafs, confined walls, and high temperature points). However, its disadvantages, such as susceptibility to airflow dilution interference, strong timeliness, point sampling limitations, safety risks, and dependence on sampling points require careful selection of sampling points, standardized operation, timely inspection, and often need to be combined with other monitoring methods (such as beam tube monitoring and sensor networks) to more comprehensively and accurately assess the coal fire situation [32].

(2) Adsorption sampling method.

Adsorption sampling is a gas harvesting technology based on the principle of selective enrichment of adsorbents, which mainly targets early landmark gases with very low concentrations (ppb level) in coal fire areas. This method uses high-performance adsorption materials such as activated carbon, silica gel or Tenax TA to actively capture trace target components through physical/chemical adsorption, and then combines laboratory thermal desorption technology to achieve gas resolution and high-sensitivity analysis. Its core is to significantly improve the enrichment efficiency of trace gases: activated carbon has strong adsorption capacity for non-polar gases (such as methane CH_4 and ethane C_2H_6) with its developed microporous structure and high specific surface area. Silicone is suitable for polar gases (such as oxygen and sulfur compounds). Tenax TA (polymer adsorbent) has outstanding adsorption efficiency and selectivity for volatile organic compounds (VOCs), especially olefins (such as ethylene C_2H_4, propylene C_3H_6) and alkanes (such as propane C_3H_8) due to its strong hydrophobicity, low background interference and excellent thermal stability, which can effectively avoid sample degradation.

Sampling operations need to be strictly regulated: the adsorption tube (usually a glass or stainless steel tube filled with a specific sorbent) needs to be connected to the constant flow pump through a sealed interface to precisely control the sampling flow rate (typical range of 50–200 mL/min) and continue to pump for hours or even longer. This long-term, low-flow cumulative sampling mode is the key to achieving trace gas enrichment, which is suitable for two typical scenarios: First, the very low concentration of early warning gas released in the early oxidation stage of coal fire (such as ethylene C_2H_4 concentration is often less than 1ppm) is targeted capture, which provides a key basis for the identification of fire budding stage. Second, in remote or non-powered areas (such as abandoned mines and surface fissure areas), battery-powered portable pumps can be used to achieve long-term passive enrichment, overcoming the equipment limitations of on-site real-time analysis.

The core advantage of this method is its ultra-high detection sensitivity—the enrichment action increases the concentration of trace gases that are difficult to detect in situ to the measurable range of laboratory instruments (such as GC—MS), which solves the problem of missed detection caused by environmental dilution or insufficient detection limit of the direct sampling method. However, its application also has significant limitations: Firstly, it is highly dependent on the back-end equipment of the laboratory, and the adsorption tube must be analyzed by a sophisticated thermal desorption instrument (usually heated to 250–300 °C) in the laboratory, and directly combined with chromatography or mass spectrometry analysis, and the results cannot be obtained immediately on site. Secondly, the sampling period is long, and a single sampling takes several hours, which is difficult to reflect the instantaneous dynamic change of gas concentration. In addition, adsorbents are sensitive to temperature and humidity, and high humidity environments may reduce the efficiency of some adsorbents (such as activated carbon), and they need to be strictly sealed and refrigerated before and after sampling to prevent contamination or desorption loss. In addition, the differences in selectivity of different sorbents require the optimization of the sampling scheme according to the characteristics of the target gas in advance, which increases the complexity of operation. Therefore, although this method is irreplaceable in terms of sensitivity and trace gas capture, it is usually used as an important supplement to direct sampling method or online monitoring technology, especially for early warning and in-depth traceability of coal fires.

(3) Online monitoring method.

Online monitoring is a dynamic monitoring method based on real-time sensing and data transmission technology, which continuously collects, transmits and analyzes gas concentration data in the coal fire area by deploying fixed or mobile sensor nodes to achieve immediate response to fire. In fixed monitoring mode, explosion-proof gas sensors (usually detect key parameters, such as CO, CH_4, O_2, CO_2, and temperature) are permanently installed in high-risk areas of underground fires, such as near the confined wall of goafs, return air roadways, contact lanes or cutting holes in the working face. These nodes transmit data to the ground monitoring center in real time via RS485 wired communication or low-power wireless networks (such as ZigBee, LoRa, 4G/5G mining intrinsic safety systems), forming a continuous time series database that provides the basis for fire evolution analysis. The mobile monitoring mode relies on explosion-proof robots, inspection drones or portable detection equipment, equipped with multi-parameter sensors (which can be extended to trace gases such as H_2S, ethylene C_2H_4 and other trace gases) and precise positioning modules (such as GPS, UWB downhole positioning or lidar SLAM mapping) to dynamically scan hidden fire areas that are difficult to reach manually (such as high-level collapse cracks and high-temperature points on open-pit mine slopes), and accurately locate abnormal leakage sources and combustion core areas through spatial heat maps and gas concentration cloud maps [33].

The core value of this method is to break through the limitations of time and space monitoring: on the one hand, its data update frequency from seconds to minutes can capture the sudden trend of gas concentration (such as the sudden rise of CO), significantly improve the timeliness of fire warning, and win a key window for underground personnel evacuation and emergency decision-making. On the other hand, through integration with smart mine systems (such as digital twin platforms and fire prediction models), multi-source data fusion analysis can be realized, and hierarchical alarms can be automatically triggered, fire extinguishing facilities can be linked, or ventilation plans can be optimized. Therefore, this method is widely used in all-weather risk control in active coal fire areas, especially in the monitoring of large-area fire areas in open-pit mines, the fire prevention system of underground intelligent working faces, and the reignition monitoring of closed fire areas.

However, the main challenges faced by online monitoring methods are data reliability and system operation and maintenance complexity:

Sensor accuracy attenuation: Electrochemical sensors (such as CO detection units) are susceptible to temperature, humidity, and cross-gas interference, with an average monthly drift of up to $\pm 5\%$, and need to be calibrated on site or automatically zero point every month. Although the infrared sensor has better stability, it is sensitive to dust pollution and needs to be cleaned regularly.

Communication and power supply constraints: the cost of underground wired deployment is high, the scalability is poor, and the wireless signal is susceptible to metal equipment and roadway bending and attenuation. The limited battery life of mobile devices (drones are often less than 30 min), which restricts the efficiency of large-scale inspections.

Environmental adaptability requirements: Underground equipment needs to meet the mining intrinsic safety standards of explosion-proof, moisture-proof, and impact resistance, and open-air scenarios need to cope with extreme temperatures and wind and sand erosion.

Data analysis complexity: Massive real-time data needs to be dynamically interpreted in combination with geological models and ventilation networks to avoid false positives (such as CO false positive signals caused by diesel vehicle exhaust).

In summary, the online monitoring method has become the pillar technology of modern coalmine fire prevention and control system with real-time data and spatial coverage as its core advantages. However, its reliability is highly dependent on strict sensor calibration mechanism, robust communication architecture and intelligent data analysis algorithms, so it often forms a technical closed loop with the laboratory verification of direct sampling method and trace gas supplementary monitoring of adsorption sampling method, and jointly builds a multi-scale and high-confidence coal fire early warning network [34].

(4) Passive diffusion sampling method.

Passive diffusion sampling is a non-powered gas collection technology based on the principle of free diffusion of gas molecules, which accumulates target gas through physical adsorption or chemical absorption medium for a long time to obtain the average concentration value during the monitoring period [35]. Its core component is a passive gas sampler (e.g., CO detection tube, hydrocarbon diffusion tube), which is filled with specific adsorbents (e.g., activated carbon, silica) or reaction reagents (e.g., iodine pentoxide for CO oxidation). When the sampler is exposed to the monitoring environment, the gas molecules are naturally permeated into the medium by selective diffusion membranes such as PTFE microporous membranes, which are dustproof and moisture-proof, and are enriched in situ through adsorption/chemical reactions. Sampling cycles typically last from a few days to several weeks and culminate in laboratory analysis (e.g., weighing, chromatography, or colorimetry) to determine the total cumulative gas volume, which is converted to time-weighted average concentration (TWA).

The core application scenarios of this method focus on areas without power supply, difficult to access, or long-term monitoring:

(1) Abandoned roadways and closed fire areas: Long-term deployment in isolated areas without power supply and communication to monitor the slow release trend of CO or hydrocarbon gases in confined spaces.
(2) Surface fissures and open-pit mine slopes: deployed at scattered surface air leakage points to assess the long-term impact of coal fire on shallow soil gases.
(3) Ventilation blind spots and high roofs: hidden positions that are difficult to cover by manual inspection, supplementing the blind spots of the fixed monitoring network.
(4) Evaluation of fire prevention and control effect: Quantify the effectiveness of treatment by comparing the average concentration changes before and after the implementation of fire extinguishing measures.

Its significant advantages are:

(1) Zero-energy operation: no power supply or pump drive equipment is required, greatly reducing deployment costs and maintenance difficulties.
(2) Extreme environmental adaptability: high-temperature resistance and explosion-proof structure design suitable for flammable and explosive environments.
(3) Ultra-long-term monitoring capability: a single sampling can last for several weeks, capturing slowly evolving fire signs.
(4) Spatial deployment flexibility: The lightweight design supports dense distribution to form a high-resolution monitoring grid.

However, this method has inherent limitations:

(1) Only the average concentration is provided: cannot reflect the instantaneous fluctuations or peaks of gas concentration (such as sudden CO outbursts), which may mask short-term risk events.
(2) Poor timeliness: laboratory analysis is required after sampling, and the data lags several days to weeks.
(3) Sensitive to environmental interference: strong airflow will disrupt the diffusion balance, resulting in low results, Temperature and humidity changes may affect the adsorption/reaction rate.
(4) Risk of media saturation: ultra-long period sampling may saturate the adsorbent and cause underestimation of concentration.
(5) Cross-interference: In a complex gas environment, the selectivity of the diffusion film on molecular weight may lead to component deviation.

In summary, the passive diffusion sampling method is an irreplaceable tool for long-term gas trend analysis in unpowered areas, especially for the evaluation of abandoned mine management, tracking of hidden fire areas and large-scale screening. However, the data lag and dynamic response defects must be complemented by real-time early warning of online monitoring and transient verification of direct sampling, so as to jointly build a coal fire gas monitoring system covering the whole cycle of "instantaneous, short-term, and long-term". During actual deployment, it is necessary to optimize the sampling point (avoid strong wind flow areas), calibrate environmental parameters, and strictly control the exposure time to improve data reliability [36].

6.1.4 Main Detection Technologies

(1) Gas chromatography (GC).

Gas chromatography is the core laboratory technology for coal fire gas detection. The principle is to use a carrier gas (such as helium or nitrogen) to bring a gas sample into the column, separate different gas molecules according to the difference

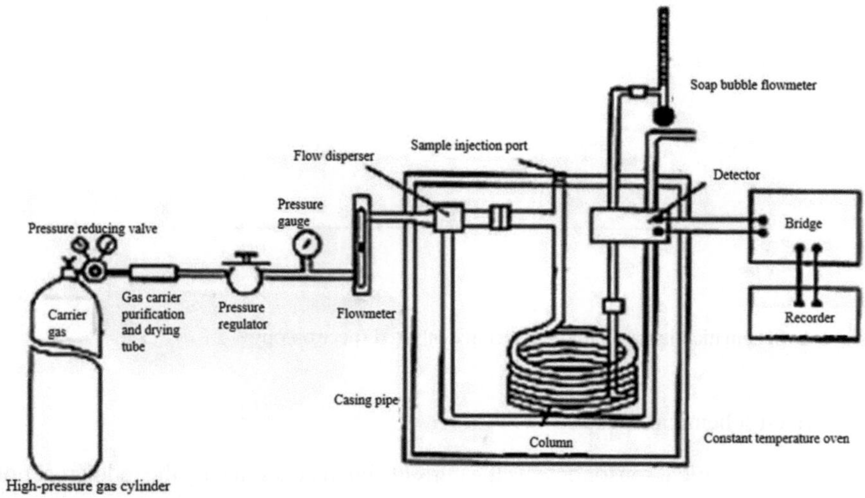

Fig. 6.3 Schematic diagram of a gas chromatograph

in adsorption/desorption capacity in the stationary phase, and then quantify them by thermal conductivity detector (TCD) or flame ionization detector (FID). This method can accurately measure more than ten index gases such as CO, CO_2, CH_4, C_2H_4, and C_2H_2 at the same time, with detection limits as low as ppm (parts per million) or even ppb (parts per billion), which is especially suitable for the identification of trace ethylene (C_2H_4) and acetylene (C_2H_2), which is very important for early warning of coal spontaneous combustion [37, 38]. However, GC relies on the laboratory environment, complex sample preprocessing, and long analysis cycle (about 30–60 min at a time), which is difficult to meet the needs of real-time monitoring, and is mostly used for fire area diagnosis and scientific research verification (Fig. 6.3).

(2) Non-dispersive infrared spectroscopy (NDIR).

Non-dispersive infrared spectroscopy technology is based on the principle of characteristic absorption of gas molecules into a specific infrared band. When infrared light passes through the gas to be measured, molecules such as CO_2 (4.26μm) and CH_4 (3.3μm) absorb specific wavelength energy, and the gas concentration can be inverted by detecting the degree of light intensity attenuation. NDIR sensors have the characteristics of fast response (<10 s), high stability, long life (usually >5 years), and are not subject to cross-interference with combustible gases, making them very suitable for continuous online monitoring systems in coalmines. For example, NDIR probes can be deployed in mine return air lanes or goaf buried pipe monitoring points to transmit CO_2 and CH_4 data in real time. However, its detection sensitivity for diatomic molecules (such as CO) is low, and water vapor may interfere with the measurement, so it needs to be equipped with dehumidification or compensation algorithms (Fig. 6.4).

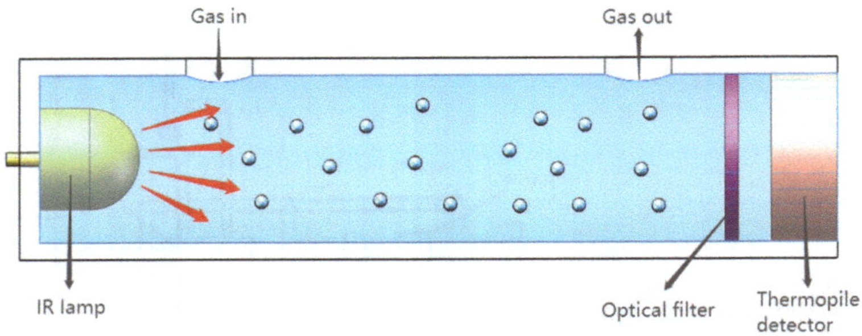

Fig. 6.4 Schematic diagram of non-dispersive infrared spectroscopy

(3) Electrochemical sensor.

The electrochemical sensor detects the current signal generated by the redox reaction of the gas on the electrode surface. Take CO detection as an example: the gas diffuses to the working electrode to react ($CO + H_2O \rightarrow CO_2 + 2H^+ + 2e^-$), and the electron transfer forms an electric current, the strength of which is directly proportional to the concentration. This type of sensor is small in size, low in power consumption, low in cost (unit price is about 100 yuan RMB), and is widely integrated in portable detectors and handheld alarms, suitable for inspectors to quickly screen hidden danger points in fire areas. However, electrochemical sensors have obvious shortcomings: they are susceptible to acid gas poisoning such as H_2S and SO_2, temperature and humidity fluctuations affect accuracy, and have a short lifespan (usually 1 and 2 years), so they need to be calibrated every 3–6 months. The reliability is significantly reduced in the coal-fire environment with high temperature and humidity (Fig. 6.5).

(4) Tunable diode laser absorption spectroscopy (TDLAS).

TDLAS technology uses a tunable laser to emit a narrow linewidth laser that scans specific absorption spectral lines of gas molecules (e.g., CH_4 at 1653nm, C_2H_2 at 1530nm), and achieves high-selectivity, high-sensitivity (ppb-level) gas quantification by detecting laser transmission intensity or second harmonic signals. Its advantage is non-contact measurement: the laser can penetrate several kilometers for telemetry (e.g., emission to the ground to the downhole), avoiding the sensor being placed in the high-temperature danger zone. At the same time, it has the ability to resist dust interference, and is suitable for the detection of high-temperature pyrolysis gases (such as H_2, C_2H_2) in the fire area. In recent years, TDLAS has been used in combination with open optical paths (such as reflective prism arrays) to monitor the boundary of open-pit mine fire areas and three-dimensional scanning of gas in underground goafs, becoming a cutting-edge technology for coal fire warning (Fig. 6.6).

Fig. 6.5 Carbon monoxide sensor

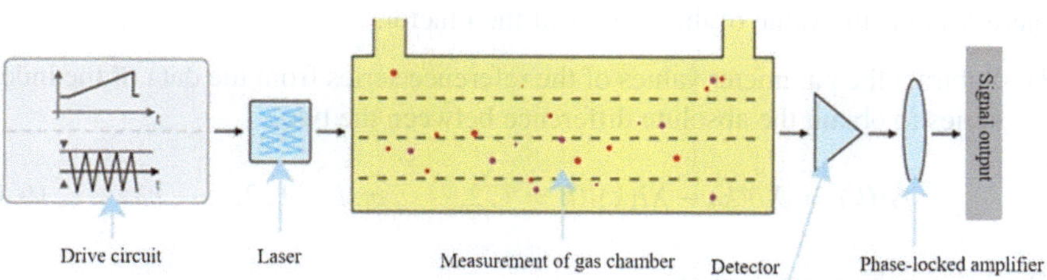

Fig. 6.6 TDLAS schematic diagram

6.1.5 Index Gas Selection and Mechanism of Action

In the process of coal fire development, characteristic gases will be released at different temperature stages, and the variation of their types, concentrations and proportions have key guiding significance for fire warning and combustion state determination [39].

(1) Optimization of index gas based on gray correlation analysis.

The basic idea of gray correlation analysis is to determine whether the connection is close by determining the geometric similarity between the reference data column and several comparison data columns, reflecting the degree of correlation between parameters.

(1) Determine the reference series and the comparison series. In the gas system for determining coal spontaneous combustion indexes, various gas indicators are comparative series, and the expression is shown in Eq. (6.1). Generally, temperature is used as a reference sequence, and the expression is shown in Eq. (6.2).

$$X_1X_2\cdots X_n = \begin{bmatrix} x_1(1) & x_2(1) & \cdots & x_n(1) \\ x_1(2) & x_2(2) & \cdots & x_n(2) \\ \vdots & \vdots & \vdots & \vdots \\ x_1(m) & x_2(m) & \cdots & x_n(m) \end{bmatrix} \tag{6.1}$$

$$X_0 = [x_0(1), x_0(2), \cdots x_0(m)]^T \tag{6.2}$$

where n is the number of gas indicators; m is the number of temperatures in each stage.

(2) Dimensional unification of raw data. In order to accurately reflect the actual situation, this paper adopts the averaging method.

$$X_i(k) = \frac{X_i(k)}{\overline{X_i}}, \overline{X_i} = \frac{1}{n}\sum_{1}^{n} X_i(k) \tag{6.3}$$

where $X_i(k)$ is the value of the k index of the i factor.

(3) Subtract the parameter values of the reference series from the data of the index series to obtain the absolute difference between the two.

$$\Delta i(k) = |Xo(k) - Xi(k)|(k = 1, 2, \cdots, m; i = 1, 2, \cdots, n) \tag{6.4}$$

(4) Ask for the contact number.

$$\zeta i(k) = \frac{\Delta i(\min) + \rho \times \Delta i(\max)}{|Xo(k) - Xi(k)| = \rho \times \Delta i(\max)} \tag{6.5}$$

where ρ is the resolution coefficient, the value is taken within (0,1), the smaller the coefficient, the greater the difference between the correlation coefficients, and the stronger the discrimination ability, usually 0.5. For the four stages of coal spontaneous combustion oxidation, the correlation degree between temperature and index gas was calculated respectively. Taking 100–186 °C as an example, the reference eigensequence t = {100, 110, 120, 130, 140, 150, 160, 173, 186}. A total of 13 parameters were compared: gas index values C_2H_4, C_2H_6, C_3H_8, CO, CO/CO_2, C_2H_6/CH_4, C_3H_8/CH_4, C_2H_4/CH_4, C_2H_4/CO, CO/CH_4, C_3H_8/C_2H_6, C_2H_4/C_2H_6, and C_2H_4/C_3H_8. Among the index parameters, CO_2 is greatly affected by the atmospheric environment under the mine, and CH_4 is the gas released normally at room temperature during the mining of the working face, so these two gases are generally not directly used as index gases for coal spontaneous combustion. According to the analysis steps of the gray correlation analysis method, the correlation degree ranking of index gases at different temperature stages is obtained.

(2) Optimization of index gas based on Kendall τ_B correlation coefficient.

τ_B correlation coefficient is suitable for evaluating the relationship between two ordered categorical variables. The working principle is to check whether the two sets of values are ranked consistently by sorting the two sets of measured values separately and performing rank conversion. If X and Y are positively correlated, then the two variables change in the same direction. If X and Y are negatively correlated, then the two variables change in opposite directions. If at least 1 variable does not change in the change, it is called an invariant pair. Kendall τ_B calculates the correlation based on the number of these pairs. When using this non-parametric method to detect the correlation between ordinal variables, the value range is between -1–1. The formula for calculating τ_B correlation coefficient is:

$$\tau_B = \frac{n_c - n_d}{\sqrt{(n_0 - n_1)(n_0 - n_2)}} \tag{6.6}$$

$$n_0 = \frac{n(n-1)}{2} n_1 = \sum_i \frac{ti(ti-1)}{2} n_2 = \sum_j \frac{\mu i(\mu i - 1)}{2} \tag{6.7}$$

where n_c is the number of pairs of consistent changes, n_d is the number of pairs of inconsistent changes, and ti and μi are the numbers of values of the i group of two variables, respectively.

$\tau > 0.80$ is a strong correlation. $0.61 < \tau < 0.80$ is a strong correlation, $0.41 < \tau < 0.60$ is a moderate correlation, $0.21 < \tau < 0.40$ is a weak correlation, and $\tau < 0.20$ is a very weak or no correlation. The Kendall τ_B correlation coefficient between each temperature stage of the coal sample and the index gas is calculated by Eq. (6.7), and the gas parameter of $\tau > 0.80$ is obtained, and the correlation coefficient of the index gas in each temperature interval is shown in Table 6.1.

Combined with the correlation results of gray correlation analysis and τ_B correlation coefficient, when the temperature is in the range of 30–100 °C, CO/CH_4 is the primary index for predicting coal spontaneous combustion, and CO/CO_2 and CO are auxiliary indicators. When the temperature is in the range of 100–186 °C, CO/CO_2 is the primary index for predicting coal spontaneous combustion, and C_2H_4/CO is the auxiliary index. When the temperature was 186–264 °C, CO/CO_2 was the primary index for predicting coal spontaneous combustion, and C_2H_4/C_3H_8 and C_3H_8/C_2H_6 were the auxiliary indicators. When the temperature is 264–342 °C, C_2H_4 is the primary index for predicting coal spontaneous combustion, and C3H8 and CO are the auxiliary indexes [40].

Carbon monoxide (CO) is the core indicator of early warning of coal fires. When the coal temperature rises to 60–80 °C, coal oxidation reaction can produce CO, and its concentration usually precedes a significant increase in temperature. In on-site monitoring, if the CO concentration continues to exceed 5–10ppm, it indicates that

Table 6.1 Correlation coefficient of the index gas in each temperature interval [41]

Temperature range	Gas indicators	Correlation coefficient
30–100 C	CO	1.000
	CO/CO_2	1.000
	CO/CH_4	1.000
100–186 C	CO/CO_2	0.929
	C_2H_4/CO	0.929
	C_2H_4/C_3H_8	0.810
186–264 C	CO/CO_2	0.905
	C_3H_8/C_2H_6	0.810
264–342 C	C_2H_4	1.000
	C_3H_8	0.867
	CO	1.000

the coal body has entered the accelerated oxidation stage. If it reaches more than 24ppm and shows an upward trend, emergency intervention needs to be initiated. The sensitivity of CO makes it the preferred target for manual detection and sensor deployment and control in downhole.

Ethylene (C_2H_4) is regarded as a "turning signal" for coal temperature to break through the critical point. The gas begins to be formed at a coal temperature of 110–130°C, and the formation reaction is irreversible. Its appearance marks the transition from slow oxidation to intense oxidation of coal, and even trace amounts (0.1ppm) have a very high warning value. In practice, the ratio of CO to C_2H_4 concentration is often used as the basis for predicting the development rate of fire: a sharp drop in the ratio indicates an intensification of oxidation reaction.

The determination of the high-temperature stage depends on hydrogen (H_2) and acetylene (C_2H_2). The formation of H_2 accelerated above 130 °C, and the concentration increased exponentially after 200 °C. C_2H_2 needs to be produced by high-temperature cracking reaction above 300 °C, and its detection directly confirms the existence of open flame combustion in the fire area. The collaborative monitoring of the two can accurately locate the core area of high temperature.

The application of composite indicators improves the reliability of judgment.

CO/CO_2 ratio: ratio >0.3 reflects the deepening of coal oxidation degree, and >1.0 indicates entering the combustion stage. Graham coefficient ($ICO/\Delta O_2$): CO generation is corrected by oxygen consumption to avoid ventilation interference. Chain ratio (C_2H_6/CH_4): An increase in the ratio indicates an intensification of the pyrolysis reaction [42].

6.2 Indicator Gas Detection Equipment

The accurate detection of coal fire indicator gas relies on specialized equipment, which can be divided into three categories: portable, fixed and intelligent unmanned according to the application scenario, and its technical characteristics and applicable conditions are as follows:

(1) Portable testing equipment.

Portable gas detection equipment is the core tool for rapid response to coal fire sites, and its design takes into account mobility, reliability and multi-functional integration, which is mainly divided into two types of technical routes:

1) Multi-parameter gas detector.

This type of equipment is based on electrochemical sensors (detecting CO, H_2, H_2S, etc.) and catalytic combustion sensors (detecting hydrocarbon gases such as CH_4), and real-time multi-gas synchronous analysis is achieved through multi-channel parallel processing technology. Taking the German Dräger X-am® 8000 as an example, its weight is less than 300g, but it can monitor 6 gases at the same time, the lower limit of CO detection is 0.1ppm, the response time is as short as 10 s, and a 95dB sound and visual alarm is triggered when the concentration exceeds the limit [43]. To adapt to the harsh underground environment, the equipment must be certified as intrinsically safe for mining (e.g. Ex ib I Mb), and the housing has an IP68 protection rating to withstand a 1.5 m drop. In practical application, inspectors need to record key data (such as CO mutation point, CH_4 gradient change) every 15 min, and combine GPS positioning to generate a gas diffusion path map in the fire area (Fig. 6.7).

2) Portable Gas Chromatograph (GC).

For the needs of complex trace gas analysis, miniaturized GC adopts fast thermal desorption technology and high-sensitivity detectors (such as photoionization detector PID, hydrogen flame ionization detector FID) to complete the qualitative and quantitative analysis of C_1–C_5 hydrocarbon gases under field conditions. Taking INFICON HAPSITE ER, for example, it has a built-in miniature capillary column with a detection limit of 0.05ppm for ethylene (C_2H_4), and the analysis cycle is compressed to 8 min. The operation of the equipment needs to be cooperated with a pre-concentration adsorption tube: after the gas is adsorbed on site, it is instantaneously released by resistance heating to 300 °C, and then pushed into the separation system by the carrier gas. This type of equipment is especially suitable for delineating the boundaries of fire zones, and when C_2H_4 concentrations > 0.1ppm or traces of C_2H_2 are detected, the range of influence of high-temperature heat sources can be determined (Fig. 6.8).

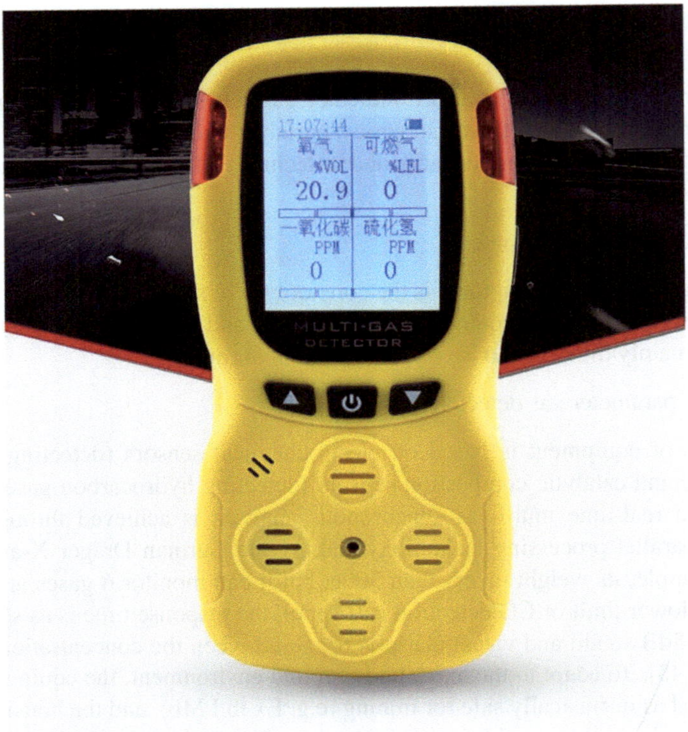

Fig. 6.7 Multi-parameter gas detector

(2) Fixed online monitoring system.

The fixed online monitoring system is an important part of the coalmine fire early warning system, which can realize continuous and real-time monitoring of the index gas. These systems are typically installed in critical locations underground in coalmines and provide reliable monitoring data through a variety of advanced sensing technologies.

1) Infrared spectroscopy multi-gas analysis system.

Infrared spectroscopy systems use non-dispersive infrared (NDIR) technology to measure concentration by detecting the absorption characteristics of gas molecules in specific infrared bands. This type of system is particularly suitable for monitoring key index gases such as CO, CO_2 and CH_4, and has the characteristics of high detection accuracy and good stability [44, 45]. The system is often equipped with an automatic calibration function that allows for regular zero and range calibration, ensuring accuracy in long-term monitoring. In terms of deployment, it can be installed on the wall to monitor the gas changes in the air flow in the roadway, and can also be drilled into the goaf to directly monitor the gas release of the hidden fire source. Modern infrared monitoring systems also have the function of data remote

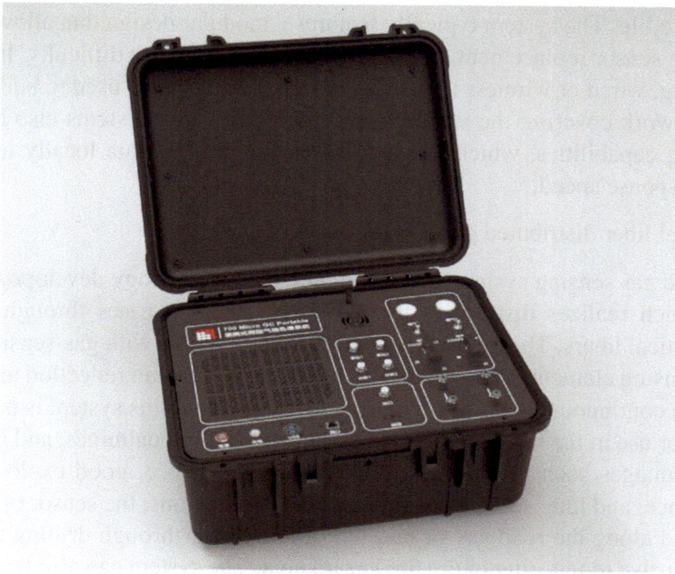

Fig. 6.8 Portable gas chromatograph

transmission, which can transmit monitoring data to the ground monitoring center in real time via industrial Ethernet or wirelessly.

2) Laser gas monitoring system.

Tunable diode laser absorption spectroscopy (TDLAS) systems represent the advanced level of current gas monitoring technology. The system uses a laser beam to scan the target area and analyze the spectral characteristics of the laser absorbed by gas molecules to achieve high-precision and long-distance gas concentration measurement. The main advantage of the TDLAS system is its extremely high detection sensitivity, which can detect gas concentrations at the ppb level with virtually no interference from other gas components. In coal mining applications, this system is particularly suitable for monitoring characteristic gases such as C_2H_2 and H_2 in high-temperature fire areas. The system can be installed at a fixed base station and realize gas monitoring within a range of hundreds of meters through a reflection device, providing an important basis for the delineation of the fire area and the assessment of fire development.

3) Electrochemical sensor array system.

The electrochemical sensor array system consists of multiple specialized gas sensors capable of monitoring multiple index gases simultaneously. This type of system has the characteristics of relatively low cost and flexible installation, and is suitable for large-scale deployment in coalmines. Modern electrochemical sensors use nanomaterials and intelligent algorithms to significantly improve detection accuracy and service life. The system typically features a modular design that allows for hot-swappable sensor replacement, greatly reducing maintenance difficulty. In terms of networking, wired or wireless transmission technology can be used to build a monitoring network covering the whole mine. Some advanced systems also have edge computing capabilities, which can preprocess monitoring data locally to improve system response speed.

4) Optical fiber distributed gas sensing system.

Fiber optic gas sensing system is a new monitoring technology developed in recent years, which realizes distributed monitoring of specific gases through specially treated optical fibers. The system uses optical fibers coated with gas-sensitive materials as sensing elements, combined with optical time-domain reflection technology, to achieve continuous monitoring for several kilometers. This system is particularly suitable for use in the harsh underground environment of coalmines, and has significant advantages such as anti-electromagnetic interference, good explosion-proof performance, and long service life. In practical applications, the sensor optical fiber can be laid along the roadway or laid out into the goaf through drilling to achieve comprehensive monitoring of the fire hazard area. The system can also be combined with distributed temperature monitoring to provide more comprehensive fire warning information.

(3) Intelligent unmanned detection platform.

1) Explosion-proof gas detection robot.

Explosion-proof gas detection robots are revolutionary equipment for high-risk underground environments. Its core is composed of a high-temperature resistant mobile chassis, a multispectral sensor array, an autonomous navigation system and a real-time communication module, which is specially designed for complex coal fire environments. The robot uses LiDAR and vision-inertial odometer (VIO) fusion technology to achieve centimeter-level positioning and obstacle avoidance in collapse risk areas or high-temperature roadways (>60 °C) [46]. The gas detection unit adopts a three-stage sensing architecture: Electrochemical sensors monitor the basic concentration of CO/O_2 in real time, NDIR module tracks CO_2/CH_4 trends, and TDLAS laser components specialize in ppm C_2H_2/H_2 trace gases. After the data is preprocessed by the edge computing chip, it automatically generates a heat map and marks the overrun area (such as CO > 24ppm area in red) through the explosion-proof 5G gateway backhaul control center. Key technological breakthroughs include ceramic matrix composite explosion-proof enclosures (withstanding instantaneous high temperatures of 120°C) and adaptive air intake systems (dynamically adjusting sampling flow based on dust concentration), significantly improving reliability under complex operating conditions. According to the application case of Jincheng Coalmine in Shanxi, China in 2023, the robot took 8 h to complete the scanning of the 5km^2 fire area, accurately locating the hidden oxidation area, and improving the efficiency of manual inspection by 15 times [47].

Table 6.2 Comparison table of technical parameters of testing equipment [48]

Equipment type	Detect gases	Precision	Response time	Applicable scenarios
Portable multi-parameter detector	$CO/CH_4/O_2$ etc	1 ppm	<30 s	Manual inspection and emergency investigation
Stationary NDIR system	$CO/CO_2/CH_4$	0.5 ppm	15–60 s	Long-term underground monitoring
Laser spectrometer	$C_2H_2/H_2/CH_4$	0.01 ppm	1–10 min	Remote detection during high temperature open flame period
UAV GC platform	C_1–C_3 hydrocarbons	0.1 ppm	3–5 min	Quick scanning of open fire areas

2) Unmanned aerial vehicle (UAV) aerial monitoring platform.

The UAV platform has overcome the efficiency bottleneck of large-scale coal fire monitoring. The system uses a six-rotor UAV as the carrier, mounts a lightweight gas payload (<3kg), and integrates quantum cascade laser spectrometer (QCL) and micro gas chromatography (μ-GC) dual-mode detection technology. The QCL module realizes high-altitude anti-interference detection by tuning the mid-infrared band, with a sensitivity of 0.05ppm, and the μ-GC completes the on-site separation and analysis of C_1–C_5 hydrocarbon gases. The flight control system is equipped with a gas concentration gradient navigation algorithm, which automatically generates a spiral approximation path when an abnormal CO concentration is detected, and locks the maximum concentration point error < 2m. In the practice of the Ulanqab open-pit mine fire area in Inner Mongolia, the UAV swarm operated collaboratively through Mesh ad hoc network technology, covering 18km^2 in 12 sorties, constructing a contour map of CO/CH_4 concentration in the whole region, and successfully identifying 7 fracture leakage sources. The range is broken through to 150 min through the hydrogen fuel cell expansion package, supporting − 20°C low-temperature environment operation (Table 6.2).

6.3 Application Cases of Index Gas Detection

Taking Tangjiahui Coalmine as the object, the gas generation law in the process of coal spontaneous combustion was systematically studied through the program heating experiment and coal spontaneous combustion experiment, and a set of coal spontaneous combustion grading early warning method based on index gas was proposed, which was successfully applied to on-site monitoring and prevention and control.

6.3.1 Experimental Equipment and Conditions

(1) Temperature program experiment.

Samples are collected from the coal seam and transported to the laboratory. Table 6.3 shows an overview of coal sample information.

The spontaneous combustion characteristics of coal were studied using the coal spontaneous combustion temperature programming experimental device (Fig. 6.9). The raw coal is crushed into five particle sizes: 0–0.9 mm, 0.9–3 mm, 3–5 mm, 5–7 mm and 7–10 mm. Take 200 g of coal samples from each particle size and mix them to make experimental samples, and put them in a cool place for later use. During the experiment, the temperature rose from 30 °C to 400 °C, the inlet flow rate was 60–120 mL/min, and the inlet oxygen concentration was 9%, 12%, 15%, 18% and 21% respectively. Whenever the temperature of the coal sample rises by 10°C, the gas composition and its content are measured once.

(2) Coal spontaneous combustion experiment.

A large-scale coal spontaneous combustion experimental device (Fig. 6.10) was used to study the spontaneous combustion characteristics of coal. The mass of the coal sample used in the experiment was 2198.31 kg, the air supply was set at $0.1 - 0.8\ m^3/h$, the oxygen concentration at the inlet was controlled at 21%, and the temperature

Table 6.3 Basic information of coal samples [49]

Coal sample	Industrial analysis (%)				Era	Coal type
	Moisture content	Ash content	Volatiles	Fixed carbon		
Tangjiahui	10.02	7.14	35.87	46.97	Jurassic	Long flame coal

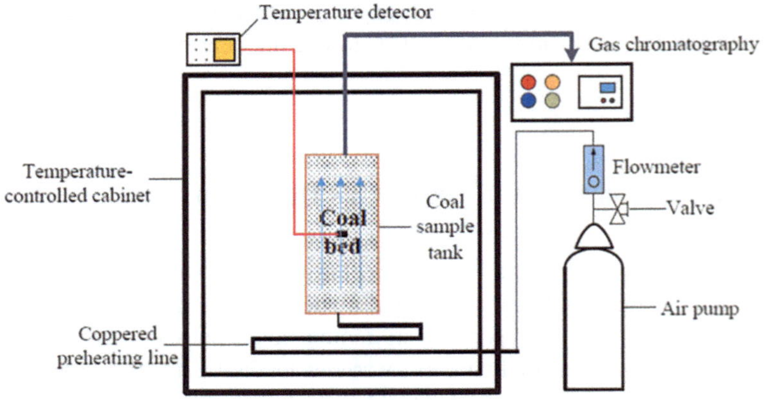

Fig. 6.9 Coal spontaneous combustion temperature programming experimental device [49]

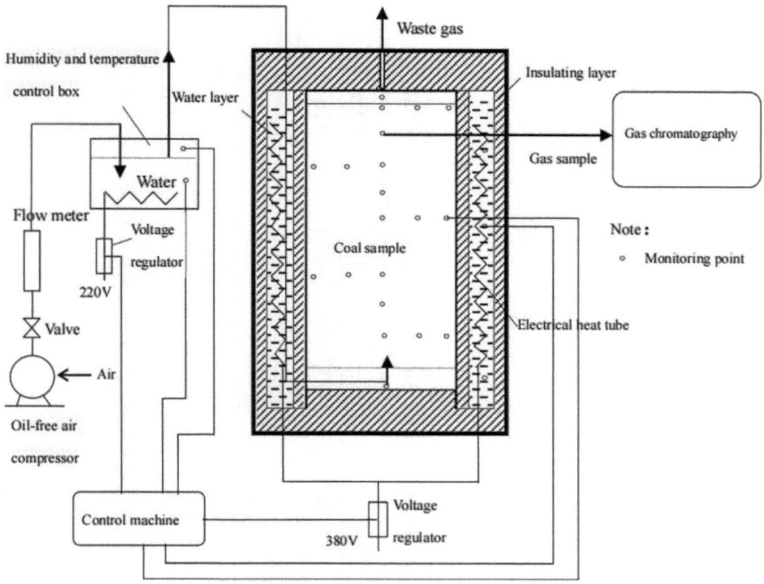

Fig. 6.10 Schematic diagram of the experimental setup [50]

gradually increased from 27.3 to 170 °C. When the temperature of the coal sample rises by 10°C, the gas composition and its content are tested once. The specific experimental parameters are detailed in Table 6.4.

6.3.2 Gas Change Law in the Process of Spontaneous Combustion of Coal

(1) Analysis of temperature programmed coal spontaneous combustion experimental data.

From Fig. 6.11, it is concluded that the O_2 concentration gradually decreases while the CO concentration continues to increase with the increase of temperature. The higher the oxygen concentration at the air inlet, the faster it is consumed and the more CO is generated during the process of spontaneous combustion of coal. When the temperature was in the range of 140–200 °C, the decline rate of O_2 concentration accelerated significantly. After exceeding 200 °C, the O_2 concentration was basically stable at about 1.0%, and the lowest was only 0.3%. The whole process shows that the oxidation reaction of coal is always in a state of high oxygen consumption.

Below 120 °C, the formation rate of CO is slow, because in the low-temperature oxidation stage, the reaction between coal and oxygen is relatively slow, and a large amount of water and oxygen complexes are generated, and some products are

Table 6.4 Experimental parameters [49]

Coal sample	Reference particle size (mm)	Height (cm)	Quality (kg)	Volume (cm^3)	Clump density(g/ cm^3)	Block density (g/ cm^3)	Porosity (%)	Air volume (m^3/h)	Initial temperature (°C)	Final temperature (°C)
Tangjiahui	3.25	185	2198.31	2,091,240	1.4	1.0512	0.2491	0.1 – 0.8	27.3	170

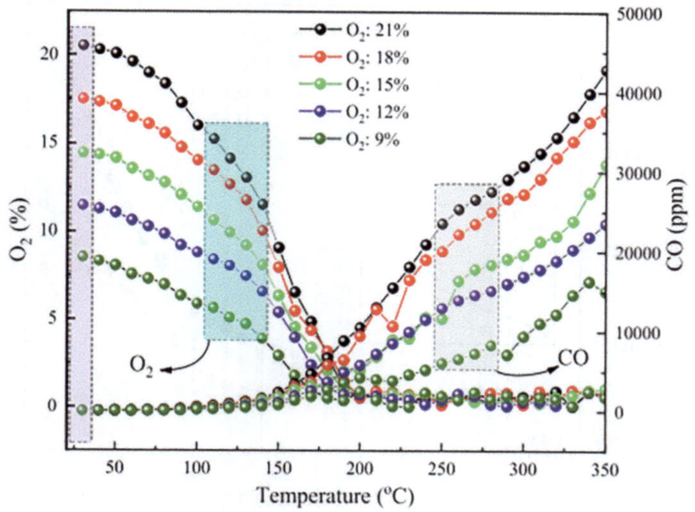

Fig. 6.11 Changes of O_2 and CO during spontaneous combustion of temperature-programmed coal under different oxygen conditions [49]

converted into carbon monoxide (CO) and carbon dioxide (CO_2). When the temperature rises to 140 °C, the carbon monoxide content increases significantly because the large amount of hydrogen and oxygen complexes generated in the early stage complete the transformation. In a high-temperature environment with lack of oxygen, incomplete oxidation of carbon elements will produce carbon monoxide, and the gas will also be produced by the cracking and decomposition process of carboxyl groups, ether bonds, and certain oxygen-containing heterocyclic compounds.

Figure 6.12 shows that a small amount of CO_2 and CH_4 were observed at the beginning of the experiment, indicating that the raw coal seam may contain CO_2 and CH_4, so it is not appropriate to use CO_2 and CH_4 as early warning indicators for low-temperature oxidation. Below 60 °C, the increase of CO_2 is relatively slow, as CO_2 is mainly derived from the conversion of water − oxygen complexes. The presence of this complex indicates that there is a chemical oxygen absorption phenomenon on the coal surface below 60°C. Above 70 °C, CO increases rapidly because CO_2 mainly comes from the decomposition of carboxylic acids, which indicates that the chemical reaction has reached a critical temperature.

Before 90 °C, the increase of CH_4 is relatively slow, and then increases rapidly, because CH_4 mainly comes from the desorption of CH_4 in coal at low temperature. As the temperature increases, thermal decomposition begins to occur in the coal sample, and the methyl side chain and fatty chain break to form CH_4. At 60 °C, C_2H_6 began to appear, and at 90 °C, C_2H_4 began to appear, indicating that C_2H_6 and C_2H_4 did not exist in the original coal seam of Tangjiahui mine.

C_2H_4 is a gaseous product of coal thermal decomposition, which indicates that the thermal decomposition process begins at 90 °C. With the increase of temperature, the lipid side chains on the aliphatic hydrocarbon ring and aromatic ring in the coal

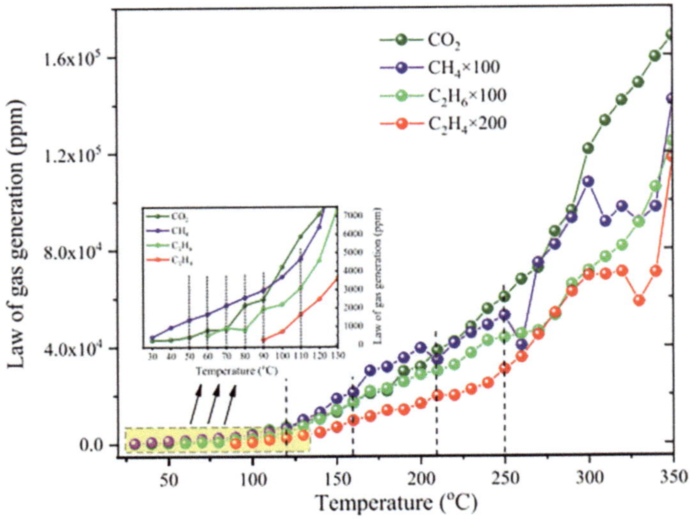

Fig. 6.12 Gas change law during spontaneous combustion of temperature-programmed coal [49]

free phase begin to form C_2H_6 and C_2H_4. Since the stability of C_2H_6 is higher than that of C_2H_4, the volume of C_2H_6 is significantly larger than that of C_2H_4 at the same temperature. When the temperature exceeded 250 °C, the content of C_2H_6 and C_2H_4 increased significantly again, indicating that the combustion reaction had begun in the coal sample.

Figure 6.13 shows that $CO/\Delta O_2$ gradually increases with increasing temperature, and the increase of $CO/\Delta O_2$ is greater between 60 °C and 250 °C. As the temperature increased, C_2H_6/CH_4 and CO/CO_2 increased first and then decreased, and a peak of C_2H_6/CH_4 and CO/CO_2 was observed in the temperature range of 60 °C to 80 °C. At 140 °C, C_2H_6/CH_4 reaches its maximum value, At 250 °C, CO/CO_2 is maximum. In the range of 90 °C to 120 °C, C_2H_4/CH_4 and C_2H_4/C_2H_6 gradually increased. In the range of 120 °C to 290 °C, C_2H_4/CH_4 fluctuated slightly around 0.18. In general, C_2H_4/C_2H_6 showed an upward trend, but the change was small before 250 °C.

(2) Experimental analysis of coal spontaneous combustion.

Figure 6.14 shows that O_2 continues to decrease as the temperature increases. When the temperature exceeds 120 °C, the decrease of O_2 increases significantly due to the slow reaction rate of coal oxygen in the low temperature stage, while the later reaction is more intense. The peak of CH_4 occurred before 40 °C, which was mainly due to the separation process of adsorbed and free CH_4 in coal. This phenomenon indicates that below 40 °C is the transition stage between CH_4 desorption and physical O_2 adsorption. In the range of 40 °C to 120 °C, the increase of CH_4 and CO was relatively slow. When the temperature exceeded 120 °C, the contents of CO, CO_2, CH_4, C_2H_4 and C_2H_6 showed significant increases. At 60 °C, the coal was observed to contain C_2H_6, and at 90 °C, the coal was observed to contain C_2H_4, indicating

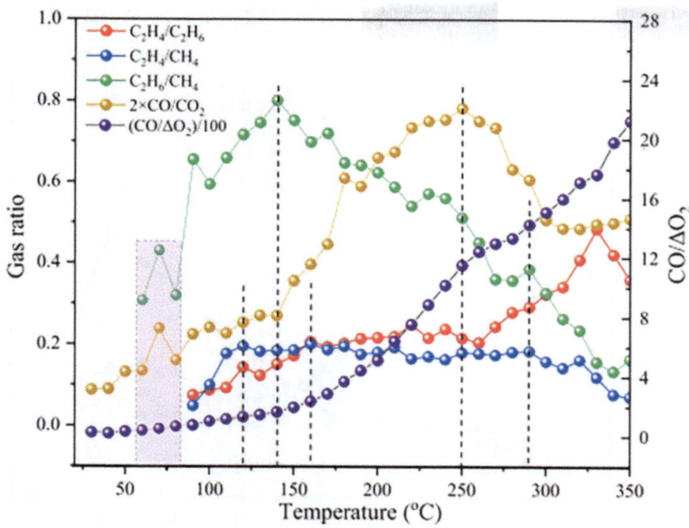

Fig. 6.13 Variation of gas ratio during spontaneous combustion of temperature-programmed coal [49]

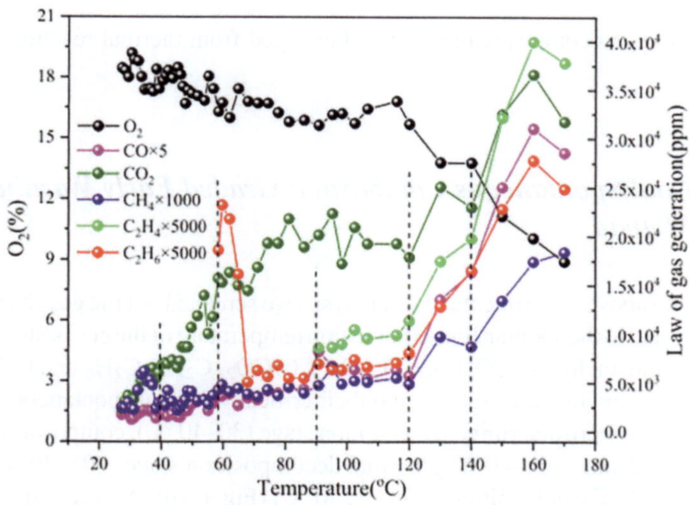

Fig. 6.14 Gas change law during coal spontaneous combustion [49]

that the spontaneous combustion process of coal reached the thermal decomposition temperature.

Figure 6.15 shows that $CO/\Delta O_2$ shows an overall upward trend with the increase of temperature. After 90 °C, the increase of $CO/\Delta O_2$ and CO/CO_2 was significant. At 140 °C, it was observed that C_2H_4/C_2H_6 reached the maximum value, indicating that

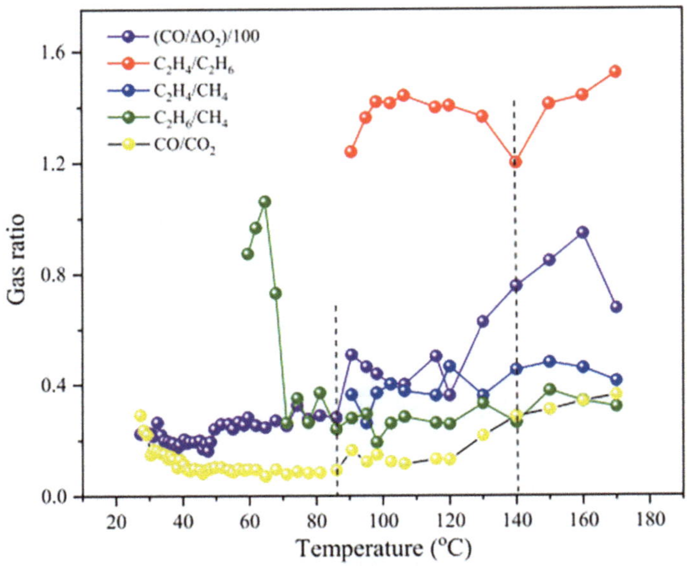

Fig. 6.15 Variation of gas ratio in the process of spontaneous combustion of coal [49]

the coal—oxygen composite reaction had changed from thermal reaction to fission reaction.

6.3.3 Coal Spontaneous Combustion Graded Early Warning Method

Based on the above experimental data analysis, we screened out the gas characteristic indexes related to the spontaneous combustion temperature of the coal seam in Tangjiahui Coalmine, including CO, O_2, $CO/\Delta O_2$, CO/CO_2, C_2H_4, C_2H_6 and C_2H_4/C_2H_6. The six characteristic temperatures and their critical values of spontaneous combustion in the mine were determined: adsorption stage (30–40 °C), composite stage (50–60 °C), critical stage (60–80 °C), thermal decomposition stage (90–120 °C), fission stage (140–160 °C) and ignition point (≥250 °C) (Fig. 6.16). Accordingly, the spontaneous combustion process of coal seam is divided into seven stages: incubation period, compound stage, self-heating stage, activation stage, thermal decomposition stage, fission stage and combustion stage. The method of determining each risk level according to the gas index is given, and the warning names of each risk level are clarified (Table 6.5). It solves the problems of criticality, danger, and difficulty in quantitatively determining the hazards of fire in the Coal Mine Safety Regulations, and lays the foundation for follow-up work.

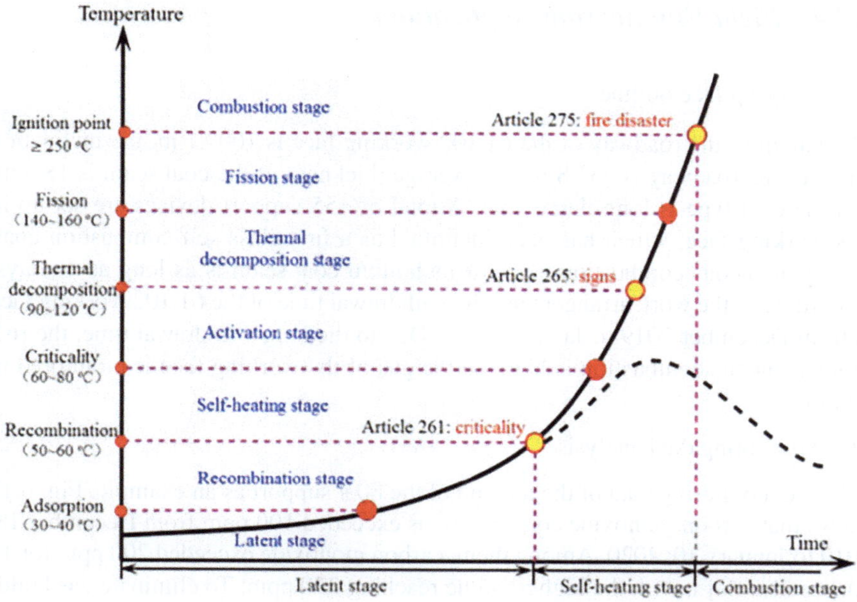

Fig. 6.16 Schematic diagram of the detailed classification of the seven levels of coal spontaneous combustion process [49]

Table 6.5 Temperature, gas index and their critical values in coal spontaneous combustion grading warning [49]

Stage name	Warning name	Temperature range/°C	Warning threshold values of gas indexes
Latent stage	Initial value for warning	$T \le 30$	$R_0 = \{O_2 > 18\% \cap CO > 50\ ppm\} \cup \{O_2 \in (15\%, 18\%) \cap CO > 80\ ppm\} \cup \{O_2 \in (12\%, 15\%) \cap CO > 150\ ppm\} \cup \{O_2 < 12\% \cap CO > 200\ ppm\}$
Recombination stage	Gray warning	$T \in [30, 50)$	$R_1 = R_0 \cap \{100 \times CO/\Delta O_2 \in (0.2, 0.3)\}$
Self-heating stage	Blue warning	$T \in [50,70)$	$R_2 = R_0 \cap \{100 \times CO/\Delta O_2 \in (0.3, 0.5)\} \cap \{C_2H_6>0\}$
Critical stage	Yellow warning	$T \in [70, 110)$	$R_3 = R_0 \cap \{100 \times CO/\Delta O_2>0.5)\} \cap \{C_2H_4>0\}$
Thermal decomposition stage	Orange warning	$T \in [110, 150)$	$R_4 = R_3 \cap \{C_2H_4>0\} \cap \{C_2H_4/C_2H_6\}_{min}$
Fission stage	Red warning	$T \in [150, 250)$	$R_5 = R_4 \cap \{CO/CO_2\}_{max}$
Combustion stage	Black warning	$T \in [250, 400)$	$R6 = R5 \cap \{(C_2H_4/C_2H_6)_{max}$, visible fumeand fire$\}$

6.3.4 Field Engineering Applications

(1) Working face outline.

The length of the roadway of the 61,102 working face is 1007.1 m, the inclination angle of the roadway is 233.8 m, the average thickness of the coal seam is 18.3 m, and the coal type is long-flame coal. A total of 155 support devices are set up in this working face, which has been identified as a first-class self-combustion coal. The spontaneous combustion cycle of its bottom coal seam is as long as 46 days. According to the work arrangement, the withdrawal time of the 61,102 working face is from December 2019 to January 2020. Due to the long withdrawal time, the risk of spontaneous combustion is high, so the gas of the working face is monitored in real time.

(2) Monitoring data analysis.

Take the monitoring data of the tail end of the 80 # support as an example, Fig. 6.17 shows that carbon monoxide concentrations exceeded 100 ppm from December 18, 2019 to January 10, 2020. Among them, carbon monoxide exceeded 200 ppm for 12 consecutive days, with the highest value reaching 370 ppm. To eliminate gas build-up and wind influences, $CO/\Delta O_2$ and C_2H_6 were analyzed (Fig. 6.18). C_2H_6 was observed between 18 and 21 December 2019, so the coal seam was determined to be in the stage of spontaneous combustion of floating coal, and a blue warning was issued. Fire prevention measures were taken to inject nitrogen into the coal seam, and a total of 76,800 cubic meters of nitrogen were injected. No C_2H_6 was observed between 22 and 24 December 2019, indicating that the injected nitrogen effectively inhibited the spontaneous combustion of coal.

However, during this period, $CO/\Delta O_2$ values ranged from 0.2 to 0.3, indicating that the floating coal in the orebody was in the recombination phase, hence the gray warning. Between 25 and 28 December 2019, C_2H_6 was observed again, and the maximum value of $CO/\Delta O_2$ reached 0.57, indicating that the floating coal in the orebody had entered a critical stage, so a yellow warning was issued. The ventilation and safety department of the coalmine has taken fire prevention and extinguishing measures, including colloidal injection and flame retardant injection. On 30 December 2019, the $CO/\Delta O_2$ value was still as high as 0.49 and C_2H_6 was present, indicating that the injected colloids were insufficient and the high temperature was not controlled, so colloids, flame retardants and nitrogen were continuously injected into the ore body.

With the implementation of various fire prevention and extinguishing measures, the concentrations of carbon monoxide and $CO/\Delta O_2$ began to decrease, and the mine temperature also decreased. Since January 6, the injection of colloids and flame retardants has been stopped, and only nitrogen gas has been continuously injected to prevent re-ignition. During the withdrawal period, a total of 10,381 kg of colloids and 6,400 kg of flame retardants were injected. The graded early warning mechanism of coal seam spontaneous combustion proposed in this paper ensures the safe completion of the evacuation operation of the No. 61102 working face.

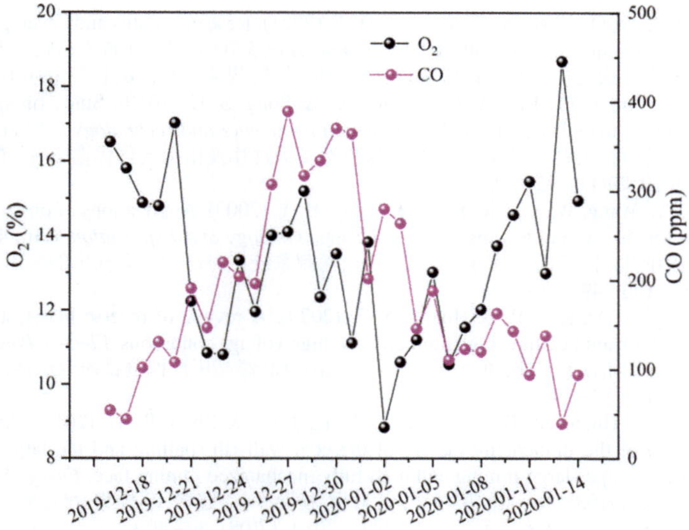

Fig. 6.17 Monitoring data of the tail end of the 80 # support [49]

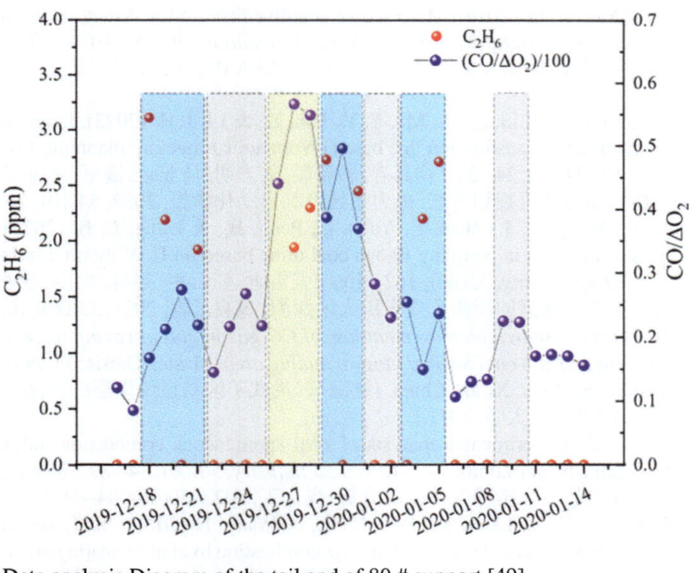

Fig. 6.18 Data analysis Diagram of the tail end of 80 # support [49]

References

1. Deng, J., Zhang, Q., Chen, W. L., & Bai, Z. J. (2024). Coal spontaneous combustion disaster monitoring and early warning technologies and development trend for coal mines. *Safety in Coal Mines*, *55*(03), 99–110. (邓军, 张琦, 陈炜乐, 白祖锦. 矿井煤自燃灾害监测预警技术及发展趋势. 煤矿安全, 2024, 55(03): 99–110.)

2. Wen, H., Tian, Q., Zhao, X. T., & Wang, W. F. (2021). Research status and development trend of coal fire monitoring and control. *Coal Engineering*, *53*(09), 101−106. (文虎, 田晴, 赵向涛, 王伟峰. 煤层火灾监测与治理研究现状及发展趋势. 煤炭工程, 2021, 53(09): 101−106.)
3. Zhan, W., An, Z. G., Ren, Y. D., Huang, X., & Song, S. L. (2010). Study on spontaneous combustion and indicators of coal gas system. *Fire Science and Technology*, *29*(11), 954−957. (占伟, 安志刚, 任砚东, 黄旭, 宋双林. 煤自然发火规律及指标气体优选研究. 消防科学与技术, 2010, 29(11): 954−957.)
4. Sun, C. Y., Wang, W. X., Niu, B. H., & Wang, H. Y. (2003). Applications of on-site detection system for the oil and gas resources in oceans. *Geology and Exploration*, (06), 49−52. (孙春岩, 王维熙, 牛滨华, 王宏宇. 气态烃快速探测系统在海洋油气勘探中的应用. 地质与勘探, 2003, (06): 49−52.)
5. Wang, Y. H., Deng, E. P., & Huang, Y. Z. (2020). A review of reverse bias test for power device in high temperature, high humidity and high voltage conditions. *Electric Power*, *53*(12), 18−29. (王延浩, 邓二平, 黄永章. 功率器件高温高湿高压反偏测试研究综述. 中国电力, 2020, 53(12): 18−29.)
6. Kang, C. Y., Huang, Z. B., Yang, H. J., Wang, J. G., & Zhao, W. H. (2019). Analysis and prevention of the influencing factors of the coal wall rib spalling and the large lump coal (Gangue) in super-large-mining and high fully mechanized mining face. *Energy Science and Technology*, *17*(08), 36−40. (康从勇, 黄志彬, 杨会军, 王建国, 赵维华. 平凉新安煤业5#煤层自燃指标气体优选实验研究. 神华科技, 2019, 17(08): 36−40.)
7. Tian, F. C., Ling, Y. T., Zhu, H. Q., Chen, M. Y., & Wang, J. C. (2022). Application of a novel detection approach based on non-dispersive infrared theory to the in-situ analysis on indicator gases from underground coal fire. *Journal of Central South University, 29*, 1840–1855.
8. Ma, A. Q., & Yao, W. Q. (2024). Multi sensor adaptive fusion SLAM method for underground mobile robots in coal mines. *Journal of Mine Automation*, *50*(05), 107−117. (马艾强, 姚顽强. 煤矿井下移动机器人多传感器自适应融合SLAM方法. 工矿自动化, 2024, 50(05): 107−117.)
9. Xu, Q. C., Guo, C. C., Liu, Z. M., Ma, Y. Y., Pan, Z., & Li, J. H. (2022). Drone autonomous inspection method for transmission line based on dynamic trajectory planning. *Proceedings of the CSU−EPSA*, *34*(10), 24−31. (徐其春, 郭晨晨, 刘志明, 马亚运, 潘卓, 李建辉. 无人机线路自主巡检的动态轨迹规划方法. 电力系统及其自动化学报, 2022, 34(10): 24−31.)
10. Zhang, L. X., Wang, Y. J., Zhao, F., Yuan, g., Peng, B., & Dang, L. B. (2021). Detection and analysis of fire area in Xinjiang Baoan coal mine based on UAV thermal infrared remote sensing. *Coal Engineering*, *53*(06), 162−166. (张雷昕, 汪云甲, 赵峰, 原刚, 彭斌, 党立波. 基于无人机热红外遥感的新疆宝安煤矿火区探测. 煤炭工程, 2021, 53(06): 162−166.)
11. Xia, H. B. (2011) *Research on movement law of CO gas in gob of caving workface in easily spontaneous combustion coal Seam o Lingwu mining area.* Master Thesis, Xi'an University of Science and Technology, Xi'an, China. (夏海斌. 灵武矿区易自燃煤层综放面采空区CO运移规律研究. 西安科技大学, 2011.)
12. Jia, X. L. (2022). Experimental analysis of coal spontaneous combustion index gas under different oxygen concentrations. *Coal Chemical Industry*, *50*(06), 84−89. (贾晓亮. 不同氧浓度环境下的煤自然发火指标气体试验分析. 煤化工, 2022, 50(06): 84−89.)
13. Wang, F., Zhong, C. Y., Li, Y. H., Ran, X. C., & Wang, F. (2021). Study on characteristic parameters and mark gases of coal spontaneous combustion oxidation heating process. *Shaanxi Coal*, *40*(02), 34−37+47. (王飞, 钟传义, 李毅恒, 冉学超, 王帆. 煤自燃氧化升温过程特性参数及标志气体研究. 陕西煤炭, 2021, 40(02): 34−37+47.)
14. Wang, Q. H., Jin, S. L., Luo, Z. M., & Li, Z. H. (2022). Influence of metamorphism degree on kinetic parameters of coal combustion. *Safety and Environmental Engineering*, *29*(03), 62−70. (王秋红, 靳松灵, 罗振敏, 李州昊. 煤变质程度对煤燃烧动力学参数的影响. 安全与环境工程, 2022, 29(03): 62−70.)
15. Chaika, L., Sokhatskii, A. A., Vasil'ev, L. E., Sushchenko, A. V., Lebed', V. V., & Moskalina, A. A., et al (2019). Investigation of the influence of the technology of blast-furnace smelting with the use of pulverized coal fuel and natural gas on the performance indicators of blast furnaces. *Metallurgist*, *62*, 1201−1212.

16. Liang, Y. T., Zhang, Y. T., Ren, M. G., Wang, L. J., Tian, F. C., & Li, Y. Q. (2025). Gas-active group correlation mechanism of spontaneous combustion indicators and hierarchical early warning system of oil-impregnated coal samples based on the reference method. *Journal of Thermal Analysis and Calorimetry., 150*(14), 1–18.
17. Huo, X. W., Chen, H. D., Xu, Y. L., Wang, L. Y., & Li, L. (2025). Coal spontaneous combustion early warning methods based on slope grey relation analysis. *Natural Resources Research., 34*(05), 1–14.
18. Lei, C. K., Feng, Q. C., Zhu, Y. Q., Cui, C. B., Bao, R. Y., & Deng, C. B. (2025). Multiple indicator gases and temperature prediction of coal spontaneous combustion oxidation process. *Fuel, 393*, Article 134991.
19. Li, D. F., Zang, Y. Y., & Shi, J. L. (2022). Intelligent mine fire early warning system. *Journal of Mine Automation, 48*(S1), 112−115,120. (李东发, 臧燕杰, 师吉林. 矿井火灾智能预警系统. 工矿自动化, 2022, 48(S1): 112−115+120.)
20. Wang, Z. Y., Bai, G. X., Zhang, Q., & Jia, M. S. (2024). Application of intelligent monitoring system for spontaneous combustion fire in coal mines based on micro-chromatography technology. *Journal of Intelligent Mine, 5*(02), 84−88. (王泽瑶, 白光星, 张琦, 贾明铄. 基于微色谱技术的煤矿自燃火灾智能监测系统应用.智能矿山, 2024, 5(02): 84−88.)
21. Lu, W., Li, J. L., Wang, K. S., Ma, H. P., & Ye, Z. L. (2015). Study on monitoring and measuring method of positive pressure bundle tube for prediction and forecast of coal spontaneous combustion. *Coal Science and Technology, 43*(10), 77−80. (陆伟, 李金亮, 王开胜, 马汉鹏, 叶正亮. 煤自燃预测预报的正压束管监测方法研究. 煤炭科学技术, 2015, 43(10):77−80.)
22. Wang, D., Lu, W., Li, J. L., & Xu, S. (2019). Study on key technology of positive pressure bundle tube monitoring system for predicting coal spontaneous combustion in mine. *Mining Safety & Environmental Protection, 46*(06), 57−60. (王栋, 陆伟, 李金亮, 徐帅. 预报矿井煤自燃火灾的正压束管监测系统关键技术研究. 矿业安全与环保, 2019, 46(6): 57−60.)
23. Wang, D., Lu, W., Li, J. L., Xu, S., & Qin, C. R. (2019). Study on high positive pressure beam tube monitoring system of sharing pipeline for gas transmission and pump control. *Coal Science and Technology, 47*(12), 141−144. (王栋, 陆伟, 李金亮, 徐帅, 秦传睿. 煤矿输气与控制共用管线的高正压束管监测系统研究. 煤炭科学技术, 2019, 47(12):141−144.)
24. Liu, Y. H., Liu, X., Zhang, S., & Zhang, Y. X. (2024). Research on multi-point measurement system of mine gas based on TDLA. *Journal of Physics: Conference Series, 2704*, Article 012033.
25. Ma, L., Fan, X. L., Zhang, S., Wang, W. F., & Wei, G. M. (2021). Research on CH_4 gas detection and temperature correction based on TDLAS technology. *Spectroscopy and Spectral Analysis, 41*(11), 3632−3638. (马砺, 范新丽, 张朔, 王伟峰, 魏高明. 基于TDLAS技术的CH_4气体检测与温度补偿方法. 光谱学与光谱分析, 2021, 41(11): 3632−3638.)
26. Shen, X. L., Hu, L. X., Gao, Y. X., & Lin, L. (2024). Monitoring of multi-component gas concentration in coal mine based on TDLAS technology (Invited). *Acta Photonica Sinica, 53*(10), 114−123. (申晓良, 胡澜夕, 高炎旭, 林列. 基于TDLAS技术的井下多组分气体浓度监测(特邀).光子学报, 2024, 53(10):114−123.)
27. Wang, Q. J., Sun, P. S., Zhang, Z. R., Zhang, L. W., Yang, X., & Wu, B., et al. (2021). Separation and analysis method of overlapping absorption spectra with cross interference in gas mixture measurement. *Acta Physica Sinica, 70*(14), 233−242. (王前进, 孙鹏帅, 张志荣, 张乐文, 杨曦, 吴边, 等. 混合气体测量中重叠吸收谱线交叉干扰的分离解析方法. 物理学报, 2021, 70(14): 233−242.)
28. Deng, J., Yang, C., Ren, L. F., Li, X., Wang, C. P., & Bai, Z. J. (2025). A review on coal spontaneous combustion prediction based on machine learning. *Journal of China Coal Society, 50*(S1), 336−360. (邓军, 杨成, 任立峰, 李鑫, 王彩萍, 白祖锦. 基于机器学习的煤自燃预测研究进展及展望. 煤炭学报, 2025, 50(S1): 336−360.)
29. Wang, Y. H., Hu, S. P., Zhu, Y. X., Dou, J. W., Zhang, X. L., & Zhang, S. J. (2024). Application status and prospect of machine learning in coal spontaneous combustion prediction and early warning. *China Coal, 50*(10), 98−103. (王玉怀, 胡硕鹏, 朱永兴, 窦静文, 张烜乐, 张思佳. 机器学习在煤自燃预测预报中的应用现状及展望. 中国煤炭, 2024, 50(10):98−103.)

30. Liu, B., Mu, K., Ye, F., Wang, F., & Wang, J. T. (2020). Prediction method of coal spontaneous combustion based on relevance vector machine. *Journal of Mine Automation, 46*(09), 104–108. (刘宝, 穆坤, 叶飞, 王帆, 王静婷. 基于相关向量机的煤自燃预测方法. 工矿自动化, 2020, 46(09):104–108.)
31. Zhai, X. W., Luo, J. L., Zhang Y. C., Song, B. B., Hao, L., & Zhou, Y. J. (2023). Prediction model of coal spontaneous combustion temperature based on data filling. *Journal of Mine Automation, 49*(01), 28–35+98. (翟小伟, 罗金雷, 张羽琛, 宋波波, 郝乐, 周妤婕. 基于数据填补的煤自燃温度预测模型. 工矿自动化, 2023, 49(01):28–35+98.)
32. Gao, F., Liang N., Jia Z., & Hou, Q. (2024). Prediction model of coal spontaneous combustion based on SSA–RBF neural network. *China Safety Science Journal, 34*(08), 128–137. (高飞, 梁宁, 贾喆, 侯青. 基于SSA-RBF神经网络的煤自然发火预测模型. 中国安全科学学报, 2024, 34(08):128–137.)
33. Zhang, L. D., Song, Z. Y., Luo, Z. M., & Zhao, S. S. (2022). Prediction of coal spontaneous combustion period based on machine learning. *China Safety Science Journal, 32*(12), 118–124. (张利冬, 宋泽阳, 罗振敏, 赵姗姗. 基于机器学习的煤自然发火期预测. 中国安全科学学报, 2022, 32(12):118–124.)
34. Xiao, Y., Cao, Y., Zhong, K. Q., Yin, L., & Deng, J. (2022). Optimized neural network to predict the experimental minimum period of coal spontaneous combustion. *Environmental Science and Pollution Research, 29*(19), 28070–28082.
35. Wang, W., Liang, R., Qi, Y., Jia, B. S., & Wu, Z. W. (2023). Prediction model of coal spontaneous combustion risk based on PSO–BPNN. *China Safety Science Journal, 33*(07), 127–132 (汪伟, 梁然, 祁云, 贾宝山, 武泽伟. 基于PSO-BPNN的煤自燃危险性预测模型. 中国安全科学学报, 2023, 33(07):127–132).
36. Zhang, T. Y., Lu, Y., Shi, S. L., Wang, T., Yang, F., & Niu, H. Y. (2019). Prediction on coal spontaneous combustion risk for multi—coal based on SVM. *Journal of Hunan University of Science and Technology (Natural Science Edition), 34*(02), 11–17. (张天宇, 鲁义, 施式亮, 王涛, 杨帆, 牛会永. 基于支持向量机分类算法的多煤种煤自燃危险性预测. 湖南科技大学学报(自然科学版), 2019, 34(02):11–17.)
37. Zou, P. Z., Ye, Y. X., Liang, X. Y., & Han, C. (2024). A Comparative analysis of coal spontaneous combustion tendency prediction based on machine learning. *Journal of Chongqing University*. https://link.cnki.net/urlid/50.1044.N.20240923.1046.002. (邹佩喆, 叶于欣, 梁晓瑜, 韩超. 基于机器学习的煤自燃倾向性预测比较分析. 重庆大学学报, 2024, https://link.cnki.net/urlid/50.1044.N.20240923.1046.002.)
38. Shukla, U. S., Mishra, D. P., & Mishra, A. (2023). Prediction of spontaneous combustion susceptibility of coal seams based on coal intrinsic properties using various machine learning tools. *Environmental Science and Pollution Research, 30*(26), 69564–69579.
39. Ma, L., Zou, L., Ren, L. F., Chung, Y. H., Zhang, P. Y., & Shu, C. M. (2020). Prediction indices and limiting parameters of coal spontaneous combustion in the Huainan mining area in China. *Fuel, 264*, Article 116883.
40. Li, Z. X., Zhang, M. Q., Yang, Z. B., Yu, J. X., Liu, Y., & Wang, H. W. (2022). Division of coal spontaneous combustion stages and selection of indicator gases. *PLoS One, 17*(4).
41. Zhu, H. Q., Sheng, K., Zhang, Y. L, Fang, S. H., & Wu, Y. L. (2018). The stage analysis and countermeasures of coal spontaneous combustion based on "five stages" division. *PLoS One, 13*(8).
42. Yao, J. Q., Yuan, W. Y., Gong, P., & Xie, L. (2022). Frequency division multiplexing and normalized harmonic method for WMS detection system in multi-gas detection. *Photonics, 9*(11), 850.
43. Zhang, Z. X. (2022). *Dynamic information fusion determination and prediction of coal spontaneous combustion and oxidation state based on FIG–SVR optimization model.* Master Thesis, China University of Mining and Technology, Xuzhou China. (张志旭. 基于FIG-SVR优化模型的煤自燃氧化状态的动态信息融合判定与预测研究.中国矿业大学, 2022.)
44. Santonab, C., Nirjhar, H. D., & Shankar, C. (2023). Grey relational analysis-based optimization of machining processes: A comprehensive review. *Process Integration and Optimization for Sustainability, 7*(4), 609–639.

45. Munoz-Pichardo, J. M., Lozano-Aguilera, E. D., Pascual-Acosta, A., & Muñoz-Reyes, A. M. (2021). Multiple ordinal correlation based on kendall's tau measure: A proposal. *Mathematics, 9*, 1616.
46. Yang, Z. J., Yu, X. X., Huang, Y., Meng, X. T., Ren, L. F., & Zhai, X. W. (2020). Research on the characteristics of characteristic gases of spontaneous combustion of No.2 coal in Zaoquan coal mine. *Energy Technology and Management*, *50*(02), 4−7. (杨致军, 余行贤, 黄楹, 孟祥涛, 任立峰, 翟小伟. 枣泉煤矿2#煤自燃标志性气体特征研究. 能源技术与管理, 2025, 50(02): 4–7.)
47. Wu, X. R. (2024). Research on the index gas and limit parameters of spontaneous combustion of coking coal in Huainan mining area. *Shandong Coal Science and Technology*, *42*(09), 77−81. (吴晓茹. 淮南矿区焦煤自燃指标气体及极限参数的研究. 山东煤炭科技, 2024, 42(09):77–81.)
48. Ma, H. S., Ma, W., Zhou, X., Zhang, Y. J., Liu, L., & Zhai, X. W. (2024). Coal spontaneous combustion characteristics and hierarchical early warning system in Meihuajing coal mine No. 2. *Shaanxi Coal*, *43*(10), 39−43+59. (马海生, 马魏, 周翔, 张彦吉, 刘龙, 翟小伟. 梅花井煤矿2号煤自燃特征及分级预警.陕西煤炭, 2024, 43(10):39–43+59.)
49. Zhang, D., Cen, X., Wang, W., Deng, J., Wen, H., Xiao, Y., et al. (2021). The graded warning method of coal spontaneous combustion in Tangjiahui mine. *Fuel, 288*, 11963.
50. Deng, J., Xiao, Y., Li, Q., Lu, J., & Wen, H. (2015). Experimental studies of spontaneous combustion and anaerobic cooling of coal. *Fuel, 157*, 261–269.

Chapter 7
Electromagnetic Detection Method

As a major disaster that threatens the safety of mining areas and damages the ecological environment for a long time, underground coal fires have strong concealment and complex spread mechanisms, making it difficult for traditional detection methods to achieve accurate positioning and dynamic monitoring of fire areas. In recent years, electromagnetic detection has gradually become the core technology for revealing the spatial distribution and combustion state of underground fire sources by virtue of its high sensitivity response to the magnetic and electrical parameters of coal and rock masses. The physical basis is that the high temperature environment caused by the spontaneous combustion of the coal seam can significantly change the phase state of the magnetic minerals in the surrounding rock, and induce the measurable magnetic anomalies on the surface. At the same time, the changes in pore structure, water migration and carrier activation caused by coal and rock pyrolysis will reshape the electrical structure of the formation and form a unique electromagnetic response signal. By capturing the magnetoelectric anomalies closely related to the evolution of the temperature field, the electromagnetic detection method provides irreplaceable technical support for the boundary delineation of the fire zone, the evaluation of combustion intensity and the early warning of rekindling.

7.1 Principle of Electromagnetic Detection

7.1.1 Magnetic Detection Principle

Magnetic exploration, referred to as magnetic method, is a geophysical method that analyzes and studies the distribution of detected objects in strata by observing and recording the magnetic differences between the detected object and the surrounding rock mass or minerals. The high temperature generated by coalmine fires will significantly affect the magnetic characteristics of the surrounding rocks, thus causing

Y. Xiao et al., *Detection of Fire Source in Underground Coal Mine and its Applications*,
https://doi.org/10.1007/978-981-95-7064-5_7

changes in the magnetic field of the fire area, which has become an extremely important geophysical indicator in coalmine fires. By observing the changes in the magnetic field on the ground and performing inversion analysis of the collected data, the spatial distribution of underground fire areas can be revealed.

(1) Magnetism of matter

The essence of the magnetism shown by matter is the movement of charged particles, matter is composed of atoms, and the movement of atomic nuclei and electrons outside the nucleus will produce a magnetic moment, and the sum of the vectors of these two magnetic moments superimposed is the reason why the material shows magnetism externally. The physical quantities included in the magnetism of matter are: magnetic intensity (M), magnetic susceptibility (χ), magnetic induction strength (B), and magnetic permeability (μ) [1–3].

Magnetization strength M is a magnetic physical parameter that characterizes the degree of magnetization of a substance, and for a uniform infinite magnetic medium, the relationship between magnetization strength and magnetic field strength can be expressed as under the action of the external magnetic field H:

$$M = \chi H \tag{7.1}$$

where, χ is the magnetization coefficient of the matter, that is, the volumetric magnetic susceptibility, dimensionless. M is the magnetization strength of the object, A/m. H is the strength of the external magnetic field, A/m.

The magnetic induction strength (B) characterizes the direction and strength of the magnetic field, and for the isotropic magnetic medium, the magnetic induction strength generated by the external magnetic field H at any point inside it, that is, the magnetic flux density:

$$B = \mu H \tag{7.2}$$

where, B represents the magnetic induction strength, T. μ is the permeability of the medium, N/A^2. H is the strength of the external magnetic field, A/m.

When the medium is vacuum, the formula can be expressed as:

$$B_0 = \mu_0 H \tag{7.3}$$

In the equation, μ_0 is the permeability of the vacuum, that is $\mu_0 = 4\pi \times 10^{-7} H \cdot m^{-1}$, under general conditions, the permeability is the relative permeability μr, which is defined as, $\mu r = \mu/\mu_0$, and can be obtained by substituting Eq. (7.3).

$$B = \mu_0 \mu_r H = \mu_0 H + \mu_0(\mu_r - 1)H = \mu_0(1 + \chi)H = \mu_0(H + M) \tag{7.4}$$

where $\chi = \mu_r - 1$. Equation (7.4) shows that compared with the vacuum medium, the magnetic medium is magnetized by the applied magnetic field and generates an additional magnetic field χH, which is directly proportional to its magnetic susceptibility χ.

According to the degree of magnetization of materials by external magnetic fields, materials can be divided into three categories: retro magnet, paramagnetic and ferromagnetic. Demagnetism, also known as diamagnetism, has a negative magnetic susceptibility of substances of this type of property, usually very small, and is often negligible in magnetic exploration. The magnetic susceptibility of paramagnetic and ferromagnetic substances is positive, the latter is much larger than the former, and the vast majority of common minerals in coal and rock are reverse magnetic and paramagnetic substances due to the influence of temperature and magnetism.

(2) Magnetic characteristics of coal and rock minerals

1) Magnetic classification of minerals

Minerals in nature can be divided into three categories according to their magnetic properties.

Reverse magnetic minerals (such as quartz, graphite): The magnetic susceptibility is low negative ($\chi < 0$). Paramagnetic minerals (accounting for the majority of minerals): The magnetic susceptibility is low positive ($\chi > 0$).

Ferromagnetic minerals (iron, cobalt, nickel, etc.): the magnetic susceptibility is significantly positive ($\chi \gg 0$), and changes with the change of temperature, magnetic field and other conditions.

Rocks usually refer to mixtures containing multiple minerals, so the type and content of minerals inside the rock determine the magnetism of its macroscopic performance. At the same time, it is also affected by the geological process after diagenesis. Normally, olivine, gabbro, basalt and other basic and ultramafic magmatic rocks have the strongest magnetic properties. Metamorphic rocks are second, Sedimentary rocks are the weakest. The magnetic characteristics of different types of rocks in terms of magnetism are shown in Table 7.1 [4].

In addition to being affected by the mineral composition of its interior, rock magnetism also has an important physical property, namely magnetic anisotropy. Magnetic anisotropy refers to the difference in magnetic response of rocks in different magnetic field directions or measurement directions, which is mainly attributed to the influence of bedding structure, directional arrangement of minerals and metamorphic deformation history inside rocks. Normally, rocks with high magnetic susceptibility and deep metamorphism have more significant magnetic anisotropy. For example, the sedimentary rocks in the fold tectonic development area have complex internal structures and orderly arrangement of mineral particles, so the degree of magnetic anisotropy is significantly higher than that of the platform area with simple structure and relatively stable rock strata. In sedimentary rocks, coal-bearing strata, as a special geological unit, have weak magnetic characteristics as a whole, which are mainly controlled by the following factors: First, the type and abundance of magnetic

Table 7.1 Comparison of magnetic characteristics of different types of rocks

Rock type	Magnetic determinants	Common rock-forming minerals and magnetic characteristics	Magnetic susceptibility	Magnetic strength
Igneous rocks	The content of ferromagnetic minerals	It is rich in ferromagnetic minerals, such as magnetite and titanium magnetite	When the rock structure is the same, the higher the content, the greater the magnetic susceptibility	The natural residual magnetization strength of ferromagnetic immersion rocks gradually increases in the order of acidic, neutral, basic and ultrabasic
Sedimentary rock	The content of ferromagnetic minerals	Most of them are quartz, calcite, feldspar, gypsum, etc. (diamagnetic or weakly magnetic minerals), The pure minerals such as siderite, ilmenite, and biotite are paramagnetic minerals, Those containing ferromagnetic mineral impurities have strong paramagnetism	The overall magnetic susceptibility is low, and the remaining magnetization strength is weak	
Metamorphic rock	Original ingredients and deterioration process	Metamorphosis from magmatic rock, magnetic enhancement, Metamorphosis from sedimentary rocks, magnetic weakening	The metamorphic conditions had a significant impact and large differences in performance	

minerals (such as pyrite, hematite, and ilmenite) in the rock. Second, the physical structure and particle orientation of the coal body. Third, the mineral transformation process caused by external environmental factors, especially temperature changes [5–9].

In the most simplified combustion reaction, coal is often simplified to carbon element (C), which is essentially a complex mixture of organic matter and inorganic minerals. Among them, carbon itself is an antimagnetic substance, and its magnetic susceptibility is negative, Inorganic mineral components mixed in coal, especially some iron-containing minerals (such as pyrite, magnetite, and hematite), are paramagnetic and even ferromagnetic [10–14]. Therefore, the overall magnetic properties of coal should be regarded as the magnetic response of the multi-component coupling system. The macroscopic magnetic susceptibility can be regarded as the weighted comprehensive result of organic components and inorganic magnetic minerals under the action of their own integrals and magnetic properties, showing a high degree of non-uniformity and environmental dependence, so the magnetic susceptibility of coal in general can be expressed by the following formula (7.5).

$$\chi_{\text{coal}} = \frac{\alpha_1 \chi_1 + \alpha_2 \chi_2 + \cdots\cdots + \alpha_n \chi_n}{\alpha_1 + \alpha_2 + \cdots\cdots + \alpha_n} = \sum_{i=1}^{n} \alpha_i \chi_i \tag{7.5}$$

where, χ_{coal} is the magnetic susceptibility of the coal as a whole. α_i is the content of each component in the coal.

Therefore, the magnetism of coal seams in a normal occurrence state under normal temperature conditions is mainly determined by the following factors: First, the type and content of paramagnetic minerals contained in coal, and the response ability of these minerals under the action of external magnetic field directly affects the overall magnetic susceptibility. Second, the microstructural characteristics of coal body, including the spatial distribution and interaction between organic matter and inorganic minerals. The third is the degree of fragmentation and structural integrity of the coal seam, and the higher degree of fragmentation is usually accompanied by the uneven enrichment or migration of magnetic minerals, which in turn affects its magnetic response characteristics. For coal seams that have undergone spontaneous combustion, pyrolysis, or are in a high-temperature combustion environment, their magnetic characteristics are significantly dominated by temperature field changes [16, 25].

(3) Magnetic characteristics of coal and rock in the fire area

The evolution process of coalfield fire areas can usually be divided into five basic stages: the initial stage, the acceleration stage, the severe stage, the slow burnout stage, and the final extinguishing stage [17]. According to the characteristics of coal seam combustion, the coal seam combustion stage can be divided into natural initial stage, center formation stage, combustion development stage and extinguishing and cooling stage. In the actual coal seam burning area, different locations may be in different combustion stages within the same fire area, and the magnetic anomalies in the combustion areas of different combustion stages will also be different, as shown in Table 7.2.

In the coalfield fire area, the high temperature environment generated during coal combustion will significantly change the magnetic properties of coal and surrounding rock, and the mineral components in the coal and surrounding rock undergo pyrolysis and phase change, resulting in changes in their mass magnetic susceptibility and the composition of magnetic minerals, and then reshaping the magnetic field in the fire area. Field geomagnetic observations show that the fire area shows obvious magnetic anomalies at different development stages: in the initial stage of spontaneous combustion, the coal rock has not undergone significant thermal metamorphism, and the magnetic changes are limited, making it difficult to obtain effective magnetic anomaly signals. During the formation period of the combustion center,

Table 7.2 Magnetic characteristics of coal and rock at different combustion stages in coal fire area [17]

Combustion stage	Combustion characteristics	Surface ejection temperature (°C)	Underground combustion body temperature (°C)	Magnetic features
Spontaneous combustion begins	Weathering period	Normal	Normal	None or magnetic
Spontaneous combustion begins	Oxidation period	30–60	100–250	None or magnetic
Spontaneous combustion begins	Spontaneous combustion period	60–100	250–400	None or magnetic
Form the center	Formation period of the combustion center	100–200	400–600	Enhancement and great change
Form the center	Formation period of the combustion center	200–300	600–900	Lower
Combustion development	Burning period	>300	>900	No magnetism
Combustion development	Formation period of the oxerence rock	>800–200	>1500–400	Enhancement, change
Extinguish and cool down	Precipitation period of sulfur, mirabilite and other minerals	200–100	400–100	Intensifier
Extinguish and cool down	Dormant period	100–60	200–100	Intensifier
Extinguish and cool down	Extinguishing period	60–room temperature	100–room temperature	Intensifier

the coal and rock are modified by heat and the magnetic anomaly is enhanced, but most of them are distributed in the form of points or lines, and the abnormal strength is uneven. From the later stage of combustion to the extinguishing and cooling stage, the coal rock undergoes significant thermal alteration, forming a burned rock, and the magnetic anomaly distribution tends to be in the form of pieces or bands, and the strength is significantly enhanced, with a wide range, basically covering the coal fire area. Therefore, the influence of coal fire on coal and rock magnetism has clear stage and spatial distribution characteristics, and the fundamental reason lies in the transformation and enrichment process of magnetic minerals induced by high temperature.

7.1.2 *Influence of Temperature on the Magnetic Parameters of Coal and Rock Mass*

The magnetic properties of coal and rock are usually determined by the type and content of their magnetic minerals. Magnetic parameters such as susceptibility and permeability are very sensitive to temperature, especially near the Curie point (magnetic transition temperature). For most magnetic materials, the change of magnetic susceptibility χ with temperature can be described by the Curie—Weiss Law [18].

$$\chi(T) = \frac{\mathrm{C}}{\mathrm{T} - T_c} \tag{7.6}$$

where, C is the Curie constant, which depends on the magnetic minerals and types of the material. T is the absolute temperature, K. T_c is the Curie temperature, which is the critical temperature at which the phase transition of the magnetic phase occurs, corresponding to the temperature point at which the material changes from ferromagnetic to paramagnetic.

The derivation of Curie—Weiss law is based on the molecular field theory in statistical physics, which assumes that the interaction between magnetic particles can be simulated as an efficient molecular field. Magnetic particles exhibit inverse magnetic behavior when the temperature exceeds Tc, and their magnetization process follows Boltzmann statistics. The magnetization intensity M can be expressed as the response of the corresponding molecular field H_{eff} [19].

$$M = \frac{N\mu B_J\left(\beta H_{eff}\right)}{V} \tag{7.7}$$

where, N is the total number of particles, μ is the magnetic moment of a single particle, A·m^2, B_J is the Brillouin function, $\beta = 1/\chi_B T$ is the Boltzmann factor.

Effect field, H_{eff} is contributed by the internal molecular field caused by the external magnetic field H and the magnetization strength M.

$$H_{eff} = H + \lambda \tag{7.8}$$

where, λ is the molecular field coefficient. At temperatures above T_c, the Brillouin function can be approximated as a linear form of its parameters to derive the expression of magnetic susceptibility [20].

$$\chi(T) = \frac{N\mu^2}{3\chi_B V} \cdot \frac{1}{T - \lambda N\mu^2/3\chi_B V} \tag{7.9}$$

Let $\frac{N\mu^2}{3\chi_B V} = C$ and $\frac{\lambda N\mu^2}{3\chi_B V} = C$, That is, the simplified form can be obtained as shown in Eq. (7.9).

During the heating process, the magnetic susceptibility of the coal sample showed a four-stage response: The initial stage (room temperature to anti-Curie temperature point): the magnetic susceptibility remained stable and there was no spontaneous magnetization. Mesothermal stage: the magnetic susceptibility shows an inverse magnetic response at Hopkinson Peak. Medium and high temperature transition stage: magnetic susceptibility decays rapidly with the increase of temperature. High temperature stage: The magnetic susceptibility first decreases, then rises to a peak, and then slowly decreases, and the final value is higher than the initial room temperature magnetic susceptibility. When the magnetic susceptibility of rocks heats up, the inverse magnetism increases below the Curie temperature to Hopkinson peak, and then changes to paramagnetism after exceeding the Curie point. When the temperature is cooled, the demagnetism first increases to the peak and then gradually decreases, reflecting the regulatory effect of temperature on the order of magnetic moment of magnetic minerals. The stable and reversible value of the magnetic susceptibility extreme value indicates that the phenomenon is determined by the intrinsic properties of magnetic minerals (non-thermal cycle damage), which confirms the reliable record of the magnetic field in geological history. The thermal hysteresis effect shown in some rock samples is due to the coupling effect of multiple factors such as Curie point offset and magnetic domain recombination. The magnetic evolution in the high temperature environment indicates the phase transition of the magnetic object and the redistribution of the electron energy level, while the structural stress during the cooling process affects the arrangement of the magnetic domains, resulting in the change of magnetic susceptibility. The temperature dependence of magnetic susceptibility reveals the phase evolution characteristics of magnetic objects, and its thermomagnetic behavior highlights the mechanism of inverse magnetic enhancement near the anti-Curie point [21–23].

7.1.3 Principle of Electrical Detection

Electromagnetic exploration is a geophysical exploration method that finds different types of useful deposits, identifies geological structures, and solves geological problems through the observation and study of the spatial distribution and temporal characteristics of artificial or natural electric fields, electromagnetic fields, or electrochemical fields according to the differences in electromagnetic properties (such as conductivity, conductivity, and dielectric properties) and electrochemical properties of various types of rocks or orebodies in the earth's crust [24]. The high temperature generated by underground coal fire will cause changes in the internal structure and humidity of coal and rock, which in turn affects its electrical characteristics. In coal fire detection, electrical methods are widely used, mainly including transient electromagnetic method and high-density electric method. Based on the high temperature of coal fire, the electrical characteristics of coal and rock can be accurately identified and dynamically monitored by systematically collecting surface and underground electric field/electromagnetic field response data, combined with multi-parameter inversion.

(1) Basic principle of transient electromagnetic method

Transient electromagnetic method, also known as time-domain electromagnetic method (TEM), is a method of using an ungrounded loop or grounding wire source to emit a pulsed magnetic field to the ground, and using a coil or grounding electrode to observe the secondary eddy current field during the interval of a pulsed magnetic field. By studying the spatio—temporal distribution characteristics of secondary vortex electromagnetic fields, geological problems such as finding underground minerals, detecting geological structures, and dividing underground water-rich areas can be solved. Its basic working method is to set up a transmitting coil on the ground or in the air, so as to generate an electromagnetic field in the space around it, and generate an induced current in the underground conductive rock ore body. After a power outage, the induced current attenuates over time due to heat loss. The decay process is generally divided into early, medium and late stages. The early electromagnetic field is equivalent to the high-frequency component in the frequency domain, with fast attenuation and small skin depth. The late component is equivalent to the low-frequency component in the frequency domain, with slow attenuation and large skin depth. By measuring the change of the secondary field in each time period after the power outage, the geoelectric characteristics of different depths can be obtained [25].

The basic principle of mine transient electromagnetic method is basically the same as that of ground transient electromagnetic method. The difference is that the mine and transient electromagnetic method is carried out in a roadway several hundred meters deep underground, and the transient electromagnetic field is distributed in full space, and the transient response at this time is the whole space response. From the perspective of the "smoke ring" theory, the early transient electromagnetic field is generated by the induced current of the surrounding rock medium of the roadway close to the emitting wire frame, which can reflect the electrical distribution at close range. Therefore, the induced EMF of different transient delay t can be obtained by using mine transient electromagnetic meters to record the electrical distribution of the surrounding rock of the roadway from near to far [26–28], as shown in Fig. 7.1.

(2) High-density electricity method

The high-density electrical method sets high-density observation points in the observation, which is an array exploration method. During field measurement, all electrodes are placed on the profile, and then the programmable electrode transfer switch and microcomputer engineering electrometer are used to realize the rapid and automatic collection of data on different electrode distances and electrode arrangements. Then, the data is processed by microcomputer, and the results of various physical interpretations of the distribution of geoelectric sections are obtained: the application

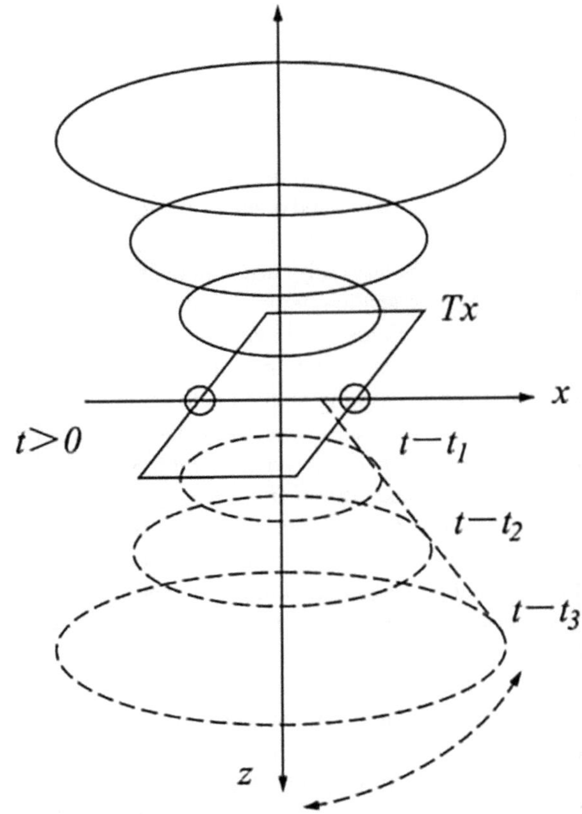

Fig. 7.1 Propagation of transient electromagnetic fields in all spaces [25]

and development of high-density resistivity exploration technology greatly improve the intelligence of electrical exploration. Compared with the traditional resistivity method, the high-density resistivity method has many advantages such as high degree of automation, rich data, low cost, high efficiency, easy interpretation, and strong exploration ability.

(3) Electrical characteristics of coal and rock mass

1) Conductive mechanism of rock ore

The mechanism of rock conductivity may be ionic conductivity, electronic conductivity and mixed conductivity, which are related to the different lithofacies composition and the conductivity of rock-forming minerals, secondary minerals and metal minerals. The resistivity of the solid phase (mineral skeleton) exceeds the liquid phase by 6–8 orders of magnitude, The gas phase is a dielectric. Therefore, the influence and spatial structure of various phases that fill the pore space of rocks are factors that determine the resistivity of most rocks. Clay and clayey rocks and gypsum are exceptions. The structure of the inclusion is important when electronically conductive accessory minerals or metallic minerals develop in rocks. According to the conductivity of rock ores, they can be divided into: conductors (electronic conductors and ionic conductors), semiconductors, and dielectrics [29].

2) Conductive characteristics of high-temperature coal and rock mass

High-temperature coal— rock mass (for the corresponding types of rocks in the sedimentary environment of the coalfield and the high-temperature environment generated by the spontaneous combustion of coal). The process and electrical changes are mainly divided into the following three states:

① Natural state. In the natural state, coal and sedimentary rocks contain pore water, interlayer water, fissure water, karst water, etc. Under the mineralization of water, the resistivity of coal rock mass is mainly ionic conductivity, and its resistivity depends on the hydrogenicity of the formation and the degree of water mineralization.

② Initial stage of coal spontaneous combustion. When the temperature rises due to the natural oxidation of coal in the goaf, the coal and surrounding rock will gradually dehydrate, and the conductivity will transition from ionic conductivity to electronic conductivity, and the resistivity will rise rapidly.

③ Coal spontaneous combustion high temperature period. The high temperature effect causes the organic matter in the coal to oxidize and decompose, the carbon content increases and the resistivity decreases. Microscopic particles in rocks break the energy barrier and become free-moving carriers due to the increase in temperature, resulting in a decrease in rock resistivity. When the temperature reaches a certain stage, the rock will differentiate according to its resistivity according to its chemical and mineral composition until it melts [31].

7.1.4 Influence of Temperature on the Electrical Parameters of Coal and Rock Mass

(1) Theory of resistivity properties of coal and rock mass

1) Theoretical basis of resistivity

Resistivity is a measurement parameter representing the ability of the substance table to block current, defined as the resistance (Ω) of the unit length (m) and unit area (m^2) inside the substance, and its unit is ohm-meter (Ω·m). In the study of coal and rocks, their resistivity is not only affected by the properties of the material itself, but also by external conditions such as temperature and pressure.

$$\rho = \frac{1}{\sigma} \tag{7.10}$$

For a uniform, isotropic material, its resistance R can be expressed by its resistivity ρ, length L and cross-sectional area S The conductivity of the underground coal ore can be expressed as resistivity [32] and conductivity.

$$R = \rho L/S \tag{7.11}$$

where, ρ is the specific electrical resistance, Ω·m.

2) Resistivity model

Archie Law [33] is a classical model describing the relationship between the resistivity of saturated rocks and the resistivity of pore liquids, which is mainly used in the exploration of oilfield geology. Its basic form is:

$$\frac{\rho_{rock}}{\rho_{fluid}} = \alpha \emptyset^{-m} \tag{7.12}$$

where, ρ_{rock} is the resistivity of the rock, Ω·m. ρ_{fluid} is the resistivity of the pore fluid, Ω·m. Ø is the porosity of the rock. α is the tortuous factor. m is the cementation index.

The derivation of Archie Law is based on geometric and physical assumptions: Rocks are made up of fully saturated pores (filled with fluids) and solid matrices. Current is conducted only through the fluid in the pores. The pore structure of rocks can be reduced to a series of continuous capillaries. According to the above model, the ratio of resistivity reflects the tortuosity and complexity of the fluid path in the pores, and the α and m parameters characterize the influence of the complexity of pore shape and distribution on resistivity.

(2) Theory of dielectric properties of coal and rock mass

1) Theoretical basis of dielectric properties

Vacuum dielectric constant, also known as vacuum capacitance, can also be called dielectric constant, is a common physical constant in electricity, with the symbol, ε_0 In the international system of units, the value of the vacuum dielectric constant is: $\varepsilon_0 = 8.854187817 \times 10^{-12}$F/m.

In electrical physics, the vacuum dielectric constant is an important constant that assumes that two charges, one positive and one negative, (charge e) are "at rest" at a distance of r. Its potential energy [34] is:

$$E_p = \frac{e^2}{4\pi\varepsilon_0 r} \tag{7.13}$$

The "stationary" state of the two charges does not mean that they are absolutely stationary between them, but also moves slowly at the temperature with the vacuum as the background. Motion under such conditions can be regarded as simple harmonic motion (usually circular motion can also be used to describe this behavior) with a period of t. Therefore, when the current to be generated by a charge is I, I = e/t.

Since the state of the electric field changes with the change of the state of charge motion (that is, the change of the electric field strength at a certain point in the space stationary state), and because the essence of the current migration is the change of the strength of the electric field, the electric field will be transferred between the two charges when this change occurs, resulting in the generation of current. The period of the displacement current at this position is also t, and the magnitude is also expressed as I = e/t. The electric current displacement formula [35] is:

$$I_D = \frac{\mathrm{d}}{\mathrm{dt}} \int_S \varepsilon_0 E ds \tag{7.14}$$

where, E is the electric field strength, V/m; S is the cross-sectional area, m^2; $\int_S$ is the area division.

The integral term on the right side of the equation is the amount of electricity Q, so the formula satisfies the relationship I = dQ/dt. The I = e/t. here is derived from the definition formula I = dQ/dt., so I = e/t. can represent the characteristics of the distribution of the displacement current in space at this position [36].

The dielectric properties of substances change with temperature and microwave frequency. It is usually expressed as a complex dielectric constant ε [37, 38]:

$$\varepsilon = \varepsilon_0 \varepsilon_r = \varepsilon_0 \left(\varepsilon' - j\varepsilon''\right) \tag{7.15}$$

where, ε_r is the relative dielectric constant or relative capacitance rate, ε is the capacitance of the dielectric, ε_0 vacuum capacitance, j is the imaginary unit.

In practical applications, due to the small absolute value of the dielectric constant of most materials, the complex relative dielectric constant is usually used to quantify the dielectric response for convenience. It can be observed that the complex relative dielectric constant is determined by the relative dielectric constant, ε' and relative dielectric loss factor ε''. ε' is a measure of the ability of a dielectric material to store electrical energy, while ε'' is a quantification of the loss of electrical energy in a material.

Dielectric loss tangent (tanδ) is a parameter for evaluating the change of dielectric constant and dielectric loss factor at the same time. Defined as the ratio between the relative dielectric loss factor and the relative dielectric constant [37, 38].

$$tan\delta = \frac{\varepsilon''}{\varepsilon'} \tag{7.16}$$

2) Polarization nature of dielectrics

The dielectric response characteristics of materials under different conditions are mainly caused by different polarization response mechanisms, including oriented polarization (also known as dipole polarization), induced polarization (both atomic and electron polarization), and interfacial polarization [39–41]. Under the action of the external electric field of time change, the charged particles inside the medium will produce displacement. However, this displacement reaction is not completed immediately, and with the adjustment of the external electric field, the displacement reaction of the particles shows a certain delay.

(3) Effect of temperature on the resistivity of coal and rock

Temperature is one of the key factors affecting the abnormal change of resistivity of coal and rock. The change of temperature directly affects the migration ability of electrons and ions inside the coal, which in turn changes the resistivity of the coal. The resistivity dependence of coal and rock on temperature can be described by the Arrhenius equation [42]. The Arrhenius equation was originally used to describe the dependence of chemical reaction rates on temperature, but it is also widely used to describe how the resistivity of a substance changes with temperature.

The Arrhenius equation expression is as follows:

$$\rho(T) = \rho_0 \exp\left(\frac{\mathrm{E}_a}{\mathrm{k}_B(T + 273.15)}\right) \tag{7.17}$$

where, ρ(T) is the resistivity when the temperature is T, Ω·m. ρ_0 is the resistivity at the reference temperature, Ω·m. Ea is the activation energy, which represents the energy required for electrons (ions) to cross obstacles, J/mol. k_B is the Boltzmann constant with a value of about 8.617×10^{-5} eV/K.

In the process of normal temperature and temperature change, the resistivity of coal shows a trend of high in the middle and low on both sides with the increase of metamorphism. For low-rank coal, the higher the moisture content, the lower the resistivity. The resistivity of coal with a higher degree of metamorphosis decreases with the increase of deterioration due to its large, orderly aromatic structure and less moisture. During the heating process, the carriers in the coal are gradually activated, promoting the migration of electrons and ions, and reducing the resistivity. When the temperature is close to the pyrolysis temperature, the rate of decline of resistivity slows down. After pyrolysis, the molecular structure of coal was destroyed and the crystal structure changed, resulting in a decrease in pores, closure of fractures, and an exponential decrease in resistivity. The resistivity response characteristics of variable temperature rocks are as follows: in the early stage, thermal expansion and water evaporation cause the increase of cracks and the increase of charge flow obstacles, resulting in a sharp increase in resistivity to the peak of pyrolysis temperature, and in the middle stage, the temperature rise leads to the activation of carriers, changes in microstructure, and rapid decrease of resistivity. At the end of the period, as the temperature increases, fracture closure or melting increases electron migration and resistivity decreases, revealing differences in the sensitivity of different rocks to temperature changes [21, 30].

(4) Influence of temperature on the dielectric parameters of coal rock

The relative dielectric constant of coal is ε'. at room temperature, it shows a trend of low in the middle and high on both sides, and ε'' and $tan\delta$ increase step by step, and the dielectric properties of the characteristic temperature points in the process of temperature change basically follow this characteristic. Under different electric field frequencies, the ε' and ε'' of coal change slightly, and the dominant polarization types are electron and atomic (ion) polarization. In the middle stage, with the increase of temperature, the thermal motion of coal molecules strengthens, resulting in the disintegration of macromolecular structure and rapid establishment of dipole polarization, reaching a peak of ε'. At the end of the stage, the internal structure failure caused by the pyrolysis of coal decreased the dielectric parameters and increased ε'', reflecting the cumulative effect of medium polarization and loss at high temperature. $tan\delta$ increases with the increase of temperature, indicating the dominant role of the loss factor in the dielectric response of coal. The dielectric response characteristics of rocks undergo three stages of change: In the first stage, the temperature rise leads to water evaporation and crack development, at which time electron polarization is the main and the dielectric parameters decrease. In the second stage, the rocks exhibit thermal stability, reflecting the interaction between weak charge holding capacity and lower polarization rate, and the dielectric parameters change little. In the third stage, the temperature rises further to strengthen the polarization effect and improve the dielectric parameters. In addition, the dispersion phenomenon indicates enhanced polarization at low frequencies and decreased dielectric constant at high frequencies, suggesting that polarization does not keep up with electric field changes. The dielectric loss tangent (tanδ) reveals the effects of frequency and temperature on energy dissipation, which are related to the conductivity process and internal polarization mechanism of the rock, and reflect the association between dielectric relaxation and the microstructure of the rock [21, 31].

7.2 Electromagnetic Detection Characteristics

7.2.1 Advantages and Limitations of Magnetic Detection

(1) Advantages of magnetic detection

The magnetic method is one of the earliest geophysical technologies applied to underground coal fire detection, which uses the magnetic enhancement of rocks in the fire area after high temperature pyrolysis, and then produces measurable magnetic anomalies on the surface, so as to realize the identification and boundary delineation of the fire area, and the specific advantages are as follows:

1) Non-contact measurement technology

One of the advantages of magnetic exploration is the means of non-contact measurement. Compared with the detection methods that require physical disturbances on the surface, such as drilling or seismic exploration, magnetic exploration only needs to deploy high-precision magnetometers to obtain the magnetic characteristics information that characterizes the underground ore body, so as to effectively avoid damage to the surface environment. This non-intrusive nature significantly enhances the application value of magnetic exploration, especially in environmentally sensitive areas where large-scale surface excavation is prohibited or restricted, such as nature reserves and urban built-up areas. In addition, magnetic exploration has outstanding adaptability to complex terrain conditions. Whether in rugged mountains, densely forested areas or water environments, magnetic instruments maintain stable performance. Therefore, the environmental adaptability and operational flexibility shown by magnetic exploration together constitute the core technical advantages of this method [43, 44].

2) Detection depth and resolution

Large detection depth and high resolution are the second advantages of high-precision magnetic exploration. Under ideal geological conditions, the method has the ability to detect ore bodies at a depth of thousands of meters, which provides strong support for the exploration of deep mineral resources. The increase in detection depth mainly depends on the magnetization strength of the magnetic minerals in the target geological body, the sensitivity of the magnetic instrument, and the background noise level of the geomagnetic field. In addition to the detection depth, the spatial resolution of magnetic exploration continues to improve. Modern high-precision magnetometers can accurately detect weak changes in the magnetic field, with spatial resolution up to the sub-meter level. This high-resolution feature significantly reduces the uncertainty and blindness of subsequent drilling projects, and is of great guiding significance for accurately delineating the boundaries, morphology, and scale of the orebody [45, 46].

3) Economy and efficiency

Compared with other geophysical exploration methods, magnetic exploration has significant advantages in terms of economy. Its instrument system has a simple structure and light weight, and the exploration operation process does not require complex equipment installation and maintenance, mainly involving data collection and processing [47]. Especially in the large-scale census stage in the early stage of regional mineral exploration, aerial magnetic survey can efficiently achieve regional coverage and significantly improve work efficiency. In addition, this method can complete the exploration of a large area in a short time, effectively reducing the overall exploration cost. Based on the above factors, magnetic exploration is recognized as a cost-effective technical means in the early stage of mineral resource exploration [48, 49].

4) Applicable scenarios

① Delineation of the boundary of the fire area

Magnetic exploration accurately defines the coal fire boundary by identifying the magnetic anomalies of the incendiary rocks (Magnetic susceptibility 0.5–5SI).

Mineral phase transformation occurs in the incendiary rock under high temperature conditions, and hematite is transformed into ferromagnetic magnetite, forming a significant magnetic marker layer [50].

② Quantitative evaluation of combustion intensity

The magnetic anomaly strength is positively correlated with the combustion temperature (the magnetic susceptibility peaks at 400–600 °C), and the combustion intensity can be inverted by ΔT outliers [50, 51].

③ Early warning of recurrence

Periodic magnetic measurements can detect sudden changes in magnetic susceptibility caused by re-ignition in the fire area (increase > 20%), such as the monitoring of re-ignition in the Wuda fire area in Inner Mongolia [52].

(2) Limitations of magnetic detection

1) Constraints of the complexity of geological background

The application effect of magnetic detection in underground coal fire monitoring is limited by the complex geological background interference. The rock formations around the fire zone often contain highly magnetic ferromagnetic minerals (such as magnetite and hematite), which generate background magnetic field strength of hundreds of nanotesla (nT), which overlaps with the thermal magnetic field caused by coal fires (usually only a few tens of nT) [53]. Especially in the marginal area of the fire zone, the thermal alteration of the surrounding rock will induce the formation of secondary magnetic minerals, resulting in the distortion of magnetic anomalies. As shown in Fig. 7.2, the spectral characteristics of thermogenic minerals in the shortwave infrared band (SWIR) indicate that the absorption peak shift of kaolinite minerals around 2200 nm is directly related to the degree of thermal metamorphism, and the phase transition of this mineral significantly changes the magnetic susceptibility of the rock mass [54]. In addition, the heterogeneity of the magnetic susceptibility of the overlying strata in the fire zone (such as basalt interlayer or iron-bearing sandstone) can form high-frequency magnetic noise, which needs to be separated by matching filtering technology, but the filtering process may lead to the attenuation of the real ignition signal [55].

(2) Interference of dynamic characteristics of the fire area

The unsteady thermodynamic process of coal-fire system poses a fundamental challenge to magnetic detection. The spatiotemporal migration of the temperature field

Fig. 7.2 Hyperion data obtained in the Panorama region of Western Australia [54]. *Note* The image on the left is a hyperspectral image cube that visualizes the three-dimensional properties of hyperspectral data.) The top image of the image cube is a false color composite image, and its red (R), green (G), and blue (B) channels correspond to bands 205 (2204 nm), 47 (824 nm), and 22 (570 nm), respectively. The middle image is another false color composite image with RGB channels corresponding to the fourth, fifth, and sixth components of the MNF (Minimum Noise Separation) transformation, respectively. Lithological differences in the target area can be clearly identified. The image on the right shows the distribution of muscovite with an absorption characteristic wavelength of less than 2200 nm (blue) and the distribution of pixels with an absorption characteristic wavelength greater than 2200 nm (red)

in the fire zone triggers the phase transition reaction of magnetic minerals: When the temperature exceeds the Curie point, the ferromagnetic material demagnetizes to paramagnetism [56]. This thermal demagnetization effect leads to a nonlinear decay of the magnitude of magnetic anomalies with the development of fire, and the difference in reflectance between short-wavelength and long-wavelength leuka directly reflects the degree of thermal deterioration, which in turn affects the magnetization strength [54]. At the same time, reducing gases (CO, CH_4) produced by combustion promote the conversion of hematite into magnetite, inducing local magnetic enhancement, which is closely related to the oxygen concentration gradient in the fire zone [57]. Such time-varying characteristics of magnetic susceptibility make it difficult to distinguish the active state of the fire zone from a single magnetic measurement data, and it is necessary to establish a dynamic model through time series monitoring, but it significantly increases the detection cost [58].

(3) Method and technical bottlenecks

① Insufficient vertical resolution

The magnetic method has a high sensitivity to shallow ignition sources, but the detection ability of deep combustion zones is sharply reduced. When the buried depth of the ignition source exceeds a certain value, the amplitude of the magnetic anomaly is usually equivalent to the noise level of the instrument [55]. Although gradient magnetic measurement can improve vertical resolution, the high temperature in the fire zone makes it impossible to deploy the probe near the surface, which limits the effective detection depth [53].

② Problem of multi-solution is prominent

There is serious non-uniqueness in magnetic anomaly inversion. The deviation between the direction of thermal residual magnetism (TRM) in the fire zone and the inclination angle of the geomagnetic field may cause the magnetic anomalous form to be "negative polarity", which is easily misjudged as a fire-free zone [57].

(4) Application scenario limitations

① Humanistic interference is sensitive

Mining facilities (rails, gas pipelines) are often present around the fire area, and the magnetic noise amplitude generated by ferromagnetic materials can reach thousands of nT [56]. While gradient measurements suppress regional noise, they have limited elimination of local point source interference, such as abandoned mine equipment.

② Topographic effect is significant

Coal fires are mostly developed in erosive landform areas (such as open-pit mine slopes), and the terrain undulations lead to spatial distortion of the magnetic field. When the height difference between the measuring point and the ignition source is greater than 50 m, the gradient change of the magnetic anomaly caused by the terrain can reach 20 nT/m, which is highly similar to the characteristics of the ignition anomaly [53]. Although the equivalent source method can be used for terrain correction, the meshing process in complex terrain can introduce edge effects and blur the boundaries of the fire zone [56].

7.2.2 *Advantages and Limitations of Electrical Detection*

(1) Advantages of electrical detection

1) Strong deep detection ability

Transient electromagnetic (TEM) and controllable source audio magneto telluric (CSAMT) have the ability to penetrate high-resistance cover layers and are suitable for deep fire zone location. TEM uses the attenuation characteristics of the secondary magnetic field to detect depths of up to 500–800 m, and the late signal attenuation rate is positively correlated with the conductivity of the formation [24]. CSAMT achieves imaging of fire zones at a depth of > 1000 m in complex tectonic areas through artificial field source excitation [59].

2) High spatial resolution and cost-effectiveness

High-density electrical (ERT) and mine transient electromagnetic methods provide sub-meter resolution with low cost and efficiency in the field. ERT uses a multi-electrode array to generate two-dimensional/three-dimensional resistivity sections through regularization inversion algorithm, with a vertical resolution of 0.5 m, which can finely depict the boundary morphology of the fire zone.

3) Widely applicable to scenarios and supports dynamic monitoring

Electrical detection is suitable for surface, open-pit mines and underground roadways, and has the ability to monitor the dynamic evolution of fire areas. Ground ERT is suitable for shallow fire areas, and the mine transient electromagnetic method can locate the hidden fire source on the roof of the roadway, and the full-space correction technology eliminates the interference of metal support [60].

4) Non-destructive and environmental adaptability

Electrical detection does not require surface excavation, reduces disturbance to sensitive environments, and is suitable for complex terrain. Non-invasive measurement avoids damage to surface ecology (such as nature reserves, urban built-up areas), and has strong adaptability to terrain undulation. Anti-jamming technology effectively suppresses power frequency noise.

(2) Limitations of electrical detection

1) Accuracy of deep detection

The resolution ability of transient electromagnetic (TEM) for deep low-resistance bodies decreases sharply with the increase of depth. When the detection depth is greater than 800 m, the secondary field signal strength is lower than the instrument noise level, resulting in the loss of valid data.

2) Shallow detection blind spot effect

The early signal is disturbed by the self-sensing effect, forming a shallow detection blind zone. Inductive electrical methods such as transient electromagnetic method have significant blind spots in shallow detection. Its technical roots lie in the fact that at the moment of the transmitted current shutdown, the strong transient self-inducting electromagnetic field generated by the coil or wire itself will drown out the early secondary field response signal from the shallow underground medium. This signal "submersion period" constitutes a theoretical detection dead zone (or "dead zone time"), which can usually reach a few microseconds to milliseconds. This effect results in limited detection capabilities for shallow critical regions.

3) Multi-source interference coupling effect

Humanistic and geological noise significantly reduces the signal-to-noise ratio of data. Humanistic noise: Strong electromagnetic interference is generated by the operation of large mechanical and electrical equipment, high-voltage transmission lines, rail transit systems, etc., and its intensity often far exceeds the weak abnormal signal caused by the target coal fire, Geological noise: Resistivity changes caused by non-ignition sources are widespread, such as shallow heterogeneities, water-rich formations, high conductive bodies, fault fracture zones, etc., and their response is easily confused with coal fire anomalies in amplitude and morphology. Coupling effect: Noise does not exist independently, but is often coupled and superimposed on each other in the temporal domain, frequency domain and space, forming a complex background field, which makes the target signal easily overwhelmed, resulting in a significant reduction in the signal-to-noise ratio of the original data.

7.3 Electromagnetic Detection Equipment

7.3.1 Magnetic Detection Equipment

As the core equipment of geophysical exploration, the function of magnetic exploration instruments is to accurately measure the geomagnetic field and its spatial variation characteristics. Based on the law of electromagnetic induction, this technology provides key data support for geological structure analysis, mineral resource assessment and geodynamic research by detecting magnetic field anomalies caused by magnetic differences in underground rock ores. According to the detection accuracy and operation mode, modern high-precision magnetic detection equipment can be divided into two categories: aviation magnetic measurement system and ground magnetic measurement system.

(1) Instrument composition and principle

The standard magnetic exploration instrument consists of a magnetometer, an inclinometer, an azimuth and a data recording device to form an integrated measurement unit. As the core sensing component, the magnetometer directly measures the total strength of the geomagnetic field. The inclinometer and azimuth meter obtain the geomagnetic field vector inclination angle and direction parameters respectively. The data logging device realizes real-time storage and processing of magnetic field information [61]. Each component needs to complete the three-dimensional feature capture of the underground magnetic anomaly field through the collaborative working mode, and then invert the occurrence state and spatial distribution law of the hidden ore body, which has significant theoretical and practical value for mineral resource exploration.

(2) Detection system and application scenarios

1) Aerial magnetic survey system: Mounted on fixed-wing aircraft or unmanned aerial vehicle platforms, it has the ability to quickly scan large areas, suitable for regional geological survey and metallogenic prospect zoning, and its core advantages are reflected in detection efficiency and data coverage density.

2) Ground magnetic measurement system: Implement high-resolution detailed investigation of the target area, which can accurately describe the boundary morphology of the ore body and the vertical change of magnetic parameters, and is suitable for ore body positioning and resource assessment in the exploration stage of the mining area.

(3) Instrument type and performance indicators

Table 7.3 shows the magnetic detection equipment and its performance of the current mature technology.

7.3.2 Electrical Detection Equipment

(1) Geological radar technology

It realizes multi-band, narrow-band pulse, high-speed, nanosecond-level sampling technology. It can not only be used for conventional geological exploration, but also to complete non-destructive testing of highways, non-destructive testing of various concrete castings, etc. For example, the two-dimensional imaging method of high-density resistance method and its application, and the application of geophysical methods in geological disasters [63].

Table 7.3 Magnetic detection equipment

Probe type	Equipment classification		Applicable scenarios	Advantage
Ground magnetic equipment	Proton magnetometer		Shallow fire area (<200 m), small-range high-precision detection	Low cost, high portability, suitable for complex terrain (e.g. mountains, open pits)
	Optical pump magnetometer		Shallow to medium to deep fire zones (<500 m)	High sensitivity, no zero drift, strong anti-interference ability
	Fluxgate magnetometer		Scenarios that require three-component information of the magnetic field	Magnetic field vector information is available in a small volume (160–920 g)
Aeromagnetic equipment	Flying platform	Fixed-wing drones	Large-scale census (>100 km^2), deep fire areas	Long endurance, high efficiency, adapt to bad weather
		Multi-rotor drone	High-precision detection of small areas (<10 km^2), complex terrain	Vertical take-off and landing, ultra-low-altitude flight, high positioning accuracy
	Magnetometer	cesium light pump magnetometer	High-precision aeromagnetic measurement	It is suitable for hard-rod connections and has a noise < 10 pT
		Helium Pump Magnetometer	Medium and low precision aeromagnetic census	Localized equipment with lower cost

(2) Transient electromagnetic method

The ground devices commonly used in the transient electromagnetic method can be divided into two categories: magnetic source and electrical source according to the emission source, the magnetic source emits a pulsed electromagnetic field through an ungrounded coil, and the electrical source emits a pulsed current through the grounding wire. The devices using magnetic emission sources include central circuits, overlapping circuits, large fixed source circuits, separation circuit devices, etc., and devices using electrical emission sources include grounding short offset devices and grounding long offset devices. The main difference between the magnetic source and the transient electromagnetic method is their emission source and the components of detection, and the similarities and differences between the two types of methods are illustrated by the center loop device in the magnetic source transient

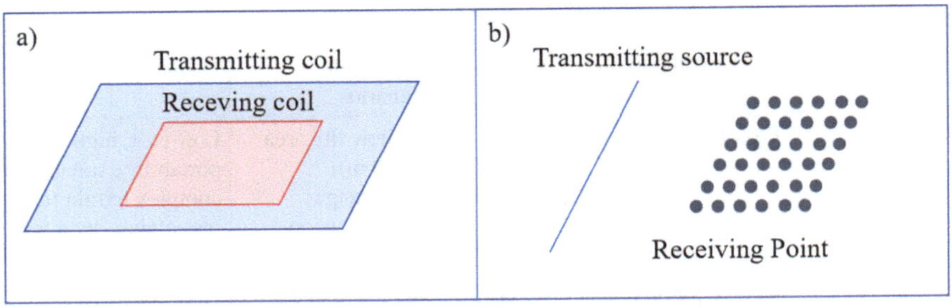

Fig. 7.3 Transient electromagnetic device diagram

electromagnetic method and the grounding long offset distance device in the transient electromagnetic method of the electrical source. The two devices are shown in Fig. 7.3.

The transient electromagnetic transient electromagnetic transmitter of the magnetic source has two calculation methods in the forward simulation, one is to integrate the area of the response of the vertical magnetic dipole source within the area of the transmitting coil, and the other is to integrate the response of the horizontal electric dipole along the four sides of the transmitting coil. The transmitter device of the transient electromagnetic source is a grounding wire, and its forward simulation calculation method is to integrate the response to the horizontal electric dipole along the emission wire. The receiving device of the magnetic source transient electromagnetic method is a closed coil, and its observed component is the vertical magnetic field change rate (dBz/dt), which is calculated by measuring the induced voltage generated by electromagnetic induction in the receiving coil. The measured value of the electrical source transient electromagnetic method is the electric field component in the direction parallel to the emission source.

(3) Ground film image technology: It realizes multi-track, multi-arrangement, continuous rolling observation and image reconstruction technology. It not only realizes the automatic implementation of long sections in field observation, but also realizes the true reproduction of the electrical distribution of formations in terms of results display.

(4) Controllable source geotelluric technology: the use of time-domain frequency-division high-effective bit massive acquisition technology, frequency-domain digital frequency-division processing technology, and auxiliary frequency band controllable source supplementary technology to realize the exploration requirements of electromagnetic broadband dense frequency point exploration. It has made breakthroughs in solving shallow geological problems and improved in fine exploration at medium and deep levels, such as CSAMT and high-density electrical methods for groundwater resources [63].

7.4 Electromagnetic Detection Application Cases

7.4.1 Application Cases of Magnetic Detection

(1) Dynamic monitoring of spontaneous combustion fire area of coal residual coal in northern Shaanxi and time-shifting aeromagnetic application

1) Background and challenges

A coalmine at the junction of Shaanxi and Inner Mongolia left a large amount of coal left over from top coal mining, causing spontaneous combustion of the coal seam to form an underground fire area. Although the traditional aeromagnetic method can locate the static fire area, it cannot capture the dynamic expansion trend of the fire area, and it is difficult to guide accurate fire extinguishing. The spontaneous combustion depth of the coalmine is about 100 m, and the high temperature converts stratigraphic hematite and siderite into strong magnetic magnetite (Fe_3O_4), forming a significant magnetic anomaly (>500 nT) [64].

2) Technical methods

The time-shifted aeromagnetic method involves two aeromagnetic flights at an interval of 15 days, and the expansion trend of the burning area is inverted by the magnetic anomaly difference.

3) Detection system

The DN20-G4 UAV and the aerospace optical pump magnetometer were used for data acquisition, as shown in Fig. 7.4. The data acquisition time is no less than 2 h, the sampling interval is 0.5s, and the static noise is 0.02nT.

4) Data processing

For the separation of the potential field, the local magnetic anomaly between the local field and the ignition area was separated by the Butterworth filter. Regularization inversion, the magnetic susceptibility distribution is solved by Tikhonov algorithm. The displacement comparison, the results of the two inversions are subtracted, and the change in the burning area is extracted.

5) Application effect

In the first flight, two burning areas were located, with a buried depth of 76m/83m and an area of about 0.12 km^2. In the second flight (15 days later), the center of the burning area sank to 78 m/84 m, and the boundary expanded by 11.3%, as shown in Fig. 7.5.

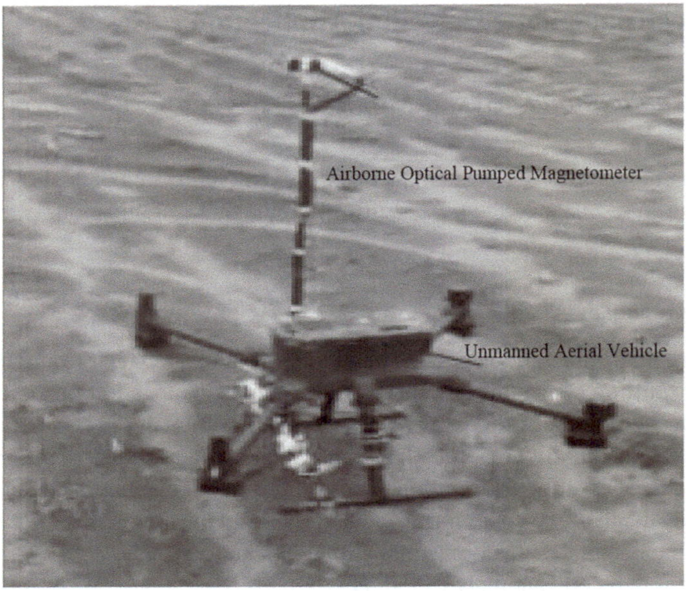

Fig. 7.4 Drone and aviation optical pump magnetometers [64]

Fig. 7.5 Time-shifting aeromagnetic method detection results of burning area [64]

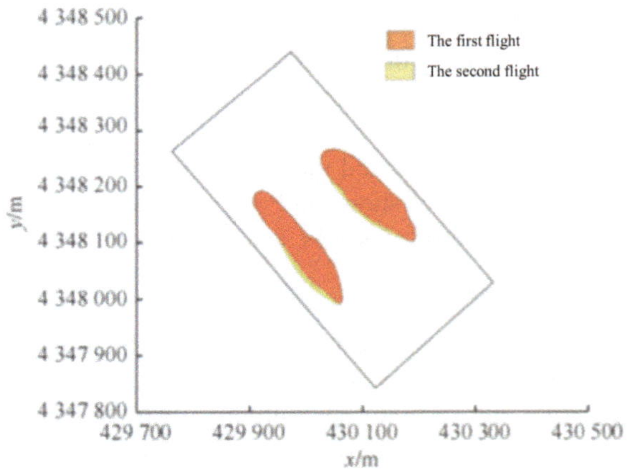

Aiming at the physical property premise of high magnetic anomalies caused by the formation of magnetic minerals (Fe_3O_4) in coal seams, the time-shifted aeromagnetic method is used to break through the limitations of traditional aeromagnetic static detection. Through two time intervals of aeromagnetic detection (UAV equipped with optical pump magnetometer), the dynamic expansion trend of the burning area is accurately captured based on the difference of the inversion results. At the same time, the regular-irregular composite grid segmentation technology is used to ensure the accuracy of terrain fitting with tetrahedral grids in the undulating areas, and hexahedral grids are used to improve the calculation efficiency in the underground homogeneous areas, which increases the inversion efficiency by nearly 6 times compared with pure tetrahedral sectioning, and meets the accuracy requirements of complex terrain. Numerical models and measured verifications show that the technology can simultaneously realize the spatial distribution and positioning of the burning area and the analysis of the temporal evolution trend, and provide key dynamic data support for coalmine fire prevention and extinguishing projects.

(2) Application of high-resolution magnetic detection in the North Colorado coal fire (NCF)

1) Background and challenges

The target area is located in the Hogback monoclinic tectonic belt of the San Juan Basin, and the coal seam belongs to the Fruitland Formation (8m thick, dip angle 12°), and is overlaid with 15 m shale/sandstone. The coal fire originated in an outcrop and spread about 200 m to the southeast. Surface fissures, thermally altered rocks and snowmelt boundaries show continuous activity in the fire area, but the traditional method (satellite thermal imaging resolution is > 100 m, The temperature measurement of the borehole is limited by high temperature) cannot accurately delineate the combustion boundary, as shown in Fig. 7.6 [65].

2) Data collection and processing

In this study, a backpack-type cesium vapor magnetometer (Geometrics G859) combined with a built-in GPS unit (positioning accuracy < 0.5 m) was used to carry out high-resolution magnetic detection of the North Coal Fire (NCF) area of the South Ute Reservation in Colorado. The survey area covers an area of 600 m × 200 m, and about 363,000 measurement points are collected through non-uniform survey line traversal, with a sampling frequency of 5 Hz, achieving sub-meter spatial resolution. In order to eliminate the diurnal interference of the geomagnetic field (the amplitude can reach 60 nT), the proton rotation magnetometer (G856) base station is deployed in the unburned coal seam area, and the background field data is continuously recorded at 30 s intervals, and the data is deducted from the mobile terminal data. The data processing methods of diurnal correction, peak filtering, polarization processing and anomaly threshold screening significantly improve the signal-to-noise ratio of the data, and finally generate a magnetic anomaly distribution with a spatial resolution of up to the sub-meter level, which provides a reliable basis for the partitioning of the combustion state.

3) Detection results and verification

The burned cooling area is blue, and the magnetic anomaly is + 50–100nT, corresponding to the outcropping thermally altered rock. The active combustion zone is red, and the magnetic anomaly is weakened, which coincides with the position of the

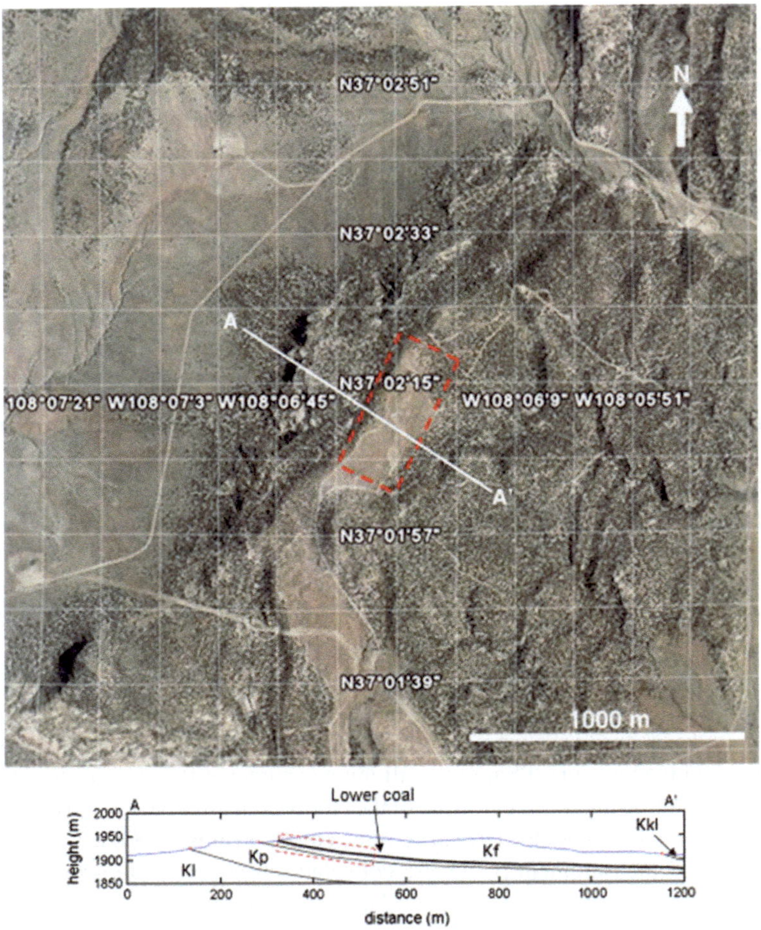

Fig. 7.6 (Above) Satellite image map around the NCF (North Coal Fire) area, (Bottom figure) Geological profile and lithology description of the NCF (North Coal Fire) area [65]

high-temperature fracture. The unburned area is green, and the background magnetic anomaly (0–20nT) is shown in Fig. 7.7.

As a major disaster that threatens environmental safety, public health and the economy of mining areas, its precise management urgently needs to clarify the spatial boundaries of fire areas. In this case, the coal fire near Durango, Colorado, USA is used as the object to reveal the thermo— chemical mechanism of the overlying rock layer of the coal-burning system: under high temperature (>200 °C) and low oxygen conditions, the alteration of ferrous minerals in the rock layer forms magnetite (Fe_3O_4), and its magnetic moment is arranged along the geomagnetic field

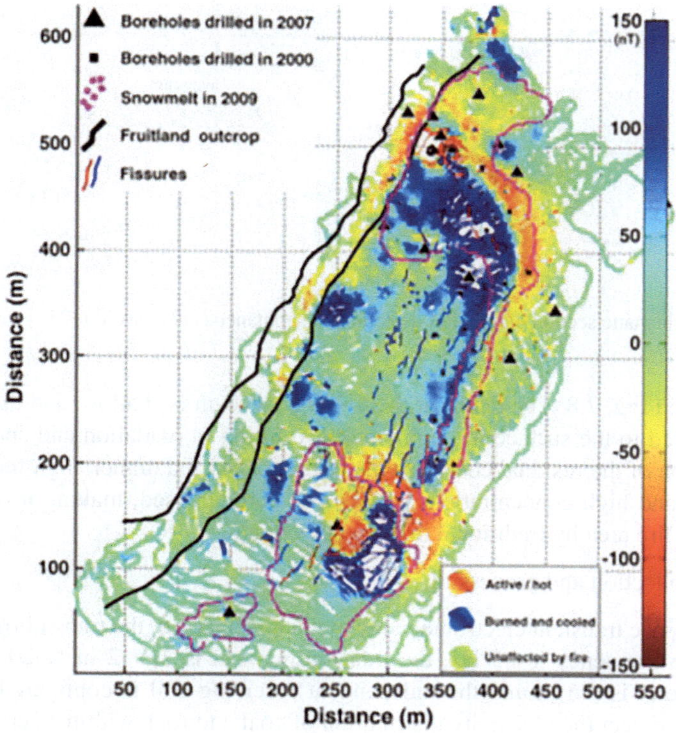

Fig. 7.7 Magnetic anomaly distribution after diurnal correction, background magnetic field rejection, burst value removal and data asymmetry correction. Color scale unit: nat (nT) [65]

to form a detectable magnetic anomaly. Based on this physical response, the magnetometer detection successfully realizes the refined distinction of combustion state and the mutual verification of multi-source data. A quantifiable and dynamic monitoring (sub-meter-level fire front migration tracking) coal fire detection technology paradigm has been established to provide a scientific basis for fire extinguishing engineering.

7.4.2 Application Cases of Electrical Detection

(1) Application of transient electromagnetic method in the detection of Beizu mining area

1) Background and challenges

Beizu Coalmine is a resource integration mine, and a large number of intricate old roadways are left by the early small coal mining using the roadway mining method,

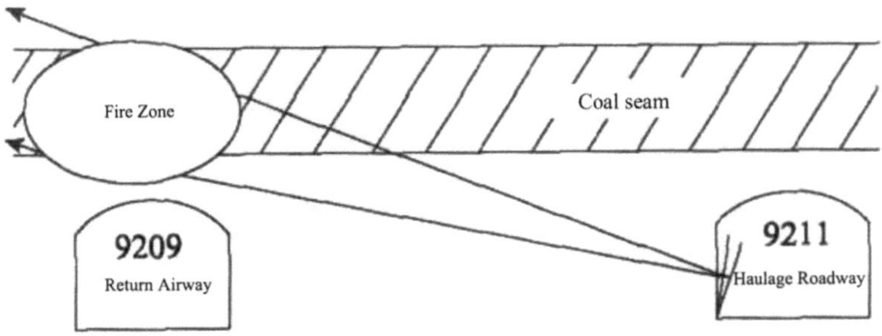

Fig. 7.8 Schematic section diagram of detection on spontaneous combustion [25]

as shown in Fig. 7.8. The depth of the 4 + 9 coal seam is shallow, and the old lane is connected to the surface through cracks, resulting in oxidation and spontaneous combustion of the residual coal. Driving in the 9209 wind tunnel, high-temperature (100 °C) and high concentrations of CO gas were exposed, making it difficult to locate the fire area by traditional drilling detection [25].

2) Data collection and processing

The full-space transient electromagnetic method is adopted, the transmitting current is 2A, the frequency is 25 Hz, the transmitting coil is 2 × 2 m × 20 turns, the coil diameter is 0.6 m and the independent receiving coil is configured, so as to accurately detect the resistivity distribution of coal and rock within a certain range. Data processing is carried out through least squares method and anomaly extraction, so as to improve the accuracy of full-space response and accurately separate fire source anomalies.

3) Detection results and verification

The detection results show that the combustion of the coal seam in the ignition source area also has a great impact on the nearby non-ignition coal seam, so there is also an abnormal high resistance area and the junction of high and low resistance in the detected non-ignition area apparent resistivity Fig. 7.9, but the high resistance value is significantly smaller than that of the ignition source area Fig. 7.10. By capturing the electrical structure of the fire area and inverting the resistivity, the transient electromagnetic method can accurately locate the fire area of the underground coal seam, and successfully solve the problem of locating the hidden fire area of Beizu Coalmine. This case provides a scientific basis for the management of mine fire areas.

(2) Application of high-density electricity method in coal fire detection in live chicken rabbit coalmine

1) Background and challenges

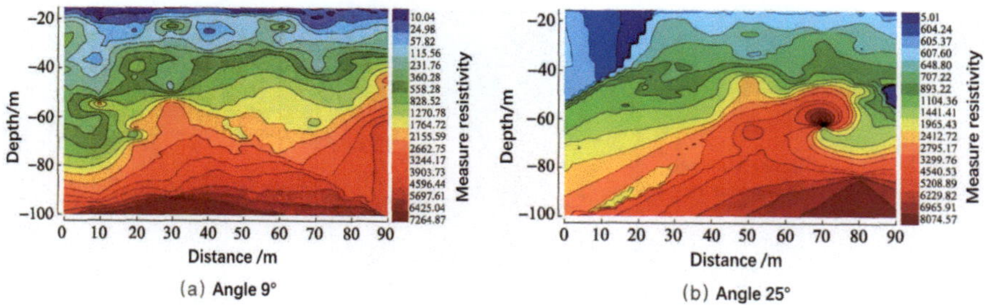

Fig. 7.9 Apparent resistivity of the detection of the fire source area profile [25]

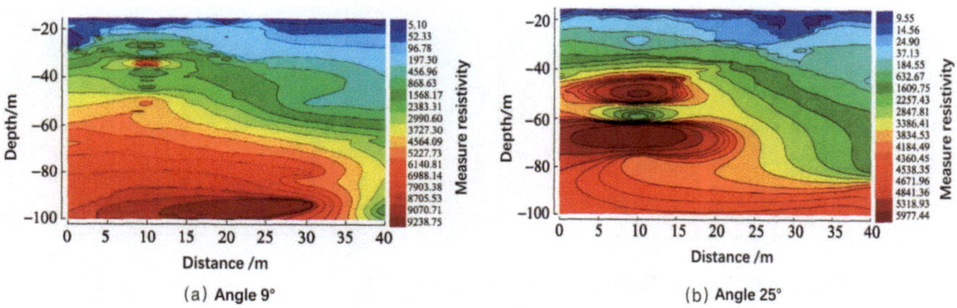

Fig. 7.10 Apparent resistivity of detection in the non-ignition area profile [25]

The Huojitu Coalmine is located in Daliuta, Shenmu, Shaanxi, China with an area of 2.71 km^2, which belongs to the Jurassic coalfield in northern Shaanxi. After years of mining, a large area of goaf has been formed, and the oxidation and spontaneous combustion of the residual coal seam cause underground coal fires, resulting in surface cracks, collapse and escape of high-temperature gases, as shown in Fig. 7.11. It is urgent to accurately detect the scope of the fire area and the distribution of goafs to guide the fire extinguishing project. Although thermal infrared remote sensing technology can identify fires, as shown in Fig. 7.12, it is ineffective for deep underground coal fires, so the high-density electrical method is used for deep detection [66].

2) Data collection and processing

This detection adopts the EDJD-1A high-density resistivity measurement system produced by Chongqing Geological Instrument Factory, equipped with 120–288 electrodes, arranged along the measurement line with a fixed electrode distance of 5m, supporting 26 kinds of arrangement and combination devices, with a maximum power supply voltage of 900V and a current of 5A. For the complex terrain of the live chicken and rabbit coalmine (steep cliff, stripping pit, etc.), four survey lines

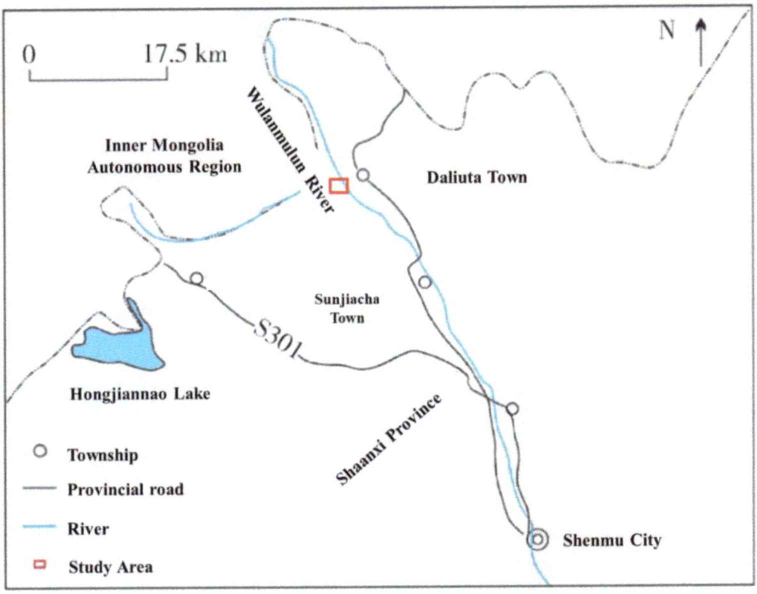

Fig. 7.11 Division of Yan'an Formation coal seam [66]

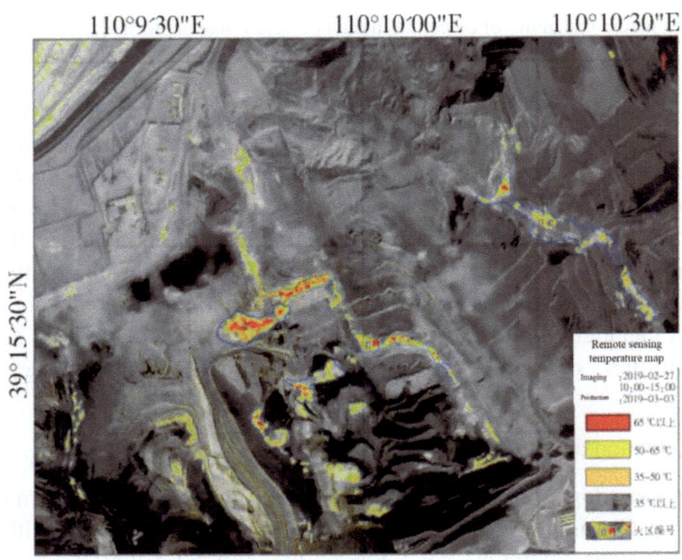

Fig. 7.12 Thermal infrared remote sensing image of the mining area [66]

with a total length of 3.48 km are laid to avoid the high drop area, as shown in Fig. 7.13. The working principle is based on the difference in the electrical properties of the rock layer: the resistivity of the coal seam is significantly higher than that of the surrounding rock. The formation conduction current is excited by the artificial electric field, and the resistivity section is generated by three-dimensional inversion, so as to accurately identify the nature of the goaf and the distribution of coal fire. The collected high-density electrical data is inverted by the damping least squares method for two-dimensional inversion, and the electrical response characteristics of rock strata at different depths are systematically analyzed, and the spatial distribution of goafs and burning areas is accurately located.

3) Detection results and verification

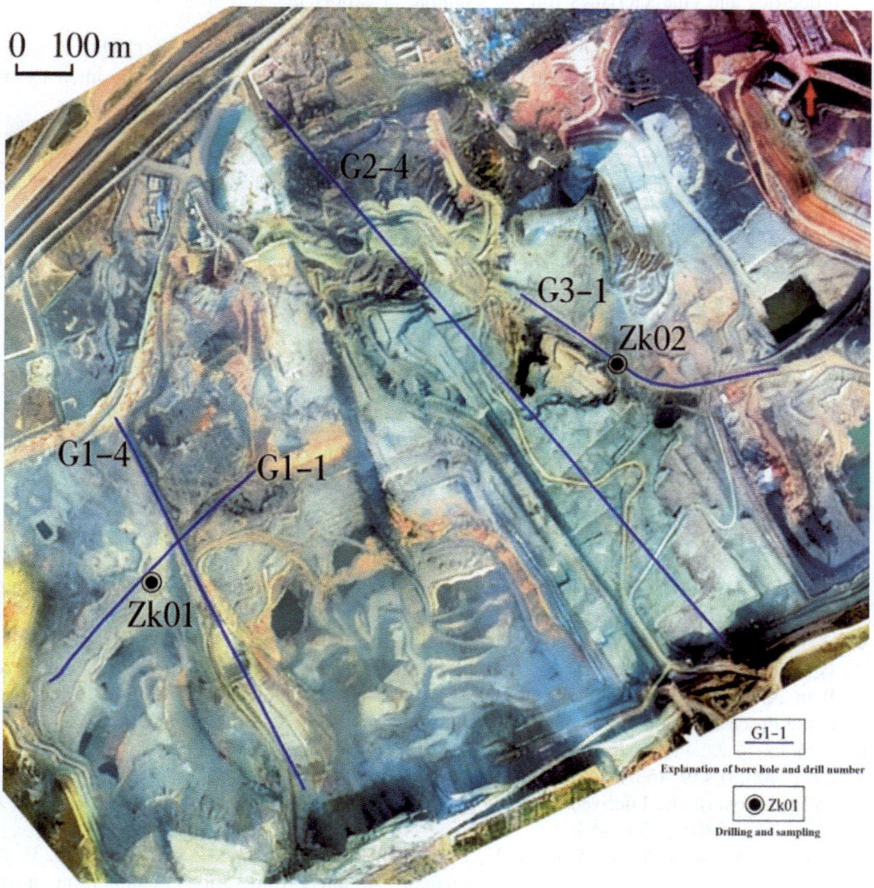

Fig. 7.13 High-density electrical method measurement line and drilling position [66]

After a small number of boreholes, the high-density electrical method accurately locates the underground goaf and the spatial distribution range of coal fire in the Huojitu Coalmine, which can provide a certain reference for the later coal fire control.

References

1. 辛邈. 煤自燃分段电学特性及瞬变电磁探测技术研究. 中国矿业大学(北京), 2017. Xin, M. (2017). *Segmented electrical characteristic of coal spontaneous combustion and transient electromagnetic detection technology on localization of combustion area in coal mine*. Ph.D. dissertation, China University of Mining and Technology (Beijing), Beijing China.
2. Sandulyak, A. A., Sandulyak, A. V., Shitikova, M. V., Tkachenko, R. Y., Sandulyak, D. A., & Gorpinenko, Y. O. (2022). Concentrated dispersed magnetwith different relative lengths: Basic properties and characteristics (on the example of a polyspherical structure). *Mechanics of Advanced Materials and Structures, 29*(26), 4972–4978.
3. Zhang, L., Xiao, L. Z., Wu, W. S., Liao, G. Z., Zhang, Y., Lu, S. H., et al. (2024). A study on the temperature sensitivity of NMR porosity inporous media based on the intensity of magnetization. *Magnetic Resonance Letters, 4*(01), Article 100085.
4. 王慧琳, 陈超, 张昌达. 最新的岩石圈磁场模型及中国磁异常特征的几点认识. 地学前缘, 2008, (03):64–71. Wang, H.L., Chen, C., & Zhang, C.D. (2008). The latest lithospheric magnetic models and some understanding of magnetic anomalies in China. *Earth Science Frontiers*, (03):64−71.
5. 曾鸣, 郑建中, 邵绪新, 郭梦熊. 高梯度磁选煤脱硫的理论基础及实验研究. 洁净煤技术, 1996, 2(01): 20–24. Zeng, M., Zheng, J. Z., Shao, X. X., & Guo, M. X. (1996). Theoretical bassis and experimental study of coal desulfurization using HGMS. *Clean Coal Technology*, *2*(01): 20−24.
6. 崔敬媛. 煤粉磁性质及磁选可选性的研究. 焦作:河南理工大学, 2009. Cui, J. Y. (2009). *Study on magnetic properties and separability of fine coal*, Master Thesis, Henan Polytechnic University,Jiaozuo China.
7. 韩文彬, 吴晗, 秦政, 任仲九. 升温过程中弱黏煤磁性变化关键因素实验研究. 煤矿安全, 2022, 53(06): 50–56. Han, W.B., Wu, H., Qin, Z., & Ren, Z.J. (2022). Experimental study on key factors of magnetic change of weakly caking coal during heating . *Safety in Coal Mines*, 53(06): 50−56.
8. 张辛亥, 苗于惠, 王伟峰, 白亚娥, 万旭, 丁峰. 高温加热对神东5–2煤磁性影响的实验研究. 煤炭技术, 2016,35(11):100–102. Zhang, X. H., Miao, Y. H., Wang, W. F., Bai, Y. E., Wan, X., & Ding, F. (2016). Experimental study of impact of heated high temperature on Shendong 5^{-2} coal magnetic. *Coal Technology, 35*(11):100−102.
9. 张辛亥, 苗于惠, 张铎, 秦政, 丁峰, 万旭. 煤自燃过程特征气体与磁性变化规律实验研究. 煤矿安全, 2017, 48(03): 36–39. Zhang, X. H., Miao, Y. H., Zhang, D., Qin, Z., Ding, F., & Wan, X. (2017). Experimental Study on Change Laws of Characteristic Gases and Magnetic in the Process of Coal Spontaneous Combustion. *Safety in Coal Mines, 48*(03): 36–39.
10. 焦红光, 李沙, 王灿, 陈清如. 热处理强化煤系黄铁矿磁性的研究. 中国矿业大学学报, 2011, 40(03):459–463. Jiao, H. G., Li, S., Wang, C., & Chen, Q. R. (2011). Study of magnetic enhancement of coa-l derived pyrite by pyrolysis. *Journal of China University of Mining & T echnology, 40*(03): 459−463.
11. Chen, Y. Y., Shen, Y. X., Xiao, S. Y., Liu, Z. J., Li, M. Y., Peng, Z. X., et al. (2023). A detailed magnetic characterization of combustion products from various metamorphic grade coals. *Journal of Applied Geophysics, 217*, Article 105168.
12. 熊盛青, 于长春. 地下煤层自燃区岩石磁性增强特征及机理研究——以内蒙古乌达和宁夏汝其沟煤矿为例. 地球物理学报, 2013, 56(08):2827–2836. Xiong, S. Q., & Yu, C. C.

(2013). Characteristics and mechanisms of rock magnetic increasing in underground coal spontaneous combustion area—take Wuda coal mine of Inner Mongolia and Ruqigou coal mine in Ningxia as examples. *Chinese Journal of Geophysics, 56*(08): 2827−2836.
13. Zhao, Y. Y., Sun, Q., Li, W. Z., Wang, S. F., Meng, Y., & Wang, X. F. (2022). Effect of high temperatures on the magnetic susceptibility of loess. *Springer Science and Business Media Llc, 29*(36), 54309–54317.
14. Wang, S. F., Sun, Q., Wang, N. Q., Luo, T., & Zhang, H. (2021). Responses of the magnetic susceptibility and chromaticity of loess to temperature in a coal fire area. *Acta Geodaetica Et Geophysica, 56*(03), 425–437.
15. 周姚秀, 董金明, 卢晓东. 煤田自燃区上磁异常的成因. 物探与化探, 1992, (02):98–106. Zhou, Y. X., Dong, J. M., & Lu, X. D. (1992). The causes of magnetic anomalies in the spontaneous combustion zone of coalfield. *Geophysical and Geochemical Exploration*, (02): 98−106.
16. 唐华风, 田志文, 赵志刚, 鲁宝亮, 唐武, 何凯伦, 等. 婆罗洲北部岩石密度和磁化率特征及其对南海南部前新生界岩性识别的约束. 地球物理学报, 2021, 64(06):2153–2169. Tang, H.F., Tian, Z. W., Zhao, Z. G., Lu, B. L., Tang, W., & He, K. L., et al. (2021). Characteristics of density and magnetic susceptibility of lithology in northern Borneo and its significance of constraining on identification of pre-Cenozoic rocks in the southern South China Sea .*Chinese Journal of Geophysics, 64*(06): 2153−2169.
17. Kuenzer, C., & Stracher, G. B. (2012). Geomorphology of coal seam fires. *Geomorphology, 138*(01), 209–222.
18. Li, Y., Winter, S. M., Kaib, D. A. S., Riedl, K., & Valentí, R. (2021). Modified Curie−Weiss law for j eff magnets. *Physical Review B, 103*(22), L220408.
19. Melnikov, N. B., Paradezhenko, G. V., & Reser, B. I. (2021). Magnetic susceptibility in metals above the curie temperature. *Journal of Magnetism and Magnetic Materials, 525*(36), Article 167559.
20. Zhu, Y. H., Wu, S., Tu, B., Jin, S. J., Huq, A., Persson, J., et al. (2020). High-temperature magnetism and crystallography of a $YCrO_3$ single crystal. *Physical Review B, 101*(01), 1–15.
21. 张桉. 变温煤岩体磁电异常响应特征及隐蔽火区定位研究. 辽宁工程技术大学, 2024. Zhang, A. (2024) *Characteristics of magneto−electric anomaly response of variable-temperature coal rocks and localisation of hidden fire zones.* Ph.D. dissertation, Liaoning Technical University, Huludao China.
22. Robert, L. H. (1987). Factors affecting the magnetic susceptibility of baked rocks above a burned coal seam. *International Journal of Coal Geology, 9*(02), 157–169.
23. Rob, S., & Caitlin, L. (2004). Magnetic surveys over clinkers and coal seam fires in western north dakota. *Presentation at the Denver Annual Meeting of the Geological Society:* 2103−2107.
24. 李金铭. 电法勘探方法发展概况. 物探与化探, 1996, (04):250–258+249. Li, J.M. (1996) Overview of the development of electrical exploration methods . *Geophysical and Geochemical Exploration*, (04):250−258+249.
25. 朱红青, 杨成轶, 秦晓峰, 和超楠. 基于瞬变电磁法的整合矿井火区分布探测方法. 科技导报, 2014, 32(25):15–19. Zhu, H. Q., Yang, C. Y., Qin, X. F., & He, C. N. (2014). Integrated coal mine fire district detecting method based on transient electromagnetic method. *Science & Technology Review, 32*(25):15−19.
26. 刘树才, 岳建华, 刘志新. 煤矿水文物探技术与应用. 徐州:中国矿业大学出版社, 2005. Liu, S. C., Yue, J. H., & Liu, Z. X. (2005). *Hydrogeophysical exploration technology and application in coal mines.* China University of Mining and Technology Press, Xuzhou China.
27. 章永风. 复杂地质条件下防治老窑水害的综合分析方法. 煤炭技术, 2000, (05):34–36. Zhang, Y. F. (2000) Comprehensive analysis method for controlling water hazards in old kilns under complex geological conditions. *Coal Technology,* (05):34−36.
28. 傅佩河, 祝仰民, 周长根, 项顺怀, 高清涛. 断层富水性的矿井瞬变电磁法探测. 煤矿开采, 2006, (02):13–16. Fu, P. H., Zhu, Y. M., Zhou, C. G., Xiang, S. H., & Gao, Q. T. (2006). Detection of yielding property of faults in mines by transient electromagnetic method. *Coal Mining Technology,* (02):13−16.

29. Torukian, Y. S., Judd, W. R., & Roy, R. F. (1990). *Physical Properties of Rocks and Minerals.* Petroleum Industry Press.
30. 黄小刚, 黄晓葛, 白武明. 高温高压下矿物岩石电导率的实验研究进展. 地球物理学进展, 2010, 25(04):1247–1258. Huang, X. G., Huang, X. G., & Bai, W. M. (2010). Progress of high temperature and high pressure experimental study on the electrical conductivity of the minerals and rocks. *Progress in Geophysics, 25* (04) : 1247−1258.
31. 葛欢, 韩猛, 刘磊, 卢君实, 赵维俊, 张桉. 高温煤岩体电性特征及地球物理数值模拟研究. 煤矿安全, 2023, 54(11):92–98. Ge, H., Han, M., Liu, L., Lu, J. S., Zhao, W. J., & Zhang, A. (2023). Study on electrical characteristics and geophysical numerical simulation of high temperature coal and rock mass. *Safety in Coal Mines, 54*(11): 92−98.
32. He, X. T., Zheng, X. C., Pu, Z. G., Qin, W., Duan, D. L., Xie, P., et al. (2024). Experimental study of the influence of saturation on the resistivity response of sandstone. *Frontiers in Earth Science, 12*(29), 1–10.
33. Zhao, J. H., Liu, C. L., Li, C. F., Zhang, Y. C., Bu, Q. T., Wu, N. Y., et al. (2022). E-scale investigation of the electrical property and saturation exponent of archie's law in hydrate-bearing sediments. *Journal of Marine Science and Engineering, 10*(01), 1–13.
34. Zhang, A., Jia, B. S., Wang, H. Y., & Zhou, Q. A. (2024). Experimental study on the dielectric properties of coal in situ with variable temperatures. *ACS Omega, 9*(17), 19481–19492.
35. Lovric, D., Vujevic, S., & Krolo, L. (2020). Computation of per-unit-length internal impedance of a multilayercylindrical conductor with possible dielectric layers11. *Facta Universitatis—Series: Electronics andEnergetics. 33*(04):605−616.
36. 武岳山, 于利亚. 介电常数的概念研究. 现代电子技术, 2007, (02): 177–179+185. Wu, Y. S., & Yu, L. Y. (2007). Research on the concept of dielectric constant. *Modern Electronic Technique*, (02):177−179+185.
37. Fan, W., Jia, C. Y., Hu, W., Yang, C. F., Liu, L. Y., Zhang, X. S., et al. (2015). Dielectric properties of coals in the low-terahertz frequency region. *Fuel, 162*(2015), 294–304.
38. Xu, G., Huang, J. X., Hu, G. Z., Yang, N., Zhu, J. Q., & Chang, P. (2020). Experimental study on effective microwave heating/fracturing of coal with various dielectric property and water saturation. *Fuel Processing Technology, 20*, Article 106378.
39. 王晴东, 王光华, 陈彪, 王国成, 王世杰. 东胜褐煤与Yallourn褐煤热解过程中微波吸收特性. 煤炭学报, 2016, 41(06):1540–1545. Wang, Q. D., Wang, G. H., & Chen, B. (2016). Microwave absorption characteristics of Dongsheng lignite and Yallourn lignite during pyrolysis process.*Journal of China Coal Society, 41*(06): 1540−1545.
40. Liu, H. Y., Xu, L., Jin, L., Fan, B. G., Qiao, X. L., & Yang, Y. X. (2015). Effect of coal rank on structure and dielectric properties of chars. *Fuel, 153*(2015), 249–256.
41. Xu, L., Liu, H. Y., Jin, Y., Fan, B. G., Qiao, X. L., & Jing, B. (2014). Structural order and dielectric properties of coal chars. *Fuel, 137*(08), 164–171.
42. Hazra, B., Wood, D. A., Chakraborty, P., Singh, D. P., Sahu, S. G., & Sarkar, P. (2023). Comparative thermal behavior, RockEval signature and kinetics of distinct thermally mature vitrain, fusain lithotypes and coals from India. *International Journal of Coal Preparation and Utilization, 43*(08), 1428–1450.
43. Gerlinde, S., Bernhard, S., & Yu, C. C. (2008). Geophysical investigation of Wuda coal mining area, inner mongolia: Electromagnetics and magnetics for coal fire detection. *International Research for Sustainable Control and Management, 4*(24), 336–350.
44. 刘双, 胡祥云, 郭宁, 蔡红柱, 张恒磊, 李永涛. 无人机航磁测量技术综述. 武汉大学学报(信息科学版), 2023, 48(06):823–840. Liu, S., Hu, X. Y., Guo, N., Cai, H. Z., Zhang, H. L., & Li, Y. T. (2023). Overview on UAV aeromagnetic survey technology. *Geomatics and Information Science of Wuhan University, 48*(06): 823−840.
45. 张登奎, 张津源, 王志刚, 王肃良, 孙强. 磁法勘探在矿产资源调查中的应用研究. 物探装备, 2024, 34(05):296–299. Zhang, D. K., Zhang, J. Y., Wang, Z. G., Wang, S. L., & Sun, Q. (2024). Research on the application of magnetic exploration in mineral resource survey. *Geophysical Prospecting Equipment, 34*(05): 296−299.
46. 兰超歌. 磁法勘探技术在复杂地质构造区矿产资源调查中的应用研究. 世界有色金属, 2025, (01):37–39. Lan, C. G. (2025). Application of magnetic exploration technology in

mineral resource investigation in complex geological structure areas. *World Nonferrous Metals,* (01):37−39.
47. 张静, 胡尊平, 李延清, 肖艳东, 刘双, 刘尧. 磁法勘探在富蕴县矿产地质调查中的应用效果. 西部探矿工程, 2017, 29(07):172–175. Zhang, J., Hu, Z. P., Li, Y. Q., Xiao, Y. D., Liu, S., & Liu, Y. (2017). Application effect of magnetic prospecting in mineral geological survey of fuyun county. *Western Prospecting Engineering, 29*(07): 172−175.
48. 赵禹深, 许献磊, 曹仕龙, 朱鹏桥. 地下煤火探测技术及工程应用综述. 矿业科学学报, 2025, 10(01):24–39. Zhao, Y. S., Xu, X. L., Cao, S. L., & Zhu, P. Q. (2025). A review of underground coal fire detection technology and engineering application. *International Journal of Mining Science and Technology, 10* (01): 24−39.
49. 吴璋, 张振振, 李雄伟, 郭建磊, 侯彦威. 基于不同高度异常交叉约束反演的航空磁法探测遗煤火区. 煤炭技术, 2023, 42(06):133–136. Wu, Z., Zhang, Z. Z., Li, X. W., Guo, J. L., & Hou, Y. W. (2023). Aeromagnetic method for detecting residual coal fire area based on cross constraint inversion of different height. *Coal Technology, 42*(06):133−136.
50. 张秀山. 磁法探测煤层自燃火区.煤田地质与勘探, 1980, 12(06):43–48. Zhang, X. S. (1980). Magnetic detection of spontaneous combustion fire zones in coal seams. *Coal Geology and Exploration, 12*(06): 43−48.
51. 朱晓颖, 于长春, 熊盛青, 陈斌. 磁法在煤火探测中的应用. 物探与化探, 2007, (02):115–119. Zhu, X. Y., Yu, C. C., Xiong, S. Q., & Chen, B. (2007). Application of magnetic method in coal fire detection. *Geophysical and Geochemical Exploration, 31*(02): 115−119.
52. 曹代勇, 时孝磊, 樊新杰, 吴查查. 煤田火区环境效应分析. 中国矿业, 2007, (07):40–42. Cao, D. Y., Shi, X. L., Fan, X. J., & Wu, C. C. (2007). Environmental effect analysis of coalfield fire. *China Mining,* (07): 40−42.
53. Sharma, P. V. (1987). Magnetic method applied to mineral exploration. *Oregeology reviews, 2*(04), 323–357.
54. Okada, K. (2022). Breakthrough technologies for mineral exploration. *Mineral Economics, 35*(3–4), 429.
55. Paterson, N. R., & Reeves, C. V. (1985). Applications of gravity and magnetic surveys: The state-of-the-art in 1985. *Geophysics, 50*(12), 2558–2594.
56. Adagunodo, T. A., Sunmonu, L. A., & Adeniji, A. A. (2015). An overview of magnetic method in mineral exploration.
57. Nabighian, M. N., Grauch, V. J. S., Hansen, R. O., Lafehr, T. R., Pearson, W. C., & Phillips, J. D. (2005). The historical development of the magnetic method in exploration. *Geophysics, 70*(06): 33ND−61ND.
58. Bandelow, F. K., & Gielisch, H. F. (2004). Modern exploration methods as key to fighting of uncontrolled coal fires in China. Denver Annual Meeting, Denver, Co: 173−177.
59. 张莹莹, 李貅. 地空瞬变电磁法研究进展. 地球物理学进展, 2017, 32(04):1735–1741. Zhang, Y. Y., & Li, X. (2017). Research progress on ground−airborne transient electromagnetic method. *Progress in Geophysics, 32*(04):1735−1741.
60. 陈载林, 黄临平, 陈玉梁. 我国瞬变电磁法应用综述. 铀矿地质, 2010, 26(01):51–54. Chen, Z. L., Huang, L. P. & Chen, Y. L. (2010). Review on the application of transient electromagnetic method in China. *Uranium Geology, 26*(01):51−54.
61. 魏万鸿. 高精度磁法在矿产资源勘查中的效果分析. 世界有色金属, 2020, (19):97–98. Wei, W. H. (2020). Effect analysis of high precision magnetic method in mineral resources exploration. *World Nonferrous Metals,* (19):97−98.
62. 杨进, 武炜. 地球物理方法在地质灾害勘查中的应用. 物探与化探, 2003, 27(05):333–337. Yang, J., & Wu, W. (2003). The application of geophysical methods in geological disaster investigation. *Geophysical and Geochemical Exploration, 27*(05): 333−337.
63. 底青云, 石昆法, 王妙乐, 王若, 李英贤, 张庚利, 等. CSAMT法和高密度电法探测地下水资源. 地球物理学进展, 2001, 16(03):53–56. Di, Q. Y., Shi, K. F., Wang, M. Y., Wang, R., Li, Y. X., & Zhang, G. L., et al. (2001). The CSAMT method and high-density electrical method are used to detect groundwater resources. *Advances in Geophysics, 16*(03): 53−56.
64. 于永宁, 李雄伟, 石磊, 柳凯元, 郭建磊, 马国庆. 时移航空磁法在煤矿火烧区探测中的应用研究.工矿自动化, 2023, 49(08):114–120. Yu, Y. N., Li, X. W., Shi, L., Liu, K. Y., Guo, J.

L., & Ma, G. Q. (2023). Research on the application of time shifting aeromagnetic method in detecting coal mine burning areas .*Industrial and Mining Automation, 49*(08):114−120.
65. Ide, T. S., Crook, N., & OrrJr, F. M. (2011). Magnetometer measurements to characterize a subsurface coal fire. *International Journal of Coal Geology, 87*(3–4), 190–196.
66. 贺小元, 王峰, 雒铮, 管涛, 李聪聪, 崔胜, 等. 高密度电法在煤火勘查中的应用: 以活鸡兔煤矿为例. 中国矿业, 2022, 31(01):168–174. He, X.Y., Wang, F., Luo, Z., Guan, T., Li, C. C., & Cui, S., et al (2022). Application of high-density electrical method to prospect coal fire: taking Huojitu coal mine as an example.*China Mining, 31*(01):168−174.

Chapter 8
Isotope Radon Measurement Method

Accurate identification of concealed fire sources within coal seams represents a critical challenge in mitigating coal fire disasters and ensuring the safety of mining operations as well as ecological restoration efforts. Conventional detection techniques are constrained by limitations in accessing deep-seated information, challenges posed by complex topography, and high operational costs. Therefore, there is an urgent need to adopt highly sensitive, minimally invasive, and reversible geochemical tracer technologies. Radon-222, due to its relatively long half-life, high diffusivity, and its sensitivity to coupled variations in temperature, stress, and porosity, serves as an effective natural tracer for delineating the three-dimensional spatial distribution of subsurface fire sources [1, 2].

8.1 Principle of Isotope Detection

The application of isotope-based radon measurement in the field of underground coal fire detection is based on its high-precision capability to detect the radioactive isotope Rn-222 [3]. This approach integrates principles from radiochemistry, seepage mechanics, and the coupled physical —chemical processes associated with coal fire dynamics, thereby establishing a three-stage theoretical framework of "radon source—migration—anomaly". The three principal isotopes of radon—Rn-222, Rn-220, and Rn-219—originate from the radioactive decay chains of three naturally occurring series: the uranium series, the thorium series, and the actinium series. The uranium decay series is illustrated in Fig. 8.1, while the corresponding physical radioactive characteristics are summarized in Table 8.1. Due to their short half-lives, Rn-220 and Rn-219 exhibit very low atmospheric concentrations. In contrast, Rn-222, with its relatively long half-life and gaseous mobility, is widely utilized as a tracer for conveying information from deep geological formations in geophysical exploration projects. Rn-222 is produced via alpha decay of U-238, possessing a

Y. Xiao et al., *Detection of Fire Source in Underground Coal Mine and its Applications*,
https://doi.org/10.1007/978-981-95-7064-5_8

half-life of 3.825 days and an alpha particle energy of 5.489 MeV. Its distinct physical properties allow for vertical migration from depths of several hundred meters to the surface, where it can be reliably detected using appropriate instrumentation [4]. During the spontaneous combustion of coal seams, the oxidation-induced temperature rise significantly enhances the emission of radon gas [5]. As illustrated in Fig. 8.2, this process can be categorized into three distinct stages. In the initial stage (30–100 °C), the evaporation of pore water within the coal matrix, driven by increasing temperature, facilitates the release of radon gas, leading to a gradual rise in radon concentration. In the intermediate stage (100–350 °C), thermal decomposition of the coal results in the rapid release of radon from closed pores, causing a sharp increase in its surface concentration. In the final stage (350–450 °C), the depletion of radon atoms due to extensive release in the earlier stages leads to a rapid decline in radon concentration [6]. This characteristic variation of radon concentration with temperature serves as a critical geochemical indicator for the detection and localization of subsurface fire sources [7].

In practical applications, the isotope radon measurement method involves establishing measurement points on the ground surface and utilizing specialized instruments, such as the RAD7 professional electronic radon detector or the CD-1 α cup radon detector, to measure surface radon gas concentrations. Figure 8.3 presents a

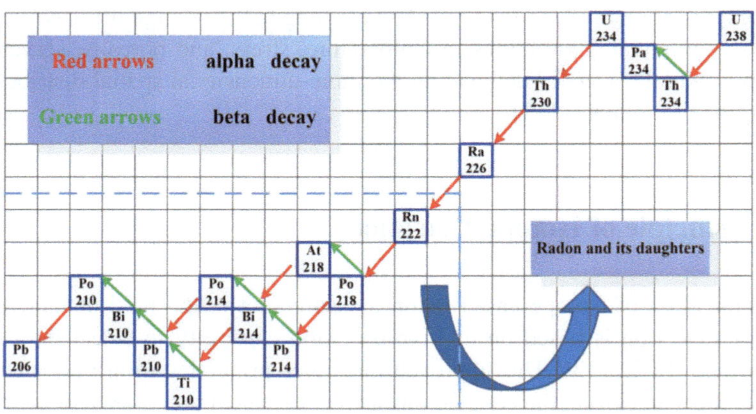

Fig. 8.1 Decay scheme of the uranium series [8]

Table 8.1 Radioactive characteristics of Rn-222, Rn-220 and Rn-219 [4]

Name	Atomic number	Mass number	Direct radiogenic parent	Direct radiogenic daughter	Half-life	Decay types	Radiation energy / Mev
Rn-222	86	222	Ra-236	Po-218	3.825 d		5.489
Rn-220	86	220	Ra-224	Po-216	55.6 s		6.288
Rn-219	86	219	Ra-237	Po-215	3.965 s		6.819

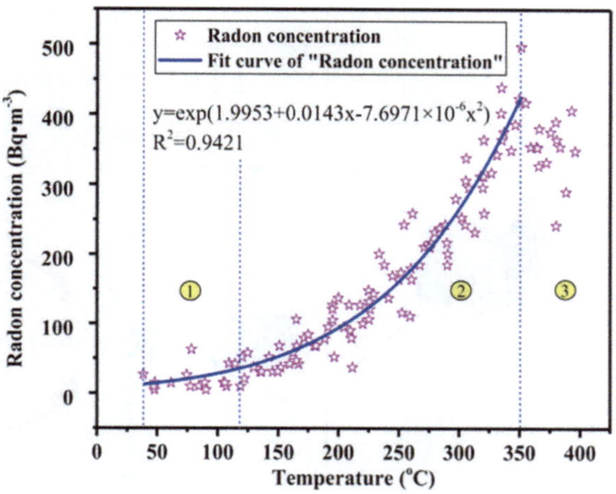

Fig. 8.2 Relationship between radon exhalation rate and coal body temperature [8]

schematic diagram illustrating the underground placement of the α cup. Figure 8.4 outlines the operational principle of the radon detector. Using the RAD7 radon detector as a representative example, the device functions by detecting α particles emitted during the decay of radon [9]. When radon gas enters the detection chamber, the α particles generated through radioactive decay are collected by the electrode under the influence of an electric field, producing an electrical signal that is directly proportional to the radon concentration. This enables the quantitative determination of radon levels. Following data processing with specialized software (e.g., Surfer), contour maps and three-dimensional visualizations can be generated to depict the potential location and extent of subsurface fire sources. The advantages of this technique include operational simplicity, cost-effectiveness, adaptability to various terrains, and the capacity to deliver relatively precise fire source localization, thereby offering robust technical support for the detection and management of coalmine fire sources [9, 10].

The isotope radon measurement method fundamentally relies on a precise understanding of the radon release mechanism during coal seam spontaneous combustion. During this process, the physical and chemical transformations of the coal body significantly enhance radon emission. As temperature increases, the coal body's

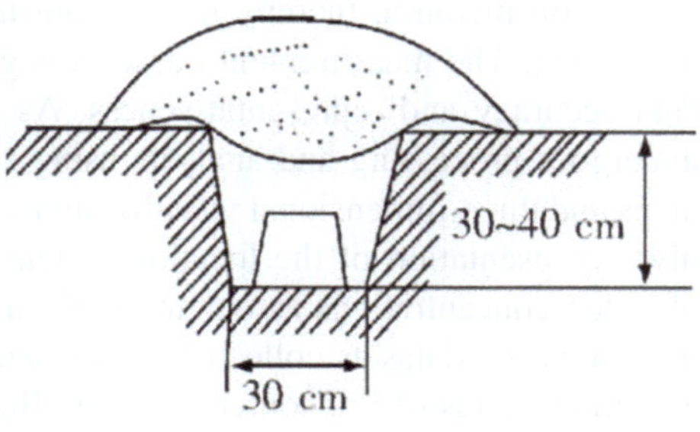

Fig. 8.3 Schematic of the α-cup Buried underground [11]

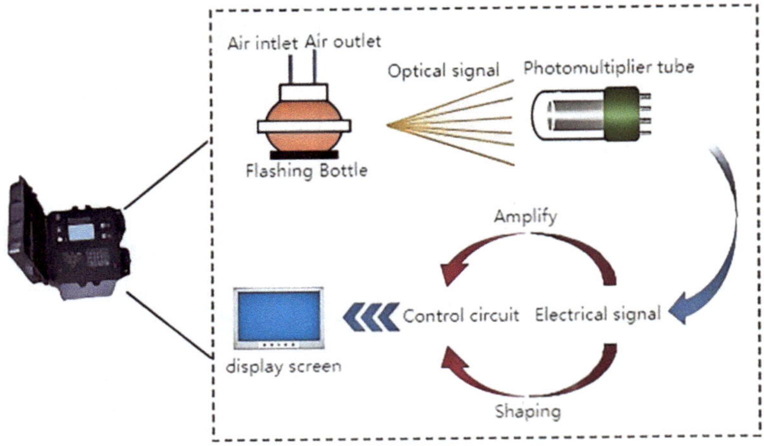

Fig. 8.4 Working principle of the radon detector [9]

lattice structure becomes loosened and distorted, leading to a marked rise in radon exhalation rate. Concurrently, micro-cracks within the coal seam expand under high-temperature conditions, forming pathways that facilitate the vertical migration of radon gas. Driven by mechanisms such as diffusion, convection, and thermal gradients, radon migrates along micro-fractures in the subsurface rock mass toward the surface, as illustrated in Fig. 8.5. Consequently, high-concentration radon anomaly zones are formed at the surface [12]. Surface radon concentration measurements can thus be utilized to infer the location and extent of underground coal seam spontaneous combustion sources [13, 14]. This process involves not only radon generation and migration but is also closely influenced by the geological structure of coal seam, oxidation level, and hydrological conditions.

From a detection technology perspective, the isotope radon measurement method entails the establishment of a surface measurement point network, with data collected using specialized radon detection instruments. The spacing between measurement points is typically determined based on geological conditions and required detection precision, ranging from 10–20 m. The operational procedure of the radon detector is illustrated in Fig. 8.6. During the measurement process, a pit of specified depth is excavated at each measurement point to accommodate the detector probe or adsorption medium, thereby minimizing interference from surface environmental factors [8]. The measurement duration is generally set between 3–5 min to ensure data accuracy and representativeness. As depicted in Fig. 8.7, the acquired data undergoes processing and analysis using specialized software to generate contour maps and three-dimensional visualizations. These graphical outputs provide an intuitive representation of the location, extent, and spatial distribution characteristics of radon concentrations associated with subsurface fire sources [15]. Furthermore, by comparing datasets collected at different time intervals, the dynamic status and evolution trends of fire sources can be effectively monitored, offering timely early

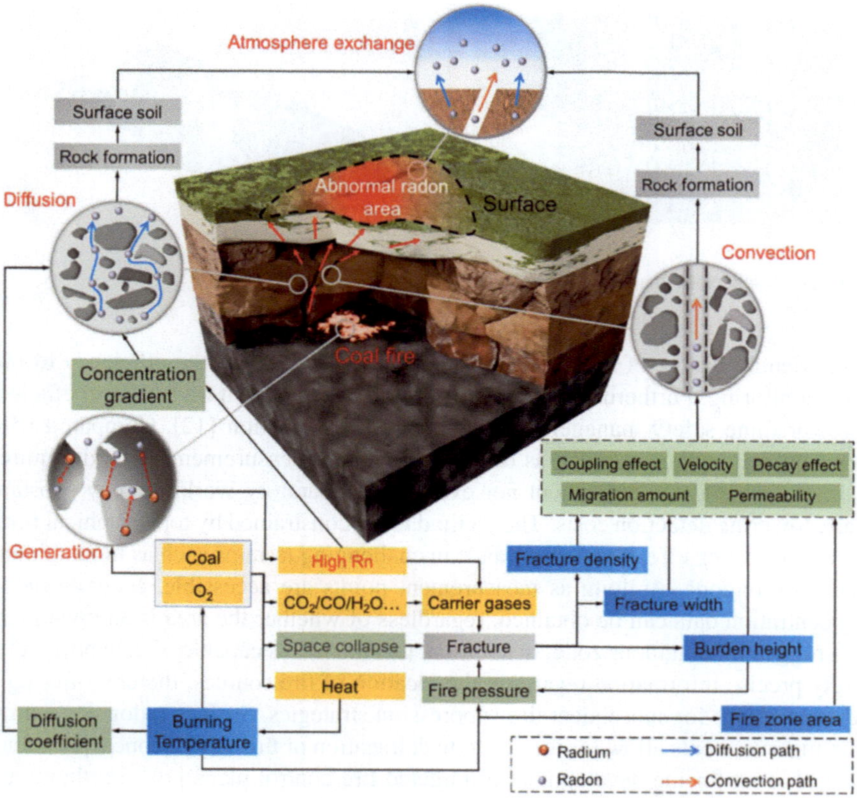

Fig. 8.5 Generation—diffusion—convection process mechanism of radon [14]

warning capabilities and supporting informed decision-making for coalmine safety management.

In the complex and dynamic environment of coal mining operations, the isotope radon measurement method exhibits distinct advantages. Its operational simplicity facilitates rapid deployment across diverse terrain conditions. The portability and

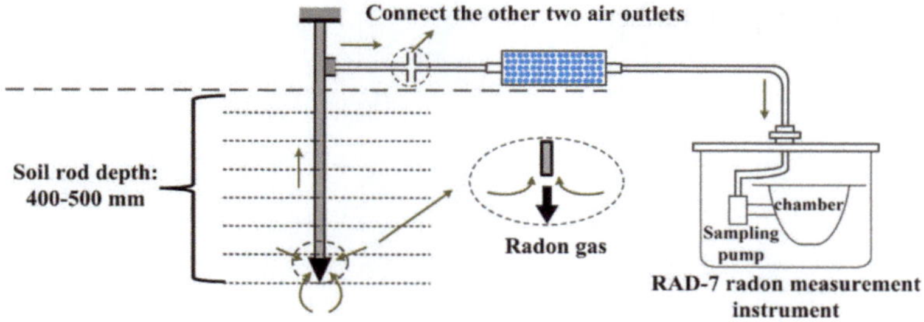

Fig. 8.6 Testing procedure of the radon detector [8]

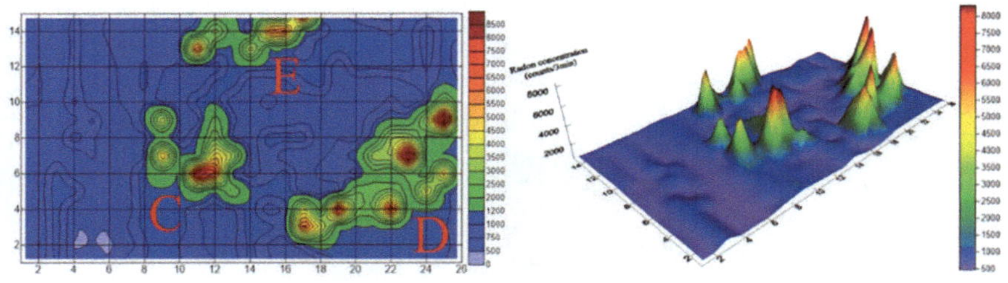

Fig. 8.7 A schematic map of radon concentration in the study area [15]

user-friendly design of radon detectors significantly improve the efficiency of on-site monitoring. Furthermore, its cost-effectiveness renders it a viable and efficient tool for mine safety management and fire source detection [12]. Compared with conventional techniques such as drilling, the radon measurement method requires neither sophisticated equipment nor extensive preparatory work, thereby substantially lowering detection costs. The method is not constrained by topographical limitations, enabling effective performance in challenging terrains such as mountainous and hilly regions. As long as measurement points are accessible, accurate radon concentration data can be obtained, regardless of whether the area is an open plain or a rugged mountainous zone. In addition, this method is capable of delivering relatively precise information regarding the location of fire sources, thereby offering a scientific basis for subsequent fire suppression strategies. Surface radon concentration measurements allow for the accurate delineation of fire source zones, providing critical input for the development of targeted fire control plans [16]. Furthermore, the isotope radon measurement method can reflect the progression and thermal variations of underground fires. Through comparative analysis of data collected across different time intervals, the activity level of the fire source can be assessed, and potential development trends can be predicted, thus offering timely early warning and decision-making support for coalmine safety production.

In conclusion, the isotope radon measurement technology not only serves as a critical technical safeguard for coalmine safety, but also contributes significantly to the sustainable development of coal resources and environmental protection. As the technology continues to advance and mature, the application prospects of the isotope radon measurement method in the field of underground coal fire detection are expected to expand further.

8.2 Isotope Detection Methods and Techniques

8.2.1 Static Radon Measurement Method

The static radon measurement method is a technique that relies on the accumulation of radon gas within a sealed environment. Its theoretical basis originates from the radioactive decay properties of radon gas and its diffusion behavior in various environmental media. Radon gas, primarily Rn-222, is generated through the decay of radium-226 present in soil, rock, or coal formations. It exhibits alpha radioactivity and has a half-life of 3.82 days. Under static conditions, radon is gradually released from subsurface materials and accumulates within a closed container. The quantity accumulated is determined by the ambient radon concentration, the diffusion rate, and the duration of the accumulation period.

The design of radon measurement containers plays a critical role in ensuring the accuracy of measurement results. The container material must exhibit low radon adsorption capacity and chemical inertness in order to minimize interference caused by radon adsorption and desorption on the container walls. Materials such as polytetrafluoroethylene (PTFE) and quartz glass are commonly employed due to their superior chemical stability and minimal inherent radon content. The geometric configuration and volume of the container should be optimized based on the measurement objectives and required precision. A low and wide shape is typically preferred to increase surface area, thereby facilitating radon gas release and diffusion. The interior surfaces are frequently coated with low-adsorption materials such as Teflon to further reduce potential radon loss through adsorption. Additionally, the container's sealing integrity must be validated through pressure testing to ensure that the radon leakage rate remains below 1% throughout the measurement period.

During the measurement process, the radon collection container is positioned within the target environment to capture radon gas via diffusion. The duration of the measurement is determined by the ambient radon concentration and the sensitivity of the detection system, typically spanning from several hours to multiple days. Radon concentration within the container increases exponentially over time, governed by the following equation:

$$C(t) = C_0\left(1 - e^{-\lambda t}\right) \tag{8.1}$$

where, C(t) denotes the radon concentration at time t, represents the ambient radon concentration, λ is the decay constant of radon $\left(\lambda = \ln 2/T_{1/2}\right)$, and $T_1/_2$ denotes the half-life of radon. As the accumulation time increases, the radon concentration within the container asymptotically approaches an equilibrium level dictated by the ambient radon concentration and the leakage characteristics of the container.

Upon completion of the measurement process, the radon content within the container is subjected to quantitative analysis using either a high-sensitivity alpha spectrometer or a low-background multi-channel gamma spectrometer. The alpha spectrometer enables precise determination of radon concentration by identifying the characteristic energy peaks of alpha particles emitted during radon decay, achieving a detection limit as low as 1 Bq/m^3. To ensure data comparability and measurement accuracy, the results must be standardized by incorporating the container's geometric factor, radon decay correction, and relevant environmental parameters (e.g., temperature and pressure) [15, 17]. The static radon measurement method, owing to its operational simplicity and cost-effectiveness, is well-suited for large-scale environmental radon surveys and long-term monitoring applications. However, its applicability in dynamically changing environments is constrained by the inherent requirement for a defined accumulation period.

8.2.2 Dynamic Radon Measurement Method

Dynamic radon measurement is a methodology employed for the real-time and continuous monitoring of environmental radon concentrations. The fundamental principle involves the use of an active air sampling system to continuously draw environmental air samples into the detection chamber at a controlled flow rate, while highly sensitive detection instruments record the radioactive decay events of radon in real time. This enables uninterrupted tracking of radon concentration levels [15, 18].

The dynamic radon measurement system is composed of four primary components: a gas extraction unit, a flow control system, a detection unit, and a data processing unit. The gas extraction unit employs a diaphragm pump or vortex pump to generate a consistent aspiration force, ensuring continuous airflow through the detection chamber at a predefined flow rate, typically ranging from 0.5 to 5 L/min. The flow control system utilizes a mass flow meter in conjunction with a needle valve to achieve precise regulation and stability of the flow rate. The flow rate fluctuation is maintained within $\pm$ 0.5%, thereby minimizing measurement inaccuracies caused by variations in flow conditions.

The detection unit constitutes a critical component of the dynamic radon measurement system. At present, two primary detection methodologies are employed: electrostatic collection and semiconductor-based detection. The electrostatic collection technique utilizes an electric field to direct radon progeny (e.g., polonium-218) onto a Faraday cup, where the accumulated charge is measured to determine the corresponding radon concentration. The semiconductor detection approach employs high-purity silicon or germanium detectors, which exploit the material band structure to achieve a sensitive response to alpha particles, thereby enabling direct acquisition of the alpha energy spectrum resulting from radon decay. Semiconductor detectors can achieve an energy resolution of up to 15 keV, allowing for effective differentiation between alpha decay peaks of radon progeny and those of other radionuclides, thereby significantly improving measurement selectivity and accuracy. Modern dynamic radon instruments integrate pulsed ionization chambers with advanced digital signal processing techniques, enabling real-time monitoring of radon concentrations with a temporal resolution of 1 min and a detection threshold as low as 0.1 Bq/m^3.

To ensure measurement accuracy, dynamic radon detectors must undergo rigorous calibration procedures. This process entails multi-point calibration using certified radon sources with known activity concentrations, thereby establishing a quantitative correlation between the instrument signal output and the corresponding radon concentration. Additionally, long-term environmental radon monitoring data are utilized to perform drift correction, maintaining the stability and reliability of measurement results over extended periods. Owing to their rapid response characteristics and high temporal resolution, dynamic radon measurement techniques are particularly well-suited for monitoring transient variations in environmental radon levels, conducting real-time assessments of indoor radon contamination, and tracking dynamic radon emission sources such as coal seam fires [11]. Nevertheless, these instruments are associated with relatively high acquisition and maintenance costs, and require controlled environmental conditions, particularly with regard to temperature and humidity stability.

8.2.3 Cumulative Radon Measurement Method

The cumulative radon measurement method is a passive sampling technique designed to quantify the total accumulation of radon gas over extended periods. It is especially suitable for monitoring environments characterized by low and steady radon emission rates. This methodology relies on the principle of radon retention within a designated adsorbing medium, and is implemented through the deployment of passive radon detectors at predetermined sampling locations. These detectors are typically fabricated from materials with high adsorption capacity, enabling efficient capture and retention of radon atoms.

In cumulative radon measurement, two widely employed detection devices are solid-state nuclear track detectors and electrochemical deposition thin-film detectors. Solid-state nuclear track detectors determine radon concentration by capturing and recording the alpha particle tracks generated during radon decay on the detector surface. These detectors are commonly fabricated from materials such as polycarbonate or Lexan, which exhibit excellent chemical stability and mechanical durability. Prior to deployment, these detectors must undergo rigorous surface preparation and precise thickness calibration to ensure measurement accuracy. In contrast, electrochemical deposition thin-film detectors operate based on electrochemical reaction principles, wherein radon gas is selectively deposited onto an electrode surface. The quantity of the deposited material is then quantified to estimate the radon concentration. This methodology demonstrates superior environmental adaptability and is especially suitable for monitoring in high-humidity environments or locations with significant temperature fluctuations [19].

The cumulative radon measurement method typically requires an extended exposure period, ranging from several days to several months, depending on the measurement objectives and the anticipated radon concentration levels. Upon completion of the exposure phase, the detectors must undergo processing and subsequent analysis.

For solid-state nuclear track detectors, alpha particle tracks are typically revealed through chemical etching, followed by microscopic or automated counting and analysis. In the case of electrochemical deposition thin-film detectors, the quantity of deposited material is determined using gravimetric or spectroscopic techniques. It is important to note that the measurement outcomes obtained through this method are susceptible to environmental influences. Variations in temperature and humidity, for instance, can significantly alter the diffusion and adsorption behavior of radon gas, thereby compromising measurement accuracy [10]. Consequently, continuous monitoring of environmental parameters is essential during the measurement period, and appropriate corrections must be applied during data processing to ensure reliable results.

8.2.4 Detector Technology

Detectors represent the critical core component in isotope-based radon measurement systems, with their operational performance directly influencing the accuracy and reliability of the obtained measurement data. Based on detection mechanisms and specific application requirements, detector technologies can be broadly categorized into solid-state nuclear track detectors, semiconductor detectors, and low-temperature liquid scintillation detectors [18, 20].

(1) Solid-state nuclear track detector

Solid nuclear track detectors constitute a class of radiation detection technology based on the formation of latent particle tracks induced by alpha particles in solid-state materials. Common detector materials include polycarbonate, Lexan, and mica, which exhibit high alpha particle sensitivity and minimal background interference, enabling effective recording of alpha tracks generated during radon decay processes.

The fabrication process of these detectors involves multiple stages, including material selection, surface preparation, thickness calibration, and sensitivity characterization. Prior to deployment, rigorous performance evaluation must be conducted to ensure measurement reliability and reproducibility. For instance, the track recording resolution of such detectors can achieve sub-micron precision, allowing differentiation of alpha particle tracks with varying energy levels. Furthermore, enhancements in measurement accuracy and stability can be attained through optimization of surface treatment protocols and calibration methodologies [21, 22].

In the data processing phase, the recorded latent tracks undergo chemical etching to enhance visibility, followed by image analysis using high-resolution optical microscopy or automated counting systems. The integration of advanced image recognition algorithms and digital data processing techniques significantly improves both the precision and efficiency of track quantification, enabling measurement errors to be maintained within $\pm$ 5%.

(2) Semiconductor detector

Semiconductor detectors operate based on the sensitive response of semiconductor materials to alpha radiation, enabling radon concentration measurement through the detection of induced electrical signals. Widely utilized materials include high-purity silicon (Si) and high-purity germanium (Ge), both of which possess superior crystalline structures and electronic properties, facilitating high-resolution energy measurements.

The development and fabrication of semiconductor detectors involve sophisticated technological processes, such as crystal growth, wafer slicing, surface polishing, dopant introduction, and electrode fabrication. Key performance metrics of these detectors encompass energy resolution, detection efficiency, background noise level, and long-term stability. Among these, energy resolution serves as a critical parameter for evaluating detector performance, as it determines the ability to distinguish alpha particles with differing energy levels. Contemporary semiconductor detectors typically achieve energy resolutions within the range of 10–15 keV, allowing for clear differentiation between alpha decay peaks of radon progeny (e.g., Po-218, Po-214) and those of other radionuclides, thereby enhancing measurement selectivity and accuracy.

In signal processing, semiconductor detectors are commonly integrated with low-noise amplifiers and multi-channel analyzers. These components amplify weak signals while suppressing noise interference, and categorize detected events according to their energy levels, enabling precise alpha energy spectroscopy. Accurate radon concentration and associated radioactive parameters can be derived by analyzing characteristic spectral peaks. The compact design and real-time measurement capability of semiconductor detectors confer distinct advantages in field applications and emergency monitoring scenarios. Modern portable semiconductor-based radon instruments incorporate miniaturized detectors, low-power electronics, and rechargeable power sources, enabling rapid and precise on-site radon measurements, thus providing timely data support for environmental radon surveillance and coal seam fire detection.

(3) Low-temperature liquid scintillation detector

Low-temperature liquid scintillation detectors represent a class of highly sensitive instrumentation for radon measurement. These detectors enable high-precision quantification of radon activity by exploiting the enhanced scintillation efficiency of organic liquid media under low-temperature conditions in response to alpha particle irradiation. The core components of the system include the scintillation liquid, detection chamber, cryogenic cooling system, and photomultiplier tube assembly [23, 12].

In low-temperature liquid scintillation detectors, detection efficiency and background level are two critical parameters that significantly influence overall detector performance. Detection efficiency is primarily determined by the scintillation liquid's response efficiency to alpha particles and the quantum efficiency of the photomultiplier tube. By refining the formulation and compositional ratios of the scintillation liquid, both its luminescence efficiency and response kinetics to alpha radiation can be enhanced. Concurrently, the cooling system must be engineered to maintain the detection chamber at a stable low temperature, thereby ensuring optimal operational conditions for the low-temperature scintillation process. To simultaneously improve detection efficiency and suppress background noise, multiple technical strategies are implemented. These include the selection of ultra-high-purity scintillation components and the optimization of the chamber's optical geometry to minimize background contributions from impurities and surface interactions. Furthermore, advanced signal processing techniques such as pulse shape discrimination and energy window gating enable effective differentiation of alpha particle signals from interfering radiation types, such as beta particles and gamma rays. These measures collectively enhance the signal-to-noise ratio and measurement precision of the detector.

In summary, advancements in detector technology serve as a pivotal driving force behind the evolution of isotope-based radon detection methodologies. As materials science, electronics, and data processing techniques continue to progress, detector performance is expected to achieve further enhancements, thereby offering increasingly robust technical support for the widespread application of isotope radon detection in diverse fields such as environmental monitoring, coal fire detection, and resource exploration.

8.2.5 Technical Optimization and Development Trends

(1) Deepened application of data fusion and intelligent analysis technologies

Within the domain of isotope-based radon measurement technology, the integration of data fusion and intelligent analysis methodologies continues to advance, offering substantial support for improving the accuracy and reliability of radon concentration assessments. Multi-source data fusion techniques are designed to consolidate information acquired from diverse detector systems, including the synergistic integration of data from static, dynamic, and cumulative radon measurement approaches. Each measurement methodology exhibits distinct error characteristics and environmental sensitivities. Through the implementation of joint probability distribution models or Bayesian fusion algorithms, the inherent uncertainties associated with individual methods can be systematically mitigated, thereby yielding fused radon concentration estimates that more closely approximate the true values. For instance, in complex geological settings such as mined-out areas, the coordinated application of multiple measurement techniques combined with data fusion can markedly enhance the precision of fire source localization while minimizing measurement discrepancies induced by variations in subsurface geological structures.

The advancement of intelligent analysis technology centers on refining machine learning and pattern recognition methodologies. As radon measurement data continues to grow exponentially, conventional manual analysis and basic algorithmic approaches are no longer sufficient to fulfill the requirements of high-efficiency and high-precision detection. Deep learning techniques, including Convolutional Neural Networks (CNN) and Long Short-Term Memory networks (LSTM), have been increasingly employed for time series forecasting of radon concentration data and for identifying spatial distribution patterns [11, 24]. These models are capable of autonomously extracting intricate data features—such as periodic variations, long-term trends, and abrupt anomalies—thus enabling predictive insights into the evolving dynamics of coal fires. Concurrently, data mining approaches grounded in clustering and classification algorithms, such as K-Means and Support Vector Machines (SVM), facilitate the identification of distinct environmental patterns across diverse radon measurement scenarios, offering a scientific foundation for optimizing detection strategies and resource deployment.

(2) Innovations regarding the miniaturization and integration of detectors

Miniaturization and integration represent key developmental trends in isotope-based radon detection technology [21]. Advances in micro-nano fabrication techniques have established a robust foundation for detector miniaturization, enabling the realization of compact radon detection devices with high sensitivity and low power consumption. For example, semiconductor detectors fabricated using Micro-Electro-Mechanical Systems (MEMS) technology can be scaled down to millimeter or even micrometer dimensions while preserving superior detection performance. These miniaturized detectors offer significant advantages, including ease of deployment in confined spaces or complex environments, reduced power requirements and operational costs, as well as enhanced system reliability and maintainability.

Integrated design consolidates multiple micro-detectors, signal processing circuits, data storage units, and wireless communication modules into a compact package, resulting in a miniaturized radon detection system equipped with intelligent sensing, real-time processing, and wireless transmission capabilities. This highly integrated system demonstrates broad applicability in distributed radon monitoring networks, enabling real-time and synchronized surveillance over large geographic areas. In complex environments such as abandoned coal mining zones or coal gangue heaps, the deployment of a wireless sensor network (WSN) composed of numerous micro-sensing nodes allows for the acquisition of high spatial resolution radon concentration data in real time, thereby enhancing the accuracy of coal fire localization and extent determination.

(3) Evolution towards high performance and intelligence of portable radon detection instruments

Portable radon detection instruments are gaining increasing prominence in the field of isotope-based radon measurement, with the pursuit of high performance and intelligent capabilities emerging as critical developmental directions. Enhanced performance is primarily characterized by improvements in detection sensitivity, reduction of the detection limit, and expansion of the measurable concentration range. The latest generation of portable radon detectors incorporates advanced detector materials and fabrication techniques, integrated with optimized electronic architectures, thereby enabling the reliable detection of weak radon signals at extremely low concentrations while maintaining measurement accuracy under high-concentration conditions.

The advancement of intelligent technologies has significantly enhanced the automation and adaptability of portable radon measurement instruments. Modern devices are equipped with functionalities such as automatic calibration, fault diagnosis, and intelligent alarm systems, enabling rapid on-site identification of anomalies and the implementation of appropriate corrective actions. Integrated with built-in Global Positioning System (GPS) and Geographic Information System (GIS) modules, these instruments can record real-time location data, correlate measurement results with spatial information, and generate intuitive radon concentration distribution maps. Furthermore, the support for wireless connectivity with mobile devices or cloud platforms facilitates remote data transmission and sharing, allowing users to access measurement results and perform analysis and decision-making at any time and from any location [8].

(4) Technological breakthroughs propelled by interdisciplinary convergence

The deep integration of isotope radon measurement technology with multiple disciplines, including geochemistry, geomechanics, and computational mathematics, is driving significant technological breakthroughs that expand the application scope and enhance the effectiveness of radon detection. Collaborative research with geochemistry enables precise identification of radon sources, detailed analysis of the occurrence of radioactive elements in coal seams and surrounding rocks, and the activation and release mechanisms of radium. This integration facilitates the construction of element geochemical behavior models, which can prospectively identify potential high-radon production zones, thereby optimizing targeted detection strategies and fundamentally improving detection efficiency. Coupling with geomechanics elucidates the influence of geological structures on radon migration pathways, quantifies the effects of fault and joint connectivity as well as stress field distribution on radon flow, and employs the discrete element method to simulate fracture surface undulations and permeability variations. These capabilities allow for the predictive identification of preferential migration channels and support the scientific planning of detection routes and sampling density, thereby enhancing adaptability in complex geological settings [25, 26]. Integration with computational mathematics leads to the development of efficient numerical simulation tools that address inversion challenges, enabling the inference of coal seam temperature and stress from radon measurement data. Combined with forward modeling, this approach reconstructs the characteristics of coal fire systems and provides robust algorithmic support for accurate fire source localization and dynamic monitoring of fire progression. The interdisciplinary integration permeates the entire R&D process of isotope radon measurement technology, where the convergence and interaction of diverse theoretical frameworks and methodologies give rise to comprehensive technical solutions. This multidisciplinary synergy significantly strengthens the core competitiveness of radon measurement technology, promotes its continuous expansion and enhancement, and sustains its leadership in driving industry-wide technological innovation.

8.3 Isotope Detection Equipment

Isotope radon measurement technology is founded on the detection and quantification of alpha particles emitted by Rn-222 and its characteristic short-lived decay progenies, primarily Po-218 and Po-214, enabling quantitative assessment of radon activity concentration in environmental or material media. The implementation of this technology depends on highly precise physical detection mechanisms and advanced electronic systems, with equipment performance critically influencing the accuracy and reliability of the acquired data.

8.3.1 Classification and Underlying Principles of Radon Detection Equipment

Radon isotope detection equipment is primarily classified into two distinct categories—active and passive—based on fundamental differences in sampling and measurement methodologies. Active radon detectors employ mechanical devices, such as air pumps, to actively extract air or water samples, followed by degassing processes that introduce the sample into specially engineered detection chambers. Within these chambers, alpha particles generated from the decay of radon and its progeny are measured in real-time or near real-time. The resulting measurements are typically reported in units of Bq/m^3 for gaseous environments or Bq/L for aqueous media. Prominent active detection technologies include the electrostatic collection method, pulse ionization chamber method, scintillation chamber method, and the active sampling mode of semiconductor alpha spectrometry [27, 18].

The electrostatic collection method (exemplified by the RAD7 detector) operates by introducing radon-laden air into an ionization chamber. Positively charged Po-218 ions, generated through the radioactive decay of Rn-222, are directed under the influence of a high-voltage electrostatic field toward the surface of a solid-state alpha detector—commonly a gold–silicon surface barrier detector or a passivated ion-implanted planar silicon detector [28, 29]. The subsequent decay of Po-218 (half-life of 3.1 min) emits 6.00 MeV alpha particles, which are then detected and quantified. Advanced instruments may also detect the 7.69 MeV alpha particles emitted during the decay of Po-214 (half-life of 164.3μs), thereby enabling differentiation between radon (Rn-222) and thoron (Rn-220). This method offers high sensitivity and rapid response, with stable measurement results typically achieved within tens of minutes to several hours, and a detection limit as low as 2 Bq/m^3. The pulsed ionization chamber method employs alpha particles emitted by radon and its decay progeny within a sealed chamber to ionize the internal gas medium. Under the influence of an applied electric field, the resulting electron–ion pairs migrate toward the respective electrodes, generating transient current pulses. These signals are amplified and discriminated using a high-gain, low-noise electronic system. The method is characterized by minimal background interference and exceptional measurement stability, making it particularly suitable for high-precision laboratory applications and the establishment of reference standards. In the scintillation chamber method, radon-containing air is introduced into a sealed chamber whose inner walls are coated with a ZnS (Ag) scintillation material. Alpha particles emitted during the decay of radon and its progeny strike the scintillator, inducing the emission of fluorescent photons. A photomultiplier tube (PMT) detects and converts the optical signal into electrical pulses, which are then counted. Particular attention must be given to the chamber's "memory effect" and the potential influence of ambient humidity on measurement accuracy.

The active sampling mode of semiconductor alpha spectrometry involves the continuous aspiration of radon-containing air through a high-efficiency filter membrane, which captures radon progeny such as Po-218, Pb-214, and Bi-214. The filter is subsequently transferred to a vacuum alpha spectrometer equipped with a high-resolution silicon detector, such as a PIPS (Pulsed Ionization Photodiode Spectrometer) detector, for direct measurement of the characteristic alpha energy peaks of Po-218 (6.00 MeV) and Po-214 (7.69 MeV). This method provides superior nuclide discrimination and yields comprehensive radiation data, supporting detailed isotopic analysis [30, 24, 31].

8.3.2 Calibration and Quality Control of Detection Equipment

Calibration of isotope radon detection equipment is a critical procedure to ensure the accuracy and reliability of measurement results. The calibration process comprises two fundamental components: energy calibration and efficiency calibration. Energy calibration aims to characterize the detector's response to α particles across different energy levels. This is achieved through the use of reference radioactive sources with precisely known emission energies, such as Ra-226 and Rn-222, which emit α particles at approximately 4.78 MeV and 5.49 MeV, respectively. During the calibration procedure, a standardized radon source is positioned under identical measurement conditions as the detector. The detector is then activated, and the resulting signals are recorded. The acquired data are subsequently processed using dedicated software to generate an energy-pulse height distribution curve. Characteristic peaks corresponding to specific α particle energies are identified on this curve. By correlating the positions of these peaks with the known energy values, the detector's energy response function can be accurately derived. This function is subsequently applied to correct energy-related deviations in measurement data, thereby ensuring the precision and traceability of quantitative radon activity assessments.

Efficiency calibration is performed to evaluate the detector's ability to accurately capture and register alpha particles emitted by the source. This process typically involves the use of a radon source with a precisely known activity, which is positioned under geometric conditions identical to those used during routine measurements. A series of repeated measurements are carried out using the detector, and the resulting count rates are recorded for each trial. To account for the temporal variation in source activity, the radioactive decay formula is applied:

$$N = N_0 e^{-\lambda t} \tag{8.2}$$

where, N denotes the remaining activity, N_0 represents the initial activity, λ is the decay constant, and t refers to the elapsed time. The detection efficiency is then calculated as the ratio between the recorded count rate and the true activity of the calibrated radon source [20, 15]. This derived efficiency value is subsequently used to adjust actual measurement data, compensating for alpha particles that were not detected due to instrumental limitations.

(1) Construction and implementation of quality control system

The quality control system is systematically integrated throughout the entire life-cycle of detection equipment, spanning from the manufacturing phase to on-site measurement operations. During the production stage, each detector must undergo rigorous quality inspection and performance testing, including material purity analysis, process control monitoring, and evaluation of key technical specifications. For semiconductor detectors, specific tests are conducted to assess crystal structural integrity, electrical performance stability, and the quality of surface passivation, ensuring compliance with high-precision radon measurement standards prior to deployment. In field applications, periodic on-site calibration using certified radon sources and radon gas with known activity concentrations constitutes a critical component of quality assurance. This procedure not only verifies the measurement accuracy and long-term stability of the equipment but also facilitates the early detection and correction of measurement deviations or anomalies. During on-site calibration, the standard radon source is positioned under environmental conditions identical to those of the actual measurement location, and data are collected using the detection equipment. These results are then compared against reference values to evaluate the measurement performance of the equipment. If the deviation is found to exceed the allowable range, adjustments or repairs to the detection equipment are required. Furthermore, inter-comparison of measurement data from multiple detectors at the same location enables the assessment of consistency and reliability across the instrument fleet. Such comparative analysis supports the identification of potential performance issues in individual units, thereby enabling timely corrective interventions.

The quality control system also encompasses rigorous review and management procedures for measurement data. All collected data must undergo a dual-review process: an initial examination is conducted by on-site personnel to verify data completeness and logical consistency, followed by a comprehensive evaluation performed by qualified data analysts. During the second review stage, statistical techniques are employed to detect outliers, which may then be either excluded or corrected based on established protocols. Detailed documentation of the time, location, and environmental conditions associated with each anomaly is required to support root cause analysis and facilitate targeted process improvements. Only data that has successfully passed these quality verification steps are incorporated into the final analytical reports, thereby ensuring the credibility and reliability of the measurement outcomes. Furthermore, the quality control framework mandates routine maintenance and servicing of detection equipment, including cleaning of detector surfaces, inspection of electronic connections, and updating of software systems, to ensure continuous operational reliability. Comprehensive maintenance records are systematically archived and serve as essential documentation within the quality control framework.

(2) Practical application cases and challenges of calibration and quality control

In practical applications, the calibration and quality control of isotope radon detection equipment encounter multiple technical and operational challenges. A typical example is the detection of fire sources in coalmines, where the complex and harsh environmental conditions—including high humidity, excessive dust, and strong electromagnetic interference [9, 32]—can significantly affect the performance and measurement accuracy of detection systems. To mitigate these challenges, comprehensive calibration and quality assurance protocols have been implemented. Prior to each underground measurement, portable standard radon sources are employed for on-site calibration, ensuring the reliability of the equipment under adverse environmental conditions. Additionally, repeated measurement points are established at various locations to enable cross-validation of data collected by different detectors, thereby assessing the consistency and reliability of the measurement results. In a specific fire source detection project conducted in a coalmine, the implementation of rigorous calibration and quality control procedures enabled the successful identification of multiple concealed fire sources. These findings provided precise guidance for subsequent fire suppression operations. The spatial correlation between the detected anomalies and the actual fire source locations reached over 95%, clearly demonstrating the critical role of systematic calibration and quality control in enhancing the accuracy and effectiveness of isotope radon detection equipment. Despite these achievements, several challenges remain difficult to fully eliminate. For instance, dynamic fluctuations in environmental conditions may lead to performance drift in detectors, prolonged usage can result in gradual degradation of detector sensitivity, and variations in technical proficiency and operational practices among personnel may introduce measurement uncertainties. To address these issues, continuous optimization of calibration and quality control methodologies is essential. Strengthening operator training to enhance technical competence and quality awareness is also crucial. Concurrently, the development of more robust and reliable detection systems should be prioritized to minimize the influence of environmental and human factors on measurement outcomes. Through sustained efforts and technological advancements, the calibration and quality control standards of isotope radon detection equipment are expected to continually improve, thereby supporting its broader application in critical fields such as the detection of spontaneous combustion fire sources in coal seams.

(3) Continuous improvement strategies for calibration and quality control

To ensure the long-term stability and measurement accuracy of isotope radon detection equipment, it is essential to establish a comprehensive and systematic continuous improvement mechanism. This mechanism encompasses the ongoing refinement of calibration methodologies, periodic evaluation of quality control procedures, and the timely integration of emerging technologies. With rapid advancements in science and technology, novel calibration techniques—such as laser spectroscopy-based calibration and quantum sensor calibration—are continuously being developed. These innovations offer superior precision, enhanced stability, and greater environmental adaptability, thereby significantly improving the overall calibration performance of detection systems. It is crucial to closely monitor the evolution of these technologies and incorporate them into the calibration workflow in a timely manner, based on operational requirements and technical feasibility. Periodic assessment and optimization of quality control procedures are also vital. By conducting statistical analysis of quality control data, potential vulnerabilities and systemic inefficiencies within the process can be identified and addressed through targeted corrective actions. Strengthening collaboration with research institutions and academic organizations through industry – university – research cooperative projects enables the joint exploration and resolution of key technical challenges in calibration and quality control. Through resource integration and complementary expertise, continuous innovation and technological advancement in this field can be effectively promoted. Furthermore, active participation in domestic and international academic and technical exchange activities is essential for staying informed about the latest industry developments and cutting-edge technologies. Such engagement provides valuable insights and references that support the continuous enhancement of calibration and quality control practices.

8.3.3 Enhancement of the Portability and Reliability of Detection Equipment

Portability represents a critical developmental priority for modern isotope radon measurement equipment, significantly broadening its applicability across diverse operational environments. The enhancement of portability primarily centers on three key aspects: miniaturization, weight reduction, and the integration of ergonomic design principles.

The foundation of miniaturization design is high-level integration. Modern detection systems achieve substantial size reduction by consolidating key functional modules—such as detectors, power supplies, signal processing units, and data storage—into a compact and unified enclosure. The latest generation of portable semiconductor radon detectors employs system-on-chip (SoC) technology, wherein data acquisition, processing, and storage functionalities are integrated onto a single chip. This advancement reduces the overall device volume to approximately one-tenth that of conventional systems. Concurrently, miniaturized detector technologies—such as micro-scale semiconductor detectors and compact low-temperature liquid scintillation detectors—are utilized. These detectors maintain high performance while achieving dimensions on the order of centimeters or even millimeters, thereby further advancing the miniaturization of radon detection equipment [33, 27].

Lightweight design aims to minimize equipment mass in order to enhance portability and facilitate field operations. A critical factor in achieving effective weight reduction is the innovation of shell materials. Modern detection equipment employs advanced high-strength, lightweight materials such as engineering plastics, aluminum alloys, and particularly carbon fiber composites. These materials offer superior mechanical strength and corrosion resistance while significantly decreasing overall equipment weight. For example, carbon fiber composite materials exhibit a density that is merely one-third that of conventional metallic materials, yet they demonstrate enhanced resistance to impact and pressure loads. Furthermore, optimization of the internal structural design—through the elimination of redundant components and the incorporation of lightweight connectors and fasteners—contributes substantially to weight reduction. As a result, single-person portable radon detection systems can now be engineered to weigh less than 5 kg.

Ergonomic design emphasizes the enhancement of operational convenience and user comfort. It incorporates intuitive and user-friendly interfaces, such as color touchscreens and clearly labeled control buttons, enabling rapid deployment and minimizing the time required for on-site configuration and troubleshooting. For example, certain portable radon detection systems are equipped with graphical user interfaces that visually present measurement data and real-time device status. Operators can perform tasks such as selecting measurement points and retrieving data with simple touchscreen interactions. Concurrently, the handle and strap configurations are ergonomically optimized to align with natural hand positioning, thereby reducing operator fatigue during extended use and improving maneuverability in challenging environments. Some models feature adjustable handle mechanisms specifically engineered to accommodate a range of hand sizes, thereby enhancing grip stability and overall handling performance.

Hardware optimization serves as a critical approach for enhancing the performance of portable radon detection equipment. By refining the detector design, both sensitivity to radon signals and resistance to external interference can be significantly improved. The implementation of advanced semiconductor materials and structural configurations—such as high-mobility transistor (HMT) technology—enhances the detector response speed and measurement accuracy for alpha particles. Concurrently, advancements in packaging technology, including the adoption of gas-tight and microcavity packaging methods, mitigate the adverse effects of environmental factors, thereby improving the detector long-term stability and operational reliability.

Power management constitutes another essential aspect of hardware optimization. Utilizing high-performance battery technologies, such as lithium-polymer and solid-state batteries, offers advantages in terms of higher energy density and extended service life, ensuring sustained power supply for field operations. Furthermore, optimization of the power management system through the integration of intelligent power control chips enables dynamic regulation of power output based on operational demands. This results in reduced overall power consumption and prolonged battery endurance. For instance, an intelligent power management system can automatically enter a low-power state during standby mode and rapidly restore full power output during active measurement, thereby ensuring efficient and adaptive performance across various operational modes.

The system self-diagnostic functionality leverages integrated sensors and advanced diagnostic algorithms to continuously monitor critical operational parameters, such as temperature, humidity, and voltage, in real time. Upon detecting any anomalies, the system is capable of promptly issuing alerts and initiating appropriate protective actions, including automatic shutdown or transition to a safe mode, thereby preventing continued operation under abnormal conditions that may result in data inaccuracies or hardware failure. For example, certain portable radon measurement instruments are fitted with internal temperature and humidity sensors. In the event of elevated internal temperatures or excessive humidity levels, the system can automatically activate a cooling fan or trigger a humidity alarm to notify the operator for timely intervention.

To further improve the reliability of portable radon detection equipment, modern systems are integrated with real-time monitoring and remote management capabilities. Equipped with embedded wireless communication modules, these devices can continuously transmit measurement data and operational status to a centralized remote monitoring station. Authorized personnel can access device status, receive alert notifications, and execute remote control commands through dedicated mobile applications or web-based platforms. For example, in the detection of spontaneous combustion zones within coal seams in remote locations, operators can initiate device operations, adjust measurement settings, and retrieve real-time data from their office workstations via network connectivity, thereby enabling comprehensive remote management. This remote functionality not only enhances operational efficiency but also facilitates prompt responses to equipment malfunctions, significantly reducing system downtime. Additionally, the real-time monitoring feature supports long-term performance tracking and evaluation. By collecting key operational metrics such as measurement frequency, fault incidence, and battery performance, a detailed performance profile can be constructed for each unit. Leveraging big data analytics, these datasets can be processed to identify potential failure patterns, enabling predictive maintenance and proactive servicing, which ultimately improves system reliability and extends equipment lifespan. For instance, by analyzing historical battery charge–discharge cycles, the remaining useful life of the battery can be estimated, allowing for timely replacement of degraded units to ensure uninterrupted device operation.

8.3.4 Intelligence and Integration of Detection Equipment

(1) Intelligent data collection and analysis

The intelligent data acquisition system constitutes a core component of modern isotope-based radon measurement equipment, providing the system with capabilities for autonomous learning and environmental adaptability. Leveraging advanced sensor technologies, the system is capable of real-time detection of subtle variations in environmental parameters, with rapid data processing performed by an embedded processor. Key environmental parameters monitored include temperature, humidity, atmospheric pressure, and background radiation levels, all of which exert significant influence on the emission and migration behavior of radon gas. For example, elevated temperatures generally accelerate the release of radon from geological media such as soil and coal seams, whereas fluctuations in humidity may affect the diffusion coefficient of radon in ambient air. Continuous real-time monitoring of these parameters, coupled with appropriate compensation calculations, can substantially improve the accuracy of radon measurement outcomes. Furthermore, the integration of multi-source data fusion techniques enhances the overall intelligence and analytical capabilities of radon measurement systems. By synthesizing data from diverse sensor modalities, such as radon concentration readings, ground-penetrating radar imagery, and infrared thermal imaging, the system can generate a more comprehensive and spatially detailed model of fire source distribution. Ground-penetrating radar facilitates the identification of subsurface structural anomalies, while infrared thermal imaging provides insights into thermally anomalous zones. The integration and joint analysis of these datasets with radon concentration measurements enable more precise localization of fire sources, as well as improved assessment of fire intensity and progression trends.

The application of artificial intelligence algorithms in data processing constitutes a crucial dimension of intelligent system functionality. Machine learning methodologies, such as Support Vector Machines (SVM) and Random Forest algorithms, are capable of performing automated classification and regression analysis on large-scale radon measurement datasets. These techniques enable the extraction of latent patterns and association rules within the data, thereby facilitating precise forecasting of radon concentrations and effective anomaly detection. In the domain of time-series data analysis, deep learning architectures—particularly Long Short-Term Memory networks (LSTM) and Generative Adversarial Networks (GAN)—demonstrate considerable advantages. LSTM networks are particularly effective in capturing temporal dynamics of radon concentration fluctuations, supporting predictive modeling of fire source evolution. Meanwhile, GANs can be utilized to generate synthetic datasets that enhance model generalization and robustness. In the context of detecting spontaneous combustion fire sources within coal seams under complex geological conditions, intelligent data acquisition and analysis systems integrate real-time multi-source monitoring with data fusion techniques and advanced AI algorithms. This integrated approach enables accurate localization of fire sources and reliable prediction of fire progression, thereby establishing a scientifically grounded basis for timely and effective fire suppression measures [34, 15].

(2) Hardware implementation of equipment integration

The integrated design of detection equipment is primarily manifested through a multi-functional and unified hardware architecture. Contemporary radon detection systems incorporate a variety of sensors and measurement modules—including alpha particle detectors, temperature sensors, humidity sensors, barometric pressure sensors, and GPS positioning modules—within a compact enclosure, thereby constituting a highly capable integrated detection platform. The GPS module enables precise recording of geographic coordinates for each measurement point, serving as foundational data for subsequent analysis and GIS-based modeling of fire source locations. For instance, in large-scale coalmine goaf area surveys, this integrated hardware architecture allows for concurrent acquisition of radon concentration levels, environmental parameters, and geospatial information, significantly enhancing both detection efficiency and data completeness. A key characteristic of this integration is the modular design of the hardware structure. Functional modules are interconnected via standardized interfaces, supporting plug-and-play functionality and facilitating system scalability and maintenance. As an example, the addition of gas composition analysis capability can be achieved simply by integrating the corresponding gas sensor module into the system, without necessitating extensive modifications to the entire apparatus. This modular approach not only improves the adaptability and expandability of the equipment but also contributes to reduced maintenance costs and operational downtime.

High-performance communication modules represent a critical component of integrated hardware systems. Wireless communication technologies, including 4G/5G, Wi – Fi, and Bluetooth, enable real-time data transmission between detection equipment and remote monitoring centers. In environments lacking conventional network coverage, such as remote areas or underground mining operations, the satellite communication modules serve as a dependable alternative for continuous connectivity. Furthermore, wired communication interfaces such as Ethernet and USB facilitate secure and stable data exchange with local computing devices or peripheral systems, ensuring both the integrity and confidentiality of transmitted information. Within the complex and challenging underground conditions of coalmines, an integrated hardware architecture allows radon detection systems to be seamlessly interfaced with existing mine safety monitoring infrastructure, thereby enabling data interoperability and cooperative analysis. For example, upon detecting anomalous radon concentrations, the system can activate alarms within the gas monitoring network, promptly alerting personnel to implement necessary safety protocols. Simultaneously, integration with the mine ventilation system enables dynamic adjustment of airflow parameters in response to fluctuating radon levels, effectively mitigating health risks for miners. This integrated hardware framework not only enhances the overall safety management capabilities of coal mining operations but also provides essential technical support for the advancement of intelligent and automated mine construction [35].

8.3.5 On-Site Application and Environmental Adaptability of Detection Equipment

The performance of isotope radon detection equipment in complex environmental conditions serves as a critical indicator of its operational reliability. In high-temperature environments, such as underground coalmines or near-surface geothermal zones, electronic components and detector materials may experience thermal stress, potentially leading to performance degradation or functional failure. To mitigate these risks, modern radon detection systems incorporate advanced thermal management technologies and high-temperature-resistant materials. For example, the integration of high thermal conductivity metal enclosures and heat sinks, in conjunction with internal heat pipe cooling systems, enables efficient heat dissipation and ensures stable equipment operation under elevated temperature conditions. Certain advanced detectors also feature specialized cooling mechanisms, such as thermoelectric cooling modules or liquid circulation systems, which further reduce operational temperatures and maintain functionality in extreme heat. Low-temperature environments also pose significant challenges, particularly in cold open-pit mining areas or high-altitude regions. Exposure to low temperatures may result in reduced battery efficiency, material embrittlement, and lubricant solidification. To address these issues, radon detection equipment is typically equipped with wide-temperature-range lithium-ion batteries and supercapacitors, which maintain high discharge efficiency even in sub-zero conditions, ensuring continuous and reliable power supply. Simultaneously, equipment housings are fabricated from materials with excellent low-temperature resilience, such as specialized alloys or fiber-reinforced composites, while internal mechanical components utilize low-temperature lubricants to ensure smooth operation in extreme cold. Furthermore, certain systems are designed with integrated preheating capabilities that allow for the warming of critical components prior to startup, thereby enabling rapid deployment and stable performance in low-temperature environments.

High-humidity environments present significant operational challenges, particularly in southern coalmines with inherent moisture or in abandoned mining areas adjacent to water sources. Elevated humidity levels may result in condensation forming on detector surfaces, electrical short circuits, and disruptions to signal transmission. To address these issues, radon detection equipment incorporates advanced moisture protection mechanisms, including waterproof breathable membranes, precision sealing gaskets, and fully integrated sealing structures. For example, high-performance rubber sealing rings are strategically installed at casing seams and interface points to ensure compliance with IP67 or IP68 environmental protection standards. Internal circuit boards are treated with specialized moisture-resistant coatings and encapsulated using hermetic sealing technologies to further improve humidity tolerance. Certain detector models also integrate desiccant chamber systems that periodically absorb internal moisture, maintaining a consistently dry internal environment [36]. Furthermore, signal transmission systems employ shielded cables and industrial-grade waterproof connectors to minimize humidity-induced interference, thereby ensuring the accuracy, consistency, and reliability of data acquisition under adverse environmental conditions.

In practical applications, the field deployment of radon detection equipment presents numerous technical challenges, yet has also led to the accumulation of substantial operational experience. A representative case is the detection of underground fire sources in coalmines, where the complex and demanding environmental conditions—such as high humidity, excessive dust, strong electromagnetic interference, and confined spatial layouts—impose stringent requirements on both the performance and usability of detection systems. In a fire source detection project conducted at a large-scale coalmine, portable isotope-based radon detection equipment was deployed for underground operations. To accommodate the narrow tunnels and challenging terrain typical of underground mining environments, the equipment was engineered with a modular, disassemblable structure, enabling efficient transportation via mine carts and manual handling. The system was further equipped with explosion-proof detectors and intrinsically safe power modules to ensure operational safety in high-gas-concentration environments. Throughout the detection process, routine calibration and field data validation procedures were implemented to guarantee the accuracy and reliability of measurement results. Additionally, the equipment's real-time data transmission capability enabled continuous transfer of detection data to the surface control center, facilitating remote monitoring and real-time analysis, thereby providing critical technical support for the timely development of fire suppression strategies.

Fire source detection in open-pit coalfields presents unique operational challenges, including exposure to strong winds, frequent sandstorms, and extreme diurnal temperature fluctuations. In a case study involving the detection of spontaneous combustion fire sources within a large open-pit coalfield, radon detection equipment was deployed featuring a robust protective enclosure and an advanced dust filtration system, effectively mitigating the adverse effects of airborne particulates. To address power supply limitations in remote locations, the system integrated a solar-powered charging solution, ensuring continuous operation without reliance on conventional grid infrastructure. The efficiency of on-site detection was significantly enhanced through the optimization of measurement point layouts, combined with GPS-based positioning and automated measurement protocols, which collectively reduced manual labor input and shortened data acquisition cycles. Furthermore, the system's intelligent data processing capabilities enabled real-time analysis and interpretation of collected data, facilitating rapid identification of fire source locations and predictive modeling of fire propagation trends, thereby providing critical time advantages for fire suppression efforts. To further enhance the performance of radon detection equipment in complex environmental conditions, continuous improvements in both design and functionality are essential. First, interdisciplinary collaboration with fields such as materials science and electronic engineering should be strengthened to foster the development of novel materials and technologies tailored for extreme environments [37, 34, 38]. Examples include high-temperature-resistant and corrosion-resistant detector materials, advanced high-capacity battery systems, and efficient thermal management solutions. Second, equipment design and operational workflows should be iteratively refined based on feedback from field applications. For instance, modular and customizable configurations can be developed to meet the specific requirements of diverse operational scenarios, thereby enhancing equipment flexibility and environmental adaptability. Concurrently, comprehensive training and technical support for operators should be emphasized to improve their proficiency and emergency response capabilities under adverse conditions, ensuring the reliable, efficient, and accurate deployment of radon detection systems in real-world applications.

Future on-site applications of isotope radon detection equipment are expected to evolve toward greater intelligence, automation, and network integration. One key development direction involves the integration of robotic systems and unmanned aerial vehicle (UAV) platforms to enable autonomous inspection and remote operation. For example, in hazardous or inaccessible areas, UAVs equipped with radon detection modules can perform aerial monitoring, with significantly enhancing detection efficiency as minimizing risks associated with manual fieldwork. Another critical advancement lies in the implementation of Internet of Things (IoT) technologies and big data analytics to establish an intelligent fire source monitoring system. By deploying a distributed network of radon detection nodes across high-risk zones—such as abandoned mining areas and coal gangue stockpiles—a real-time, dynamic surveillance framework can be constructed. These nodes are capable of autonomously acquiring and transmitting environmental data, which is then aggregated and analyzed on a centralized cloud computing platform. Through advanced big data analytics, the system can achieve capabilities such as early fire source detection, precise localization, and predictive trend modeling, thereby offering comprehensive technical support for coalmine safety management and production risk mitigation.

8.4 Isotope Detection Application Cases

In the treatment of spontaneous combustion fire area of coalmines, the detection of the precise location of spontaneous ignition sources in coal seams by isotope radon measurement is the key technology for fire area control, and the collection of temperature information in the fire area provides a reliable basis for fire extinguishing schemes and fire extinguishing parameters. However, the radon anomaly formed by the ignition source on the surface is quite weak, and various factors on the surface have a certain influence on the radon measurement results. The factors affecting radon measurement are studied through practical testing methods, in order to highlight useful anomalies, improve the application effect of radon measurement, and eliminate or reduce various interference factors [3, 39, 19].

8.4.1 *Overview of the Fire Area*

The No. 1 shaft of Dahuangshan Coalmine is a 150,000-ton per year that has been renovated and expanded into a 600,000-ton per year production, which is developed by inclined shafts, and the current production level is + 780 level, with a mining depth of 201 m. The whole well is now mined with two layers of coal, from top to bottom are medium and large troughs and eight-foot troughs. The coal quality is No. 45 gas fertilizer coal, the spontaneous ignition period of the coal seam is 3–6 months (the shortest is one month), and the coal dust is explosive, with an explosion index of 39%.

The 780 horizontal east wing is 2,000 m long, with a comprehensive mining area and four artillery mining areas. The strike length of the fully integrated mining area is 960 m. The western boundary of the mining area is the riverbed coal pillar stop mining line, the eastern boundary is the ninth mining area (blasting mining area), the upper part is the goaf of the medium and large groove coal seam in the six-level 4–8 mining area, and the lower part is the unmined solid coal seam.

The 780 comprehensive surface is arranged in the medium and large trough coal seam, with a strike length of 950 m, an oblique length of 75 m, an inclination angle of 120, and a mineable coal seam thickness of 23 m in the working face. A total of three main roadways are arranged, + 780 transportation along the roof of the coal seam, + 796 return air along the bottom of the coal seam, and + 810 gas discharge lane along the roof of the coal seam.

At present, the upper part of the working surface is the position of the goaf of the + 820 horizontal seventh mining area, and there is an oblique length of 21 m isolated coal pillar between the comprehensive release face. The upper horizontal coal mining method is warehousing: 10 silos are arranged in a mining area, the width of the silo is 8 m, the two silos are a group, the coal pillars between the silos are 6 m, the coal pillars between the groups are 8 m, and each group is constructed with an explosion-proof seal (explosion-proof wall thickness 0.5 m, filled with loess 3 m). According to the analysis of the detection gas concentration in the goaf, it was found that ethylene and acetylene gases had appeared in the extracted gas samples, and it was speculated that there might be a fire zone in the upper goaf.

At 11:20 on December 2, 2005, the CO probe in the + 780 horizontal gas emission lane exceeded the limit with a concentration of 26 ppm. By 15:00, the CO reached 46 ppm, and by 21:05, the CO reached 168 ppm. At 2:25 on December 3, the CO was 7200 ppm, and ethylene gas appeared, and at 15:07, the ambulance team conducted reconnaissance on + 780 haulage roadway, + 796 return air roadway, and + 810 watts of alleys, and after opening No. 5 up the mountain, it was found that the smoke in + 810 watts was very large, and the personnel could not enter, so they changed from level 5 to − 14° alley reconnaissance, and found that smoke was coming out, and CH_4 was detected to be 1.2% and CO was 1700 ppm. At 10:20 on December 7, the ambulance team reconnaissance situation: CH_4 2.6% at the entrance of the alley at − 14°, CO at 0.5%, temperature 25 °C, CH_4 at the wind measurement station at − 14° is 20%, CO is 1.7%, temperature is 30 °C, visibility is 3 m, there is white smoke.

On March 7, 2006, + 796 was ventilated, and the CO exceeded the limit at − 16° and + 796 corresponding to the uphill area of No. 5, reaching 500ppm.

8.4.2 Measuring Device and Measurement Method

The measuring device consists of KZ-D02α cup radon meter, α probe cup and global positioning system (GPS).

(1) KZ-D02α cup radon meter

The radon measurement instrument uses the KZ-D02α cup radon meter, and the physical diagram is shown in Fig. 8.8. This instrument was chosen because of its advantages:

(1) High sensitivity. It directly uses the inner surface of the α cup to collect the daughter body of radon, and the collector area is large, and the useful information is used to the fullest.
(2) The method is simple. No need to do any treatment of the α cup when burying and picking up the cup.
(3) Flexible cup burial. The upper limit of the burial time of the α cup is not limited, which is especially conducive to the work of large areas and long sections.
(4) High work efficiency. The calffar ratio with electrostatic α can be more than doubled.
(5) Multiple parameters can be obtained. Due to the high sensitivity of the α cup method, the α cup can be measured for the second time, so as to obtain four parameters: 222Rn + 220Rn, 222Rn, 220Rn, and 222Rn/220Rn, which is conducive to the interpretation of geological phenomena.
(6) Many functions. The ionization chamber α cup method can α measurement of α cups, α card, activated carbon, soil radium, 218P0, 210P0 and other methods.
(7) Simple structure. The atmospheric air pulse ionization chamber is adopted, which does not need to be used away from light, and is easy to discharge pollutants.
(8) Low cost. There is no need to consume anything other than one battery during work.
(9) Power consumption is less than 135mW, weight 2 kg, volume 250 × 140 × 170 mm.
(10) It can be connected with the computer recording system to form a micro computerized instrument to realize automatic data recording, processing and mapping.

The main technical indicators of KZ-D02α radon meter are:

(1) Instrument background: not greater than n counts/minute.

Fig. 8.8 Physical image of KZ-D02α cup radon meter

   (2) Counting capacity: 1–9999.
   (3) Counting time: 1 min, 2 min, 3 min, …… Any minute.
   (4) Stability:

      Statistical: keep the measurement conditions unchanged, and the continuous measurement count values (greater than ten readings) conform to the law of radio statistical fluctuations.

      Accuracy: 10 measurements of the same sample with a relative standard error of no more than ± 20%.

      Stability: 8 h of continuous work, the change of count value is less than ± 10% after taking the logarithm.

   (5) Working environment temperature: 0 °C– + 40 °C.
   (6) Humidity in the working environment: less than 75%.

(2) α probe the cup

   The α detection cup burial diagram is shown in Fig. 8.9, α the detection cup is made of high-adsorption material and has a collection area of 12 cm × 8 cm. When testing or measuring, dig a square pit with a width of about 300mm and a depth of about 300–400mm at each numbered measuring point, put the cup mouth down, cover it with plastic sheeting, and then bury it with soil, and write down the dot number and placement time.

   After burying the cup for 4 h, the cup can be taken out of the pit and quickly put into the radon detector ionization chamber for measurement.

(3) GPS

To effectively and accurately determine the location of the underground fire source, it is crucial to accurately compare the upper and lower mines. The

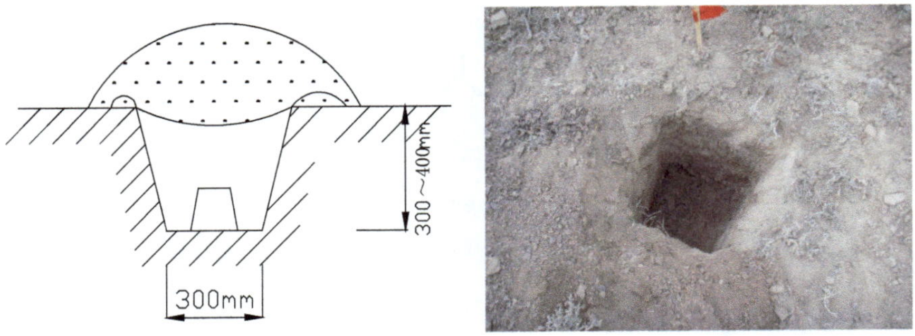

Fig. 8.9 Burial diagram of the α detection cup

control between the well and the bottom is mainly positioned by the global positioning system (GPS). At the same time, to overcome the corresponding position of the well up and down due to the terrain wave when measuring on hills or areas with large terrain fluctuations, it is generally not recommended to use the meter ruler line for fixed-point positioning, and the use of the global positioning system is more scientific and reasonable.

When positioning, first find a correction reference point according to the geological report provided by the mine, and then use the GPS receiver to measure the measurement coordinates of the correction reference point, and use the measurement coordinate measurement value to calculate the deviation of the receiver. After that, select the reference point of the measurement area, the measurement reference point is generally 3 points, and arranged in a vertical direction, and then the measurement point is arranged in this area, the shape is rectangular, square or irregular grid, and it is extended according to the measurement situation during the measurement process. According to the required accuracy, it is generally 20m × 20m, 15m × 15m, 10m × 10m, 5m × 5m. Each measurement point should be pre-numbered for easy data recording and analysis.

8.4.3 Fire Area Detection

(1) Benchmark point selection

According to the topography of Dahuangshan Coalmine (Fig. 8.10), the measurement reference points are selected from the top and bottom of the shaft on the ground, as shown in Fig. 8.11, and three reference points are selected. The reference point 1 is based on the intersection of 10 m beyond the cutting eye of the 780 mining area and the roof transportation along the groove as the reference point, the reference point 2 is 10 m outside the cutting eye of the 780 mining area and the intersection of the stone gate in the 7 mining area as the reference point, and the reference point 3 is 230 m along the extension line of the reference point 2. The three datum points are arranged vertically.

(2) Detection area

The detection area is expanded based on the three selected reference points. In the east – west direction, it extends approximately 315 m westward, bounded by a distance of 10 m beyond the cut-off point of the fully mechanized top-coal caving face in mining area 780. In the north—south direction, it extends approximately 180 m northward, bounded by a distance of 10 m beyond the

Fig. 8.10 Topographic and geomorphological map of Dahuangshan Coalmine

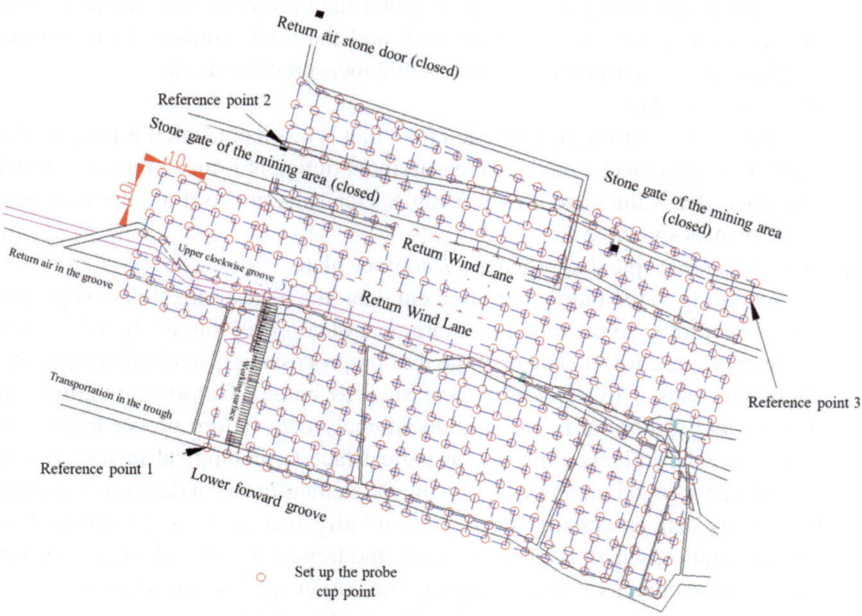

Fig. 8.11 Layout of the measurement site plan

Fig. 8.12 Physical diagram of the probe cup burial

lower chute of the fully mechanized top-coal caving face. The total detection area is 43,000 m^2. The detection area covers all the fully integrated working faces and the goaf of the seventh mining area, and covers part of the goaf of the eighth mining area.

(3) Probe the cup and bury it

According to the measurement selected measurement field, the probe cup is buried according to the grid with a detection accuracy of 10 m × 10 m in the measurement field, the probe cup buried grid diagram is shown in Fig. 8.11, and the overhead view of the probe cup on the ground is shown in Fig. 8.12. In addition, according to the step of radon data preprocessing method, some measurement points are supplemented, and the total number of normal and supplementary measurement points is 492 measurement points.

(4) Measurement plan

After the measuring cup is buried for 4 h, it is taken out, and it is measured in the CD-1α cup detector, and the measurement timing is 1 min, 2 min, and 3 min. The measured values are drawn using Surfer software to draw the measured value plan as shown in Fig. 8.13.

(5) Outlier stereoscopic diagram and its contour plan

According to the radon measurement data, the measurement data is preprocessed according to the radon measurement data processing method described in the previous chapter. The main goal of preprocessing is to eliminate pseudo-anomalies caused by random factors such as human factors or instrumental factors, and to normalize the pseudo-anomaly data before proceeding to the next step. The influence of meteorological factors, geological factors, instrumental factors and human factors on the radon measurement data was corrected by primary data processing. Then, the similarity analysis is used to find out the data anomalies, which are used to guide the trend surface analysis and profile analysis of secondary data processing. The outlier stereoscopic diagram and its contour plan are drawn with Surfer software (Fig. 8.14).

(6) Analysis of the location and cause of the fire

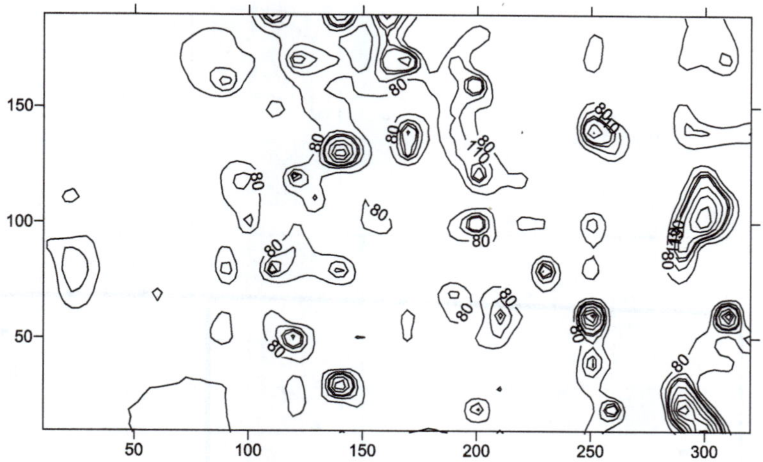

Fig. 8.13 Schematic diagram of measurement results

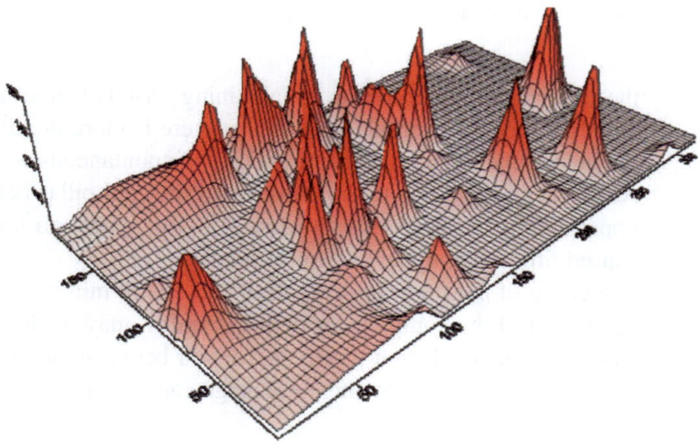

Fig. 8.14 Three-dimensional diagram of outliers

According to the analysis of the detection results, Fig. 8.15 shows that there are the following three ignition high temperature zones in the Dahuangshan No. 1 well.

(1) Comprehensive release goaf.
(2) Goaf of the upper seventh mining area.
(3) + 796 lane confined area.

For the above ignition sites, the cause of the fire may be.

(1) Causes of fire in the goaf of the comprehensive release face: Due to the long installation time of the bracket, the broken top coal is exposed to

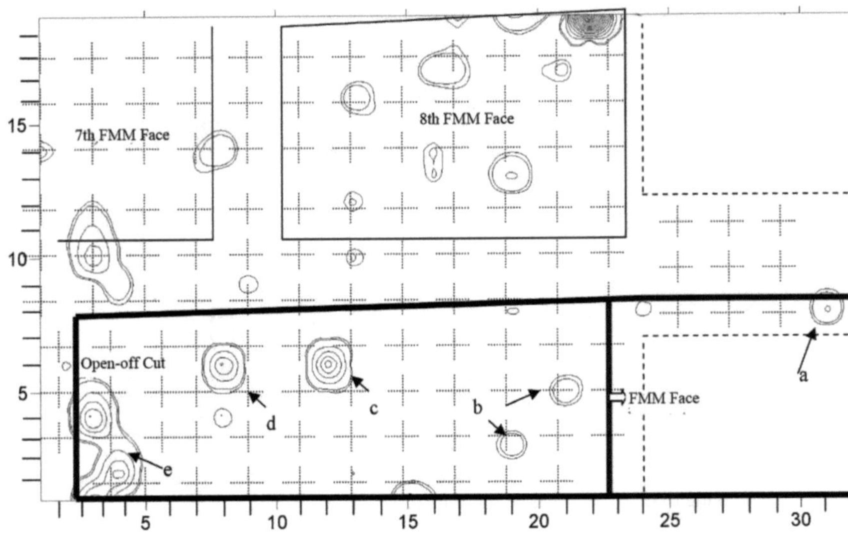

Fig. 8.15 Detection area of fire area

the air for a long time, and the initial mining period of the working face is only mined and not released, so that there is more floating coal in the goaf, which increases the possibility of spontaneous combustion. At the same time, due to the crushing of the coal pillar between the upper and lower levels, the air leakage in the upper and lower goafs caused fires in the upper and lower goafs.

(2) The cause of goaf fire in the seventh and eighth mining areas of the upper part of the comprehensive placement: it may be due to the air leakage of the explosion-proof closed wall between the warehouses, and it is also affected by the air leakage caused by the crushed coal pillars.

(3) Analysis of the cause of fire in the + 796 roadway confined area: The falling area was not carefully treated during the roadway excavation process, forming a high-temperature fire source.

References

1. 文虎,田晴, 赵向涛, 王伟峰. 煤层火灾监测与治理研究现状及发展趋势.煤炭工程, 2021, 53 (09):101–106. Wen, H., Tian, Q., Zhao, X. T., & Wang, W. F. (2021). Research status and development trend of coal fire monitoring and control. *Coal Engineering, 53*(09): 101−106.
2. Chan, Z. K., Zhou, B., Wang, Z. F., Lu, Z. F., Yang, Q. F., & Dong, Z. Y., et al (2023). Long-distance migration law of radon in overburden of abandoned goaf during coal spontaneous combustion. *Journal of environmental radioactivity, 270*:1.1–1.13.

3. 邓军, 李贝, 王凯, 王彩萍. 我国煤火灾害防治技术研究现状及展望. 煤炭科学技术, 2016, 44(10):1–7+101. Deng, J., Li, B., Wang, K., & Wang, C. P. (2016). Research status and outlook on prevention and control technology of coal fire disaster in China. *Coal Science and Technology, 44*(10): 1–7+101.
4. 邬剑明, 刘艳, 周春山. 同位素测氡法在柳湾矿自燃火源位置探测中的应用. 中国煤炭, 2006, (09):37–39+4. Wu, J. M., Liu, Y., & Zhou, C. S. (2006). Application of isotope radon measurement method in the detection of spontaneous combustion fire sources in Liuwan coal mine. *China Coal*, (09): 37–39+4.
5. 文虎,程小蛟,许延辉,马砺,郭军,姜希印,等. 松散煤体自然发火过程氡析出及运移规律. 煤炭学报, 2019, 44(09):2816–2823. Wen, H., Cheng, X. J., Xu, Y. H., Ma, L., Guo, J., & Jiang, X. Y., et al (2019). Research on the status of underground coal fire distribution and development of the detection technology. *Journal of China Coal Society, 44*(09): 2816−2823.
6. 张会见, 何毓俊. 同位素测氡技术在确定煤矿自燃火源位置中的应用. 矿业安全与环保,2005,(02):70–71. Zhang, J. H., & He, Y. J. (2005). Application of isotope radon measurement technology in determining the location of spontaneous combustion sources in coal mines. *Journal of Mine Safety & Environmental Protection*, (02): 70−71.
7. Du, B., Liang, Y. T., & Tian, F. C. (2021).Detecting concealed fire sources in coalfield fires: An application study.*Fire Safety Journal*, (01):103298.
8. Wen, H., Cheng, X. J., Fan, S. X., Xu, Y. H., Ren, S. J., & Gao, J. (2020).A method for detecting hidden fire source in deep mine goafs based on radon measurement and its experimental verification. *Applied Geochemistry*: 104603.
9. 李金良, 张友谊. 浅谈煤矿火源探测技术. 煤矿开采, 2006, (04):86–87+91. Li, J. L., & Zhang, Y. Y. (2006). Research on the application of "radon measurement-magnetic" method to detect spontaneous combustion technology in small coal mines. *Coal Mining* (04): 86–87+91.
10. 朱红青, 袁杰, 赵金龙, 朱晓曼. 地下煤火分布及探测技术现状研究. 工业安全与环保, 2019, 45(12):28–32. Zhu, H. Q., Yuan, J., Zhao, J. L., & Zhu, X. M. (2019). Analysis of detection technology of fire origin in colliery. *Industrial Safety and Environmental Protection, 45*(12): 28−32.
11. 王刚. 测氡法和红外热成像法在小窑区火源位置探测的综合应用. 西安科技大学学报, 2015, 35(05):573–578. Wang, G. (2015). Comprehensive application of radon measurement and infrared thermal imaging in the detection of fire source location in small mine areas. *Journal of Xi'an University of Science and Technology, 35*(05): 573−578.
12. Zhang, Z. Y., Zhang, H. M., Ma, T., Li, J. Y., Wang, Y. J., & Hung, V. T. (2025). Environmental hazards of coal fires: Fracture-driven radon anomalies and migration. *Process Safety and Environmental Protection, 201*, Article 107592.
13. 张俊英, 方熙杨, 王海宾, 王海燕, 姚海飞, 李琨. 基于同位素测氡的煤矿火区圈划方法对比研究. 中国安全科学学报, 2021, 31(01):38–44. Zhang, J.Y., Fang, X.Y., Wang, H.B., Wang, H. Y., Yao, H. F., & Li, K. (2021). Research on fire detection technology based on isotopic radon measurement method. *China Safety Science Journal, 31*(01): 38−44.
14. Wang, K., Dou, K., Zhao, S. C., & Zhang, S. S. (2023). Research on hidden fire source detection in goaf of Hengyi coal mine.*E3S Web of Conferences, 385*(000):5.
15. Cheng, X. J., Wen, H. Xu, Y. H., Fan, S. X., & Ren, S. J. (2021).Environmental treatment technology for complex coalfield fire zone in a close distance coal seam—a case study.*Journal of Thermal Analysis and Calorimetry, 144*(02):563–574.
16. 张俊英,方熙杨,王海宾,王海燕,姚海飞,李琨.基于同位素测氡的煤矿火区圈划方法对比研究. 中国安全科学学报, 2021, 31(01):38–44. Zhang, J. Y., Fang, X. Y., Wang, H. B., Wang, H. Y., Yao, H. F., & Li, K. (2021). Current situation and new technology prospects of hidden fire source location detection of residual coal spontaneous combustion in goaf. *China Safety Science Journal, 31*(01): 38−44.
17. Wu, S., Zhou, B., Wang, J. F., Yang, Q. F., Dong, W. Z., & Dong, Z. Y. (2023). Isotope radon measurement method to identify spontaneous combustion regions in coal gangue hills: Case study for a coal mine in China. *International Journal of Coal Preparation and Utilization, 44*(04), 468–480.

18. 吴玉国, 邬剑明, 王俊峰. 煤矿自燃火区治理温度检测技术及装备的研究应用. 煤矿安全, 2007, (09):4–6. Wu, Y. G., Wu, J. M., & Wang, J. F. (2007). Research and application of temperature detection technology and equipment for fire zone control in coal mines. *Safety in Coal Mines*, (09): 4−6.
19. 赵禹深, 许献磊, 曹仕龙, 朱鹏桥. 地下煤火探测技术及工程应用综述. 矿业科学学报, 2025, 10(01):24–39. Zhao, Y.S., Xu, X.L., Cao, S.L., & Zhu, P.Q. (2025). Application of radon-magnetic-electrical combined detection technology. *Journal of Mining Science and Technology*, *10*(01):24−39.
20. Duan, S. L., Fang, L. F., Shi, Q. L., Qi, J. S., Huang, Q. M., & Wang, H. S. (2023). Application of rapid identification technology in shallow coal seam fire detection.
21. 刘海胜, 李雄. 测氡法在郭家湾煤矿火源位置探测中的应用. 山东煤炭科技,2023, 41(03):104–106. Liu, H. S., & Li, X. (2023). A review of underground coal fire detection technology and engineering application. *Shandong Coal Science and Technology*, *41*(03): 104−106.
22. 商宇航, 刘春雨, 邰振华, 刘世明. 基于瞬变电磁法和同位素测氡法的煤层综合探测.黑龙江科技大学学报, 2020, 30(01):16–21. Shang, Y.H., Liu, C. Y., Tai, Z. H., & Liu, S. M. (2020). Comparative study of coal mine fire areas zoning methods based on isotopic radon measurement technique. *Journal of Heilongjiang University of Science and Technology*, *30*(01): 16−21.
23. 李东凯, 何启林, 李金亮. 氡-磁-电联合探测火区技术的应用研究. 安徽理工大学学报(自然科学版), 2021, 41(04):60–64. Li, D. K., He, Q. L., & Li, J. L. (2021). Detection on hidden fire source of goaf in Fengjiata Coal Mine. *Journal of Anhui University of Science and Technology (Natural Science)*, *41*(04): 60−64.
24. 安世岗, 孙勇, 王世栋, 王刚, 张德鹏. 同位素测氡技术在煤矿火源位置探测中的应用. 煤矿安全, 2014, 45(08):159–161. An, S. G., Sun, Y., Wang, S. D., Wang, G., & Zhang, D. P. (2014). Application of isotope radon measurement technology in the detection of fire source location in coal mines. *Safety in Coal Mines*, *45*(08): 159−161.
25. 胡相明, 洪海博, 董浩, 吴佰谦, 孙自超, 王伟, 等. "测氡-磁"法探测小窑自然发火技术的应用研究. 煤炭科学技术, 2025, 53(02):200–210. Hu, X. M., Hong, H. B., Dong, H., Wu, M. Q., Sun, Z. C., & Wang, W., et al (2025). Comparative study of coal mine fire areas zoning methods based on isotopic radon measurement technique. *Coal Science and Technology*, *53*(02): 200−210.
26. Lu, Z. F., Zhou, B., Wang, J. F., Chan, Z. K., Yang, Q. F., & Dong, Z. Y., et al (2023).Study on the formation mechanism of a radon source during coal spontaneous combustion in a goaf. *Fuel*, *336*:127135.
27. 鲍翔宇, 邓强. 基于同位素测氡法火区探测技术研究. 煤炭技术, 2024, 43(12):143–145. Bao, X. Y., & Deng, Q. (2024). Application of isotope radon measurement technology in the detection of mine fire area. *Coal Technology*, *43*(12): 143−145.
28. 安帮, 邸晟钧. 同位素测氡法在煤自燃火源位置探测中的应用. 山西焦煤科技, 2022, 46(08):13–15+18. An, B., & Di, S. J. (2022). Application of isotope radon measurement method in the detection of coal spontaneous combustion fire source location. *Shanxi Coking Coal Science & Technology*, *46*(08): 13–15+18.
29. 李斯昊, 张景钢. 同位素测氡法在浅煤层地质火源探测中的应用. 华北科技学院学报, 2011, 8(02):9–11. Li, S. H., & Zhang, J. G. (2011). Application of isotope radon measurement method in the detection of geological fire sources in shallow coal seams. *Journal of North China Institute of Science and Technology*, *8*(02): 9−11.
30. 迟克勇. 同位素测氡技术在老窑采空区隐蔽火源探测的应用研究. 内蒙古煤炭经济, 2021, (16):21–23. Chi, K. Y. (2021). Application of isotope radon measurement in detection of coal mine fire location. *Inner Mongolia Coal Economy*, (16): 21−23.
31. Wang, T., Wang, H. Y., Fang, X. Y., Wang, G. D., Chen, Y. Q., & Xu, Z. Y., et al (2023). Research progress and visualization of underground coal fire detection methods. *Environmental Science and Pollution Research*.
32. 张铎, 王帅豪, 文虎, 邓军, 翟小伟, 王伟峰. 采空区遗煤自燃隐蔽火源位置探查现状及新技术展望. 煤矿安全, 2024, 55(11):108–119. Zhang, D., Wang, S. H., Wen, H., Deng, J., Zhai, X. W., & Wang, W. F. (2024). Application of radon measurement and infrared thermal imaging method in the detecting fire area of small coal Kiln. *Safety in Coal Mines*, *55*(11): 108−119.

33. Wang, T., Wang, H. Y., Fang, X. Y., Wang, G. D., Chen, Y. Q., Xu, Z. Y., et al. (2023). Research progress and visualization of underground coal fire detection methods. *Environmental Science and Pollution Research International, 30*(30), 74671–74690.
34. 张立，李霞，穆效治. 煤矿火灾区域同位素测氡探测技术研究. 能源技术与管理，2020, 45(06):111–112. Zhang, L., Li, X., & Mu, X. Z. (2020). Study and application of techniques for detection and measurement of temperature information related to spontaneous combustion zone in coal mines. *Energy Technology and Management*, *45*(06): 111−112.
35. 刘轩, 王俊峰, 周斌, 刘硕. 基于普通克里金法的同位素测氡探火数据优化处理. 太原理工大学学报, 2022, 53(04):690–696. Liu, X., Wang, J. F., Zhou, B., & Liu, S. (2022). Integrated detection of coal seam based on transient electromagnetic method and isotope radon measurement technique. *Journal of Taiyuan University of Technology*, *53*(04): 690−696.
36. 张辛亥，牛恒，费金彪. 测氡法探测煤层自燃火源在巴依里矿的应用. 煤矿安全，2009, 40(07):29–30. Zhang, X. H., Niu, H., & Fei, J. B. (2009). Application of radon detection method for identifying spontaneous combustion sources in coal seams in Baili mine. *Safety in Coal Mines*, 40(07): 29−30.
37. 李智飞, 许召稳, 陈印. 冯家塔煤矿采空区隐蔽火源位置探测. 陕西煤炭, 2015, 34(02):73–75+4. Li, Z. F., Xu, Z. W., & Chen, Y. (2015). Data optimization based on ordinary kriging for radon detection to identify spontaneous combustion areas. *Shaanxi Coal*, *34*(02): 73–75+4.
38. 隋涛, 潘跃飞, 邬剑明. 同位素测氡法探测火源位置在柴里煤矿火区治理中的应用. 山西煤炭, 2006, (03):46–48. Sui, T., Pan, Y. F., & Wu, J. M. (2006). Application of isotope radon detection method in fire source location for fire area control in Chaili coal mine. *Shanxi Coal*, (03): 46−48.
39. 金永飞, 邓军, 许延辉, 孙海河. 测氡法探测地下煤火技术研究与应用. 西安科技大学学报, 2010, 30(06):647–650. Jin, Y.F., Deng, J., Xu,Y.H., & Sun, H.H. (2010). Research and application of underground coal fire detection technology based on radon measurement. *Journal of Xi'an University of Science and Technology*, *30*(06): 647−650.
40. 罗颖初，王宇. 测氡法探测自燃火源位置在万柏林煤矿的应用. 科技情报开发与经济, 2009, 19(11):190–191. Luo, Y.C., & Wang, Y. (2009). Application of radon measurement method in the detection of fire source location in Guojiawan coal mine. *Sci−Tech Information Development & Economy*, *19*(11): 190−191.
41. 吕英华, 于婷婷. 同位素测氡技术在矿井火区探测中的应用. 陕西煤炭, 2015, 34(03):80–82. Lv, Y. H., & Yu, T. T. (2015). Application of isotope radon measurement technology in the detection of fire zones in mines. *Shaanxi Coal*, *34*(03): 80−82.
42. Li, P. F., Sun, Q., Deng, Y. H., Zheng, X. C., & Ge, Z. L. (2024).Study on the characteristics of radon exhalation from fly ash filling materials in coal fire goaf based on the evolution of pore structure.*Journal of Thermal Analysis & Calorimetry*, *149* (01).

Chapter 9
Acoustic Detection Method

In recent years, acoustic temperature measurement methods have been widely applied in temperature measurement in fields such as industrial boilers, grain silos, accumulated materials, and aero engines due to their advantages of non-contact, large measurement space, high temperature measurement accuracy, simple operation, and real-time continuous operation. From the perspective of temperature measurement principles, the acoustic method can meet the requirements for detecting concealed fire sources of coal spontaneous combustion and is expected to become a highly promising approach for precisely detecting the location of concealed fire sources. This chapter aims at the rapid and accurate detection of concealed fire sources in coal spontaneous combustion, explores the spatio–temporal distribution characteristics of the sound field within loose coal, clarifies the main propagation paths of sound waves in loose coal, explains the response characteristics of sound waves to the temperature of loose coal, constructs a sound wave temperature measurement model for the spontaneous combustion process of loose coal, reveals the sound wave temperature sensing mechanism within loose coal, and realizes the acoustic reconstruction of the temperature field of coal spontaneous combustion.

9.1 Propagation Path of Acoustic Waves in Loose Coal Bodies

The basic principle of acoustic temperature measurement method is to use the physical relationship that the propagation velocity of sound waves in the medium is a single-valued function of the medium temperature [1], and then reconstruct the three-dimensional temperature field of the medium according to the physical relationship between sound velocity and temperature. In order to explore the main path of sound waves in the propagation of sound waves in loose coal bodies, taking long-flame coal, lean coal and anthracite coal as the research objects, the flight time of sound waves

Y. Xiao et al., *Detection of Fire Source in Underground Coal Mine and its Applications*,
https://doi.org/10.1007/978-981-95-7064-5_9

in different gas environments and different saturated gas environments (air, nitrogen, carbon dioxide) of loose coal bodies is calculated by using the acoustic wave flight time test system, and the propagation path of sound waves in the loose coal body is analyzed from the perspectives of theoretical analysis and experimental test. It provides a theoretical basis for the application of acoustic temperature measurement technology in loose coal bodies.

9.1.1 Loose Coal Grading Characteristics

The classification of gradation includes three types: continuous grading, intermittent grading, and continuous open grading [2], which can better reflect the voidness of the loose medium. Intermittent gradation refers to the exclusion of one or several particle sizes in the entire range of mineral particle distribution, so that the mineral particles form a discontinuous gradation, Continuous open grading refers to the narrow distribution range of the entire mineral particle size, from the largest particle size to the smallest particle size only appears in a continuous form at several particle size levels, which is called continuous open grading. In this experiment, the grading method was used to study the distribution of coal samples of each level of particle size of loose coal, reflecting the void rate of loose coal [4], and the porosity of loose coal bodies of different particle sizes studied by Deng et al. [5] is shown in Table 9.1.

In the experiment, three coal samples were selected as research objects: three samples were selected from long-flame coal in Simengou Coalmine (Shenmu, Shaanxi), lean coal in Malan Coalmine (Gujiao, Shanxi), and anthracite coal in Sihe Coalmine (Jincheng, Shanxi) as research objects, and the coal samples were transported from the underground to the laboratory: some of the coal samples with 5 particle size ranges (0.9–3 mm, 3–5 mm, 5–7 mm, 7–10 mm and greater than 10 mm) were screened out by jaw crusher, in order to exclude the influence of other factors on the experimental results. The three coal samples were made into the same five-level gradation specimen (0.9–3 mm, 3–5 mm, 5–7 mm, 7–10 mm and more than 10 mm accounted for 10%, 15%, 50%, 15% and 10% respectively) for the experimental test of the flight time of sound waves in the loose coal body. the other part of the coal sample is made into columnar briquettes with a diameter of 50 mm and a height of 100 mm, which is used for uniaxial compression experiments to test the mechanical properties of coal. The industrial analysis results of the three coal samples are shown in Table 9.2.

Table 9.1 Different gradations of loose coal void ratios [6]

Particle size	0.9–3 mm	3–5 mm	5–7 mm	7–9 mm	9–10 mm	7–10 mm	10 mm or more
Void rate	3.9	8.7	14.1	16.5	15.7	18.1	22.3

Table 9.2 Industrial analysis of coal (%)

Coal sample	M_{ad}	A_{ad}	V_{daf}	FC_{ad}
Long−flame coal	4.38	3.44	35.48	56.70
Lean coal	1.06	3.93	15.08	79.93
Anthracite coal	5.32	10.09	7.22	77.37

9.1.2 Flight Time of Acoustic Waves Under the Condition of Saturated Gas of Loose Coal

(1) Experimental device and test process

The sound velocity measurement system includes a hardware system and a software test platform, in which the hardware system includes a computer, a power amplifier, a speaker, an experimental chamber, a data collector, a microphone, etc., among which the experimental cavity includes a strong sound absorber, a blocking net and a small hole in the air inlet and outlet, and the experimental cavity has the characteristics of airtightness. The software test platform includes signal generation, testing, data acquisition, and processing analysis, as shown in Fig. 9.1. The software test platform generates a signal, which is transmitted to the speaker through the power amplifier, the speaker converts the electrical signal into an acoustic signal and transmits it to the experimental chamber, and the microphone receives the sound wave through the experimental chamber and transmits it to the data collector.

a. Hardware system

The experimental chamber is a non-standard part designed and manufactured by itself, as shown in Fig. 9.2. The experimental chamber is composed of three parts: the detachable speaker cavity, the main body of the experimental cavity and the removable back cover, with a total length of 2.3 m, an inner diameter of 0.1 m and an outer diameter of 0.12 m. The removable speaker cavity contains the speaker, air intake duct, Lotus terminal and shield net, There are two small holes for microphone insertion at a distance of 1.5 m between the upper part of the main body of the

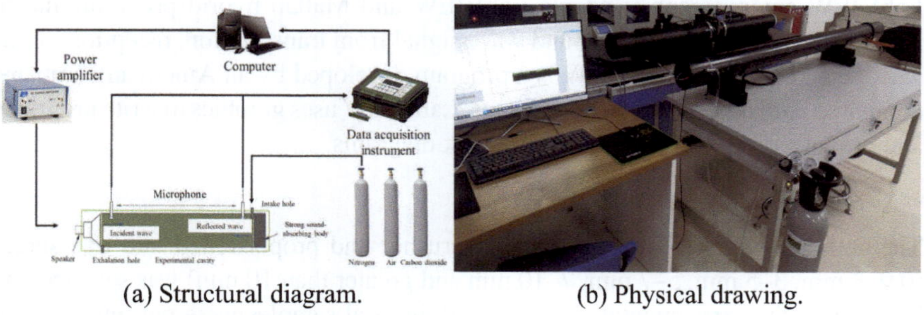

(a) Structural diagram. (b) Physical drawing.

Fig. 9.1 Sonic flight time measurement system

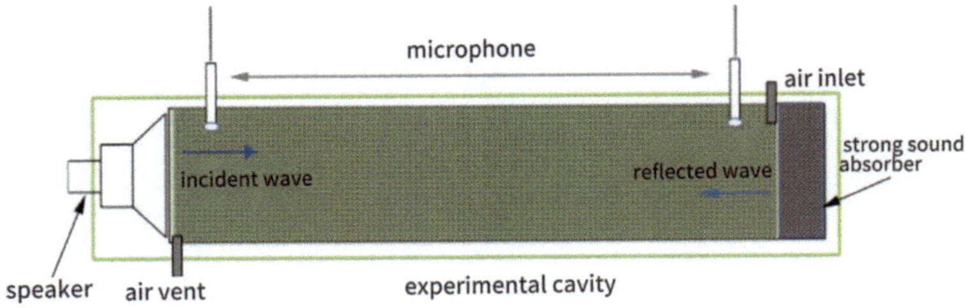

Fig. 9.2 Schematic diagram of the experimental cavity

experimental chamber, and the two small holes are on the same horizontal line. The length of the removable back cover is 0.3 m, including a conical sound-absorbing sponge, a shielding net and an outlet pipe. The experimental chamber is made of aluminum alloy as a whole, and the surface of the inner wall is smooth, which can ensure that the viscous friction of sound waves when passing through the inner wall is small.

The types of sound sources used to generate signals include electric sound sources, pneumatic sound sources, and spark discharge sound sources [7], and for temperature measurement applications, electric sound sources and pneumatic sound sources are mostly used. The speaker selected for this test system is the Huiwei M4N full-band speaker, which has a maximum power of 30W and a rated impedance of 8Ω. The microphone used is an electret microphone of model MPA201, with an open-circuit sensitivity of 40 mV/Pa (+2dB), a response frequency of 20–20 kHz, a dynamic range of 16–134 dBA, a temperature range of −30–80 °C, and a humidity range of 0–95%RH. The PA700 measurement power amplifier is adopted, and its maximum output power is 270WRMS at room temperature, the maximum output voltage is 75VPeak, and the frequency response range is 20–20 kHz. Using a 16-channel data collector with model XCQ108, the sampling rate of data acquisition is 51200 Hz, the frequency response range is 20–20 kHz, the sampling accuracy is 24 bits, and the noise floor is less than 30 dBA.

b. Software system

Based on the "sonic flythrough time" measurement software developed in the LabVIEW environment, it adopts LabVIEW and Matlab hybrid programming to achieve the whole process of sound wave signal from transmission, reception, oscilloscope and analysis. LabVIEW is a program developed by an American company, compared with other computer languages, LabVIEW uses graphics to write programs and produce programs in the form of block diagrams.

c. Experimental testing process

The anthracite coal crushed by the jaw crusher and proportioned into five stages (0.9–3 mm, 3–5 mm, 5–7 mm, 7–10 mm and greater than 10 mm) was selected and loaded into the experimental chamber, and the coal samples were put into the coal sample and compacted to ensure the compactness of the loose coal body. The sound source signal used in the experiment is 1–4000 Hz 11-level pseudo-random signal, and the sampling frequency of the microphone is 51200 Hz. When measuring the sound wave flight time before and after the reloading of the coal sample, open the back cover of the experimental cavity, place the coal sample to be measured in the experimental cavity, seal the experimental cavity, and calibrate the accuracy of the microphone before each test, and the treatment method of the remaining coal samples is consistent with the above process. The signal received by the two microphones is shown in Fig. 9.3.

(2) Calculation of acoustic flight time based on cross-correlation method

Acoustic flythrough time measurement is to calculate the delay time of the sound source signal when it propagates between two microphones, and is commonly used in acoustic temperature measurement signal delay estimation methods, including artificial neural networks, wavelet transforms, and cross-correlation function methods [9]. According to the experimental research of many scholars, the acoustic delay estimation based on the cross-correlation function is the most reliable [10]. Delay estimation is the use of cross-correlation function for homologous signals with time delay, and the abscissa corresponding to the maximum value point of the intercorrelation is the time of sound wave flight. If the signals received by microphone 1 and microphone 2 are separate.

$$x_1(k) = s(k) + \mathrm{W}_1(k) \tag{9.1}$$

$$x_2(k) = s(k - D) + W_2(k) \tag{9.2}$$

In the formula, $x_1(k)$ represents the signal received by microphone 1, $x_2(k)$ represents the signal received by microphone 2, s(k) represents the sound source signal emitted by the speaker, W_1 and W_2 are the signals mixed with background noise, and D is the signal delay time. If the sound source signal emitted through the speaker is received by two microphones, then the cross-correlation function between the two microphones is [10]:

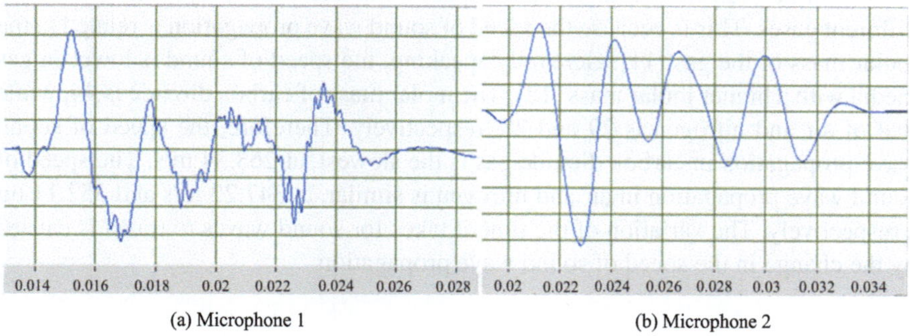

(a) Microphone 1 (b) Microphone 2

Fig. 9.3 Sonic flyover time test signal

$$\begin{aligned} R_{x1x2}(\tau) &= E\{x_1(k)x_2(k+\tau)\} = E\{(s(k)+w_1(k))(s(k-D+\tau)+w_2(k+\tau))\} \\ &= R_{ss}(\tau-D)+R_{sw1}(\tau-D)+R_{sw2}(\tau)+Rw_1w_2(\tau) \end{aligned} \tag{9.3}$$

If the signal s(k) and the noise W_1(k) and W_2(k) meet the uncorrelated assumptions, there is.

$$R_{SW1}(\tau-D)=0, \quad R_{SW2}(\tau)=0, \quad R_{W1W2}(\tau)=0$$

Get:

$$R_{x1x2}(\tau)=R_{ss}(\tau-D) \tag{9.4}$$

According to the nature of autocorrelation, obtained:

$$|R_{ss}(\tau-D)| \le R_{ss}(0) \tag{9.5}$$

$R_{ss}(\tau-D)$ at $\tau-D=0$ Achieve the maximum at that time, Then τ is the travel time of the sound wave between the two microphones, that is, by calculating the cross-correlation number of the signals received by the two microphones, the travel time of the sound wave between the two microphones is obtained.

(3) Analysis of sound wave travel time

a. Travel time of sound waves in a gaseous environment

The transit time and speed of sound waves in different gases are shown in Fig. 9.4. The results show that the flight time of sound waves varies under different gas conditions. The flight time is the longest in carbon dioxide gas and the shortest in nitrogen gas (that is, the propagation speed of sound waves is the fastest in nitrogen gas and the slowest in carbon dioxide gas). The transit time of sound waves in air and nitrogen is very different. Generally speaking, the order of sound speed among different gases is nitrogen > air > carbon dioxide, and the speed of sound wave propagation is between 268.34 and 352.11 m/s.

As can be seen from Fig. 9.4b, the speed of sound wave propagation varies in different gases. This is because the speed of sound wave propagation is related to the molar mass of the gas [11]. Generally speaking, the speed of sound is lower in gas media with a higher molar mass [12]. The molar mass of carbon dioxide is 44, while that of air and nitrogen is 29 and 28, respectively. Therefore, the speed of sound wave propagation in carbon dioxide gas is the slowest, at 268.34 m/s. The speed of sound wave propagation in air and nitrogen is similar, at 347.22 m/s and 352.11 m/s respectively. The variation of the time it takes for sound waves to travel is caused by the change in the speed of sound wave propagation.

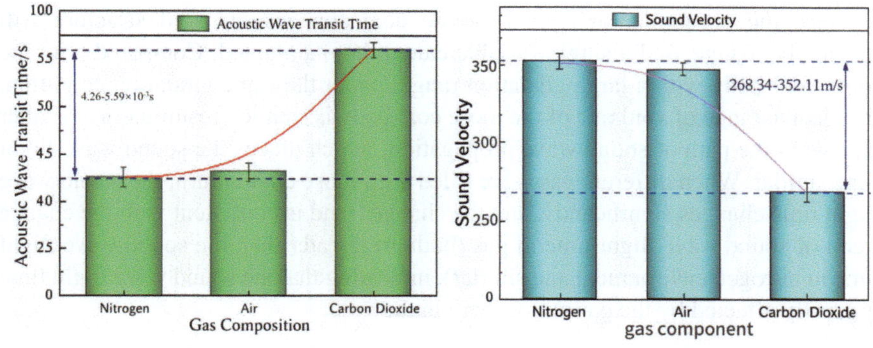

(a) Sonic Flying Time. (b) Sound wave velocity

Fig. 9.4 Flight time and sound velocity of sound waves in different gas environments

b. Acoustic wave flight time of the three coal samples under different atmospheric conditions

The flight time of sound waves in three coal samples under different atmospheric conditions is shown in Fig. 9.5a, and the sound velocity calculated by the total distance between the two microphones is 1.5 m as shown in Fig. 9.5b. The results show that the acoustic wave flight time of coal samples under different atmosphere conditions is different, the sound wave flight time is the longest in the carbon dioxide atmosphere, and the shortest in the nitrogen atmosphere. The difference between the acoustic wave flight time and the degree of deterioration of coal showed no obvious relationship between the sound wave flight time and the degree of deterioration of coal, and the values were between 5.37×10^{-3}–6.81×10^{-3}m/s.

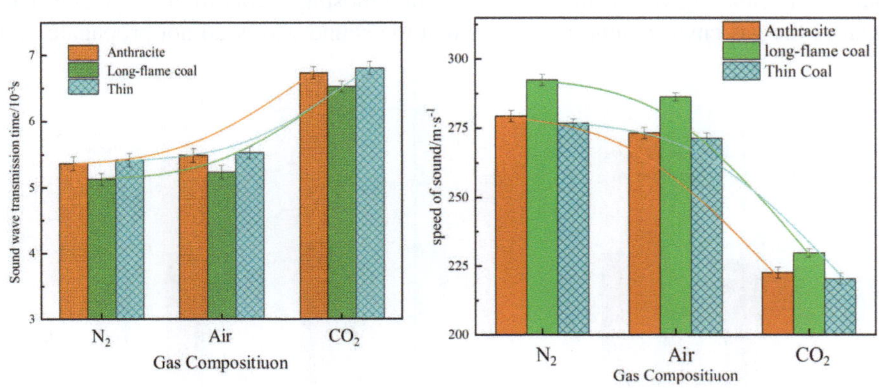

(a) Sonic Flying Time. (b) Sound wave propagation velocity.

Fig. 9.5 Sound wave flight time and sound velocity of the three coal samples under different atmospheric conditions

Once the accumulation form of loose coal changes, its void structure will inevitably change, and a single variable cannot be maintained. Compared with the loose coal body with a large gradation range, under the same gradation condition, the change range of void rate of the loose coal body is smaller, resulting in a smaller change in the path of sound wave propagation, which makes the sound wave flight time similar. When different gases are filled with loose coal in turn, the sound wave flight time changes significantly, and the change trend is consistent with the change trend of sound wave flight time in gas (both are greater than the sound wave flight time in nitrogen environment and similar), indicating that the sound wave flight time is greatly affected by the gas atmosphere in the void.

c. Analysis of the propagation path of sound waves in loose coal bodies

To exclude the influence of other factors on the acoustic crossing time, anthracite coal was selected as the experimental coal sample, and the acoustic wave flying time and sound velocity were compared before and after the resampling of anthracite loose coal samples under gas environment and different atmosphere conditions, as shown in Fig. 9.6. The results show that the acoustic wave flight time changes significantly before and after sampling under the same atmosphere condition, and the acoustic wave flying time becomes shorter after reloading the coal sample. Compared with the flight time of sound waves in the gas environment, the flight time before and after the reloaded coal samples under different atmospheric conditions became longer. The variation law of sound wave flight time in different atmospheres in the three states is the same, all of which are carbon dioxide > air > nitrogen.

The anthracite coal sample is poured out and reloaded, which changes the accumulation form of the loose coal body, and then changes the original void structure, resulting in a change in the path in the process of sound wave propagation, and shortening the flight time of sound waves. The flight time of sound waves before and after reloading anthracite under different atmosphere conditions is greater than that in the gas environment, indicating that the sound waves do not propagate in a

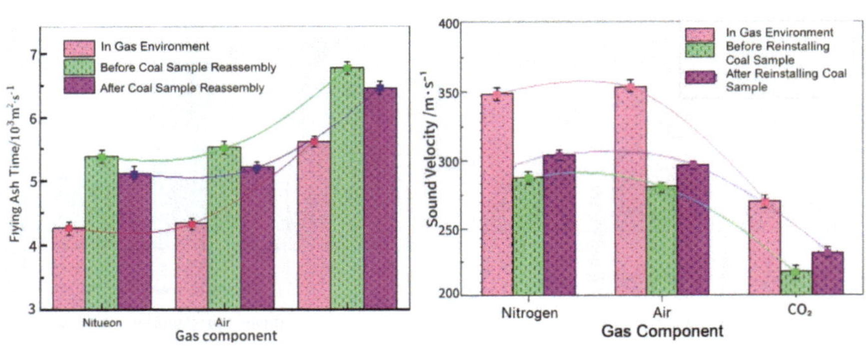

(a) Sonic Flying Time (b) Sound wave velocity

Fig. 9.6 Flight time and sound velocity of acoustic waves in different gases before and after anthracite reloading straight line, but in a zigzag way in the loose coal body, and the total propagation path becomes longer. The variation law of acoustic wave passage time in different gas conditions in the gas environment, before and after reloading coal sample is the same (all are carbon dioxide > air > nitrogen, and the difference in the flight time of acoustic waves in air and nitrogen is small), indicating that the medium propagating the acoustic wave is all gaseous medium, and the acoustic wave propagates along the void gas in the loose coal body.

d. Analysis of the change rate of acoustic flight time in three atmospheric environments

The rate of change of flight time when the sound waves propagate in different coal samples in the three atmospheric environments is shown in Fig. 9.7. It can be seen from the figure that the change rate of flight time of sound waves in the nitrogen−air atmosphere is the smallest in different states. The change rate of flight time in the nitrogen−carbon dioxide atmosphere was the largest. The change rate of acoustic crossing time in different environments has the same trend. The change rate of sound wave flight time before and after anthracite reloading is small in different atmospheres. The molar mass of nitrogen and air is similar, and the speed difference between sound waves propagating in it is small, so the change rate of flight time between nitrogen and air atmosphere is the smallest. The molar mass difference between nitrogen and carbon dioxide is the largest, and the difference in sound velocity is the largest, so the flight time of nitrogen and carbon dioxide in the atmosphere is the largest. The accumulation mode of anthracite coal changed before and after reloading, but under the same gradation condition and the type of coal did not change, the difference in the change rate of flight time was small, which further confirmed that the sound waves propagated through the gas medium in different coal samples.

(4) Acoustic wave propagation path in loose coal body

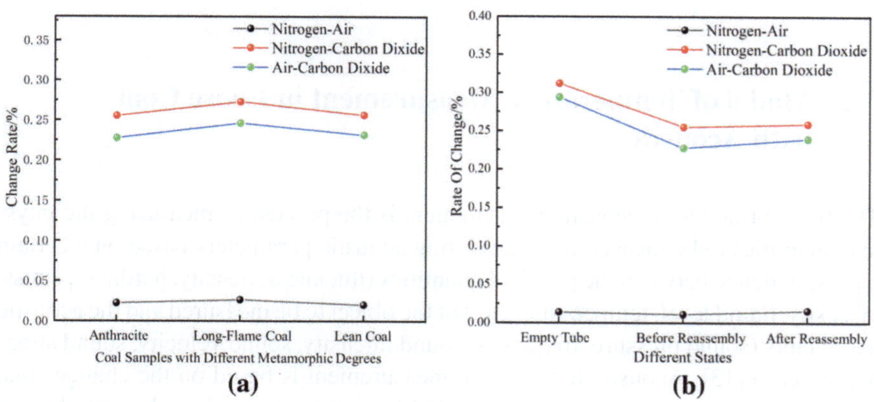

Fig. 9.7 Variation of sound wave flight time under different atmospheres. **a** In three coal samples with different degrees of metamorphosis, **b** Before and after the bituminous coal reload

By calculating the mechanical parameters of coal, it is found that coal belongs to a rigid porous medium, in which sound waves propagate through pores, but due to the uneven distribution of holes in raw coal and poor pore connectivity, the attenuation of sound waves in coal is large, and sound waves may propagate through other easier paths.

The flight time of sound waves in different gas environments is measured, and it is found that the flight time of sound waves propagating in different gases is different, and the law is carbon dioxide > air > nitrogen. By studying the acoustic flight time of three coal samples under different saturated gases (nitrogen, air, carbon dioxide), it is found that the sound wave flight time is close to the difference under the same gradation, indicating that the coal quality has little effect on the sound wave flight time. When different gases are filled into the loose coal body in turn, the sound wave flight time changes significantly, and the change trend is consistent with the change trend of sound wave flight time in gas, indicating that the sound wave flight time is more related to the gas atmosphere in the void. After resampling the anthracite coal sample, the acoustic wave flight time changes, indicating that the acoustic wave propagation is related to the accumulation form of the loose coal body, and the change trend of the acoustic wave flying time in the loose coal body of different atmospheres is the same, and the difference between the acoustic wave flying time in the nitrogen atmosphere and the air atmosphere is small, which further indicates that the sound wave propagates along the air in the void of the loose coal body.

This also explains why the sound wave flight time of the three coal samples under different atmosphere conditions becomes longer than the flight time of sound waves in different gas environments, which is due to the fact that the sound waves do not propagate along a straight line in the loose coal body, and the total path of sound wave propagation becomes longer, and the change of sound wave propagation path leads to the change of sound wave flight time. The sound velocity of sound waves is the slowest in the carbon dioxide atmosphere and the fastest in the nitrogen atmosphere, which is consistent with the results of the sonic wave velocity experiment under different gas conditions.

9.2 Model of Temperature Measurement in Loose Coal with Acoustic

The basic principle of acoustic measurement is the process of measuring the physical quantities to be measured by measuring acoustic parameters based on a certain correspondence between the physical quantities (thickness, density, hardness, stress, viscosity, liquid level, temperature, etc.) of the object to be measured and the acoustic parameters (sound pressure, frequency, sound intensity, sound velocity, sound attenuation, etc.) [13]. Acoustic temperature measurement is based on the changes that occur in some acoustic parameters affected by temperature when the sound wave propagates in the object to be measured, and the temperature of the measured object is derived according to the correspondence between these changes and temperature. There are two types of acoustic temperature measurement methods: one is to measure the temperature of an object based on the change of sound wave propagation speed, and the other is to measure the temperature of an object based on the change of sound wave frequency [14]. This paper mainly focuses on the study of sound velocity measurement, first analyzes the basic principle of acoustic wave temperature measurement, obtains the physical relationship between temperature and sound velocity in the atmosphere in free space, and establishes the propagation model of acoustic waves in loose coal. According to the void distribution characteristics in loose coal and combined with the fluctuation equation, the theoretical model of acoustic temperature measurement of loose coal body is derived, and the main factors affecting acoustic temperature measurement are clarified, which provides theoretical support for subsequent experimental research on acoustic temperature measurement.

9.2.1 Basic Principles of Temperature Measurement with Acoustic

The basic principle of sound velocity temperature measurement technology is that there is a single-valued function relationship between the propagation velocity of sound waves in the gas medium and the temperature of the gas medium [15–17], and the process of establishing this functional relationship requires the knowledge of acoustic basics, thermodynamics, mechanics and other related disciplines.

If the length of the ideal medium is dx, the area is S, and the density is ρ, then the mass of the medium is ρSdx, and when the planar sound wave propagates along the direction perpendicular to the medium section, the left section of the medium will be subject to the force F exerted by the left medium, and the right section will be subjected to the force exerted by the right medium: $F + \frac{\partial F}{\partial x}dx$, As shown in Fig. 9.8.

When the vibration displacement ζ the above medium is subjected to the force F, assuming that the interface of the medium does not change, the strain generated by the medium with a length of dx after being subjected to F is $\partial\xi/\partial x$. Assuming that the elastic modulus of the medium is E, the relationship between stress and strain can be obtained according to Hooke law, as shown below [18]:

$$F = -ES\frac{\partial\xi}{\partial x} \tag{9.6}$$

The resultant force dF on the microelements of the medium is calculated according to Eq. (9.6):

$$dF = F - (F + \frac{\partial F}{\partial x}dx) = -\frac{\partial F}{\partial x}dx = ES\frac{\partial^2\xi}{\partial x^2}dx \tag{9.7}$$

It can be obtained according to the equation of motion [18]:

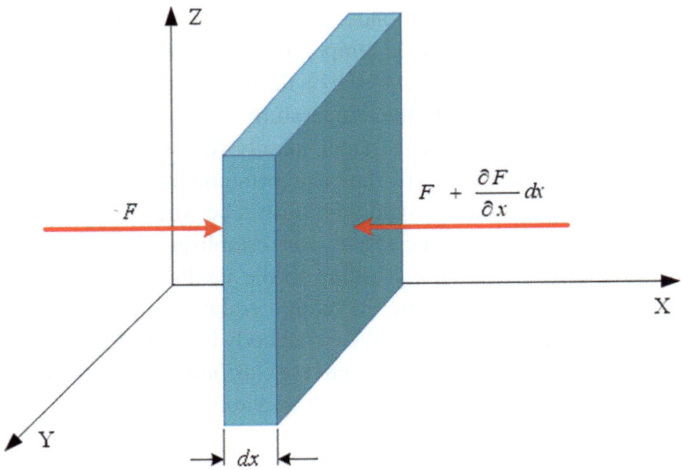

Fig. 9.8 Schematic diagram of fluctuations

$$ES\frac{\partial^2\xi}{\partial x^2}dx = \rho Sdx\frac{\partial^2\xi}{\partial t^2} \tag{9.8}$$

Further sorted out:

$$\frac{\partial^2\xi}{\partial t^2} = \frac{E}{\rho}\cdot\frac{\partial^2\xi}{\partial x^2} \tag{9.9}$$

From the wave theory, it can be seen that the wave equation for sound wave propagation along the x-axis is [18]:

$$\frac{\partial^2\xi}{\partial t^2} = u^2\frac{\partial^2\xi}{\partial x^2} \tag{9.10}$$

It can be obtained by formula (9.9) and formula (9.10):

$$u = \sqrt{\frac{E}{\rho}} \tag{9.11}$$

From the principle of acoustics, when the medium propagating by sound waves is regarded as a homogeneous, isotropic, infinitely large, and non-attenuating fluid, the sound waves only propagate in the form of longitudinal waves, in which case the stress on the medium is the sound pressure p [13]. Combined with formula (9.6):

$$p = \frac{F}{S} = -E\frac{\partial\xi}{\partial x} \tag{9.12}$$

For an ideal gas of a certain mass, the propagation of sound waves in it can be regarded as a process of rapid adiabacy, and the equation of state can be obtained according to the ideal adiabatic conditions:

$$PV^{\gamma} = const \tag{9.13}$$

where, P is static pressure, Pa; γ is the ratio of constant pressure specific heat and constant volume specific heat of gas; V is the volume of the gas, m^3, which can be obtained by deriving both sides of the formula (9.13):

$$dP = -\gamma P \frac{dV}{V} \tag{9.14}$$

Since the cross-sectional area of the medium remains unchanged, the volume changes, dV/V is actually equivalent to the change in the length of the medium, dξ/x. At the same time, for small amplitude sound waves, the differential of the pressure P is the sound pressure p [18], so the formula (9.14) can be changed to:

$$dP = p = -\gamma P \frac{dV}{V} = -\gamma P \frac{d\xi}{x} \tag{9.15}$$

Compare the formula (9.12) and the formula (9.15) to derive:

$$E = \gamma P \tag{9.16}$$

The speed of sound wave propagation represents the compression characteristics of the medium when it is disturbed by sound, and for the medium with high compressibility, the speed of sound disturbance propagation is relatively slow, and for the medium with low compressibility, the speed of sound disturbance propagation is relatively fast, according to the physical meaning of the above sound velocity, the sound velocity expression of the adiabatic process is [18]:

$$c = u = \sqrt{\frac{\gamma P}{\rho}} \tag{9.17}$$

The loose coal body will release its own moisture during oxidation and heating, so the influence of moisture in the air on the speed of sound needs to be considered, and the density of wet air is expressed as follows [1, 19]:

$$\rho = \frac{PM_a}{ZR(T + 273.15)}\left[1 - x_v\left(1 - \frac{M_v}{M_a}\right)\right] \tag{9.18}$$

where, Z is the compression factor; T is the temperature of the gas, °C; R is the universal constant of the ideal gas, and the value is 8.3144598 J/(mol·K). Ma is the molar mass of air, $28.9644424 \times 10^{-3}$ kg/mol; xv is the molar fraction of water vapor;

Mv is the molar mass of water vapor, 18.01534×10^{-3} kg/mol [20]. Substituting the formula (9.18) into the formula (9.17) can be obtained.

$$c = \sqrt{\frac{\gamma ZR(T + 273.15)}{M_a\left[1 - x_v\left(1 - \frac{M_v}{M_a}\right)\right]}} \tag{9.19}$$

The Molar fraction of water vapor in Eq. (9.19) is [21]:

$$x_v = 0.01RHf_z\frac{P_{sv}}{P} \tag{9.20}$$

where, RH is the relative humidity of the air, %; Psv is saturated vapor pressure, Pa; fz is the enhancement factor. When the temperature is between 0–100 °C, Psv can be calculated by the formula proposed by Hardy [22], and fz can be calculated with reference to the formula proposed by Greenspan [23].

$$P_{sv} = exp\left[\sum_{i=0}^{6} g_i(T + 273.15)^{i-2} + g_7 ln(T + 273.15)\right] \tag{9.21}$$

$$f_z = exp\left[\xi_1\left(1 - \frac{P_{sv}}{P}\right) + \xi_2\left(\frac{P_{sv}}{P} - 1\right)\right] \tag{9.22}$$

$$\xi_1 = \sum_{i=0}^{3} A_i \cdot T^i \tag{9.23}$$

$$\xi_2 = exp\left[\sum_{i=0}^{3} B_i \cdot T^i\right] \tag{9.24}$$

The values from g0 to g7, A0 to A3, and B0 to B3 are shown in Table 9.3.

Table 9.3 Value of each parameter

Parameter	Numeric value	Parameter	Numeric value
g_0	-2.9912729×10^3	A_0	3.53624×10^{-4}
g_1	-6.0170128×10^3	A_1	2.93228×10^{-5}
g_2	1.887643854×10^1	A_2	2.61474×10^{-7}
g_3	$-2.8354721 \times 10^{-2}$	A_3	8.57538×10^{-9}
g_4	1.7838301×10^{-5}	B_0	-1.07588×10^1
g_5	$-8.4150417 \times 10^{-10}$	B_1	6.32529×10^{-2}
g_6	$4.4412543 \times 10^{-13}$	B_2	-2.53591×10^{-4}
g_7	2.858487	B_3	6.33784×10^{-7}

The specific heat ratio in the formula (9.19) is calculated γ the formula proposed by Bohn [24]:

$$\gamma = \frac{7+x_v}{5+x_v} \tag{9.25}$$

The compression factor Z in the formula (9.19) can be calculated by the formula proposed by Hyland and Wexler [25]:

$$Z = 1 + AP_{sv} + BP_{sv}^2 \tag{9.26}$$

$$A = C_1 + C_2 \cdot e^{\frac{C_3}{T+273.15}} \tag{9.27}$$

$$B = K_1 + K_2 \cdot e^{\frac{K_3}{T+273.15}} \tag{9.28}$$

In the formula, C_1, C_2, and C_3 are 0.7×10^{-8} Pa^{-1}, -0.147184×10^{-8} Pa^{-1}, 1734.29 K^{-1}, respectively; K_1, K_2, K_3 are 0.104×10^{-14} Pa^{-2}, $-0.335297 \times 10^{-17}$ Pa^{-2}, 3645.09 K^{-1}, respectively.

Through the derivation of the above temperature measurement principle formula, it can be seen that the propagation velocity of sound waves in the gas medium is a function of the temperature of the gas, and is also related to the humidity, atmospheric pressure and composition of the gas medium, and in most cases, the speed of sound wave propagation is less affected by the gas composition [26], so the temperature of the gas medium can be inverted according to the function relationship between sound velocity and temperature by measuring the speed of sound waves in the gas medium.

For the free atmosphere environment, when the measured temperature field is determined, the distance between the two microphones can be determined first by installing a pair of acoustic transducers on both sides of the measured area, and the average velocity on the path of sound wave propagation can be further calculated by measuring the time t of the propagation of the sound wave between the two acoustic transducers:

$$c = \frac{l}{t} \tag{9.29}$$

At the same time, by measuring the humidity and air pressure values of the atmospheric environment in the measured area, the temperature of the measured area can be calculated according to the formula (9.19).

9.2.2 Acoustic Temperature Measurement Model of Loose Coal Body is Established

(1) Propagation model of sound waves in loose coal bodies

Based on the "Acoustic Fundamentals" [18, 27], the sound velocity model of sound waves in loose coal is derived. Firstly, the path of sound wave propagation in the loose coal body is simplified into an approximate cylindrical pipe, assuming that the sound wave propagates along the x direction of the cylindrical pipe with a radius of a, because the gas near the pipe wall will be affected by the friction of the pipe wall, resulting in the gas in the pipe producing a velocity gradient, as shown in Fig. 9.9.

Due to the viscosity force between the gases in each layer, the velocity gradient is generated, and the magnitude of the viscosity force is directly proportional to the velocity gradient and the contact area, which can be expressed as follows.

$$F_{\eta} = -\eta \frac{\partial v}{\partial r} d\delta \tag{9.30}$$

where, F_{η} is the viscosity force; η is the shear viscosity coefficient of the fluid; dδ is the microelement area of the contact surface; r represents radial distance; v is the speed of the movement of the particle.

To further analyze the viscosity force of each layer, the cylindrical tubular propagation channel is divided into several small cyclics, and a small cyclic element is taken for analysis, the length of the small cyclic element is dx, the inner diameter is r, and the thickness is dr, then the inner surface area $d\delta = 2\pi r\,dx$, and the volume $dV = 2\pi r\,dr\,dx$. The speed of the cyclic element is different from the adjacent gas layer, and the slower layer will produce a tensile force on the fast layer, that is, viscous resistance, assuming that the inner surface of the cyclic element is subjected to a force of $F\eta(r)$ and the outer surface is subjected to a force of $F\eta(r + dr)$, as

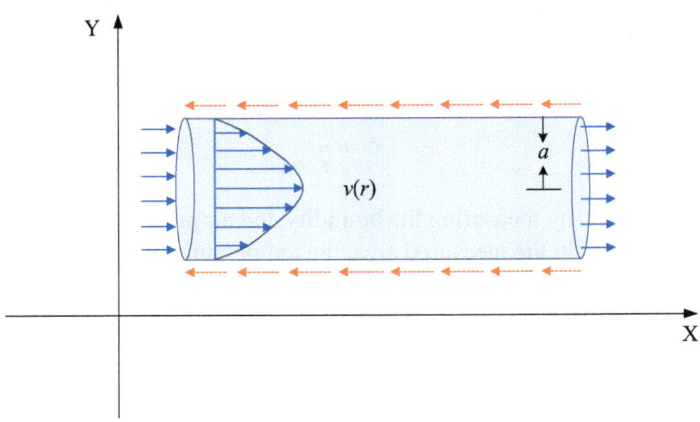

Fig. 9.9 Velocity gradient diagram

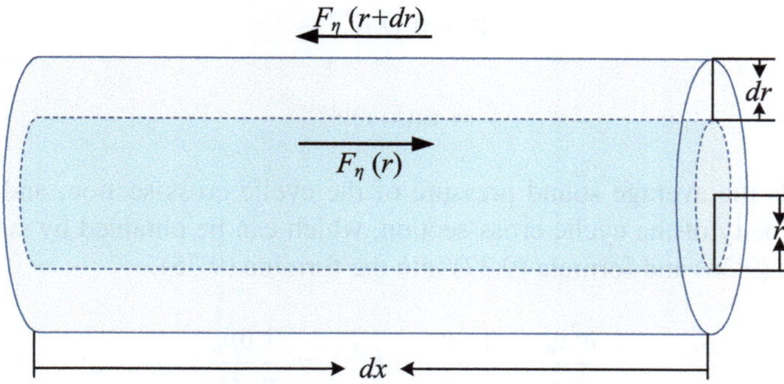

Fig. 9.10 Cylindrical tube ring element

shown in Fig. 9.10. The viscous forces acting on the inner surfaces are shown as follows.

$$F_{\eta} = -\eta \frac{\partial v}{\partial r} 2\pi r dx \tag{9.31}$$

The total viscosity force acting on the cyclic element in this segment is:

$$dF_{\eta} = F_{\eta}(r) - F_{\eta}(r + dr) = \frac{\partial\left(2\pi \eta dx \frac{\partial v}{\partial r} r\right)}{\partial r} dr \tag{9.32}$$

When sound waves travel in a tube, the cyclotron is subjected to an elastic force caused by sound pressure p, which can be expressed as:

$$\mathrm{dF_K} = -\frac{\partial \mathrm{p}}{\partial \mathrm{x}} 2\pi \mathrm{rdrdx} \tag{9.33}$$

According to Newton second law:

$$\mathrm{dF} = \mathrm{dF_K} + \mathrm{dF_\eta} = \rho_0 \mathrm{dV} \frac{\partial \mathrm{v}}{\partial \mathrm{t}} \tag{9.34}$$

where, ρ_0 is the static density of the cyclotron. Substituting the formula (9.32) and formula (9.33) into the above formula can be obtained:

$$\frac{1}{\mathrm{r}} \frac{\partial\left(\mathrm{r}\frac{\partial \mathrm{v}}{\partial \mathrm{r}}\right)}{\partial \mathrm{r}} - \frac{\rho_0}{\eta} \frac{\partial \mathrm{v}}{\partial \mathrm{t}} = \frac{1}{\eta} \frac{\partial \mathrm{p}}{\partial \mathrm{x}} \tag{9.35}$$

According to the above derivation, the sound pressure p is a function of the lateral coordinate x, and the particle velocity v is a function of the lateral coordinate x and the radial coordinate r.

$$\mathrm{p} = \mathrm{p_a(x)e^{j\omega t}} \tag{9.36}$$

$$\mathrm{v} = \mathrm{v_a(x, r)e^{j\omega t}} \tag{9.37}$$

where, p_a is the average sound pressure of the cyclic cross-section, and v_a is the average velocity of the cyclic cross-section, which can be obtained by substituting the formula (9.36) and formula (9.37) into the formula (9.35).

$$\frac{\partial^2 v_a}{\partial r^2} + \frac{1}{r}\frac{\partial v_a}{\partial r} + K^2 v_a = \frac{1}{\eta}\frac{\partial p_a}{\partial x} \tag{9.38}$$

In the formula, $K = (1 - j)\sqrt{\frac{\rho_0 \omega}{2\eta}}$.

It can be seen that the formula (9.38) is a non-homogeneous differential equation for v_a, so that z = Kr, it can be converted into the standard zero-order column Bézier equation, and the general solution of va can be further obtained.

$$v_a = AJ_0(Kr) + BN_0(Kr) + \frac{1}{\eta K^2}\frac{\partial p_a}{\partial x} \tag{9.39}$$

In the formula, A and B are the coefficients to be determined, J_0 is the Bézier function of the order 0, N_0 is the Neumann function of the order of 0, and the Bezier and Neumann functions are shown as follows:

$$\mathrm{J_n(x)} = \sum\nolimits_{\mathrm{k=0}}^{\infty} (-1)^{\mathrm{k}} \frac{\left(\frac{\mathrm{x}}{2}\right)^{\mathrm{2k+n}}}{\mathrm{k!(n+k)!}} \tag{9.40}$$

$$\mathrm{N_n(x)} = \frac{\mathrm{J_n(x)\cos nx - J_{-n}(x)}}{\mathrm{\sin nx}} \tag{9.41}$$

When r = 0, it can be seen from the formula (9.41) that the Neumann function diverges at zero point, so B = 0. Assuming that the pipe wall is rigid, so when r = a, $v_a = 0$, substitute the formula (9.39).

$$\mathrm{A} = \frac{\frac{\partial \mathrm{p_a}}{\partial \mathrm{x}}}{\eta \mathrm{K}^2}\left[-\frac{1}{\mathrm{J_0(Ka)}}\right] \tag{9.42}$$

Substitute the formula (9.42) into the formula (9.39) to get the result:

$$\mathrm{v_a} = \frac{\frac{\partial \mathrm{p_a}}{\partial \mathrm{x}}}{\eta \mathrm{K}^2}\left[1 - \frac{\mathrm{J_0(Kr)}}{\mathrm{J_0(Ka)}}\right] \tag{9.43}$$

The average velocity of the particle can be obtained by averaging the cross-section of formula (9.43):

$$\overline{\mathrm{v_a}} = \frac{1}{\pi a^2}\int_0^a 2\pi r v_a dr = \frac{\frac{\partial p_a}{\partial x}}{\eta K^2}\left[1 - \frac{2J_1(Ka)}{KaJ_0(Ka)}\right] \tag{9.44}$$

From formula (9.44), it can be seen that the average velocity of a particle is a function of Ka. Let $|Ka| = \mu = a\sqrt{\frac{\rho_0\omega}{\eta}}$, the μ value reflects the influence of the boundary layer in the cylindrical tube on the propagation of sound waves. According to the magnitude of the μ value, cylindrical tubes are classified into three categories: μ>10, namely $a > 10\sqrt{\frac{\eta}{\rho_0\omega}}$ as the thin tube, $1 < \mu < 10$, namely square root $\sqrt{\frac{\eta}{\rho_0\omega}} < a < 10\sqrt{\frac{\eta}{\rho_0\omega}}$ as the microporous tube, $\mu < 1$, namely $a < \sqrt{\frac{\eta}{\rho_0\omega}}$ for capillary.

To further derive the acoustic temperature measurement model of loose coal mass, it is first necessary to determine the type of cylindrical tube to which the voids of the loose coal mass belong. 20 °C air $\frac{\eta}{\rho_0} = 15.6 \times 10^{-6}$ m²/s, this article studies the sound wave frequency is 200–2000 Hz, each frequency under the corresponding square root $\sqrt{\frac{\eta}{\rho_0\omega}}$ of the values as shown in Table 9.4.

As can be seen from Table 9.4, the radius a of the equivalent cylindrical pipe of the loose coal body void meets the requirements of the thin pipe, taking 200 Hz as an example, the radius a needs to meet $a > 10\sqrt{\frac{\eta}{\rho_0\omega}} = 1.114 \times 10^{-3}$ m, with the gradual increase of the acoustic wave frequency, the conditions to be met by radius a are getting smaller and smaller, and it is obvious that the above conditions are fully met for the void of the loose coal, so the temperature measurement model of the loose coal body is further derived by taking the thin tube as the research object.

When $a > 10\sqrt{\frac{\eta}{\rho_0\omega}}$, it is believed that the influence of the boundary layer on sound waves is not very large, which can further prove that the $\frac{J_1(Ka)}{J_0(Ka)} \approx -j$, the formula (9.44) can be changed to:

Table 9.4 Different frequencies of $\sqrt{\frac{\eta}{\rho_0\omega}}$ Value

Frequency (Hz)	$\sqrt{\frac{\eta}{\rho_0\omega}}$ (m)	Frequency (Hz)	$\sqrt{\frac{\eta}{\rho_0\omega}}$ (m)
200	1.114×10^{-4}	1200	4.549×10^{-5}
300	9.097×10^{-5}	1300	4.370×10^{-5}
400	7.878×10^{-5}	1400	4.211×10^{-5}
500	7.047×10^{-5}	1500	4.068×10^{-5}
600	6.433×10^{-5}	1600	3.939×10^{-5}
700	5.956×10^{-5}	1700	3.822×10^{-5}
800	5.571×10^{-5}	1800	3.714×10^{-5}
900	5.252×10^{-5}	1900	3.615×10^{-5}
1000	4.983×10^{-5}	2000	3.523×10^{-5}
1100	4.751×10^{-5}		

$$\overline{v_a} \approx \frac{\frac{\partial p_a}{\partial x}}{\eta K^2}\left(1 + j\frac{2}{Ka}\right) \tag{9.45}$$

Further transform the formula (9.45) to:

$$-\frac{\partial p_a}{\partial x} \approx -\eta K^2\left(1 + j\frac{2}{Ka}\right)^{-1}\overline{v_a} \tag{9.46}$$

Because $|Ka| > 10$, therefore, the formula (9.46) can be approximated as:

$$-\frac{\partial p_a}{\partial x} \approx -\eta K^2\left(1 - j\frac{2}{Ka}\right)\overline{v_a} = \left[j\rho_0\omega + \frac{\sqrt{2\eta\rho_0\omega}}{a}(1 + j)\right]\overline{v_a} \tag{9.47}$$

Cause $\rho = \rho_0\left(1+\frac{1}{a}\sqrt{\frac{2\eta}{\rho_0\omega}}\right)$, $R = \frac{1}{a}\sqrt{2\eta\rho_0\omega}$, ρ represents the effective static density, and R is the damping coefficient of the tube, then the formula (9.47) can be changed to:

$$-\frac{\partial \mathrm{p_a}}{\partial \mathrm{x}} = (\mathrm{j}\rho\omega + \mathrm{R})\overline{\mathrm{v_a}} \tag{9.48}$$

Substituting the formula (9.46) and formula (9.47) into the formula (9.48) can be obtained.

$$-\frac{\partial \mathrm{p}}{\partial \mathrm{x}} = \rho\frac{\partial \overline{\mathrm{v}}}{\partial \mathrm{t}} + \mathrm{R}\overline{\mathrm{v}} \tag{9.49}$$

Since the viscous effect of the tubules causes an equivalent increment of static density, the equations of state of matter and continuity for the tubules remain valid, as shown as follows.

$$\mathrm{p} = \mathrm{c}_0^2\rho' \tag{9.50}$$

$$\rho_0\frac{\partial \overline{\mathrm{v}}}{\partial \mathrm{x}} = -\frac{\partial \rho'}{\partial \mathrm{t}} \tag{9.51}$$

where c_0 is the adiabatic sound velocity in a free environment; ρ' is the increment of static density; The combined formula (9.49), formula (9.50) and formula (9.51) can obtain the wave equation in the thin tube.

$$\rho_0\mathrm{c}_0^2\frac{\partial^2\overline{\mathrm{v}}}{\partial \mathrm{x}^2} = \rho\frac{\partial^2\overline{\mathrm{v}}}{\partial \mathrm{t}^2} + \mathrm{R}\frac{\partial \overline{\mathrm{v}}}{\partial \mathrm{t}} \tag{9.52}$$

Take $\overline{\mathrm{v}} = \overline{\mathrm{v_a}}\mathrm{e}^{\mathrm{j}\omega\mathrm{t}}$ to formula (9.52), get:

$$\frac{\partial^2 \overline{v_a}}{\partial x^2} = \left(-\frac{\omega^2}{c_p^2} + j\frac{\omega R}{\rho c_p^2}\right)\overline{v_a} \tag{9.53}$$

In the formula, $c_p^2 = c_0^2\frac{\rho_0}{\rho}$, $\overline{v_a} = \overline{v_0}e^{-j(k-j\alpha)x}$, $\overline{v_0}$ is the average velocity of the particles at x = 0, which can be obtained by substituting it into the formula (9.53).

$$\frac{\omega^2}{c_p^2} - j\frac{\omega R}{\rho c_p^2} = -\alpha^2 - 2j\alpha k + k^2 \tag{9.54}$$

For the general case, the α is much smaller than k, so the α^2 item can be omitted, and the α and k can be further determined, as shown below.

$$\alpha = \frac{\omega R}{2\rho c_p^2 k} \approx \frac{1}{ac_0}\sqrt{\frac{\eta\omega}{2\rho_0}} \tag{9.55}$$

$$k = \frac{\omega}{c_p} \tag{9.56}$$

From the above equation, it can be seen that α is the sound wave attenuation coefficient of the thin tube, and the formula (9.56) is the wavenumber of the thin tube, where c_p is the phase velocity of the sound wave.

$$c_p = c_0\sqrt{\frac{\rho_0}{\rho}} \approx c_0\left(1 - \frac{1}{2a}\sqrt{\frac{2\eta}{\rho_0\omega}}\right) = c_0\frac{1 - \frac{\eta}{2a^2\rho_0\omega}}{1 + \frac{1}{2a}\sqrt{\frac{2\eta}{\rho_0\omega}}} \tag{9.57}$$

η can be derived from the Tsilingiris calculation method [21].

$$\eta = \frac{(1 - x_v)\mu_a}{(1 - x_v) + x_v\Phi_{av}} + \frac{x_v\mu_v}{x_v + (1 - x_v)\Phi_{va}} \tag{9.58}$$

The expressions of Φ_{av} and Φ_{va} are shown below.

$$\Phi_{av} = \frac{\sqrt{2}}{4}\left(1 + \frac{M_a}{M_v}\right)^{-\frac{1}{2}}\left[1 + \left(\frac{\mu_a}{\mu_v}\right)^{\frac{1}{2}}\left(\frac{M_v}{M_a}\right)^{\frac{1}{4}}\right]^2 \tag{9.59}$$

$$\Phi_{va} = \frac{\sqrt{2}}{4}\left(1 + \frac{M_v}{M_a}\right)^{-\frac{1}{2}}\left[1 + \left(\frac{\mu_v}{\mu_a}\right)^{\frac{1}{2}}\left(\frac{M_a}{M_v}\right)^{\frac{1}{4}}\right]^2 \tag{9.60}$$

where μ_a is the dynamic viscosity of dry air, 10^{-6} kg/(m·s); μ_v is the dynamic viscosity of water vapor, 10^{-6} kg/(m·s); X_V is the molar fraction of water vapor. μ_a and μ_v can be calculated by the following formula [1].

$$\mu_a = MA_0 + MA_1(T + 273.15) + MA_2(T + 273.15)^2 + MA_3(T + 273.15)^3 + MA_4(T + 273.15)^4 \tag{9.61}$$

$$\mu_v = \frac{\sqrt{\frac{T+273.15}{647.27}}}{MV_0 + MV_1\frac{647.27}{T+273.15} + MV_2\left(\frac{647.27}{T+273.15}\right)^2 + MV_3\left(\frac{647.27}{T+273.15}\right)^3} \tag{9.62}$$

In the formula, MA_0 is 0.40401, MA_1 is 0.074582, MA_2 is -5.7171×10^{-5}, MA_3 is 2.9928×10^{-8}, MA_4 is -6.2524×10^{-12}; MV_0 is 0.0181583, MV_1 is 0.0177624, MV_2 is 0.0105287, MV_3 is -0.0036744.

From the formula (9.58), it can be calculated that the order of $\eta/\rho 0$ is 10^{-6} m^2/s, while the order of a is 10^{-3} m, $\omega = 2\pi f$, when f is 1000 Hz, $\frac{\eta}{2a^2\rho_0\omega}$ is of the order of 10^{-3}, so the formula (9.57) can be reduced to:

$$c_p \approx \frac{c_0}{1 + \frac{1}{2a}\sqrt{\frac{2\eta}{\rho_0\omega}}} \tag{9.63}$$

The speed of sound wave propagation is divided into two types: phase velocity and group velocity, which refers to the propagation velocity of a single frequency wave, which represents the propagation velocity of the wave formation. Group velocity refers to the propagation speed of a wave packet, which is the propagation speed of wave energy in a physical sense [1]. The phase velocity of sound waves is mainly calculated by measuring the phase difference of sound waves in different positions, and the speed of sound is further obtained according to the distance of propagation. The propagation of sound waves in the loose coal body is different from the propagation in the free atmosphere environment, because the sound wave mainly propagates in the void in the loose coal body, its propagation path is tortuous and unstable, and the sound wave will be affected by the refraction and reflection of the loose coal body particles, so that the sound wave will be attenuated and distorted in the process of propagation, resulting in the phase of the sound wave signal received by different microphones will change, thus affecting the measurement of the sound wave propagation time, so the phase velocity measurement is easily interfered with by various factors and resulting in large errors. Therefore, this paper uses the cross-correlation method to calculate the time of sound wave propagation according to the similarity of the sound waveform received by different microphones, so as to further obtain the group velocity of sound wave propagation and reduce the influence of the above interference factors.

The expression for group velocity CG is shown below.

$$c_g = \frac{\partial \omega}{\partial k} \tag{9.64}$$

Combine the formula (9.56) and the formula (9.63) to further transform the formula (9.64) into:

$$c_g = \frac{c_0}{1 + \frac{1}{2d}\sqrt{\frac{2\eta}{\rho_0 \omega}}} \tag{9.65}$$

where d is the equivalent diameter of the void, d = 2a.

For thin pipes, in addition to the sound wave attenuation caused by the viscosity force of the pipe wall, it is also necessary to consider the energy attenuation caused by the heat exchange between the gas medium and the pipe wall. Therefore, the η in the formula (9.55) and formula (9.65) are replaced by η_e, which is shown below.

$$\eta_e = \eta\left[1 + (\gamma - 1)\sqrt{\frac{\lambda}{\gamma C_p \eta}}\right] \tag{9.66}$$

where Cp is the constant pressure specific heat capacity of the gas in the tube, 10^{-3} J/(g·K); λ is the thermal conductivity of the gas, W/(m·K). Cp and λ can be obtained by the calculation method proposed by Tsilingiris [21], as shown in the formulas (9.67) and equations (9.68):

$$C_p = \frac{C_{pa}(1 - x_v)M_a + C_{pv}x_v M_v}{M_a(1 - x_v) + M_v x_v} \tag{9.67}$$

$$\lambda = \frac{(1 - x_v)k_a}{(1 - x_v) + x_v \Phi_{av}} + \frac{x_v k_v}{x_v + (1 - x_v)\Phi_{va}} \tag{9.68}$$

where Cpa and Cpv are the specific heat capacity of dry air and water vapor, J/(g·K), respectively. Ka and KV are the thermal conductivity of dry air and water vapor of 10^{-3} W/(m·K), respectively. The values of CPA, CPV, KA, and KV are shown below.

$$C_{pa} = CA_0 + CA_1(T + 273.15) + CA_2(T + 273.15)^2 + CA_3(T + 273.15)^3 + CA_4(T + 273.15)^4 \tag{9.69}$$

In the formula, CA_0 is 0.103409×10, CA_1 is -0.284887×10^{-3}, CA_2 is 0.7816818×10^{-6}, CA_3 is $-0.4970786 \times 10^{-9}$, CA_4 is $0.1077024 \times 10^{-12}$.

$$\mathrm{C_{pv}} = \mathrm{CV_0} + \mathrm{CV_1 T} + \mathrm{CV_2 T^2} \tag{9.70}$$

In the formula, CV_0 is 1.86910989, CV_1 is $-2.578421578 \times 10^{-4}$, CV_2 is $1.941058941 \times 10^{-5}$.

$$k_a = KA_0 + KA_1(T + 273.15) + KA_2(T + 273.15)^2 + KA_3(T + 273.15)^3$$

$$+ KA_4(T + 273.15)^4 + KA_5(T + 273.15)^5 \tag{9.71}$$

In the formula, KA_0 is -2.276501×10^{-3}, KA_1 is 1.2598485×10^{-4}, KA_2 is $-1.4815235 \times 10^{-7}$, KA_3 is $1.73550646 \times 10^{-10}$, KA_4 is $-1.066657 \times 10^{-13}$, KA_5 is $2.47663035 \times 10^{-17}$.

$$\mathrm{k_v} = \mathrm{KV_0} + \mathrm{KV_1 T} + \mathrm{KV_2 T^2} \tag{9.72}$$

In the formula, KV_0 is 1.761758242×10, KV_1 is $5.558941059 \times 10^{-2}$, KV_2 is $1.663336663 \times 10^{-4}$.

Substituting the formula (9.66) into the formula (9.55) and formula (9.65) respectively can be obtained.

$$\alpha = \frac{1}{ac_0}\sqrt{\frac{\eta_e \omega}{2\rho_0}} = \frac{1}{ac_0}\sqrt{\frac{\omega}{2}\left[\frac{\eta}{\rho_0} + \left(\sqrt{\gamma} - \frac{1}{\sqrt{\gamma}}\right)\sqrt{\frac{\lambda\eta}{\rho_0^2 C_p}}\right]} \tag{9.73}$$

$$\mathrm{c_g} = \frac{\mathrm{c_0}}{1 + \frac{1}{2\mathrm{d}}\sqrt{\frac{2\eta_e}{\rho_0\omega}}} = \frac{\mathrm{c_0}}{1 + \frac{1}{2\mathrm{d}}\sqrt{\frac{2}{\omega}\left[\frac{\eta}{\rho_0} + \left(\sqrt{\gamma} - \frac{1}{\sqrt{\gamma}}\right)\sqrt{\frac{\lambda\eta}{\rho_0^2 C_p}}\right]}} \tag{9.74}$$

According to the viscosity coefficient of gas motion $\nu = \frac{\eta}{\rho}$, the formula (9.55) and formula (9.56) can be further reduced to [26]:

$$\alpha = \frac{\sqrt{\pi f}}{ac_0}\sqrt{\nu + \left(\sqrt{\gamma} - \frac{1}{\sqrt{\gamma}}\right)\sqrt{\kappa\nu}} \tag{9.75}$$

$$\mathrm{c_g} = \frac{\mathrm{c_0}}{1 + \frac{1}{2\mathrm{d}\sqrt{\pi f}}\sqrt{\nu + \left(\sqrt{\gamma} - \frac{1}{\sqrt{\gamma}}\right)\sqrt{\kappa\nu}}} \tag{9.76}$$

In the formula, $\kappa = \frac{\lambda}{\rho_0 C_p}$ is the thermal diffusivity of the gas; f is the frequency of the sound wave. From Eqs. (9.57) and (9.58), it can be seen that both equations contain an identical fraction composed of gas thermodynamic parameters on the right side, and in order to further simplify the above two equations, the same fraction is defined as the gas composite thermodynamic parameter F [26]:

$$\mathrm{F} = \sqrt{\nu + \left(\sqrt{\gamma} - \frac{1}{\sqrt{\gamma}}\right)\sqrt{\kappa\nu}} \tag{9.77}$$

Based on the above equation, the formula (9.75) and formula (9.76) are further simplified.

$$\alpha = \frac{F\sqrt{\pi f}}{ac_0} \tag{9.78}$$

$$c_g = \frac{c_0}{1 + \frac{F}{2d\sqrt{\pi f}}} \tag{9.79}$$

(2) Acoustic method loose coal body temperature measurement model

The distribution of medium in loose coal is complex, including coal particles and air, and there is a gas–solid interface. Sound waves mainly propagate in the voids of loose coal bodies, and the distribution characteristics of the voids determine the propagation characteristics of sound waves. In order to further characterize the propagation path of sound waves, the length ratio of the actual path of sound waves propagating in loose coal bodies and the straight path is defined as tortuous τ, which is used to describe the zigzag progress of sound waves in loose coal bodies.

Based on the above definition, the speed at which sound waves travel between two microphones can be expressed as:

$$c_g = \frac{\tau l}{\Delta t} \tag{9.80}$$

where l is the straight-line distance between the two microphones, m; Δt is the time of sound wave propagation, s.

The relationship between the sound velocity c_0 in free atmosphere and the propagation time Δt of sound waves in loose coal bodies can be further obtained by formulas (9.79) and equations (9.80).

$$c_0 = \frac{\left(1 + \frac{F}{2d\sqrt{\pi f}}\right)\tau l}{\Delta t} \tag{9.81}$$

From Eq. (9.81), it can be seen that the sound velocity in free atmosphere is not only related to the linear distance and time of sound wave propagation, but also related to the thermodynamic parameters of the gas in the loose coal, the acoustic frequency, the equivalent diameter of the void and the tortuosity of the loose coal body φ.

$$\varphi = \left(1 + \frac{F}{2d\sqrt{\pi f}}\right)\tau \tag{9.82}$$

Then the formula (9.81) can be expressed as:

$$c_0 = \frac{\varphi l}{\Delta t} \tag{9.83}$$

Theoretically, as long as the conversion factor φ is determined, the sound velocity c_0 in the same free atmospheric space as the void environment of the loose coal can be obtained by measuring the distance l and time Δt of the sound wave propagation in the loose coal body, and the temperature of the loose coal body can be further reversed by Eq. (9.19).

9.3 Acoustic Reconstruction of the Spontaneous Combustion Temperature Field of Loose Coal

9.3.1 Basic Theory of Temperature Field Reconstruction by Acoustic Method

Acoustic temperature field reconstruction involves using multiple acoustic transducers and arranging them around the area to be measured in a certain way, then controlling each transducer to transmit and receive acoustic wave signals in sequence, and obtaining the acoustic wave flight time of each acoustic wave propagation path, and finally using appropriate reconstruction algorithms to achieve this [28, 29].

Taking the square and circular areas to be measured in the two-dimensional plane as examples, the common layout of the acoustic transducer is shown in Fig. 9.11.

In Fig. 9.11, 8 acoustic transducers are installed at equal distances on the four sides of the square area to be measured (i.e., the third equinoxes on each side), and 6 acoustic transducers are installed at equal distances on the circumference of the circular area to be measured, all of which are represented by black dots. The sound wave transducers here are all selected as transmitter and receiver homogeneous, and multiple sound wave propagation paths through the area to be measured will be formed between each transducer, which is represented by a solid black line between

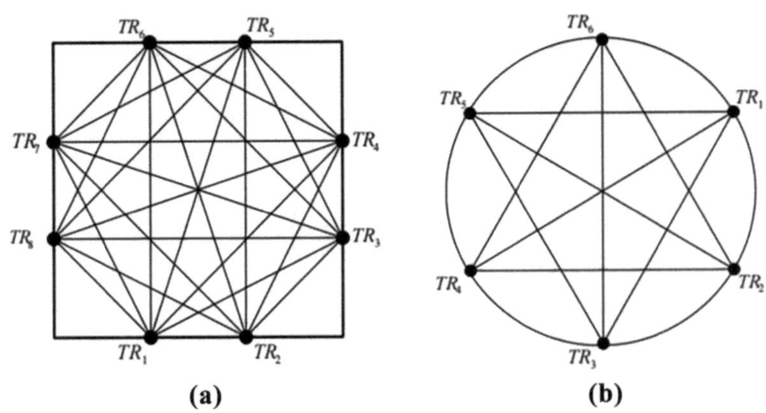

Fig. 9.11 Installation of acoustic transducer and distribution of acoustic wave path

the solid black dots (except for the solid line on the side where the dots are located in the area to be measured in the square).

The biggest advantage of the transceiver is that it can be used as a sound wave transmitter and a sound wave receiver at different times, and the conversion of its working process will be controlled by the program or related hardware equipment. For example, in Fig. 9.11a, when transducer TR1 transmits an acoustic signal as a transmitter, transducers TR2–TR8 act as a receiver to detect and receive acoustic signals sequentially, When transducer TR2 in turn transmits sound wave signals as transmitters, transducers TR1 and TR3–TR8 will detect and receive sound wave signals sequentially as receivers. In this way, the transmitting and receiving functions of each acoustic transducer are controlled sequentially in each detection cycle, and the acoustic wave signal is analyzed and processed, and then the acoustic wave flight time of each propagation path can be estimated with the help of the acoustic delay estimation algorithm. Finally, the measured set of acoustic flight time, geometric parameters of the region to be measured and its block division are substituted into the temperature field reconstruction algorithm to obtain the real-time temperature field of the area to be tested in the detection cycle. On the one hand, the real-time temperature field can be presented in the form of an intuitive image through the visualization device, and on the other hand, the specific temperature information data can be provided to the background analysis or control system as needed as a reference for formulating subsequent control strategies.

The general steps of acoustic temperature field reconstruction can be briefly summarized as follows:

(1) Considering the temperature complexity and specific application requirements in the actual environment, determine the layout mode of the acoustic transducer and select the effective acoustic wave propagation path.
(2) According to the defined transducer layout, the selected effective sound wave propagation path, the characteristics of the reconstruction algorithm, combined with the geometric characteristics of the area to be tested, the block division method is determined, and the entire area to be tested is evenly divided into several small blocks.
(3) Control the sound wave transducer to transmit and receive sound wave signals, and analyze and process the signals, and obtain the sound wave flight time corresponding to each sound wave propagation path by the sound wave delay estimation algorithm.
(4) Substitute a set of acoustic wave flight time obtained from each detection cycle with the relevant parameter information of the area to be measured into the temperature field reconstruction algorithm, calculate the temperature information of the area to be measured, draw and display the temperature field visualization image, and save or transmit the temperature field data as needed.

In the acoustic temperature measurement technology of sound velocity, in addition to the complexity of the temperature distribution of the area to be measured, the factors affecting the reconstruction effect of the acoustic temperature field mainly include the following aspects: temperature field reconstruction algorithm, layout mode of acoustic transducer and selection of effective acoustic wave propagation path, block division method of the area to be measured, sound wave signal processing and acoustic delay estimation.

(1) Temperature field reconstruction algorithm

Acoustic temperature field reconstruction derives temperature distribution information based on the propagation velocity distribution of acoustic signals in the region to be measured, which is actually an inverse process of "finding the cause from the result" [30], which belongs to the research category of inversion problems. Compared with the forward problem, the inversion problem is more complex: its solution process is often pathological, and in many cases, it is limited and constrained by objective conditions, resulting in the existence, uniqueness and stability of most reconstruction algorithms in terms of existence, uniqueness and stability of the solution. In addition, the number of acoustic transducers used in the temperature field reconstruction process is limited, resulting in limited effective acoustic measurement data. Therefore, how to reconstruct the temperature field that meets the accuracy and speed requirements under the condition of less acoustic measurement data is also a key problem that must be considered and solved by the reconstruction algorithm. In short, in the process of reconstructing the temperature field using acoustic temperature measurement technology, the reconstruction algorithm plays a crucial role in the reconstruction effect of the temperature field.

(2) Layout of acoustic transducer and selection of effective acoustic path

When reconstructing the acoustic temperature field, several acoustic transducers need to be installed in the area to be measured to form multiple acoustic wave propagation paths between the transducers and obtain the corresponding acoustic wave flight time, and finally realize it with the help of reconstruction algorithms. In this process, the layout of the acoustic transducer will be directly related to the selection of effective acoustic wave propagation path and the measurement of acoustic wave flight time, and also determine the ease of reconstruction process control, which ultimately affects the accuracy and speed of temperature field reconstruction [31].

It can be said that the reasonable layout of the sound wave transducer is actually a big reason to better obtain the effective sound wave propagation path. A reasonable and feasible layout method of acoustic transducer should fully consider the transducer performance and the constraints of the area to be measured, and strive to evenly distribute the sound wave propagation path in the area to be measured. For applications with complex temperature distributions or high temperature resolution requirements, a certain number of sound transducers must be obtained to obtain a sufficient number of effective sound wave propagation paths to reconstruct a temperature field that meets the accuracy requirements. However, the number of sound transducers is not always better, as too many transducers will increase the complexity of control and reduce the real-time reconstruction [32]. It can be seen that when determining the layout of the sound transducer, it is necessary to take into account the accuracy and speed requirements of temperature field reconstruction.

(3) Method of dividing the area to be tested

Dividing the area to be measured into several small blocks in a certain way is also an important task in the process of acoustic temperature field reconstruction. When determining the specific division method, it is necessary to comprehensively consider the characteristics of the temperature field reconstruction algorithm, the layout of the acoustic transducer, the selection of effective acoustic wave propagation path, the requirements and limitations of the application scenario, etc., and combine the characteristics of the area to be tested. For a specific temperature field reconstruction algorithm, the block division method and the sound transducer layout method should be considered and trade-offed against each other, and the following three principles should be generally followed in the specific implementation: ① In non-special cases, the small block is generally uniformly divided according to the equal area (two-dimensional situation) or the volume is equal (three-dimensional situation): ② Strive to make as many effective acoustic wave propagation paths as possible pass through each divided sub-block: ③ Strive to make each effective acoustic wave propagation path pass through as many subdivided sub-blocks as possible.

(4) Acoustic signal processing and acoustic delay estimation

Acoustic temperature measurement technology is actually the process of extracting temperature information from acoustic signals, that is, the process of using signal characteristics to study the characteristics of objects. In the actual application environment, there are various mixed noises, such as power frequency interference, line stray current, and background noise, which have different degrees of impact on the detection, analysis, processing, and delay estimation of acoustic signals [33–35]. Especially when the acoustic signal reaches the end of the propagation path after attenuation, the signal received by the receiver is difficult to discern, resulting in poor accuracy of acoustic time-of-flight measurement [36, 37], and ultimately reducing the temperature field reconstruction effect.

According to the principle of acoustic temperature measurement, the square relationship between the temperature measurement error and the acoustic wave flight time measurement error is directly proportional [38]. The accurate measurement of sound wave flight time is not only affected by the effect of sound wave signal analysis and processing. It also depends to a large extent on the acoustic delay estimation algorithm [39]. Because if the performance of the delay estimation algorithm cannot meet the requirements, even if the sound wave signal analysis and processing effect is good, the estimated sound wave flight time is difficult to be accurate. Obviously, the time of flight of the acoustic wave obtained by the delay estimation algorithm will be directly related to the accuracy of the acoustic temperature field reconstruction.

In short, the actual industrial application process is often in the application environment of large space, mixed noise and attenuation characteristics, so it poses great challenges to the analysis and processing of acoustic signals and delay estimation algorithms.

9.3.2 Common Algorithms for Temperature Field Reconstruction

(1) Acoustic CT algorithm

The least squares method is a commonly used acoustic CT temperature field reconstruction algorithm, also known as the least squares method, which is a mathematical optimization technique to find the optimal function match of the data by minimizing the sum of the squares of the error [40]. The least squares temperature field reconstruction algorithm is the earliest algorithm proposed in the field of acoustic temperature measurement, and has become a classic algorithm commonly used in subsequent acoustic temperature field reconstruction research due to its simple principle, simple calculation, and good speed [27, 41]. As an important statistical method and one of the main tools to improve the "observation combination", the least squares method provides a convenient and effective data processing method for scientific research by obtaining the "best" result or "most likely" expression based on a large number of observations of an event. From the perspective of principle, the advantages of least squares are mainly reflected in the following aspects: ① it can easily discover its internal regularity from a large number of disordered experimental measurement data, ② Pay attention to the overall trend of the data, eliminate the impact of local fluctuations, and easily fit functions or curves that can reflect the internal regularity or changing trend of the data, ③ The internal laws and overall trends between the data can be used to solve unknown data based on the actual data obtained, so as to ensure that the sum of squares of errors between the solved results and the actual measured data is minimal [42]. The following will briefly introduce its principle.

Suppose there is a dependency between the physical quantity y and l variables as follows.

$$y = f(x_1, x_2, \cdots, x_l, a_0, a_1, \cdots, a_n) \tag{9.84}$$

Among them, a0, a1,..., an are the n + 1 unknown parameters that need to be determined.

The essence of the least squares method is that first obtain m(m > n + 1) experimental observation points (xi1, xi2,..., xil, yi) (i = 1, 2,..., m), and then determine a set of parameter values (a0, a1,..., an) through certain rules, so that the sum of the squares of the error between the experimental value yi and the function value represented by the following Eq. (9.85) is minimized.

$$\overline{y} = f(x_{i1}, x_{i2}, \cdots, x_{il}; a_0, a_1, \cdots, a_n) \tag{9.85}$$

In other words, the least squares method is to obtain the minimum value of the following equation.

$$E(a_0, a_1, \cdots, a_n) = \sum_{i=1}^{m} (y_i - \overline{y_i})^2 \tag{9.86}$$

When using least squares to solve practical problems, in order to minimize random error, multi-point measurements are usually performed, so that the number of equations is greater than the number of unknowns, even if m > n + 1. From the relevant theory of differential calculus to solve the extreme value, it can be seen that the unknown needs to satisfy the following equations.

$$\frac{\partial y}{\partial a_i} = 0(i = 1, 2, \cdots, n) \tag{9.87}$$

Therefore, the result obtained by solving the equation system (9.87) is the least squares solution of the solved problem.

The algorithm divides the area to be measured into several small blocks under a certain sound transducer layout method, and then calculates the average temperature of each small block by the least squares method, and then uses the temperature as the temperature value of the geometric center of the corresponding small block for simple interpolation, and finally obtains the temperature field of the area to be measured. However, in order to ensure that the inversion problem of acoustic temperature field reconstruction has a unique solution, there is a strict limitation based on the least squares temperature field reconstruction algorithm: the number of small blocks divided into the measured area cannot exceed the number of effective acoustic wave propagation paths selected. For small-scale areas to be tested, the impact of this limitation on the reconstruction results is still acceptable. Once the area to be measured is too large and the number of small blocks is small, the temperature field obtained by simply calculating the average temperature of the finite block and simply interpolating it will have a large error, and it is difficult to accurately reflect the actual temperature distribution. In addition, this limitation also brings another serious problem that cannot be ignored and is more serious: the reconstruction results of the algorithm will have a large number of missing temperature information at the boundary of the area to be measured, so it cannot fully reflect the temperature distribution characteristics of the entire area to be measured. If the temperature value at the installation position of the acoustic transducer is known, the problem of missing boundary temperature information in the results obtained by the algorithm can be reduced to a certain extent, but the measurement of these temperature values brings additional difficulties and challenges to the reconstruction of the temperature field. The least squares method originally originated from a series of studies on the establishment of mathematical models and the fitting of observational data in astronomy, and has been widely studied and applied in mathematical statistics such as correlation regression analysis, analysis of variance and linear model theory, and has a strong theoretical foundation.

The classical least squares temperature field reconstruction algorithm was proposed by Fumio Ito of the Tokyo Electric Power Research Institute and Masayasu Sakai of the Nagasaki Research Institute of Mitsubishi Heavy Industries when

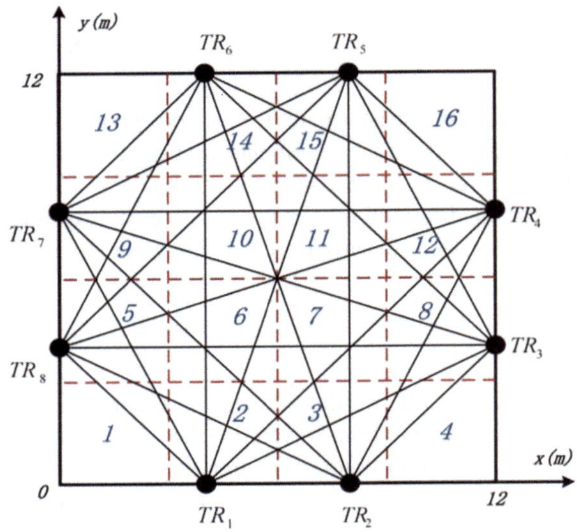

Fig. 9.12 Area to be measured, its transducer layout mode, and the selection of effective acoustic propagation path [42]

studying the attenuation characteristics of sound wave propagation, and was widely used in subsequent studies due to its simple principle and good speed [13].

The temperature field reconstruction algorithm based on least squares requires the area to be measured to be divided into several sub-blocks, and the internal temperature of each sub-block must be assumed to be uniform. In the two-dimensional square area to be tested, the layout of the sound wave transducer, the effective sound wave propagation path formed by the reconstruction algorithm, and the block division of the area to be measured are shown in Fig. 9.12, where the black solid dot is the sound wave transducer.

Assuming that the region to be measured is divided into N uniform small blocks, and M represents the number of effective acoustic propagation paths selected in the region to be measured, limited by the solution characteristics of the least squares sick matrix [46], the classical algorithm requires $N \leq M$ (as shown in Fig. 9.12, there are $N = 16$ and $M = 24$, where the solid black line between the black dots is the effective sound wave path, and the red dotted line is the block dividing line).

For a certain sound wave propagation path in the area to be tested, the theoretical value of the flight time of the sound wave signal along the path can be expressed as the following line integral.

$$t_k' = \int_0 a dl_k \tag{9.88}$$

where a represents the spatial characteristics of the acoustic wave signal in the area to be measured, that is, the reciprocal of the acoustic wave propagation velocity, lk represents the acoustic wave propagation path, and t_k' represents the theoretical value of the acoustic flight time of the path.

Since the internal temperature of each sub-block in the area to be measured is assumed to be uniform, the sound wave velocity in each sub-block is also uniform. If ai represents the reciprocal of the sound wave velocity in the ith subblock, and ΔSki represents the length of the kth acoustic wave propagation path through the ith subblock, then Eq. (9.88) can be rewritten as.

$$t'_k = \sum_{N}^{i=1} \Delta S_{ki} a_i \tag{9.89}$$

At this time, if tk is used to represent the actual measurement value of the flight time of the acoustic wave in the path, there must be a certain difference between the measured value and the theoretical value:

$$\varepsilon_k = t_k - t'_k = t_k - \sum_{N}^{i=1} \Delta S_{ki} a_i \tag{9.90}$$

For all the sound wave propagation paths in the area to be measured, the least squares method is used to minimize the sum of the squares of the error between the measured and theoretical values of the sound wave flight time of each path, even if the following equation is true:

$$\frac{\partial}{\partial a_i} \sum_{M}^{k=1} (t_k - \sum_{N}^{i=1} \Delta S_{ki} a_i)^2 = 0 \tag{9.91}$$

From Eq. (9.91), the following regular equation is easy to obtain:

$$S^T S A = S^T t \tag{9.92}$$

Thereinto, $A = \begin{bmatrix} a_1 \; a_2 \; \cdots \; a_N \end{bmatrix}^T$ is the spatial characteristic matrix of the sound wave signal, That is, the reciprocal of the propagation speed of sound waves in each subdivided block, $t = \begin{bmatrix} t_1 \; t_2 \; \cdots \; t_M \end{bmatrix}^T$ is the time-of-flight matrix, That is, the sound wave flight time corresponding to each sound wave propagation path, $S = \begin{bmatrix} \Delta S_{11} & \Delta S_{12} & \ldots & \Delta S_{1N} \\ \Delta S_{21} & \Delta S_{22} & \ldots & \Delta S_{2N} \\ \ldots & \ldots & \ldots & \ldots \\ \Delta S_{M1} & \Delta S_{M2} & \ldots & \Delta S_{MN} \end{bmatrix}$ is the distance matrix, the value of the element is the length of each sound wave propagation path through each subdivision.

To solve the regular Eq. (9.92), matrix A is obtained, that is, the reciprocal of the sound wave velocity of each small block in the area to be tested.

$$A = (S^T S)^{-1} S^T t \tag{9.93}$$

For a specific area to be tested, the distance matrix S is a constant matrix that can be calculated in advance under a certain sound transducer layout and block division. Therefore, when the acoustic wave time-of-flight matrix t is obtained by actual measurement, the acoustic wave spatial characteristic matrix A can be calculated, and then the temperature matrix can be solved according to the relationship between acoustic wave propagation velocity and temperature:

$$T = \frac{1}{A^2 B^2} \tag{9.94}$$

Of course, the temperature matrix solved by Eq. (9.94) at this time $T = [T_1, T_2, \cdots, T_N]^T$. In fact, it is only the average temperature of N sub-blocks in the area to be tested. The last step of the least squares temperature field reconstruction algorithm is to take this series of average temperatures as the temperature values of the geometric center points corresponding to each subdivision block, and then obtain the temperature field through simple interpolation operations.

The classical algorithm has simple principles, convenient calculations, and good speed. However, due to the constraints between the number of blocks divided by the area to be measured and the number of effective acoustic paths, the reconstruction results obtained by the algorithm will not only have limited accuracy, but also have serious missing temperature information at the boundary of the area to be measured.

(2) Generalized inverse matrix

In the case of the M $<$ N dimension of the A matrix, the system of equations has a large number of possible solutions, and it is mathematically necessary to find the unique solution with the best properties—this usually requires the solution to have the smallest vector norm among all possible solutions, just like choosing the spatial point closest to the origin as a representative. In order to solve such problems, it is necessary to build a solution tool that can intelligently screen out specific optimization goals in a specific system, namely the generalized inverse matrix method. It is not just an extension of ordinary matrix inverses, but also applies to non-square or irreversible matrices. Generalized inverse matrices mainly work by solving problems such as linear equations, least squares problems, data fitting, etc.

For a matrix A (whether it is a square matrix or a matrix with a dissatisfied matrix), its generalized inverse matrix usually refers to a matrix that can meet certain conditions. Its most common definition is the Moore–Penrose Inverse, which satisfies the following four conditions.

① $AA^+A = A$
② $A^+AA^+ = A^+$
③ $(AA^+)^T = AA^+$
④ $(A^+A) = A^+A$

If T represents the transpose of a matrix, then A^+ is the generalized inverse matrix of A

These conditions indicate that the generalized inverse matrix is some kind of "best" approximation of the inverse matrix, especially when solving the least squares problem, which can provide the optimal solution. Its application is very wide, especially when solving under-linear and over-linear systems (i.e., the number of equations is more or less than the number of unknowns). It plays an important role in signal processing, image processing, data science, and machine learning. It can also be used to solve least squares solutions to systems of linear equations, especially if there is no solution or infinite number of solutions to system equations.

Its generalized inverse matrix can be obtained by singular value decomposition (SVD). SVD determines whether the problem is pathological.

Suppose it is an m × n dimensional matrix, and its singular values are decomposed into.

$$A = U\Sigma V^T = \sum_{i=1}^{n} u_i\sigma_i v_i^T \tag{9.95}$$

There into $U = (u_1, u_2, \cdots u_n) \in R^{m\times \mathrm{n}}$, $V = (v1, v2, \cdots vn) \in R^{n\times \mathrm{n}}$, $\Sigma = diag(\sigma_1, \sigma_2, \cdots \sigma_n)$ is a non-negative value. $(\sigma_1, \sigma_2, \cdots \sigma_n)$ is a diagonal element, The other matrices are matrices of 0. $\sigma_1 \geq \sigma_2 \geq \cdots \geq \sigma_n \geq 0$. After the singular value is decomposed, the generalized inverse matrix of matrix A is:

$$A^+ = V\Sigma^{-1}U^T \tag{9.96}$$

In the formula, $\Sigma^{-1} = diag\left(\sigma_1^{-1}, \sigma_2^{-1}, \cdots \sigma_n^{-1}\right)$, with the help of singular value decomposition, can be represented by the condition number $cond(A) = ||A||_2\left||A^+\right||_2 = \frac{\sigma_1}{\sigma_{rank}\ (A)}$ to indicate the pathological degree of the problem.

Then the least squares solution in the sense of the minimum norm of the linear equation system is the result obtained by its generalized inverse matrix, as follows.

$$x = A^+b \tag{9.97}$$

The solution obtained by the generalized inverse matrix is shown below:

$$x = \sum_{i=1}^{n} \frac{u_i^T t}{\sigma_i} v_i \tag{9.98}$$

If A^TA is not inverse in the equation, the solution obtained by generalized inverse arithmetic is basically the same as that of least squares under full rank conditions. If singularity or approximate singularity occurs, it may not be solvable, and the least squares solution of the system of equations is not unique.

$$Ax \approx b \tag{9.99}$$

At this point, the generalized inverse will look for all x in the making of $\|Ax - b\|$ The smallest vector, And make it itself $\|x\|$ least, Its least norm solution is unique. The existing studies using radial basis function to reconstruct the temperature field have proved the effectiveness of this method. The generalized inverse matrix method is suitable for the indefinite equation, or the rank of matrix rank(A) < min{m, n} situation. However, its stability is still subject to the pathological degree of the anti-problem.

(3) Tikhonov regularization method

The Tikhonov regularization method is the most widely used approach. The main idea of this method is to incorporate a prior problem (regularization term problem) into the least squares, which can effectively reduce the pathological degree of the original problem. The regularization equation can be expressed as:

$$\min\left\{\|Ax - b\|_2^2 + \lambda^2\left\|L\left(x - x^*\right)\right\|_2^2\right\} \tag{9.100}$$

In the equation, the regularization parameter is the regularization operator, and the mass of the equation solution is measured by the sum of the squares of the residual norms, and the regular term avoids overfitting by adjusting the squareness of the norm of the solution, which is used for balance $\|Ax - b\|_2^2$ and $\|L(x - x^*)\|_2^2$ weight between. In Tikhonov's regularization algorithm, the selection of regularization parameters is the key to affecting its calculation accuracy. Generally speaking, the selection of regularization parameters should aim to minimize the norm deviation between the residual and the solution, while taking into account the stability and accuracy of the solution. At that time, it could be simplified.

$$\min\left\{\|Ax - b\|_2^2 + \lambda^2\|x\|_2^2\right\} \tag{9.101}$$

For the above equation, if the singular value decomposition method is used to look at the regularization problem, it is found that if the objective function reaches the minimum value, there are the following different forms of solutions.

$$x_{reg} = \sum_{i=1}^{n} \frac{u_i^T b}{\sigma_i} v_i \tag{9.102}$$

Compared with the least squares node, the filter factor is added:

$$f_i = \frac{\sigma_i^2}{\sigma_i^2 + \lambda^2} \tag{9.103}$$

When σ_i is much less than λ, f filters out the term of $\sigma_i < \lambda$:

$$f_i \frac{u_i^T b}{\sigma_i} v_i = \frac{\sigma_i^2}{\sigma_i^2 + \lambda^2} \frac{u_i^T b}{\sigma_i} v_i \approx \frac{\sigma_i}{\lambda^2} u_i^T b v_i \approx 0 \tag{9.104}$$

When $\sigma_i > \lambda$:

$$f_i \frac{u_i^T b}{\sigma_i} v_i = \frac{\sigma_i^2}{\sigma_i^2 + \lambda^2} \frac{u_i^T b}{\sigma_i} v_i \approx \frac{u_i^T b}{\sigma_i} v_i \tag{9.105}$$

(4) Algebraic reconstruction method

The algebraic reconstruction method is one of the more intuitive of all iterative algorithms, and its solution principle is to give the initial value of the information of the internal image of the measured area in advance [44], and calculate the image error value according to the initial value of the information and bring it into each sound wave propagation path, and perform the correction until the image error value is less than or equal to the predetermined termination condition.

In the reconstruction of the temperature field, the measured area is divided into discrete grids, and the number of grids is p = m × n (m represents the number of rows, n represents the number of columns), and the temperature values are the same in any discrete grid. Since the length of sound waves in the grid j (j = 1,...,p) can be directly measured, and the flight time of path i (I = 1,...,q) can also be directly measured, the slowness value (reciprocal value of sound velocity) in the discrete grid can be solved through calculation, and then the function relationship between temperature and sound slowness can be used to obtain the temperature value in the discrete grid to complete the temperature field reconstruction process. In summary, the matrix form of the mathematical model of temperature field reconstruction is:

$$Ax = y \tag{9.106}$$

where A—q × p dimensional distance coefficient matrix;

x—p × 1 dimensional acoustic slowness vector;

y—q × 1−dimensional acoustic TOF data vector;

q—number of sound wave propagation paths;

p—The total number of discrete meshes divided.

The execution process of the algebraic reconstruction algorithm is as follows [45]:

(1) The measured area is divided into different numbers of discrete meshes, and the initial value of sound velocity is x0.
(2) Based on the initial value of sound velocity x0 and the pre-calculated distance coefficient matrix A, the flight time of each sound wave propagation path corresponding to the initial value is obtained by using Eq. (9.102) to obtain the flight time y0 of each sound wave propagation path.
(3) Subtract the calculated flight time from the measured actual flight time to obtain the image error value.
(4) The error of each sound wave propagation path is projected back onto the corresponding path, and the error is evenly added to the discrete grid through which each sound wave propagation path passes, and the sound velocity value of the current grid is corrected.

After the above process is carried out in a cyclical manner, the image error value gradually decreases. When it is less than or equal to the termination conditions set in advance, the above execution process is completed to realize the reconstruction of the temperature field.

In the process of the kth iteration, for the jth grid through which the ith sonic propagation path passes, the iterative solution formula is.

$$x_{i,j}^{k+1} = x_{i,j}^{k} + \beta \frac{y_i - A_i x^k}{\|A_i\|^2} A_{i,j} \tag{9.107}$$

where, $x_{i,j}^{k}$ It represents the value of the sound velocity vector x of the ith sonic wave propagation path through the jth grid in the kth iteration, xk represents the value of the sound velocity vector X at the kth iteration, and β represents the relaxation factor (0 < β < 2), Ai represents the ith row of the distance coefficient matrix A, yi represents the ith row of the flying time vector y, and Ai and j represent the values of column j of row i of the distance coefficient matrix A.

The relaxation factor is a key parameter in the algebraic reconstruction algorithm, which affects the execution time and the final reconstruction quality. In general, the smaller the relaxation factor, the better the quality of the reconstruction and the more time it takes to perform. When the reconstruction error is large or the iteration speed is slow, the relaxation factor should be increased. Through the algorithm execution process, it can be seen that the final reconstruction result is obtained after several iterations and corrections, if the difference between the initial value of the set iteration and the real value is too large, it will cause too many iterations or even errors. Therefore, the setting of the initial value of the iteration is very important for the correctness of the reconstruction results.

(5) Radial basis function algorithm

The radial basis function method is also a commonly used algorithm in the field of temperature field reconstruction, which is a non-iterative method and is usually used in combination with SVD. The temperature field reconstruction algorithm based on the radial basis function and SVD combines the advantages of the two algorithms. Firstly, the mathematical model of temperature field reconstruction is obtained by relying on the radial basis function, and then the mathematical model is calculated by SVD, so as to reconstruct the temperature field of the measured area.

If the sound velocity function in the measured region is v(m, n) and the sound velocity reciprocal function is x(m, n), the following line integral can be used to represent the TOF of the sound waves along a certain propagation path in the measured region.

$$y_i = \int_{l_i} \frac{1}{v(m,n)} dl_i = \int_{l_i} x(m,n) dl_i \tag{9.108}$$

where li is the ith sound wave propagation path, and yi is the corresponding theoretical value of the sound wave flight time.

If the measured region is divided into p discrete meshes of exactly the same size, x(m, n) can be expressed as a linear combination of p radial basis function expressions with the geometric center of the divided mesh as the coefficient, as follows.

$$x(m,n) = \sum_{p}^{j=1} \varepsilon_j \phi_j(m,n) \tag{9.109}$$

In the equation, εj represents the undetermined coefficient, which depicts the spatial characteristics of the sound velocity reciprocal function x (m, n) in the measured area, and one of the main tasks of temperature field reconstruction is to calculate the value of the undetermined coefficient. (m, n) represents the position coordinates in the measured area, and (mj, nj) represents the center of the radial basis function, which is the center point of the jth mesh. φj (m, n) is the selected radial basis function, and different functions have different expression forms, and the common radial basis functions are: Gauss function, Markov function, MQ function, inverse MQ function and thin plate spline function. Table 9.5 shows the expressions of each radial basis function.

According to the characteristics of temperature field reconstruction, the Markov function is used as φj (m, n) [46, 47], where θ represents the shape parameter, and its value can usually be predetermined by numerical methods, and is generally only related to the transducer arrangement and the size information of the measured area.

After selecting the type of radial basis function φj (m, n), the expression of aij can be defined as:

$$a_{ij} = \int_{l_i} \phi_j(m,n) dl_i = \int_{l_i} \exp[-\theta \| (m,n) - (m_j, n_j) \|] dl_i \tag{9.110}$$

To obtain the undetermined coefficient εj, the Eqs. (9.104)–(9.106) are combined to obtain the following equation:

$$A = (a_{ij}), \quad i = 1, 2, \cdots, q, j = 1, 2, \cdots, p$$
$$\varepsilon = (\varepsilon_1, \varepsilon_2, \cdots, \varepsilon_p)^T$$

Table 9.5 Expressions of each radial basis function

Radial basis function name	Function expressions
Gauss	$\phi_j(m,n) = exp[-\theta^2 \| (m,n) - (m_j, n_j) \|]$
Markov	$\phi_j(m,n) = exp[-\theta \| (m,n) - (m_j, n_j) \|]$
MQ	$\phi_j(m,n) = 1 + \theta^2 \| (m,n) - (m_j, n_j) \|$
Inverse. MQ	$\phi_j(m,n) = \frac{1}{\sqrt{1+\theta^2 \| (m,n) - (m_j,n_j) \|}}$
Thin plate splines	$\phi_j(m,n) = \| (m,n) - (m_j, n_j) \| ln\left(\sqrt{\| (m,n) - (m_j, n_j) \|}\right)$

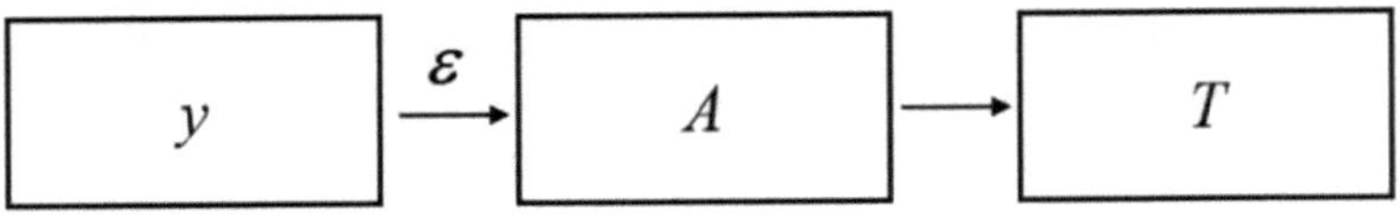

Fig. 9.13 Process of temperature field reconstruction and inversion

$$
\begin{aligned}
\mathrm{y} &= \left(y_1, y_2, \cdots, y_q\right)^T \\
y_{l_i} &= \sum_{j=1}^{p} \varepsilon_j \int_{l_i} \phi_j(m, n) dl = \sum_{j=1}^{p} a_{ij}\varepsilon_j
\end{aligned}
\tag{9.111}
$$

Then, the Eq. (9.111) can be rewritten into a matrix form:

$$
y = A\varepsilon \tag{9.112}
$$

For Eq. (9.112), if the propagation speed information of the acoustic wave signal in a certain measured area is already known, and the flyover time of a certain acoustic wave propagation path needs to be calculated, then this is a traditional forward modeling problem. Conversely, when using acoustic wave temperature measurement technology to reconstruct the temperature field, what needs to be addressed is a classic "from result to cause" inversion problem, that is, under the condition that the flyover time of all acoustic wave paths in the measured area is known, the temperature field information in that area needs to be solved. The inversion problem is shown in Fig. 9.13.

According to the SVD principle described in the previous section [48]. When we get the generalized inverse matrix A−, we can find the expression for the ε of the coefficients to be determined as follows.

$$
\varepsilon = A^{-}y \tag{9.113}
$$

In engineering, before reconstructing the temperature field, it is necessary to study the meshing method of the measured area, the arrangement of the acoustic transducer and the geometric characteristics of the measured area, and determine the shape parameter θ in advance by numerical method. From this point it can be seen that all the elements of matrix A can be found by the Eq. (9.104). The results show that the generalized inverse matrix A− is a constant matrix that can be calculated by the SVD method.

In simple terms, when the acoustic flight time matrix y is obtained by using the real measurement method, the value of the coefficient ε to be determined can be calculated using Eq. (9.113). In this case, by substituting the solved coefficients ε into Eq. (9.113), the function x (m, n) of the reciprocal sound wave velocity can be calculated, and the sound velocity function v (m, n) can be obtained.

In this case, if T (m, n) is used to represent the temperature distribution in the measured region, the temperature field reconstruction results can be obtained from the relationship between the sound velocity and the temperature of the medium as follows:

$$T(m, n) = \frac{1}{B^2} \cdot v^2(m, n) \tag{9.114}$$

The temperature field reconstruction information obtained through Eq. (9.114) can be visually displayed through graphic icons, etc., and can also be used as a reference for analysis, evaluation of operating status, or proposing control schemes when necessary.

Four commonly used algorithms for temperature field reconstruction, including acoustic CT algorithm, parabolic model algorithm, algebraic reconstruction algorithm, and radial basis function algorithm, play an important role in the field of temperature field reconstruction. The acoustic CT algorithm is based on the least squares method, which has the characteristics of simplicity and speed, and is suitable for the processing of a large amount of observation data and the reconstruction of temperature fields. The parabolic model algorithm approximately describes the temperature distribution through the parabolic function, which plays an important role in the reconstruction of the spontaneous combustion temperature field of loose coal. The algebraic reconstruction algorithm realizes the temperature field reconstruction by iteratively correcting the sound velocity value, and has an intuitive solution principle. The radial basis function algorithm is based on the radial basis function, with high accuracy and stability, and is suitable for complex temperature field reconstruction and high-dimensional data processing. These algorithms realize the reconstruction and analysis of temperature fields through different mathematical principles and models, providing effective data processing methods and technical support for scientific research and engineering practice.

The application of these algorithms provides a variety of options and methods for temperature field reconstruction, but there are still some limitations and shortcomings in practical applications. In order to further improve the accuracy and stability of temperature field reconstruction, it is necessary to introduce a temperature field reconstruction compensation algorithm. Through the compensation algorithm, new mechanisms and strategies can be introduced on the basis of existing algorithms, which can solve the limitations of existing algorithms, optimize the reconstruction results, and improve the accuracy and reliability of temperature field reconstruction.

9.4 Acoustic Temperature Measurement Results of Loose Coal

(1) Experimental conditions for reconstruction of the temperature field of loose coal

The experimental system includes 12 sets of speakers, microphones, computers, power amplifiers, data collectors, thermocouples, high-power heating rods, and thermostats. A speaker and a microphone form an acoustic transceiver, and the 12 transceivers are placed equidistantly at the four boundaries of the area to be tested. The area to be measured is a square area of 0.8 m × 0.8 m. The thermostat, thermocouple, and heating rod are formed into a set of temperature control devices for heating (Fig. 9.14).

By comparing the temperature measurement results of different sound source signals, it is found that the main peak of the interrelationship coefficient of the linear swept frequency signal in a certain frequency band range is more prominent, and the anti-interference is better than other signals, and the temperature measurement information is more accurate when the frequency sweep period is 0.1 s in the frequency band of 600–1000 Hz, so the temperature field reconstruction experiment adopts the linear swept frequency signal of 600–1000 Hz and the swept frequency period of 0.1 s. The transceiver layout used in the experiment and the path of the sound wave are shown in Fig. 9.15. The container is 0.8 m × 0.8 m, and each side is equistanically placed with a total of 46 microphones, including vertices, and a total of 42 unique sound wave propagation paths are generated. This arrangement can make each sound wave path occupy the entire area to be measured as much as possible.

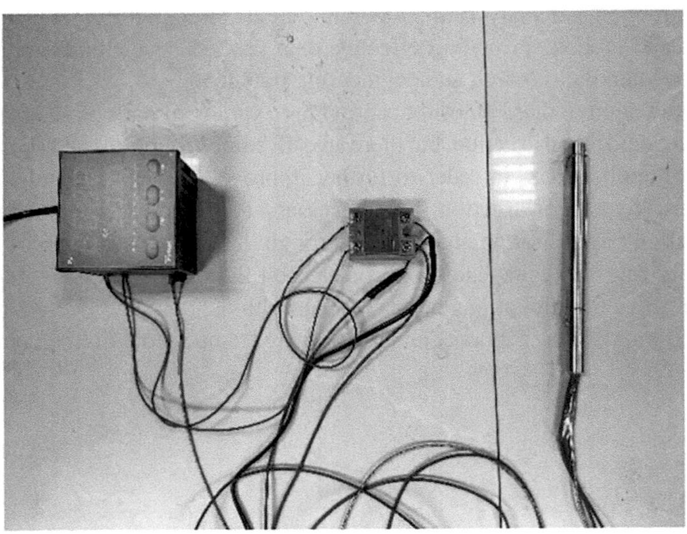

Fig. 9.14 Diagram of temperature control device

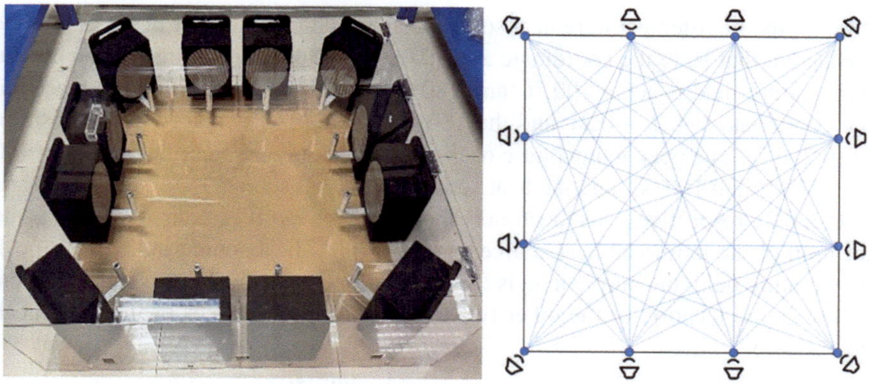

Fig. 9.15 Point location diagram of microphone arrangement on the experimental bench

Coal samples were added to acrylic containers. In goaf areas or coalmines, the particle size and stacking method of loose coal bodies are uncertain. To better conform to the actual situation, in this experiment, the particle size of the coal samples was randomly selected. The coal samples were uniformly and randomly added to the acrylic containers to completely cover the speakers and microphones. As shown in Fig. 9.16a.

Insert the heating rod into the coal pile, start the heating rod and temperature controller, heat the coal pile, because the coal is a bad conductor of heat, so it needs to be heated for a period of time, so that the coal sample is heated fully and evenly before the flight time is tested. According to the law of coal spontaneous combustion gas generation, the peak temperature of the single-peak temperature field is set to

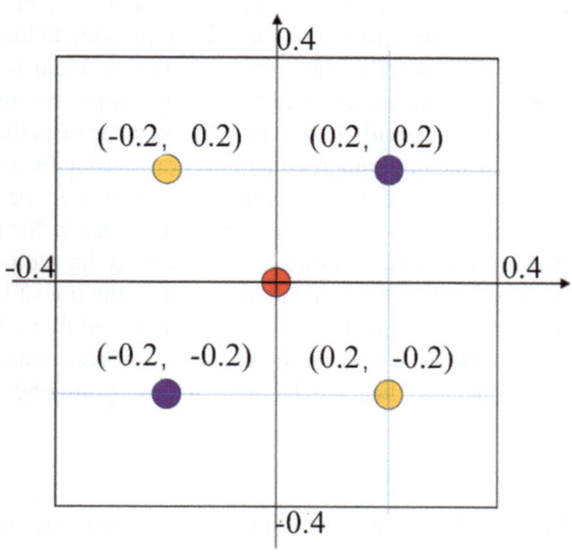

Fig. 9.16 Location of heat source and experimental process diagram

180 °C, the high-temperature heat source is arranged in the center of the area to be measured, that is, in the coordinate axis (0,0), the peak temperature of the bimodal temperature field is set at 100 °C and 180 °C, the 100 °C heating rod is placed at the coordinate (−0.2, −0.2), and the 180 °C heating rod is placed at the position of (0.2, 0.2). The peak temperature of the three-peak temperature field is set to 100 °C, 140 °C and 180 °C, and the heat source position is set to (0, 0), (0.2, −0.2) and (−0.2, 0.2), respectively. Its location is shown in Fig. 9.16b. Among them, red is the location of the unimodal heat source and the 100 °C heat source in the three-peak temperature field, the purple icon is the position of the bimodal heat source, and the yellow is the location of the other two temperature heat sources of the three-peak heat source.

(2) Reconstruction results of the temperature field of loose coal

Based on the above heat source location, temperature setting, characteristic factor selection and gas medium constant determination, the temperature field is reconstructed using the measurement data, and the reconstruction results are as follows.

It can be seen from Figs. 9.17, 9.18 and 9.19 that the SVD algorithm can be used to reconstruct the temperature field of loose coal, and the reconstructed temperature peaks are also relatively obvious. However, there are still certain errors in the reconstructed peak temperature and the position of the heat source. As can be seen from the figure, due to the resolution limitation of the mesh division during the reconstruction process, the reconstructed 2D temperature map presents the form of a square temperature field. This is because the mesh division is relatively rough, and the reconstructed sound slow value during the reconstruction process is the average sound slow value of a certain mesh. The reconstructed temperature value is also the average temperature value of a certain mesh. However, the temperature difference between the average temperature and the peak temperature of the grid where the adjacent grid is reconstructed is relatively large. Therefore, although the interpolation algorithm is used to reconstruct the temperature field, the reconstructed temperature field is not smooth. The solution to this problem is to optimize the algorithm to enable the temperature field reconstruction process to overcome the limitation that the number of grids is less than the number of paths. From the reconstructed 3D graphics and the isotherm diagrams drawn using the reconstructed temperature data, it can be seen that, except for the high-temperature peak, there are still certain errors in other areas at room temperature. The reason for the result is that the acoustic wave transit time measured experimentally has certain errors. In this experiment, the inversion method was used to deduce the transit time through temperature, and it was found that under completely ideal conditions, The flyover time derived from temperature is very accurate. In subsequent research, the reconstruction accuracy of the temperature field can be improved by optimizing the acoustic wave flyover time algorithm.

(3) Error analysis

Based on the construction of the experimental platform and the reconstruction results, a reconstruction error analysis was conducted. The reconstruction quality was verified

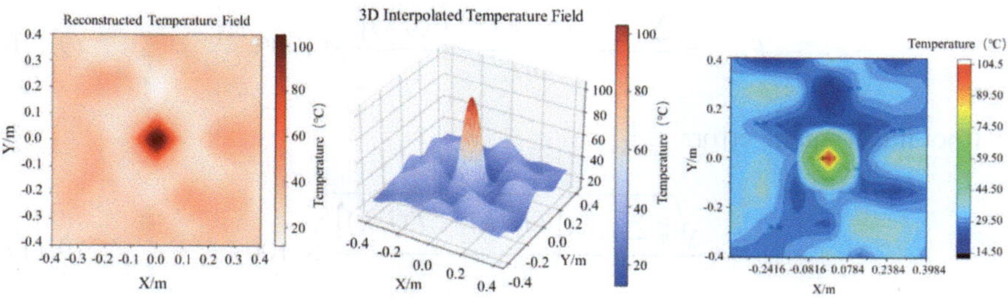

(a) 2D drawing. (b) 3D drawing. (c) Isotherm diagram.

Fig. 9.17 Reconstruction results of single peak temperature field of loose coal

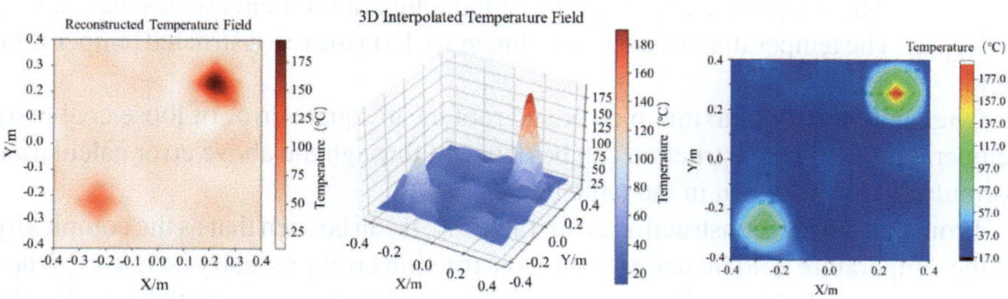

(a) 2D drawing. (b) 3D drawing. (c) Isotherm diagram.

Fig. 9.18 Reconstruction results of bimodal temperature field of loose coal

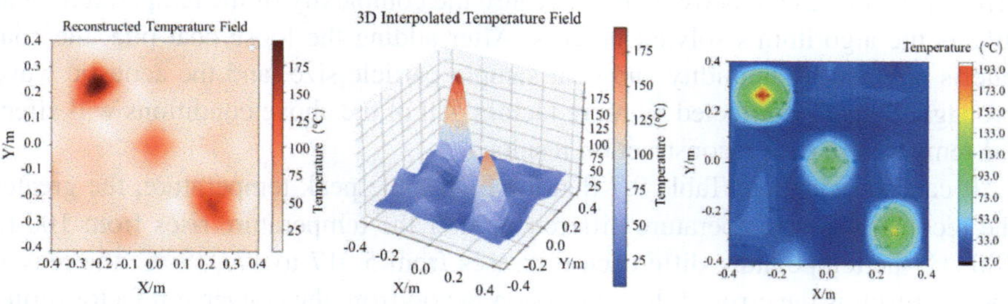

(a) 2D drawing. (b) 3D drawing. (c) Isotherm diagram.

Fig. 9.19 Reconstruction results of the three-peak temperature field of loose coal

by the maximum relative error, the relative average error, and the root mean square error. The calculation methods are as follows:

$$E\frac{max\left|T(j,k)-\hat{T}(j,k)\right|}{T_m}{}_{max} \tag{9.115}$$

Relative average error:

$$E_{ave} = \frac{\sum_{j,k=1}^{N,M} \left| T(j,k) - \hat{T}(j,k) \right|}{nT_m} \times 100\% \tag{9.116}$$

Root mean square error:

$$E_{err} = \frac{\sqrt{\frac{1}{n} \sum_{j,k=1}^{N,M} \left[T(j,k) - \hat{T}(j,k) \right]}}{T_m} \times 100\% \tag{9.117}$$

where:

n: Number of reconstruction peaks.

Tm.: The average temperature of the basal temperature field.

T(j, k): The temperature value at the actual temperature field coordinates (j, k).

T(j, k): The temperature value at coordinates (j, k) in the reconstructed temperature field.

The relative error, maximum error and root mean square error of loose coal body temperature field reconstruction can be obtained through the above error calculation formula, which is shown in the table below.

From the above reconstruction error Table 9.6, it can be seen that as the complexity of the temperature field increases, the reconstruction error gradually increases. When the temperature field transitions from a single peak temperature to three peaks, the average error increases from 7.48 to 10.73%, the maximum error increases from 6.01 to 17.03%, and the root mean square error increases from 8.25 to 13.35%. It can be observed that the more complex the temperature field, the greater the reconstruction error of the loose coal body. This is because the complexity of the temperature field affects the algorithm's solving process. After adding the loose coal pile, the coal pile is affected by humidity and coal sample particle size, and the acoustic wave propagation path is affected by more factors. All of the above conditions will affect the temperature field reconstruction results.

It can be seen from Table 9.7 that the higher the peak temperature, the greater the reconstruction temperature difference, after the temperature rises from 100 to 180 °C, the temperature difference increases from 5.017 to 17.075 °C, the reason for this result is the error of the conversion factor, from the conversion factor fitting formula, it can be seen that it rises exponentially with the increase of temperature, which causes the higher the temperature, the faster the conversion factor rises, and

Table 9.6 Error value of the temperature field

Temperature field type	Unimodal temperature field	Bimodal temperature field	Three-peak temperature field
Average error/%	7.48	8.56	10.73
Maximum error/%	9.71	12.94	17.07
Root mean square error/%	8.25	9.86	13.35

the greater the error, so the larger the error of the conversion slow matrix, the larger the error of the temperature field reconstruction.

In addition to the above influencing factors, the factors causing the error include the bending effect of the sound wave propagation path, the noise affects the sound wave propagation temperature measurement, when the sound wave signal passes through the high temperature area, its propagation path will have a bending effect, when other conditions are constant, the sound line is always bent in the direction of the decrease of sound speed, in this study it can be seen that the higher the temperature, the greater the sound velocity around it, so the sound line will bend in the direction of deviating from the high temperature area, resulting in the measurement error of the sound wave flight time.

According to the above temperature field reconstruction results, the peak coordinates of different types of temperature field reconstruction can be known, and the heat source position reconstruction error can be known compared with the actual heat source temperature. The statistical table of heat source location is shown in Table 9.8.

It can be seen from the above statistical table of the heat source location that there is also an error in the reconstruction of the heat source position by the temperature field, with a maximum difference of 0.07 m, because the initial meshing method 7×

Table 9.7 Reconstruction of peak temperature

Temperature field	Actual temperature/°C	Reconstruct temperature/°C	Difference/°C
Unimodal temperature field	100	105.017	5.017
Bimodal temperature field	100	108.418	8.418
	180	192.943	12.943
Three-peak temperature field	100	105.227	5.227
	140	153.373	13.373
	180	197.075	17.075

Table 9.8 Heat source location

Temperature field	Heat source temperature	Actual heat source coordinates	Reconstruct the heat source coordinates
Unimodal temperature field	100	(0, 0)	(1.14E^{-14}, 1.43 E^{-14})
Bimodal temperature field	100	(−0.2, −0.2)	(−0.27, −0.26)
	180	(0.2, 0.2)	(0.27, 0.27)
Three-peak temperature field	100	(0, 0)	(0, 0)
	150	(0.2, −0.2)	(0.27, −0.27)
	180	(−0.2, 0.2)	(−0.26, 0.26)

7 used in this paper is relatively rough, and the reconstructed heat source peak is in the center of a grid, so after inserting the temperature by the interpolation algorithm, the heat source position will change slightly, and the solution to this problem is to improve the grid resolution as much as possible in the follow-up research, but the problem of pathology is aggravated while increasing the resolution. May affect the reconstruction accuracy of temperature peaks. Therefore, it is necessary to improve the resolution of the grid and reduce the peak temperature error of the reconstruction to make the reconstruction results more accurate.

References

1. Guo, M. (2019). *Temperature measurement of stored biomass using acoustic method.* Ph.D. dissertation, North China Electric Power University. Beijing China (郭淼. 基于声学法的堆积生物质温度测量研究. 北京:华北电力大学, 2019)
2. Yuan, W. T., Yang, S. H., Liu, D., Zhang, W. Z., & Zhang Y. (2021). Research on optimization of aggregate gradation on self-compacting concrete. *New Building Materials, 48*(10), 70–74 (袁文韬, 杨仕海, 刘东, 张武宗, 张远. 自密实混凝土骨料级配优化研究. 新型建筑材料, 2021, 48(10):70–74)
3. Zhang, T. J., Chen, J. W., Bao, R. Y., Jiang, X. K., Zhou, A., & Ge, Di. (2018). Fractal characteristics of particle size distribution of broken coal samples with different immersion time. *Journal of Mining and Safety Engineering, 35*(03), 598–604 (张天军, 陈佳伟, 包若羽, 蒋兴科, 周敖, 葛迪. 不同浸水时间级配破碎煤样的粒度分布分形特征. 采矿与安全工程学报, 2018, 35(03):598–604)
4. Zhang, J. W., Wang, J. C., Wei, W. J., & Li, L. H. (2019). Experimental investigation on the effect of size distribution on the flow characteristics of loose top coal. *Journal of China Coal Society, 44*(04), 985–994 (张锦旺, 王家臣, 魏炜杰, 李良辉. 块度级配对散体顶煤流动特性影响的试验研究. 煤炭学报, 2019, 44(4):985–994).
5. Deng, J., Qu, G. Y., Ren, S. J., Wang, C. P., & Zhao, X. Y. (2022). Experimental study on optimization of low frequency acoustic transmission frequency in loose coal. *Safety in Coal Mines, 53*(01), 15–23 (邓军, 屈高阳, 任帅京, 王彩萍, 赵小勇. 松散煤体中低频声波传声频率优选实验研究. 煤矿安全, 2022, 53(01):15–23)
6. Zhou, H. S., Yu, Q., Zhang, H., & Wang, H. (2009). The application of acoustic pyrometry in measuring the temperature field of the furnace consuming coal. *Acoustic Technology, 28*(06), 752–756 (周红生, 喻强, 张华, 王欢. 声学测温技术在燃煤炉膛温度场测量中的应用. 声学技术, 2009, 28(06):752–56)
7. Shen, G. Q., Zhang, S. P., An, L. L., Li, G. S., Feng, Q., & Deng, Z. (2013). Research on acoustic temperature measurement sound source of power station boiler. *Electric Power Science and Engineering, 29*(02), 49–55 (沈国清, 张世平, 安连锁, 李庚生, 冯强, 邓喆. 电站锅炉声学测温声源研究. 电力科学与工程, 2013, 29(02):49–55)
8. Luo, Z., Sun, X. P., & Tian, F. (2005). Comparison of acoustic wave signal delay estimation methods in strong noise environment. *Journal of Shenyang Institute of Aeronautic Engineering*, (02), 45–48 (罗振, 孙小平, 田丰. 强噪声环境下声波信号时延估计方法的比较. 沈阳航空工业学院学报,2005,(2):45–48)
9. Liu, L. Y., Tian, F., Li, Z. Z., & Sun, E. Y. (2005). Design and implementation of acoustic flight time measurement system based on cross-correlation technology. *Journal of Shenyang Institute of Aeronautical Technology*, (1), 57–59 (刘立云, 田丰, 李治壮, 孙恩岩. 基于互相关技术的声波飞行时间测量系统的设计与实现. 沈阳航空工业学院学报, 2005,(1):57–59)
10. Liu, Z. Q. (2018). Measure of molar gas constant by sound velocity measuring apparatus. *Research and Exploration in Laboratory, 37*(5), 31–33+53 (刘竹琴. 利用声速测量仪测量摩尔气体常量. 实验室研究与探索,2018,37(5):31–33+53)

11. Zhang, K. S., & Zhang, X. Q. (2021). Method of gas concentration detection based on single-frequency sound speed. *Machinery & Electronics, 39*(10), 19−22 (张克声, 张向群. 基于单频点声速的气体浓度检测方法. 机械与电子, 2021,39(10):19–22)
12. Wang, M. J. (2012). *Acoustic measure method and technical research of gas medium temperature field.* Ph.D. dissertation, Northeast Petroleum University, Daqing China (王明吉. 气体介质温度场声学测量方法与技术研究. 大庆:东北石油大学, 2012)
13. Shen, X. H. (2016). *Research on temperature distribution reconstruction algorithm based on acoustic thermometry.* Ph.D. dissertation, Chongqing University, Chongqing China (沈雪华. 基于声学测温的温度场重建算法研究. 重庆:重庆大学, 2016)
14. Kleppe, J., Fadali, S., Lapolla, M., & Fralick, G. C. (2001). High temperature gas measurement in combustors using acoustic pyrometry method. *Proceedings of the International Instrumentation Symposium, 47*, 441–449.
15. He, Q.W., Yu, Z. Q., Li, Y. Q., Tian, G. Y., & Zhou, H. C. (2003). Acoustic detection device for furnace velocity field. *Automation & Instrumentation*, (03), 42−45 (何其伟, 於正前, 李言钦, 田国银, 周怀春. 炉膛速度场声学检测装置. 自动化与仪器仪表, 2003, (03):42–45)
16. Song, Z.Q., & Fan, X. (2005). Detection technology and application of boiler temperature field by acoustic method. *Boiler Technology, 36*(01), 20−23 (宋志强, 樊旭. 声学法锅炉温度场检测技术及应用. 锅炉技术, 2005, (01):20–23+69)
17. Du, G. H., Zhu, Z. M., & Gong, X. F. (2012). *Fundamentals of Acoustics (Third Edition).* Nanjing University Press, Nanjing China (杜功焕, 朱哲民, 龚秀芬. 声学基础(第三版). 南京: 南京大学出版社, 2012)
18. Davis, R. S. (1982). Equation for the determination of the density of moist air (1981). *Metrologia, 18*(01), 33–40.
19. Rasmussen, K. (1997). Calculation methods for the physical properties of air used in the calibration of microphones. *Technical University of Denmark Report PL−11b*
20. Tsilingiris, P. T. (2008). Thermophysical and transport properties of humid air at temperature range between 0 and 100 °C. *Energy Conversion and Management, 49*(05), 1098–1110.
21. Hardy, B. (1998). ITS-90 formulations for vapor pressure, frost point temperature, dewpoint temperature, and enhancement factors in the range −100 to +100 °C. *The Proceedings of the Third International Symposium on Humidity & Moisture*, Teddington, London, England.
22. Greenspan, L. (1976). Functional equations for the enhancement factors for CO_2-free moist air. *Journal of Research of the National Bureau of Standards Section A Physics and Chemistry, 80*(01), 41–44.
23. Bohn, D. A. (1988). Environmental effects on the speed of sound. *Journal of the Audio Engineering Society, 36*(04), 223–231.
24. Hyland, R., & Wexler, A. (1983). Formulations for the thermodynamic properties of the saturated phase of H2O from 173.15 K to 473.15 K. *Ashrae Transactions, 89*(2), 500−519
25. Chen, G. N. (2012). *The study on key technology of acoustic temperature measurement in stored grain.* Ph.D. dissertation, Shenyang University of Technology, Shenyang China (陈冠男. 声学法仓储粮食温度检测关键技术的研究.沈阳:沈阳工业大学, 2012)
26. Xu, X. M. (2003). *Fundamentals of acoustics.* Science Press, Beijing China (许肖梅. 声学基础. 北京:科学出版社, 2003)
27. Wong, K. V., Bon, B. L., Santina, V., & Samedi, S. (2012). Use of ultrasound thermometry to study natural convection in opaque nanofluids. *Combustion Science and Technology, 1567*, 154–170.
28. DeSilva, U., Bunce, R., & Claussen H. (2013). Novel gas turbine exhaust temperature measurement system. *In Proceedings of the ASME Turbo Expo*, ASME Turbo Expo 2013: Turbine Technical Conference and Exposition, 4.
29. Kim, T. K., & Lh, J. G. (2015). Inverse estimation of the temperature field within a gas-filled duct section by use of acoustic data. *Measurement Science and Technology, 26*(6).
30. Shen, G. Q., An, L. S., & Jiang, G. S. (2006). Real-time monitoring on boiler combustion based on acoustic measurement. In *Proceedings of 2006 IEEE Power India Conference*, pp 41−44
31. Pan, H., Ying, C., Jie, M., & Wu, M. (2006). Progress in developing a device for measuring heat flux from the hydrothermal vent in deep ocean using acoustic method. *In Proceedings of the OCEANS MTS/IEEE Quebec conference & Exhibition, 9*, 15–20.

32. Wu, Y. (2002). Time delay estimation of non-Gaussian signal in unknown Gaussian noises using third-order cumulants. *IEEE Electronics Letters, 38*(16), 930–931.
33. Chen, Y., Xu, K. B., & Liu, M. S. S. (2014). Application study on multiple acoustic paths intersection in temperature field measurement. *Advanced Materials Research, 902*, 167–172.
34. Viola, F., & Walke, W. (2005). A spline-based algorithm for continuous time-delay estimation using sampled data. *IEEE Transactions Ultrasonics, 52*(01), 80–93.
35. Ouadfeul, S. A., & Aliouane, L. (2010). Random seismic noise attenuation data using the discrete and the continuous wavelet transforms. *Soil Dynamics and Earthquake Engineering, 30*(11), 1168–1181.
36. Miyamoto, R., Mizutani, K., Ebihara, T., & Wakatsuki, N. (2015). Defect detection and size estimation in billet from profile of time-of-flight using ultrasonic transmission method with linear scanning. *Japanese Journal of Applied Physics, 54*(07), 07HC11
37. Kychakoff, G., Hollingshead, A. F., & Boyd S. P. (2005). Use of acoustic temperature measurements in the cement manufacturing pyro-process. In *Proceedings of 2005 IEEE Cement Industry Technical Conference*, pp 23−33
38. Zhang, S. P., Shen, G. Q., An, L. S., & Li, G. S. (2015). Ash fouling monitoring based on acoustic pyrometry in boiler furnaces. *Applied Thermal Engineering, 84*, 74–81.
39. Peng, X. J., Cai, Y., Li, Q., & Wang, K. (2016). The least-squares method based on coupling coefficients for reactor power distribution reconstruction. *Annals of Nuclear Energy, 94*, 272–278.
40. Fischer, G., Starke, M., & Ziemann, A. A. (2012). Acoustic tomography of the atmosphere: Comparison of different reconstruction algorithms. *Acta Acustica united with Acustica, 98*(04), 534–545.
41. Esmaeilbeigi, M., & Hosseini, M. M. (2013). A dynamic meshless method for the least squares problem with some noisy subdomains. *Applied Mathematical Modelling, 37*(05), 3152–3163.
42. Wang, R., An, L.S., Shen, G. Q., & Zhang, S. P. (2014). Research on simulation of furnace three-dimensional temperature field reconstruction by acoustics based on singular value decomposition. *Proceedings of the Chinese Society for Electrical Engineering, 34*(S1), 147−152 (王然, 安连锁, 沈国清, 张世平. 基于奇异值分解的炉膛三维温度场声学重建仿真研究. 中国电机工程学报, 2014, 34(S1):147–152)
43. Jin, J., Peng, S. L., Liu, Q. L., & Deng, M. (2010). A comparative study of spatial interpolation methods for meteorological elements in China. *Geotechnical Investigation & Surveying, 38*(11), 48−51 (金君, 彭思岭, 刘启亮, 邓敏. 中国陆地区域气象要素空间插值方法比较研究. 工程勘察, 2010, 38(11):48–51)
44. Gordon, R., Bender, R., & Herman, G. T. (1970). Algebraic reconstruction techniques (ART) for three-dimensional electron microscopy and x-ray photography. *Journal of theoretical biology, 29*(03), 471–81.
45. Chen, Y. S. (2023). *Algorithms and system design of temperature field reconstruction based on acoustic waves.* Master Thesis, University of Electronic Science and Technology of China, Chengdu China (陈尹胜. 基于声波的温度场重建算法与系统设计. 电子科技大学, 2023)
46. Zhu, W. H., & Yang, J. (2014). Study on the prediction of Shanghai composite index based on a fusion model of RBF neural network, Markov chain and GA. *4th International Conference on Intelligent System and Applied Material (GSAM)*, (1049), 1413−1416
47. Spivak, S. I., & Abdyusheva, S. R. (2012). Inverse problems for Markov models. *Automation and Remote Control, 73*(11), 1929–1936.
48. Minamide, A., Mizutani, K., & Wakatsuki, N. (2008). Temperature distribution measurement using reflection with acoustic computerized tomography. *Japanese Journal of Applied Physics, 47*(5S), 3967–3969.

Chapter 10
Quantum Detection Method

In recent years, the nanoscale heat transfer properties have been widely studied in crystalline materials such as semiconductors and silicon carbide. The thermal transport is essentially caused by the movement of hot carriers (electrons and phonons in coal molecular) and the energy they carry at the nanoscale, which is vector and represents the direction of heat transmission, and the frequency domain characteristics, motion characteristic parameters, and thermal motion rate of hot carriers are all related to temperature, which can meet the detection requirements of high temperature points of hidden ignition sources, and can become a promising method for accurately detecting the location and degree of concealed ignition sources.

10.1 Mechanism of Thermal Carrier Heat Transport in Coal

The effect of hot carriers on the thermal conductivity of solid materials and their contribution to the study. The main modes of thermal conduction of solids are the movement of free electrons and lattice vibration. As a solid combustible organic rock, coal has certain conductive properties and lattice structure, indicating the presence of free electrons and phonons that transport energy in coal molecules. Kang et al. [1] found that within 30–90 °C, the resistivity of coal with a high degree of metamorphism decreased linearly with the increase of temperature, the free electrons in the coal increased, and the electrical and thermal conductivity increased. Liu et al. [2] used microwave heating technology to analyze the mechanism of dielectric response during coal warming, and found that more valence electrons absorbed energy transitioned to the conduction band during coal warming, indicating that the number of sp^2-hybrid carbon atoms and delocalized electrons in coal increased with the increase of microwave radiation time. Lyu et al. [3] studied the effect of temperature on the conductivity of raw coal, and found that the resistivity of coal weakened in the

Y. Xiao et al., *Detection of Fire Source in Underground Coal Mine and its Applications*,
https://doi.org/10.1007/978-981-95-7064-5_10

range of 15–65 °C. Deng et al. [4] found that with the increase of temperature, the lattice vibration in coal intensifies, and the number of phonons and the probability of phonon collision increases, resulting in the decrease of the mean free path of phonon and the weakening of thermal diffusivity, and when the temperature exceeds 210 °C, the number of phonons in coal tends to be saturated, and the thermal diffusivity of coal begins to increase. Xiao et al. [5] used the grey correlation method to study the influence of characteristic parameters of coal crystal structure on thermal energy transport at different temperatures, and also found that the number of phonons in coal increased slowly with the increase of temperature during coal pyrolysis within 200 °C. Deng et al. [6] used the laser thermal conductivity instrument LFA457 to analyze the heat transport performance of coal in nitrogen and air atmosphere, and revealed the influence of lattice vibration in coal on its thermal diffusivity. In summary, with the increase of temperature, the conductivity and lattice vibration in the coal are enhanced, and the number of free electrons and phonons gradually increases, while when the coal is at high temperature, the number of phonons in the coal is gradually saturated, and its heat transport capacity is gradually limited. Therefore, the analysis of the relationship between the electron and phonon heat carriers and their coupling and their contribution to the heat transfer performance of coal, and the construction of an electron–phonon thermal energy transport model are helpful to understand the main heat energy transport modes of coal at different temperatures.

Multi-scale scattering characteristics of hot carriers. Under the temperature gradient, solid heat conduction is the diffusion and transport of heat energy from the region with high concentration of electrons and phonons to the region with low concentration, and there are many influencing factors hindering the diffusion during the heat transport process, resulting in the scattering of hot carriers, including between hot carriers scattering caused by collisions between hot carriers and point defects and internal impurities. Tao et al. [7] used the annealing method to adjust the thermal carrier concentration of three-dimensional AZO films, and found that when the thermal carrier concentration was low, electron–phonon scattering and electron–electron scattering jointly dominated the thermoelectric transport properties, revealing the scattering mechanism of hot carriers at low temperatures. Polshyn et al. [8] studied the electrical transport characteristics of twisted bilayer graphene with different twisting angles of 0.75°–2°, and found that electron–phonon scattering was the main factor affecting its electrical transport characteristics. Lv et al. [9] found that the resistivity of Bi2O2Se has a quadratic relationship with temperature, revealing that the electron scattering rate increases with the increase of temperature. Jeffries et al. [10] used two-dimensional electron spectroscopy to analyze the electron–electron scattering in plasma nanorods, and found that the electron–electron scattering rate and the excitation pulse energy increased quadratically. Jaćimovski et al. [11] studied the relaxation mechanism caused by the collision of electrons with impurities, phonons, vacancies, etc., and found that the collision of electrons and phonons at low temperatures is the main factor affecting the properties of thermoelectric transport. Huang and Wang [12] used the equilibrium molecular dynamics method to analyze the microscopic transport mechanism of gallium arsenide under the inelastic scattering of phonons, and found that with the increase of temperature, the lattice thermal conductance first increased and then decreased, and the maximum value appeared. Tang et al. [13] mainly discussed the scattering rate model of point defects and dislocations, and constructed a thermal conductivity model of GaN thin films based on first-principles calculations and classical Debye–Callaway models, revealing the influence of phonon scattering caused by point defects and dislocations on the heat transport parameters of GaN thin films. Nshimyimana et al. [14] studied the effect of phonon scattering caused by point defects on the lattice thermal conductivity of GeTe thermoelectric materials. He et al. [15] studied the effects of acoustic phonon, phonon-impurity and optical phonon scattering on the tunable propagation characteristics of graphene material structure, and found that acoustic phonon scattering was dominant at room temperature, while optical phonon scattering was the main influencing factor when the temperature exceeded 300 °C. In summary, the thermal carrier scattering can significantly affect the thermal conductivity and other thermal characteristic parameters, increase the relaxation time, hinder the heat conduction in solid materials, and the main scattering mechanism is inconsistent at different temperatures. Therefore, the study of the defect state characteristics of the lattice and band structure in coal, and the analysis of defects, impurities, and collisional scattering between phonon and between phonon and electron in coal are helpful to reveal the multi-scale scattering characteristics of hot carriers in the temperature field of hidden ignition sources.

Study on the electron–phonon coupled non-equilibrium diffusion transport mechanism of solid materials. From the nanoscale, it can be found that the heat conduction process is mainly the energy transfer of heat carriers through motion and collision. At present, the femtosecond–picosecond time scale laser pumping detection thermal reflectance (TDTR) and molecular dynamics analysis electron–phonon coupled non-equilibrium diffusion theory are mainly used, and in the molecular dynamics isothermal simulation, the temperature adjustment includes Berendsen thermal bath, Gaussian thermal bath and Nose–Hoover thermal bath. Zhang et al. [16] used TDTR to analyze the thermal reflectivity of single crystal silicon carbide with time at low temperature, and then the thermal transport performance of silicon carbide was studied, and it was found that the thermal conductivity of silicon carbide appeared at a maximum value at −173 °C. Wang et al. [17] explored the acoustic projectile transport phenomenon of monocrystalline silicon cells at different temperatures based on TDTR, and found that at low temperatures, the average free path of monocrystalline silicon phonon increases with the increase of temperature. Fladischer et al. [18] used TDTR combined with scanning thermomicroscopy to analyze the thermal conductivity and other thermal properties of CVD films. Lu et al. [19] used the TDTR method to detect the thermal conductivity change of transmetal-insulator transition epitaxy and polycrystalline VO_2 films, and revealed the correlation between the rate of change of thermal conductivity and the rate of change of electrical conductivity. In the molecular dynamics simulation of nanoscale non-equilibrium thermal processes, Hu et al. [20] analyzed different model-level thermal carrier excitation at the nanoscale based on the non-equilibrium molecular dynamics and phonon Boltzmann transport equation to study their thermal transport laws. Krylow et al.

[21] analyzed the time domain of the lattice structure in graphene after ultrashort laser excitation based on molecular dynamics simulation, and obtained the electron–phonon coupling characteristics. Dong et al. [22] simulated the effects of polymorphs, size, temperature and strain on the heat transport of cellulose nanocrystals through molecular dynamics, and found that the thermal conductivity depended on the polymorphs of cellulose nanocrystals, and there was significant ballistic diffusion phonon transport in cellulose nanocrystals, revealing the mechanism of ballistic diffusion phonon transport in the crystals. In summary, TDTR and molecular dynamics can be used to obtain the thermal conductivity changes of crystal materials in femtosecond–picosecond time, and analyze the non-equilibrium diffusion transport process of electrons and phonons at the nanoscale, but there are relatively few studies on the non-equilibrium thermal diffusion transport mechanism of electrons and phonons in coal molecules. Therefore, the influence of electrons, phonons and their couplings on thermal conductivity in coal is studied, and the constitutive equations of thermal characteristics of heat carriers and temperature during the thermal energy transport of hidden ignition sources are established, so as to reveal the electron–phonon coupling non-equilibrium diffusion transport mechanism in the temperature field of coal fire, and provide a theoretical basis for the exploration of thermal energy and high-temperature regional transfer of coal.

10.2 Characterization of the Basic Characteristics and Structure of Coal Samples

The thermal conductivity modes of coal at the microscopic level mainly include free electron movement and lattice vibration, and the construction of coal molecular model is to study the thermal carriers (electrons and phonons) in coal. The premise and basis of thermal movement. Therefore, this chapter first studies the basic characteristics and structure characterization of the selected coal samples from typical coal fire areas, and uses industrial analysis, elemental analysis, X-ray photoelectron spectroscopy (XPS) and solid-state NMR carbon spectroscopy (^{13}C–NMR). The structural parameters, such as the atomic ratio, carbon skeleton, and main functional groups in the coal molecule were obtained by testing and analysis, which provided a basis for the subsequent construction of the coal molecular structure model.

10.2.1 Industrial Analysis and Elemental Analysis

The experimental coal samples came from the M14 coal seam of Tunpu Coalmine, which is located in Fukang Mining Area, Changji, Xinjiang, China, with serious spontaneous combustion and a long history of fire, and the coal samples in this area can be fully used as coal samples in typical coal fire areas for experiments. The coal samples are dried in the laboratory to a constant quality, ground to below 200 mesh

Table 10.1 Industrial analysis of coal samples

Carbon fixed (%)	Ash (%)	Volatile matter (%)	Moisture (%)
53.38	7.0	40.40	3.70

Table 10.2 Elemental analysis of coal samples

C (at%)	H (at%)	N (at%)	O (at%)	H/C	N/C	O/C
72.05	3.91	0.79	21.24	0.65	0.01	0.22

(<0.08 mm), and sealed in sealed bags for storage. The 5E-MAG6700 industrial analyzer and the German Elementar vario EL cube elemental analyzer were used for industrial analysis and elemental analysis tests on coal samples.

There are two kinds of water in coal, namely chemical water and free water, free water refers to the water adsorbed in the internal pores and outer surface of the coal, and the water combined with the coal through the physical state, including internal moisture and external moisture. In industrial analysis, the moisture index mainly refers to the external moisture in the drying base after the external moisture is removed. Ash refers to the residue left after all the organic matter in the coal is burned. Volatile matter includes the substances formed by pyrolysis of coal during the heating process, and the gas produced by the decomposition of inorganic matter, which represents the degree of metamorphism of coal. Fixed carbon refers to the solid products left after the pyrolysis of organic matter in coal. The results of the industrial analysis of the coal samples shown in Table 10.1 indicate that the coal samples are a type of lignite with a low degree of metamorphism.

Table 10.2 summarizes the results of elemental analysis of coal samples. The elemental content of the samples C, H, N, O and S was determined using the Elementar vario EL cube, where the elemental analysis result was 0 because the S content was below the lower limit of the elemental analyzer detection range (0–6 mg abs. on the Elementar vario EL cube), so it is not included in the table and is not covered in the subsequent analysis. The ratio of hydrogen to carbon (H/C), the ratio of oxygen to carbon (N/C), and the ratio of nitrogen to carbon (O/C) of coal samples were calculated using industrial and elemental analysis data.

10.2.2 Solid State Nuclear Magnetic Resonance Carbon Spectrum (^{13}C-NMR)

The German Bruker Avance III. HD 600MHZ nuclear magnetic resonance instrument ^{13}C-NMR (Fig. 10.1) was performed on the coal sample. The CP/MAS broadband solid probe was used to detect a range of 31P-15N, a test temperature of 299.2 K, a relaxation delay of 0.5 s, a pulse width of 3.84 μs, a sampling time of 0.035 s, a nuclear magnetic resonance frequency of 100.63 MHz, a spectral width of 30,120.5 Hz, and a cumulative number of data acquisition times of 6000 times.

The ^{13}C-NMR spectra of the coal samples can be divided into three regions according to the chemical shift, which are the lipid carbon region of 0–100 ppm, the aromatic carbon region of 100–170 ppm, and the carbonyl carbon region of about 200 ppm. As shown in Fig. 10.2, Peakfit was used to perform peak fitting on the solid-state NMR spectrum, and the spectrum was divided into 13 peaks, all of which were Gaussian peaks. Aromatic carbon occupies the dominant position in coal, reaching 68.91%, and aromatic carbon determines the carbon skeleton and

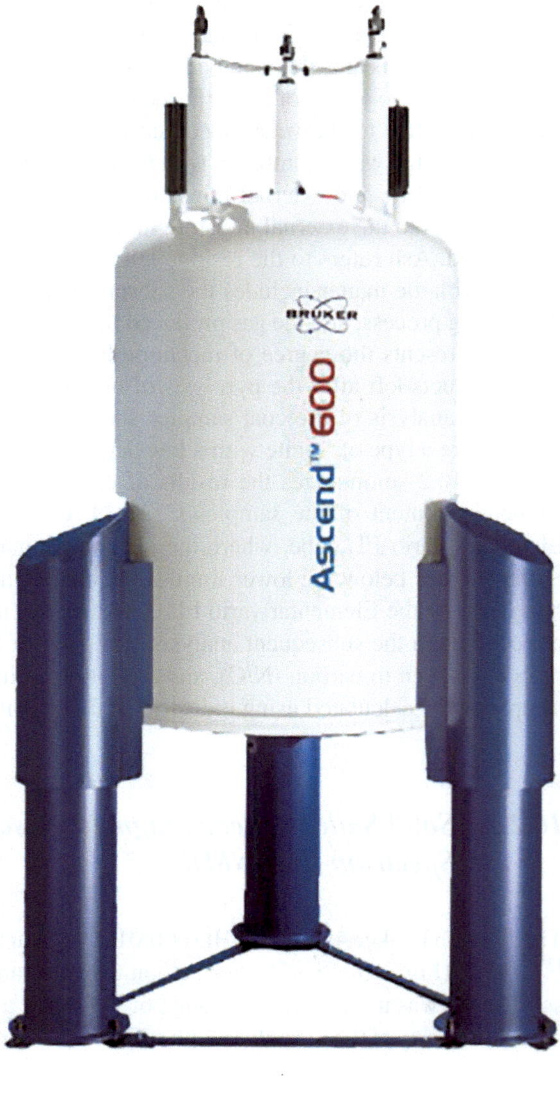

Fig. 10.1 Avance III. HD 600 MHz nuclear magnetic resonance spectrometer

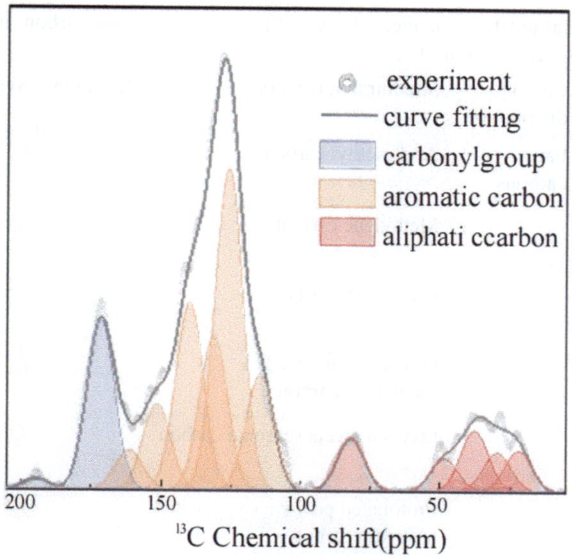

Fig. 10.2 Peak fitting diagram of ^{13}C-NMR spectra of coal sample

ring structure of coal samples. The second is fatty carbon, which determines the form of branching in the coal molecule. Carbonyl carbon accounted for the smallest proportion, only 13.88% (Table 10.3).

10.2.3 X-Ray Photoelectron Spectroscopy (XPS)

Thermo Scientific ESCALAB 250Xi photoelectron spectrometer (Fig. 10.3) was used in the United States3 X-ray photoelectron spectroscopy (XPS) test was performed on the coal sample. The X-ray source is Al–Kα, the photoelectron energy is 1486.6 eV, the line width is 0.85 eV, the obtainable spot size is 30–400 μm, the ion gun energy range is 100–4000 eV, the vacuum degree is 8, the angle of incidence is 90°, the power is 72 W, and the resolution is 0.4 eV. XPS spectra of C, N, and O elements (atoms) are obtained.

The carbon, nitrogen, and oxygen in the coal samples were present in the form of multiple functional groups, and the XPS analysis software Avanage was used for charge correction (correction C: 284.8 eV) and peak fitting, and the results are shown in Fig. 10.4 and Table 10.4. The carbon atoms in the coal samples existed in four ways, namely, aromatic graphitized carbon C = C, which peaked at 284.03 eV and accounted for 75.33%. The aliphatic carbon C–H was at 284.80 eV, accounting for 21.01%. The peak position was 285.45 eV, accounting for 3.65% hydroxycarbon C–O and carbonyl carbon C=O. The nitrogen atom comes in two forms, pyridine nitrogen and pyrrole nitrogen, where the amount of pyrrole nitrogen is twice that of

Table 10.3 Chemical shifts and relative contents of carbon of different structures in ^{13}C-NMR spectra of coal samples

Carbon distribution	Structural attribution	Location	Symbol	Chemical shift (ppm)	Relative content (%)
Fatty carbons	Aryl methyl carbon	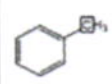	f_{al}^{a}	21.91	3.10
	Methylene carbon	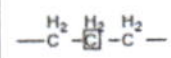	f_{al}^{1}	29.57	2.95
	Quaternary carbon	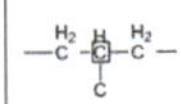	f_{al}^{2}	37.63	4.62
	Oxygen is attached to methylene carbon		f_{al}^{3}	48.38	2.48
	Oxygen meets seasonal carbon	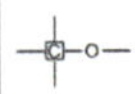	f_{al}^{O}	81.95	4.08
	Protonated positive oxygenated aromatic carbon	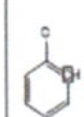	$f_{ar}^{O_1}$	114.56	8.78
	Protonation of aromatic carbon		f_{ar}^{H}	125.93	24.59
Aromatic carbon	Aromatic bridgehead carbon	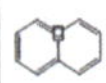	f_{ar}^{B}	131.27	11.71
	Aromatic branched chain carbon	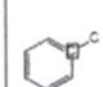	f_{ar}^{C}	139.85	14.28
	Oxygen replaces aromatic carbon	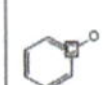	$f_{ar}^{O_2}$	151.48	6.55
				161.38	3.00
Carbonyl carbon	Carboxyl carbon		f_{a}^{C}	171.78	12.97
	Carbonyl carbon		f_{a}^{O}	194.54	0.91

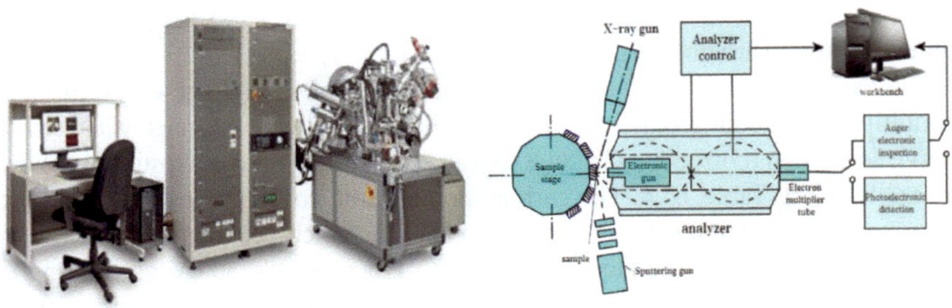

(a) Thermo Scientific ESCALAB 250Xi Photoelectron Spectrometer (b) Experimental principle

Fig. 10.3 X-ray photoelectron spectroscopy experiments

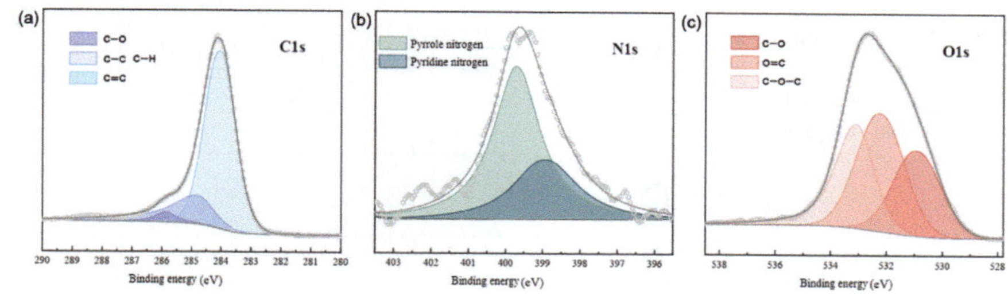

Fig. 10.4 XPS C (1 s), N (1 s), and O (1 s) peak spectra of coal sample

Table 10.4 XPS C (1 s), N (1 s), and O (1 s) peak data of coal samples

Elemental peaks	Peak position (eV)	Chemical state	Relative content (%)
C1s	284.03	C=C	75.33
	284.80	C–H	21.01
	285.45	C–O, C=O	3.65
N1s	398.91	Pyridine	32.11
	399.71	Pyrrole	67.89
O1s	530.91	C–O–C	28.65
	532.22	C=O	39.65
	533.09	C–O	31.7

pyridine nitrogen. Oxygen atoms exist in the coal sample structure in the form of ether-based oxygen C–O–C, carbonyl oxygen C=O and C–O, with peak positions of 530.91, 532.22 and 533.09 eV, and relative areas of 28.65%, 39.65% and 31.7%, respectively.

10.2.4 Coal Macromolecular Structure Analysis

Based on the above test results, heteroatom analysis, aromatic carbon and aliphatic carbon analysis were carried out to study the macromolecular structure of coal from the relatively clear nitrogen element. As shown in Table 10.5, the nitrogen element in the coal sample occurs in two forms, pyridine nitrogen and pyrrole nitrogen, where the content of pyrrole nitrogen is twice that of pyridine nitrogen. Therefore, it can be assumed that the number of nitrogen atoms in a single coal macromolecular model is 3, containing two pyrrole molecules and one pyridine molecule. According to the elemental analysis and atomic ratios of each element listed in Table 10.2, the atomic numbers of C, H and O in a single coal sample macromolecular model were calculated to be 310, 202 and 68, respectively.

The aromatic carbon ratio refers to the ratio of aromatic carbon to total carbon content, which is calculated by Eq. (10.1) and the aromatic carbon ratio of coal-like

Table 10.5 Aromatic structure in the macromolecular structure model of coal samples

Aromatic structure	Quantity (pcs)
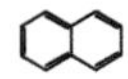	15
	2
	1
	2

macromolecules is 68.91%. The degree of aromatic ring substitution is defined as the proportion of non-hydrogen-linked aromatic carbon to total aromatic carbon, which is calculated by Eq. (10.2) and has an aromatic ring degree δ of substitution of 064. The ratio of bridge carbon to circumferential carbon is used to determine the degree of condensation of the aromatic structure in the coal, and the value obtained by Eq. (10.3) is $X_{BP} = 0.205$. In view of the ratio of bridge carbon to weekly carbon of naphthalene is 0.250, it can be seen that the aromatic structure of coal-like macromolecules is mainly bicyclic and single-ring. By changing the structure and number of aromatic rings so that their value is closer to 0.205. According to the sum of the total carbon number and aromatic carbon rate X_{BP}, it can be determined that the aromatic structure in the coal is composed of 15 bicyclic aromatic structures naphthalene and 2 monocyclic aromatic structures benzene, as shown in Table 10.5, with a X_{BP} value of 0.206.

$$f_{\mathrm{ar}} = f_{ar}^{O_1} + f_{ar}^{O_2} + f_{ar}^{H} + f_{ar}^{C} + f_{ar}^{\mathrm{B}} \tag{10.1}$$

$$\delta = \frac{\left(f_{ar}^{O_1} + f_{ar}^{O_2} + f_{ar}^{\mathrm{B}} + f_{ar}^{\mathrm{C}}\right)}{f_{\mathrm{ar}}} \tag{10.2}$$

$$X_{\mathrm{BP}} = \frac{f_{ar}^{B}}{f_{ar}^{O_1} + f_{ar}^{O_2} + f_{ar}^{H} + f_{ar}^{C}} \tag{10.3}$$

10.3 Construction of Lignite Macromolecular Structure Model

Because of the different characteristics of coal and rock composition and metamorphism, the molecular structure of coal has unique complexity, variability and heterogeneity, and it is generally believed that it has the structure of polymer, but because it does not have a unified polymeric monomer, it is generally believed that the molecular structure of coal is composed of several basic structural units with similar structures connected by bridge bonds. Combined with the structural parameters such as the atomic ratio, carbon skeleton and main functional groups in the lignite molecule, the macromolecular structure model of the coal was preliminarily constructed, and the N MR chemical shift of each C atom in the coal molecule after structural optimization was calculated. The chemical shift data of each C atom were imported into gNMR, and the peak shape was set to Gaussian peak, and the calculated spectrum of the coal macromolecular structure model was obtained. The calculated and experimental spectra are compared, and the structural model is adjusted accordingly. Subsequently, the adjusted structural model was calculated by NMR. The calculations involve several iterations to ensure that the calculations match well with the experimental spectra.

10.3.1 NMR Calculations

(1) ORCA.

ORCA [23] was developed in 1995 as a molecular spectroscopy program by a team led by Professor Frank Neese of the Max Planck Institute in Germany to study certain features of open-shell transition metal complex spectroscopy, and has since evolved into a general-purpose quantum chemistry program. ORCA uses DFT to calculate the electronic structure and spectral properties of molecules, and uses a variety of density functionals, including local LDA, GGA, and hybrid density functionals, to balance the accuracy and speed of calculation, and can calculate aperiodic macromolecular systems.

In addition, ORCA can calculate the mechanical properties and electronic structure of molecules using molecular force fields and empirical molecular orbital methods. Among them, the molecular force field method is based on the interaction potential energy between atoms in a molecule, and the empirical molecular orbital method is based on the empirical correction of the electron energy level in the molecule, which can provide higher computational speed and higher computational efficiency than DFT in some cases.

ORCA has been a molecular spectroscopy program since its early days, and in its newly published version 5.0, the ability to predict nuclear magnetic resonance (NMR) properties has been greatly optimized, with significant enhancements to chemical shielding and spin coupling, greatly improving the accuracy and timeliness of predicting NMR. In this study, all NMR calculations were done using ORCA 5.0.2.

(2) Selection of functional and base groups.

The revTPSS function and the double-hybrid functional DSD–PBEP86 [24] are much more accurate than other functions in calculating NMR. In addition, since the KT function is one of the few functions that takes into account the magnetic characteristics, the KT function is also considered as a function of NMR calculations. The base group chose def2–TZVPP [25] The base set as well as the pcSseg base set specifically for NMR calculations. The optimized geometry was calculated in a vacuum using B3LYP–D3(BJ)/def2−TZVP. In this section, Multiwfn [26] is used to process the NMR calculation output file. Table 10.6 lists the TMS at different compute levels (Fig. 10.5) C magnetic shielding value.

As shown in Fig. 10.6, the correctness of the different calculation methods was tested with benzene, naphthalene, pyridine and pyrrole. The experimental values and chemical shifts at different levels of calculation for these four molecules, as well as their mean absolute error (MAE) relative to the experimental values, are shown in Table 10.7.

From the data in Table 10.7, it can be judged that the NMR chemical shift of C calculated using revTPSS/pcSseg-1 is more accurate than that calculated using other methods. Compared with the larger double-hybrid functional, the revTPSS functional of the Meta−GGA grade shows obvious advantages in the accuracy of NMR calculations. In terms of the basis set, the def2−TZVPP basis group, which is commonly used in computational chemistry and performs well, does not show its

Table 10.6 C-magnetic shielding values of TMS (unit: ppm)

	revTPSS			KT1	KT2	DSD−PBEP86
	pcSseg-1	pcSseg-2	pcSseg-3	def2−TZVPP	def2−TZVPP	def2−TZVPP
TMS	188.68	188.99	189.34	190.93	190.49	194.32

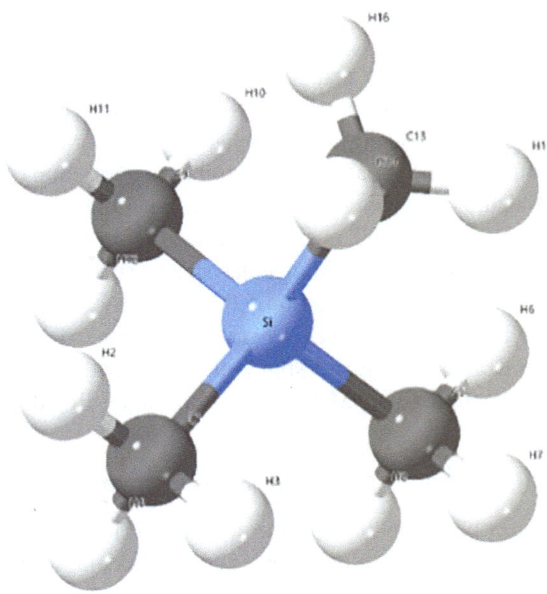

Fig. 10.5 TMS ((CH3) 4Si), tetramethyl lsilane, NMR internal standard

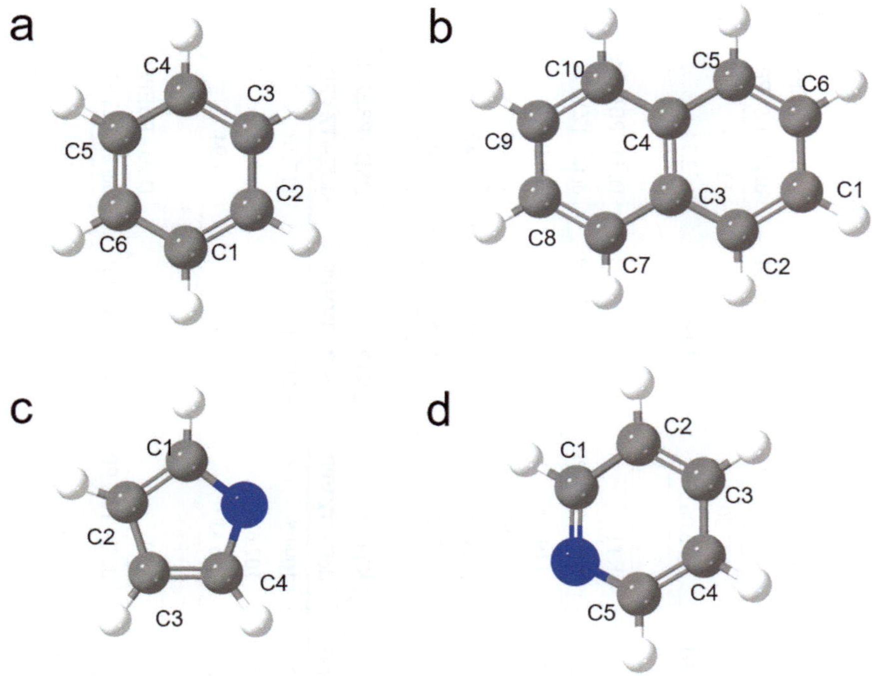

Fig. 10.6 **a** Benzene, **b** Naphthalene, **c** Pyrrole, **d** Pyridine

performance over the pcSseg base set, and the accuracy and speed of the pc Sseg base set reflect its superiority as a basis set for calculating NMR compared with the spectral base set. For the pcSseg base set, the accuracy of using the larger pcSseg-2 and pcSseg-3 base sets is not as high as that of pcSseg-1, which indicates that part of the reason for the good results of revTPSS/pcSseg-1 calculation is that the functional and base group errors are well offset, and this phenomenon is widespread.

10.3.2 Three-Dimensional Structure Model of Coal Macromolecules

(1) Geometric optimization and annealing dynamics simulation.

In Materials Studio, the coal molecular model was constructed according to the coal molecular structure parameters obtained from the coal sample structure characterization test and analysis, and the simple initial optimization was carried out after hydrosaturation to optimize the atomic arrangement of the coal molecules, and the preliminarily constructed coal molecular model was an approximate planar structure.

Table 10.7 Chemical shifts of C at different calculation levels for different aromatic structures (unit: ppm)

(a) Benzene

	Experimental values	revTPSS			KT1	KT2	DSD−PBEP86
		pcSseg-1	pcSseg-2	pcSseg-3	def2−TZVPP	def2−TZVPP	def2−TZVPP
C (1,2,3,4,5,6)	128.36	127.45	130.84	131.39	122.30	122.70	131.10
MAE (C)		0.91	2.48	3.03	6.06	5.66	2.74

(b) Naphthalene

	Experimental values	revTPSS			KT1		DSD−PBEP86
		pcSseg-1	pcSseg-2	pcSseg-3	def2−TZVPP	def2−TZVPP	def2−TZVPP
C (1,6,8,9)	125.44	124.98	128.05	128.60	119.50	119.89	128.00
C (2,5,7,10)	127.73	127.59	131.29	131.72	122.51	122.78	130.67
C (3,4)	133.25	131.04	137.34	138.08	128.06	128.66	138.42
MAE (C)		0.94	3.42	3.99	5.45	5.03	3.56

(c) Pyrrole

	Experimental values	revTPSS			KT1	KT2	DSD−PBEP86
		pcSseg-1	pcSseg-2	pcSseg-3	def2−TZVPP	def2−TZVPP	def2−TZVPP
C (1,4)	117.23	115.07	117.66	118.22	109.05	110.01	117.56
C (2,3)	107.30	108.36	110.84	111.57	102.67	103.46	110.85
MAE (C)		1.61	1.99	2.63	6.41	5.53	1.94

(d) Pyridine

	Experimental values	revTPSS			KT1	KT2	DSD−PBEP86
		pcSseg-1	pcSseg-2	pcSseg-3	def2−TZVPP	def2−TZVPP	def2−TZVPP
C (1,5)	149.90	149.93	153.79	154.70	144.21	144.75	153.31

(continued)

Table 10.7 (continued)

(d) Pyridine

	Experimental values	revTPSS			KT1	KT2	DSD–PBEP86
		pcSseg-1	pcSseg-2	pcSseg-3	def2–TZVPP	def2–TZVPP	def2–TZVPP
C (2,4)	123.29	122.36	125.42	125.93	117.45	117.81	126.21
C (3)	135.69	133.17	136.41	137.00	127.52	128.07	137.14
MAE (C)		1.16	2.25	2.92	6.57	6.08	2.59

The geometric optimization and annealing dynamics of a single coal macromolecular structure can obtain the three-dimensional optimal geometric configuration and minimum energy configuration of the molecular structure of a single coal. In the process of geometric optimization, the Smart method is used to search for the optimal configuration of coal molecules. The SMART method is a commonly used optimization method to efficiently search for the lowest energy point in a complex energy surface. The number of iterations of 5000 steps was set as the termination condition of the optimization process, and the FINE standard was used to judge whether the optimization converged, and the standard of RMS Force was 0.005 kcal/(mol·Å), the energy deviation is 0.0001 kcal/mol. In order to obtain the net charge of the atom, the charge balance method is employed to determine the molecular force field charge distribution. In molecular dynamics simulations, two interaction energies, Coulomb energy and van der Waals energy, are considered and calculated using atomic states. For annealing simulations, NVT ensembles are used to keep the number of particles, volume, and temperature constant. The nose temperature control method was used to perform high temperature relaxation under the NVT ensemble, the initial temperature was set at 300 K, the maximum temperature was set at 800 K, and the heating rate was 60 K/time. The total time for the annealing simulation was set at 1 fs and 10 cycles were performed. At the end of each cycle, the output configuration is used to perform molecular dynamics calculations. A Compass force field is used to simulate the structure, vibrational frequency, thermodynamic properties, and structure and properties of condensed matter of isolated molecules with high accuracy, which is defined as Eq. (10.4). From the output results, the configuration with the least energy is selected as the optimal geometric configuration.

$$E_{\text{total}} = E_{\text{A}} + E_{\text{B}} + E_{\text{C}} + E_{\text{I}} + E_{T} + E_{\text{van}} + E_{\text{elect}} + E_{\text{H}} \tag{10.4}$$

where E_{A} is the bond angular energy; E_{B} for the key to stretch and contract; E_{C} is the interaction energy of the cross term; E_{I} for the reversal energy; E_{T} for torsional energy; E_{van} for Van der Waalen; E_{elect} is the electrostatic interaction energy; E_{H} is hydrogen bonding energy. E_{A}, E_{B}, E_{C} E_{I}, and E_{T} are valence electron energy, E_{van}, E_{elect}, and E_{H} are nonbonding energy.

(2) Three-dimensional structure of coal macromolecules.

Figure 10.7 shows the optimized molecular structure of coal and the molecular model of coal after annealing kinetics, and the optimal structure model of coal molecules is obtained.

ORCA used the rev-TPSS/pcSseg-1 method to calculate the optimal structure model of coal molecules, and the calculated magnetic shielding value of C atom was converted into a chemical shift value. As shown in Fig. 10.8, the calculated and experimental spectra for constructing the coal macromolecular structure model are consistent as a whole. However, due to the sideband effect in the NMR carbon spectrum test experiment, the absorption peak intensity of the carbon−oxygen region in the experimental spectrum is high. Therefore, the intensity of the absorption peak

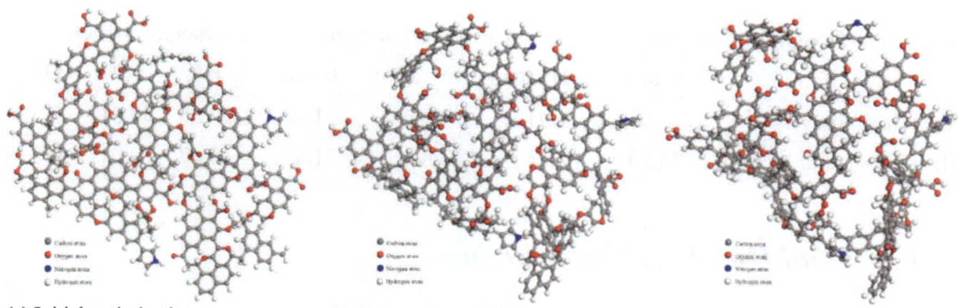

Fig. 10.7 Molecular model of coal

in the 160–180 ppm region of the calculated spectrum is significantly smaller than that of the experimental spectrum, which is common and difficult to avoid.

The chemical formula of the macromolecular structure model of Tunpu coal is $C_{311}H_{209}N_3O_{68}$, and the molecular weight is 5076.05 g/mol. The elemental mass fractions of C, H, N and O were 73.59%, 4.15%, 0.83% and 21.43%, respectively. Table 10.8 compares its results with those of elemental analysis tests with an acceptable tolerance. Therefore, the research on Tunpu coal and the established macromolecular structure model are reliable.

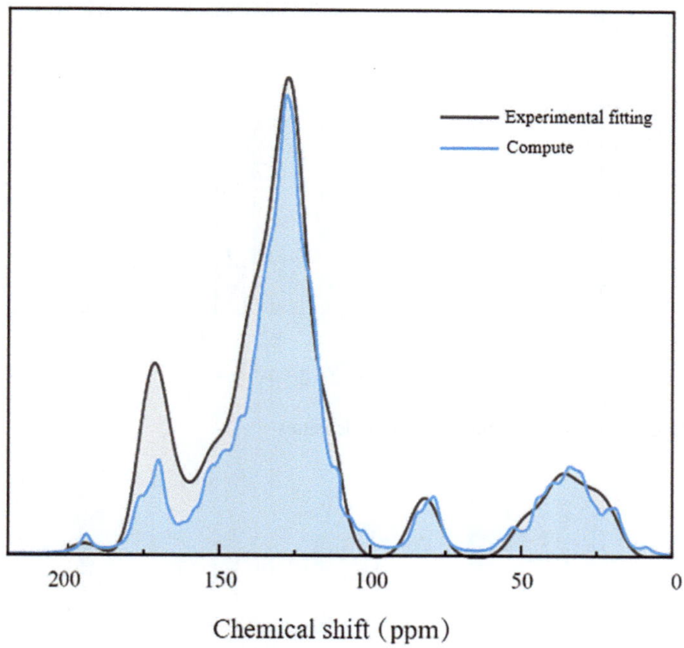

Fig. 10.8 Experimental and computational spectra of ^{13}C-NMR of coal samples

Table 10.8 Comparison of elemental analysis tests with macromolecular structure models

	C (at%)	H (at%)	N (at%)	O (at%)	H/C	N/C	O/C
Elemental analysis	72.05	3.91	0.79	21.24	0.65	0.01	0.22
Macromolecular models	73.59	4.15	0.83	21.43	0.67	0.01	0.22

10.3.3 Coal Unit Cell Construction

Amorphous Cell was used in Materials Studio to add the coal macromolecule model to the unit cell, and by changing the density of the structural model, the lowest energy configuration system was selected as the structural model for molecular dynamics simulation. As can be seen from Fig. 10.9, when the density of the structural model is 0.95 g/cm^3, the lowest energy configuration is obtained, and the unit cell size at this time is 20.7 × 20.7 × 20.7 Å (1 Å = 10–10 m), as shown in Fig. 10.10.

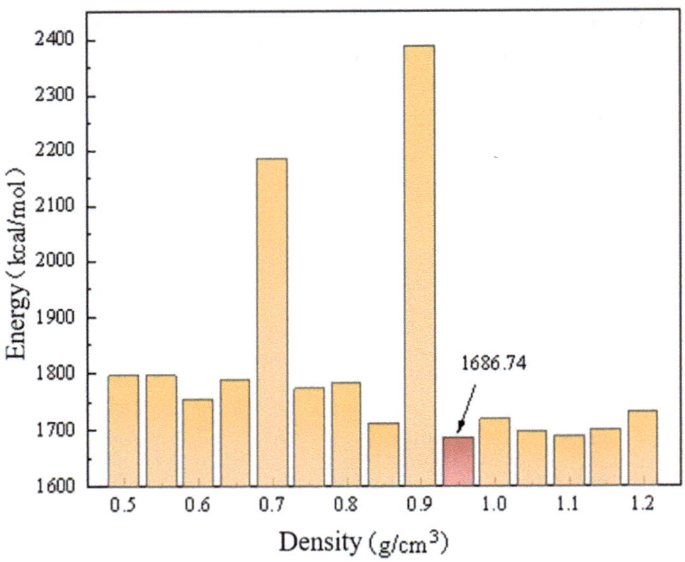

Fig. 10.9 Energy of coal unit cells at different densities

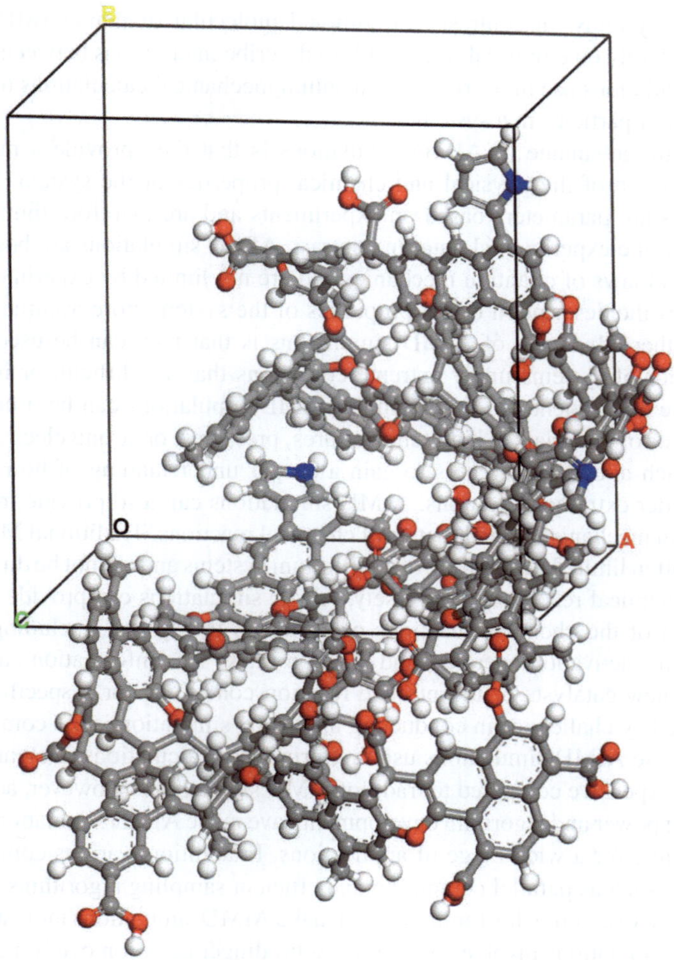

Fig. 10.10 Lowest energy structure of coal unit cells

10.4 Thermal Carrier Temperature Sensing Characteristics in Lignite

10.4.1 Ab Initio Molecular Dynamics Simulations

The Ab Initio Molecular Dynamics (AIMD) computational method in CP2K [27], which simulates the kinetic processes of chemical systems by calculating the electronic structure from scratch, is a powerful computational tool in the field of computational chemistry and materials science to study the behavior of atoms and molecules in specific systems. In contrast to traditional molecular dynamics (MD) simulations, which rely on empirical force fields to describe interactions between particles, AIMD simulations use first-principles quantum mechanical calculations to simulate the motion of particles in a system.

The main advantage of AIMD simulations is that they provide a more accurate description of the physical and chemical properties of the system. Empirical force fields are parameters based on experiments and are therefore limited by the accuracy of the experimental data. In contrast, AIMD simulations are based on the fundamental laws of quantum mechanics and are not limited by experimental data. This makes the description of the properties of the system more accurate and reliable. Another advantage of AIMD simulations is that they can be used to study the behavior of systems under extreme conditions that are difficult or impossible to replicate experimentally. For example, AIMD simulations can be used to study the behavior of materials at high temperatures, pressures, or strong electromagnetic fields, which allows researchers to gain a deeper understanding of how materials behave under extreme conditions. AIMD simulations can also provide insight into the reaction mechanisms and kinetics of chemical reactions. Traditional MD simulations are often limited to the study of equilibrium systems and cannot be directly used to study chemical reactions. Conversely, AIMD simulations can provide a detailed description of the chemical reactions occurring in the system, including reaction mechanisms, activation energies, and reaction rates. This information can be used to design new catalysts or to optimize reaction conditions for a specific process. One of the key challenges in conducting an AIMD simulation is the computational cost. Because AIMD simulations use first-principles calculations, they are computationally expensive compared to traditional MD simulations. However, advances in computing power and algorithm development have made AIMD simulations increasingly feasible for a wide range of applications. In addition, various computational techniques such as parallel computing and efficient sampling algorithms have been developed to overcome this limitation and make AIMD simulations more accessible.

AIMD simulations involve solving the Schrödinger equation over time to obtain the electronic structure of the system at each time step, and then using this electronic structure to calculate the forces acting on the atoms in the system. These forces are used to propagate the position and velocity of the atoms forward in time. Simulations can be performed in microscopic classical sets, where the total energy of the system is conserved, or in classical sets, where the temperature of the system is controlled. The accuracy of AIMD simulations depends on the quality of the electronic structure calculations used to calculate the force acting on the atom, and the most common electronic structure calculation method in simulations is DFT.

In conclusion, AIMD simulations have several significant advantages over traditional MD simulations, including a more accurate description of the physical and chemical properties of the system, the ability to study the system under extreme conditions, and the ability to directly study chemical reactions. Despite the computationally high cost of AIMD simulation, advances in computing power and algorithm development have made it increasingly feasible for a wide range of applications. In this study, AIMD was used in CP2K to study the effect of temperature on hot carriers. The Nose–Hoover thermostat was used to establish NVT ensembles with target temperatures of 300, 450, 600, 750, and 900 K, with a time step set to 1 fs and 10,000 steps (10 ps).

10.4.2 Thermal Carrier Temperature Sensing Characteristics

(1) Temperature-sensing characteristics of electronic density of states.

The results shown in Figs. 10.11 and 10.12 show that the peak position of the first peak on the conduction band of the coal molecule shifts towards the Fermi level with increasing temperature, from 1.21 eV at 300 K to 0.72 eV at 900 K. At the same time, the bandgap width was reduced from 0.59 eV at 300 K to 0.42 eV at 600 K, and then increased to 0.82 eV at 900 K. These trends reflect the effect of temperature on the electronic properties of coal molecules.

At higher temperatures, the atoms in the coal molecule have greater kinetic energy and vibrate more violently. This will lead to changes in the electronic structure of the coal molecule, which in turn will affect its conductivity and electron transport characteristics. Specifically, thermal energy can excite electrons from the valence band to the conduction band, causing the peak position of the first peak of the conduction band to shift to the left. The decrease in the band gap can be attributed to the increase in the thermal energy of the atom, which increases the probability of electrons transitioning from valence to conduction bands, allowing more electrons to participate in the conduction process, resulting in higher conductivity. These results show that the electronic properties of coal molecules are highly dependent on temperature, and changes in temperature can have a significant impact on their conductivity and electron transport properties. Changes in the bandgap width have an important impact on conductivity and electron transport. As the temperature increases, the reduction of the bandgap increases the conductivity of the material, as more electrons can cross the band structure into the conduction band. This increases electron mobility, which improves electron transport performance. On the other hand, the reduction of the band gap also increases the electron concentration of the material, which increases the conductivity of the material.

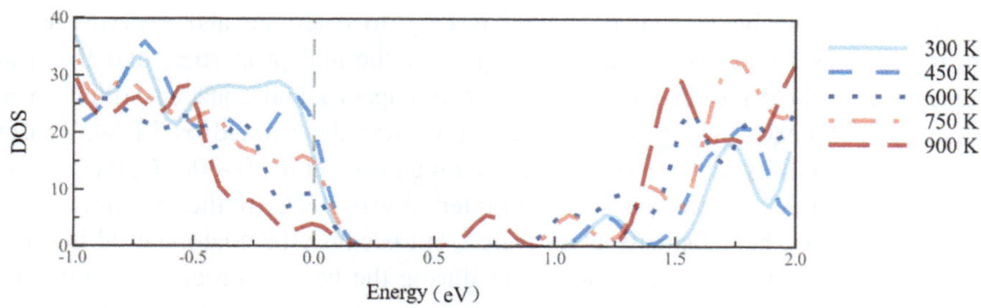

Fig. 10.11 Total electron density of states and fractional density of coal molecules at different temperatures

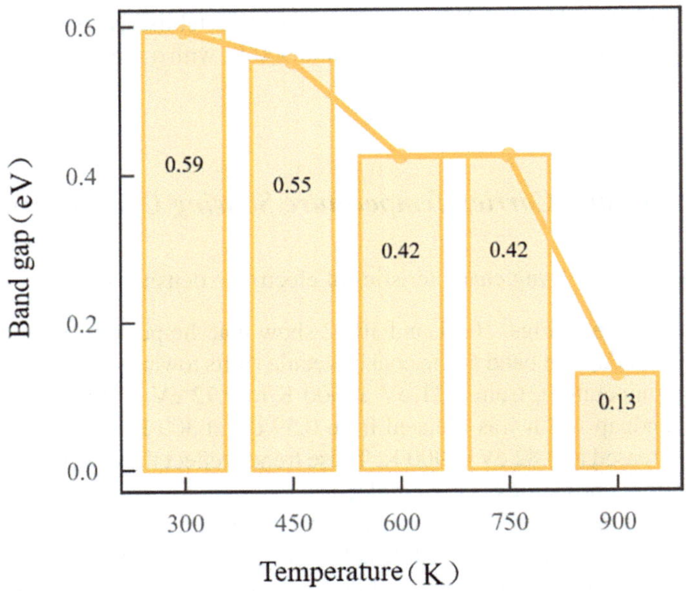

Fig. 10.12 Bandgap widths of coal molecules at different temperatures

(2) Temperature-sensing characteristics of phonon density of states.

Figure 10.13 plots plot the frequency-dependent bar graph of the total density of states of states of phonons at each temperature during the heating of coal molecules from 300 to 150 K with a gradient of 150 K and the Loess method.

The variation of the total phonon density of states of coal molecules with temperature is shown in Fig. 10.13. The total phonon density of states of coal molecules ranges from 0 to 80 THz, with slightly lower frequency maximums at 300, 750 and 900 K, and the strongest phonon density of states in the range of 10–50 THz, which corresponds to the torsional, bending and telescopic vibrations of the aromatic structure. Therefore, the aromatic structure plays a crucial role in the phononic density of states of coal molecules.

(3) Phonon electron thermal properties.

In this work, the heat capacity, which is easy to calculate and experimentally measured, is selected as the research object of thermal properties, and the heat capacity is divided into constant pressure heat capacity and constant volume heat capacity. The heat capacity of the coal sample was determined by LFA457 laser thermal conductivity meter, the temperature range was set to 30–300 °C, the heating rate was 1 °C/min, and the measurement interval was 30 °C. In the experiment, the constant pressure heat capacity is measured, but because the coal material is solid, thermal expansion almost does not occur during the heating process, so it can be considered that its constant pressure heat capacity and constant volume heat capacity are equal. The calculated heat capacity is the heat capacity of constant volume, so

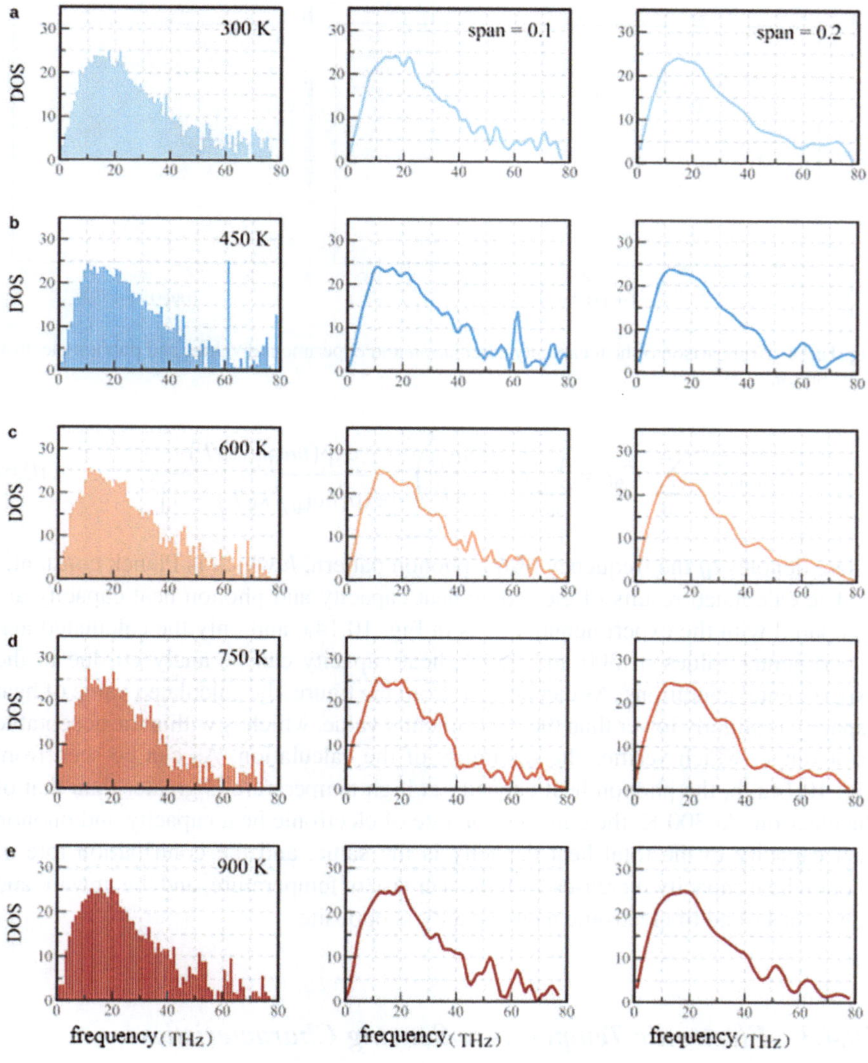

Fig. 10.13 Total density of states of states of phonons in coal molecules at different temperatures

the heat capacity mentioned in this article is the heat capacity of constant volume, which is expressed by C, and the calculation method of electronic heat capacity and phonon heat capacity is:

$$C_{\text{electron}} = \frac{1}{3}\pi^2 k_B^2 T DOS(E_F) \tag{10.5}$$

where k_B is the Boltzmann constant, T is the temperature, and $DOS(E_F)$ is the density of states at the Fermi level.

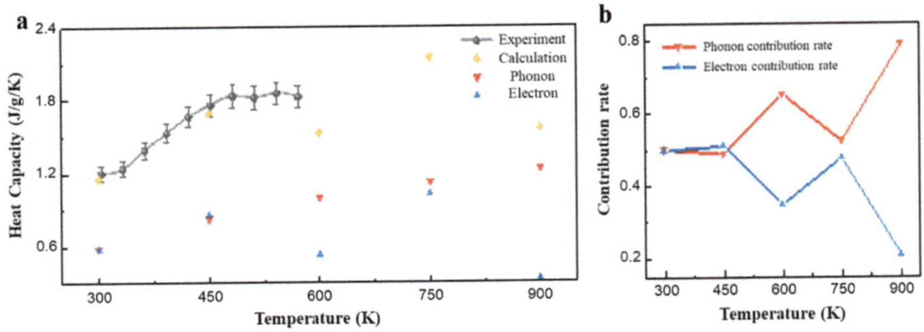

Fig. 10.14 Comparison of heat capacity calculation and experimental values and phonon electron contribution

$$C_{\text{phonon}} = \sum_{\mathbf{q}j} C_{\mathbf{q}j} = \sum_{\mathbf{q}j} k_{\text{B}} \left(\frac{\hbar\omega_{\mathbf{q}j}}{k_{\text{B}}T} \right)^2 \frac{\exp(\hbar\omega_{\mathbf{q}j}/k_{\text{B}}T)}{\left[\exp(\hbar\omega_{\mathbf{q}j}/k_{\text{B}}T) - 1\right]^2} \quad (10.6)$$

where denotes $\mathbf{q}j$ the frequency of the phonon pattern, $\hbar$ which is Planck constant.

The calculated results of electronic heat capacity and phonon heat capacity are compared with the experimental values in Fig. 10.14a, and only the calculated and experimental values of 300 and 450 K heat capacity can be analyzed due to the experimental conditions. As can be seen from the figure, the calculated value of heat capacity is slightly lower than the experimental value, which is within the acceptable error range, which verifies the reliability of the calculation. As can be seen from Fig. 10.14a, b, the phonon heat capacity at high temperature is greater than that of the electron. At 300 K, the contribution rate of electronic heat capacity and phonon heat capacity to the total heat capacity is the same, and the contribution rate of phonon heat capacity increases with the increase of temperature, and the upward and downward trend of electronic heat capacity is opposite.

10.4.3 Electronic Temperature Sensing Characteristics

(1) Temperature-sensing characteristics of electronic divider density.

Figures 10.15 and 10.16 depict the electron separation density and band gap widths C, O and H of each element in the coal molecule at different temperatures, and the band gap decreases with increasing temperature. As the temperature increases, the thermal expansion of the lattice leads to an increase in the lattice constant, which increases the distance between atoms and decreases the bond strength between atoms. Compared with other elements, the band gap of N shows a significant difference with temperature, which basically remains unchanged from 300 to 600 K, and then decreases to 750 K, but increases significantly when the temperature continues to rise to 900 K, and the electrons in N are excited by stronger and stronger thermal energy with the increase of temperature, which increases the number of excited electron states in the conduction band, but the increase of the number of excited states of electrons will also lead to an increase in the scattering of electrons and defects and impurities. This reduces the mobility and transport of electrons, and the electron–electron interaction likewise increases, partially offsetting the increase in the number of electrons involved in the conduction process excited by thermal energy, resulting in a wider band gap. For N, this characteristic temperature affecting electron transport is in the range of 750–900 K.

(2) Temperature-sensing characteristics of the electron-local density of states of carbon atoms Fig. 10.17 plots of the electron local density of states and the bandgap width bars for C211, C215, C216, C217 and C218 in coal molecules at different temperatures are plotted. These five atoms are all carbon atoms in the naphthalene molecule, and their local density of states and band gap show a certain temperature dependence. At 300 K, the band gaps of these five carbon atoms are 2.12 ± 0.04 eV, and similarly, at temperatures of 450 K and 900 K, they have band gaps of 2.00 ± 0.04 eV, and 1.79 ± 0.04 eV, respectively. The bandgap of carbon atoms at a temperature of 600 K can also be expressed as 1.95 ± 0.05 eV, and only the bandgap value of C211 is slightly larger than this range, reaching 2.04 eV, which may be because C211 is a bridge carbon atom in the naphthalene molecule, and the electron density around it is higher, and the interaction between electrons is stronger, which will lead to a larger bandgap of C211. Obviously, 600 K is a special temperature for the carbon atoms on the naphthalene molecule, and the carbon atoms at different points and structures show obvious differences at this

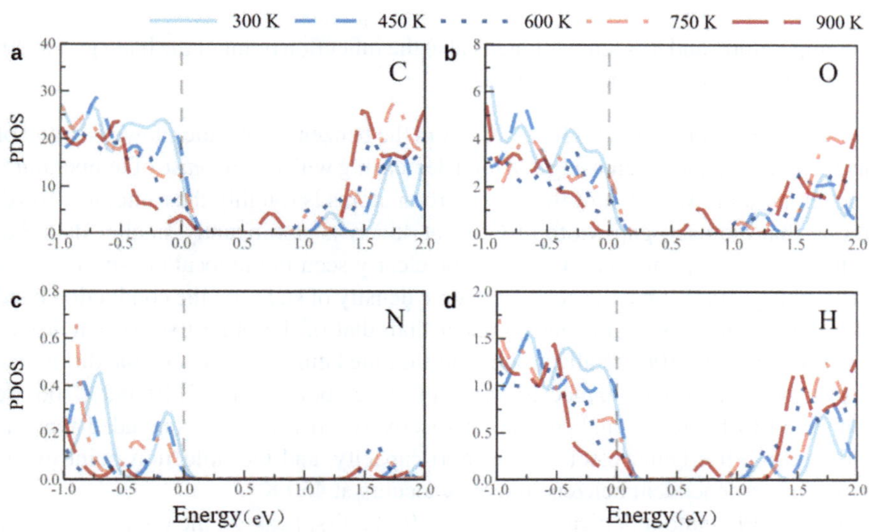

Fig. 10.15 Electron separation densities in coal molecules at different temperatures

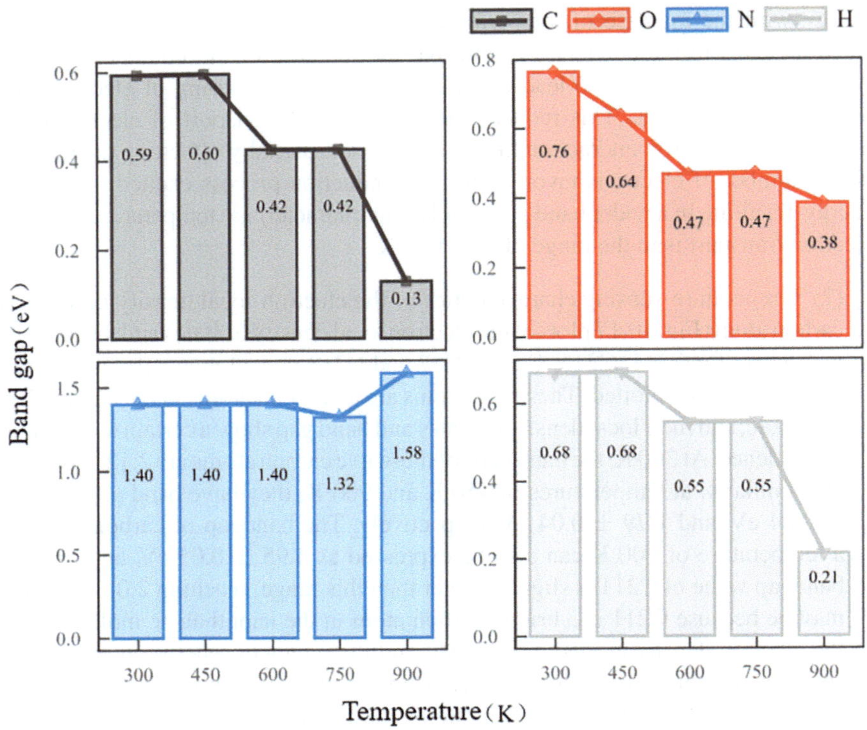

Fig. 10.16 Band gap widths of each element in the coal molecule at different temperatures

temperature, and this temperature is also the inflection point of the band gap width in the process of temperature increase.

The three atoms C106, C110 and C111 in the benzene molecule showed a trend of decreasing first, then increasing and then decreasing with the increase of temperature. The temperature variation of the three carbon atoms is roughly the same, it is worth noting that the band gap width of C111 at 450 K is significantly smaller than that of the other two atoms, and it can also be clearly seen in the local density of states curve in Fig. 10.18a that the peak electronic density of states on the conduction band of C111 at 450 K is significantly greater than that of the other two carbon atoms, C111, C110 and C106 are all carbon atoms on the benzene molecule, the difference is that the other atoms connected by C111 are carbon atoms, C110 and C106 are connected to hydrogen and oxygen, respectively, so it can be concluded that the carbon atom has better electron transport capacity, and the inflection point of the temperature-dependent benzene molecule occurs at 450 K.

In the carbon atom depicted in Fig. 10.19, the first peak on the conduction band of C51 as the carbon atom attached to the oxygen atom in the ether group at 900 K appears at a fairly low energy level, indicating that the presence of carbon atoms in the ether group optimizes the conductivity and electron transport capacity of the entire

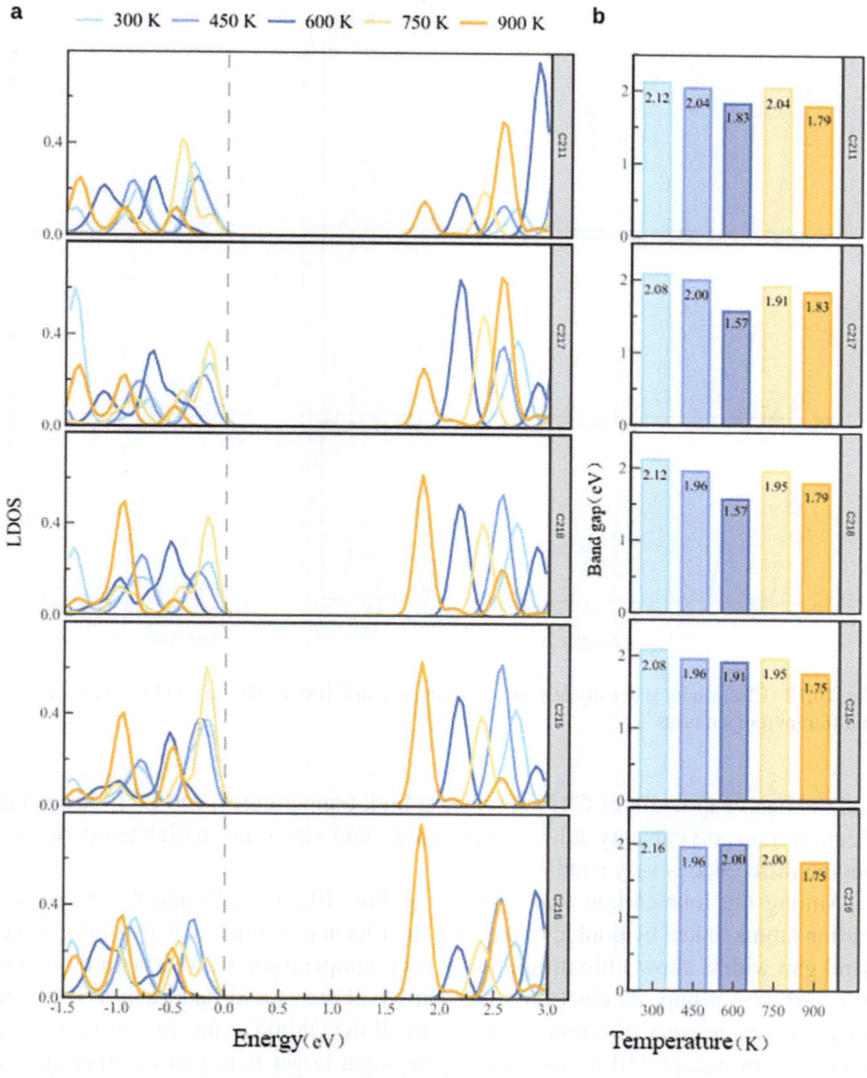

Fig. 10.17 Electron local density of states and band gap of C211, C215, C216, C217 and C218 in coal molecules at different temperatures

coal molecule at high temperatures. The band gap of C224 and C194 is consistent with that of C224 and C194 during the heating process, and the band gap of C224 is smaller than that of C194 at high temperature, indicating that the connection between fat carbon atoms has better electron transport ability than that of fat carbon atoms and aromatic carbon atoms at high temperature. C210 and C220, as carbon atoms that connect carbon atoms at different points on the naphthalene molecule, have the same temperature sensitivity. The band gap width of C220 is wider at low temperatures,

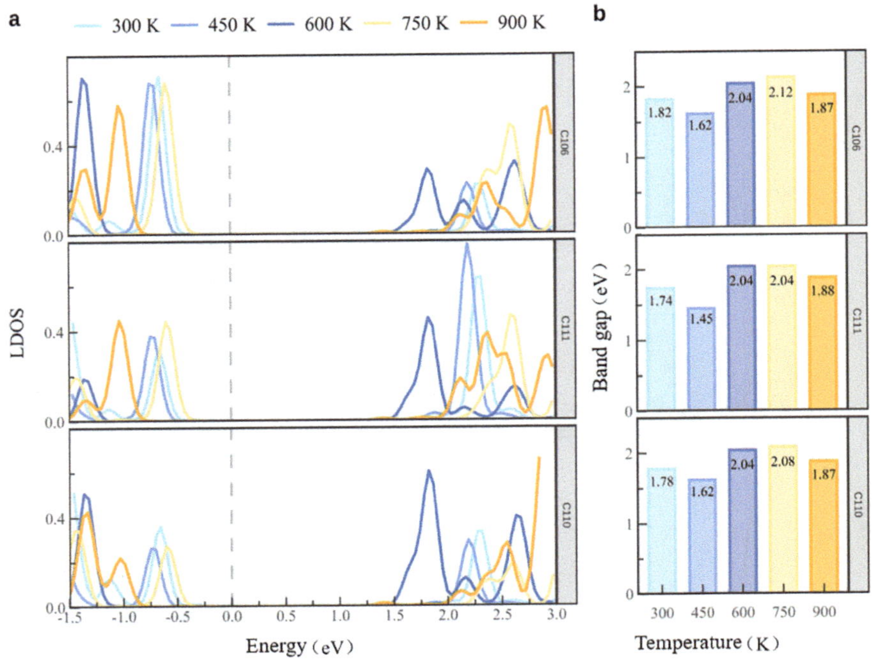

Fig. 10.18 Electron local density of states and band gap of C106, C110 and C111 in coal molecules at different temperatures

and the band gap width of C219 is wider at high temperatures, and C219 has better electron transport capacity at low temperatures, and vice versa at high temperatures, but the difference is very small.

Among the four carbon atoms shown in Fig. 10.20, C225 and C228 are two carbon atoms linked by double bonds, and their local electronic density of states and band gap widths show little difference at each temperature. C231 is a carbon atom in a carboxyl group, its electron local density of states and band gap at different temperatures are very different compared to all the carbon atoms in question, at all temperatures except 450 K, its band gap is much larger than that of other carbon atoms, it is easy to know from the local density of states diagram, this is due to the fact that the valence band is at a higher energy level, and the conduction band energy level is also relatively high, thus resulting in a larger band gap, while 450 K is obviously a special temperature for C231, at this temperature, The peaks on the conduction band of C231 are at a lower energy level, which is more conducive to electron excitation.

(3) Temperature-sensing characteristics of the electron local density of states of oxygen and nitrogen atoms. The three oxygen atoms O121, O166 and O229 depicted in Fig. 10.21 are oxygen atoms in the ether group that link the fat carbon atom, the carbon atom in the naphthalene molecule, and the carbon atom in the

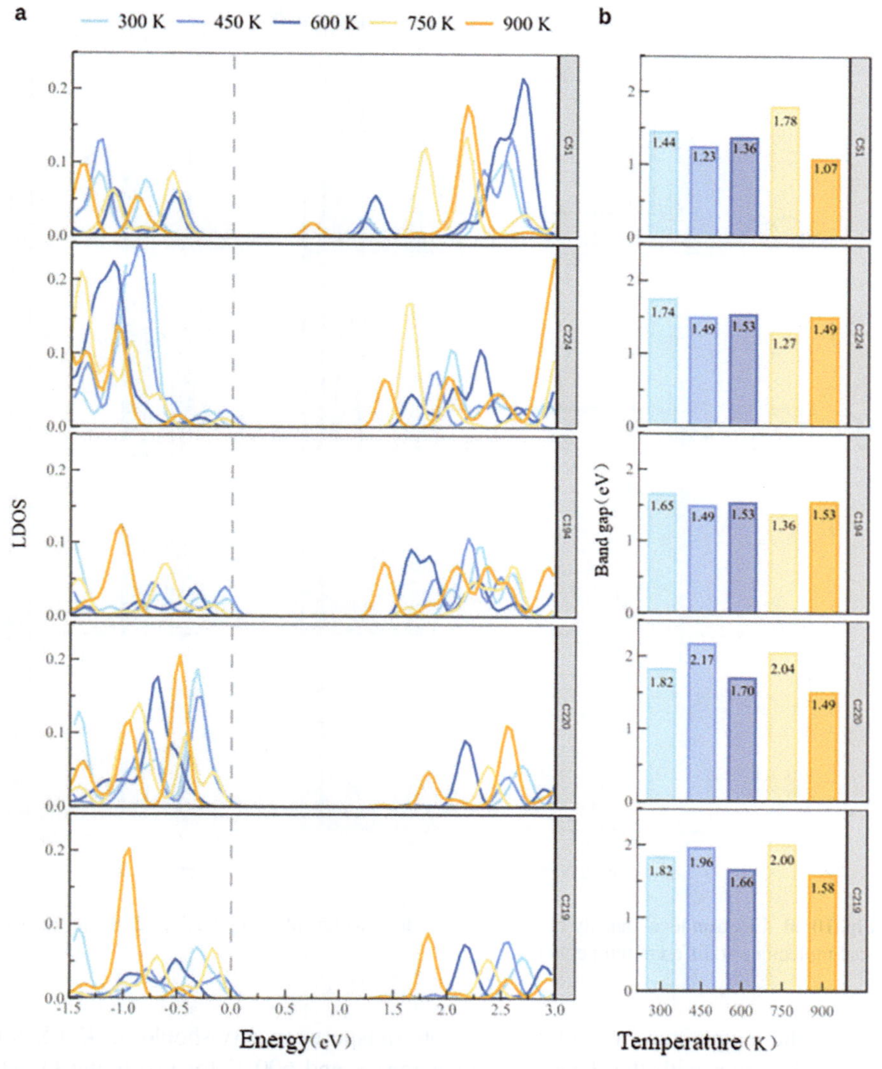

Fig. 10.19 Electron local density of states and band gap of C51, C194, C219, C220 and C224 in coal molecules at different temperatures

benzene molecule, respectively. Although the band gap of O 121 is very narrow at 900 K, the local density of states curves of nitrogen-bound electrons show that the intensity of the first peak on the conduction band is not negligible, but it is very weak, and only a few electrons can transition from the valence band to this energy level at high temperatures. As can be seen from Fig. 10.21, the first peak in the conductor band of these three oxygen atoms is the local density of states at 600 K, with 0121 being at the lowest energy level, followed by O166, and O229 being the highest. The narrowest band gap for O121 is 900 K at all temperatures,

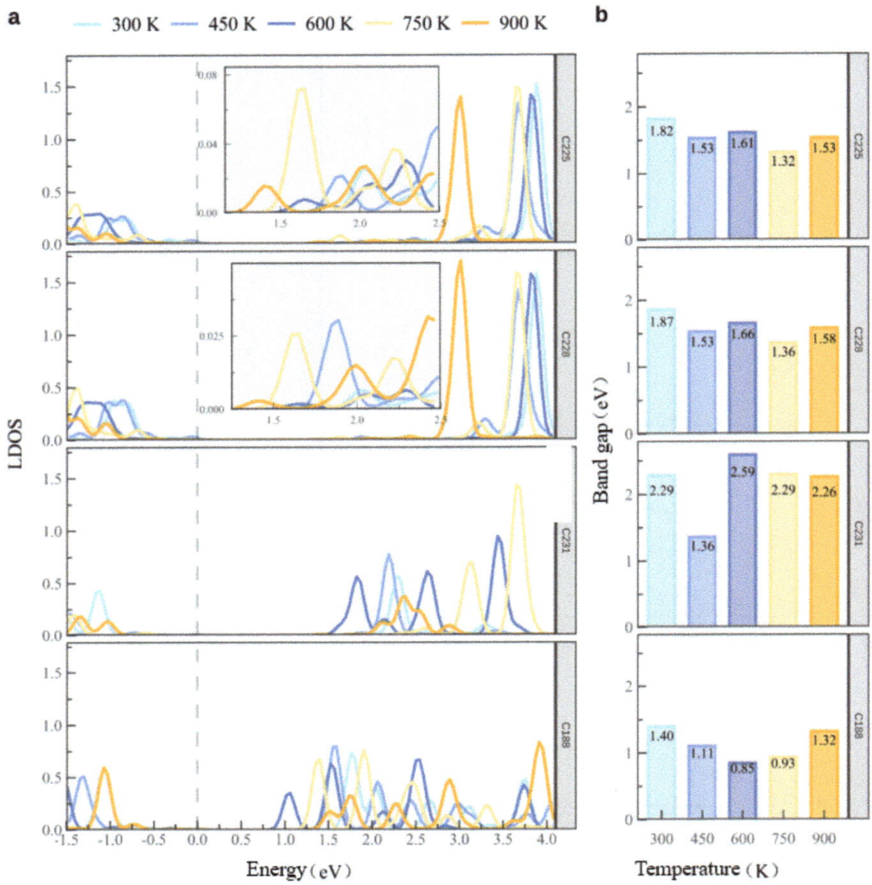

Fig. 10.20 Electron local density of states and bandgap widths of C188, C225, C228 and C231 in coal molecules at different temperatures

but the temperature for optimal electron transport capacity should be 450 K for combination with the density of states curve, and 600 K for O166 and O229. Therefore, for ether-based oxygen atoms, the electron transport capacity should be optimal at a temperature of 450–600 K.

In Fig. 10.22, O193 is the oxygen atom in the carbonyl group, and its band gap changes with temperature and the carbon atom C188 on the carbonyl group connected to it is exactly the same, showing a trend of first decreasing and then increasing, and its inflection point is 600 K, so the temperature correlation of electron transport of the carbonyl structure can be obtained, in the temperature range of 300–600 K, At 600–900 K, the electron−electron interaction caused by the increase of excited

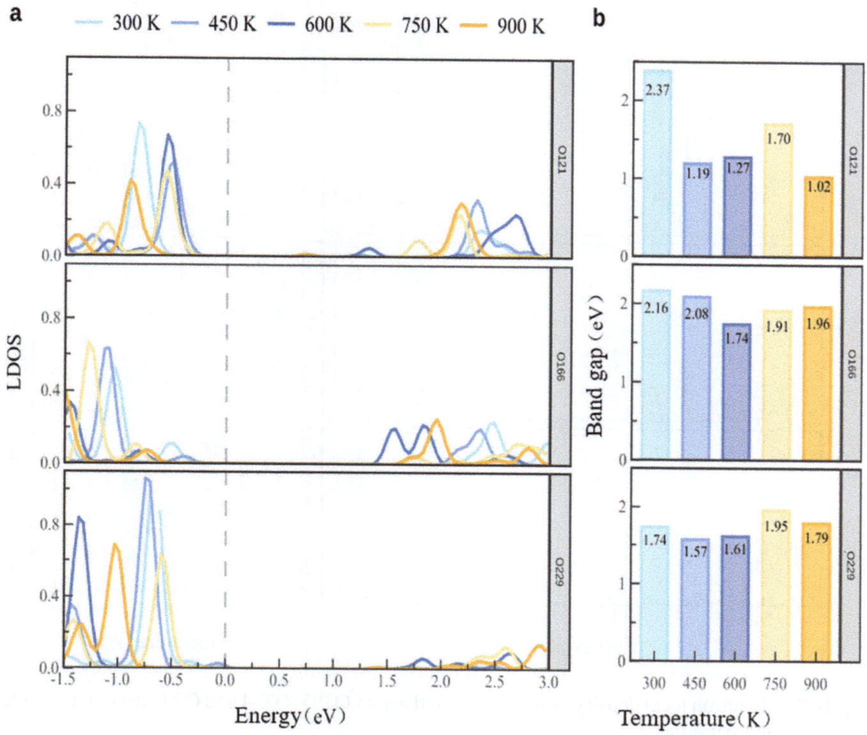

Fig. 10.21 Electron local densities of states and band gaps of O121, O166 and O229 in coal molecules at different temperatures

electrons increases the scattering of electrons and defects and impurities, which in turn affects the degradation of electron transport ability.

O234 and O235 are two carbon atoms in the carboxyl structure, and their band gap shows the same trend with the increase of temperature, the difference is that the energy level of O235 in the valence band is higher, because O235 is an oxygen atom with double bonds with carbon atoms in the structure, and its electronic structure is more stable.

N355 and N357 are nitrogen atoms on pyridine and pyrrole molecules, respectively, and according to c, it is easy to see that their electron transport capacity increases with increasing temperature, while the band gap of N355 at 450 K is significantly larger than that at other temperatures, and similarly, the energy level on the valence band of N355 is the highest at 4 50 K from the electron local density of states curve, so N355 should be the most stable at 450 K and the lowest electron transport capacity (Fig. 10.23).

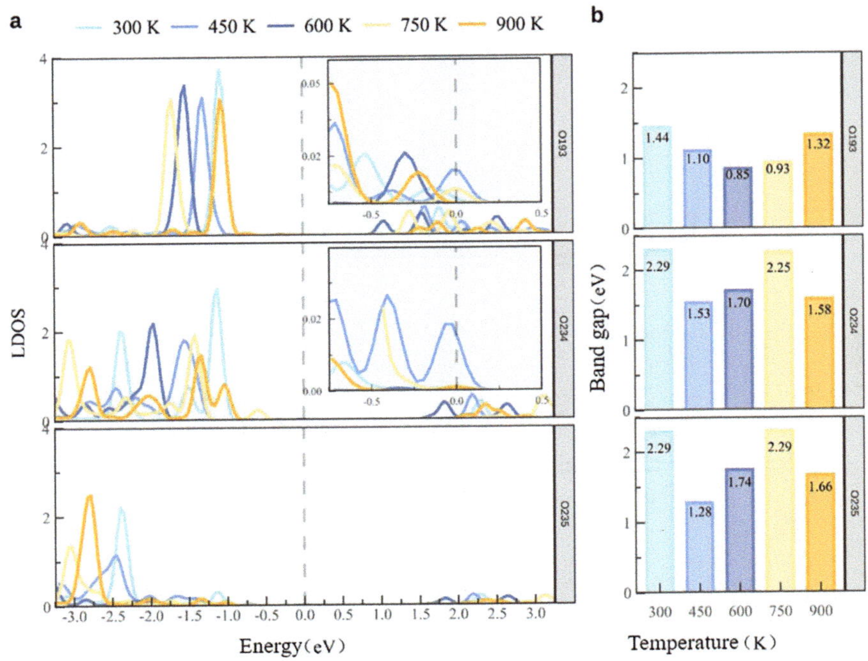

Fig. 10.22 Electron local density of states and band gap of O193, O234 and O235 in coal molecules at different temperatures

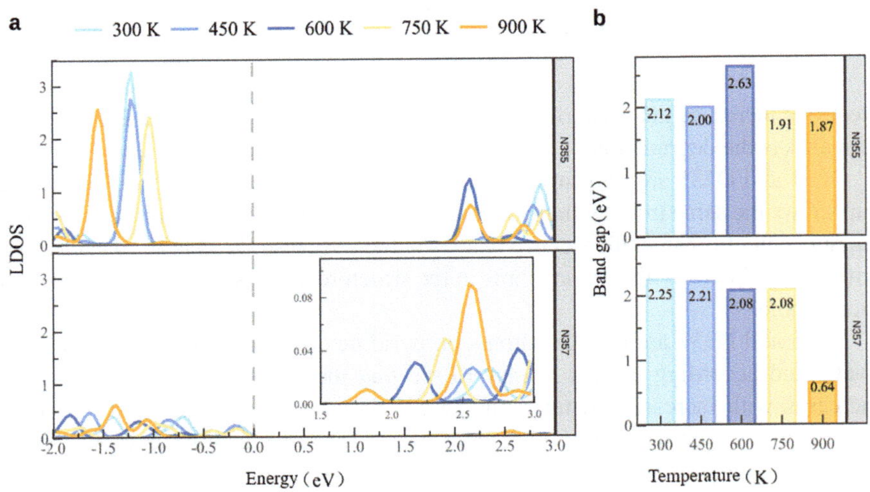

Fig. 10.23 Electron local density of states and band gap of N355 and N357 in coal molecules at different temperatures

10.4.4 Temperature Sensing Characteristics of Phonons

(1) Temperature sensing characteristics of phonon dividing density.

The change of phonon partal density of C in coal molecules with temperature is shown in Fig. 10.24. The phonon part-state density of C in coal molecules ranges from 0 to 80 THz, and in the low-frequency range of less than 40 THz, the phonon part-state density of C is highly consistent with the total phonon state density, while in the high-frequency range greater than 40 THz, it is much lower than the total phonon density of states. The results show that the phonon density of states of C controls the phonon density of the whole coal molecule in the low frequency range.

Figure 10.25 shows the variation of the phonon density of O in coal molecules with temperature, and its phonon density of states frequency ranges from 0 to 60 THz, and becomes more and more concentrated at 0–40 THz in the process of increasing temperature. The span = 0.2 curve shows that the peak with the highest intensity is around 10 THz, and this peak shifts to a lower frequency as the temperature rises.

The phonon density of states in coal molecules varies with temperature as shown in Fig. 10.26, and its phonon density of states frequency ranges from 0 to 80 THz, most of which are concentrated at 0–50 THz, and only a signal greater than 70 THz appears at low temperature, and only a small number of signals with no outstanding intensity appear at 50–70 THz as the temperature rises. This indicates that the density of phonon states in N shows a tendency to shift to lower frequencies during the process of low temperature to high temperature.

As shown in Fig. 10.27, the phonon density of states in coal molecules varies with temperature, and its phonon density of states frequency ranges from 0–80 THz, and is evenly distributed in the whole frequency range, and the intensity of the signal near 60 THz at 450 K is much higher than that in other regions and other temperatures, and at 450 K, the distribution of phonon density of states in the high-frequency range above 60 THz is significantly greater than that in other temperatures. This indicates that when the temperature of the hydrogen atom in the coal molecule is increased to 450 K, the number of phonons in the high-frequency range, especially at 60 THz, increases significantly, so the heat energy transmitted by the hydrogen atom in the coal molecule will also be relatively high, which will lead to the improvement of the thermal conductivity of the coal molecule at this temperature.

Similarly, the contribution of the phonon density of each element in the coal molecule to the total phonon density of states of the coal molecule during the growth process from 300 to 900 K is studied, as shown in Fig. 10.28. It can be seen from the figure that the contribution rate of C to the total density of states of the phonon in the low frequency range remains unchanged in the low frequency range when the temperature rises to 900 K, and fluctuates slightly in the high frequency range, but the increase and decrease of C is only 0.05, while the increase and decrease of O, N. The contribution rate of and H to the total density of states during the warming process is almost negligible. The results reveal that the influence of temperature factor on the contribution rate of each element to the total density of states is very insignificant, mainly because there are large differences in the mass fraction and number of atoms

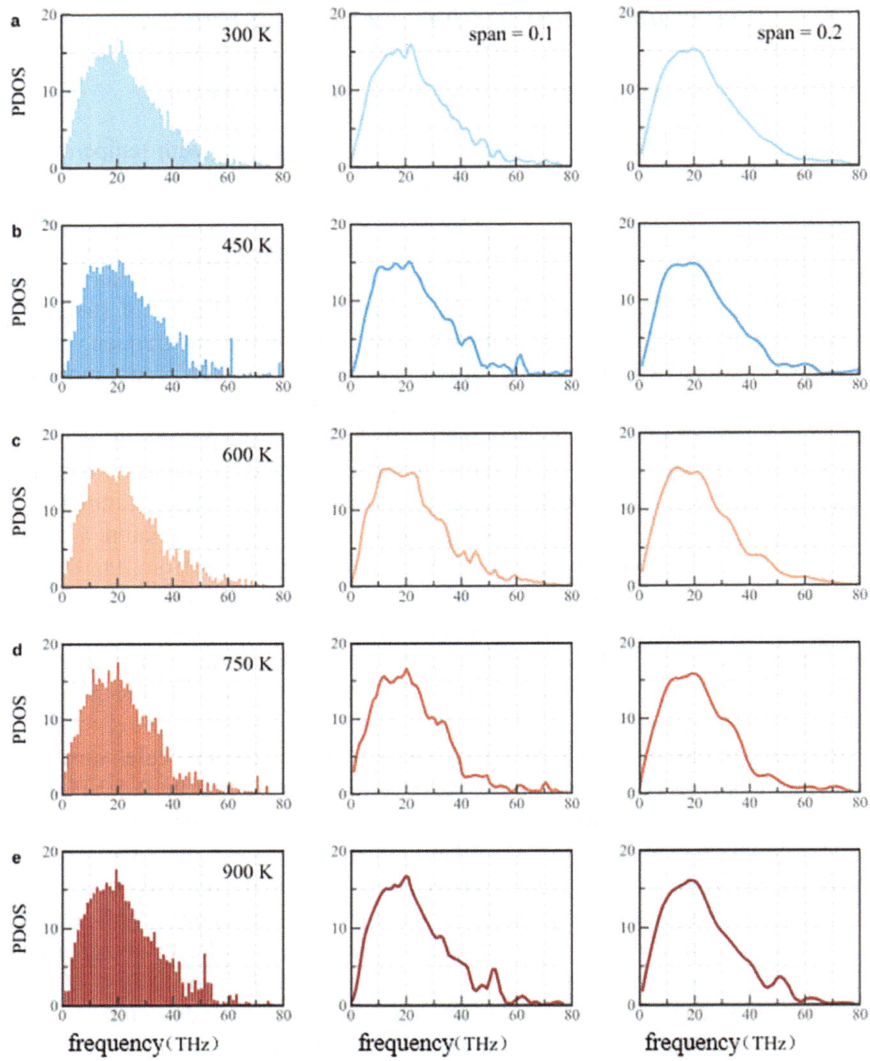

Fig. 10.24 Phonon fractional density of C in coal molecules at different temperatures

among the elements in the coal molecule, and the phonon density of states cannot be characterized at the level of different elements under this overall framework, so the phonon local density of states of carbon atoms in different structures is further explored.

(2) Temperature-sensing characteristics of the phonon local density of states of carbon atoms. Figures 10.29, 10.30 and 10.31 is a bar graph of the frequency distribution of phonon local density of states of carbon atoms at different points and different externally linked atoms in the naphthalene molecule during the

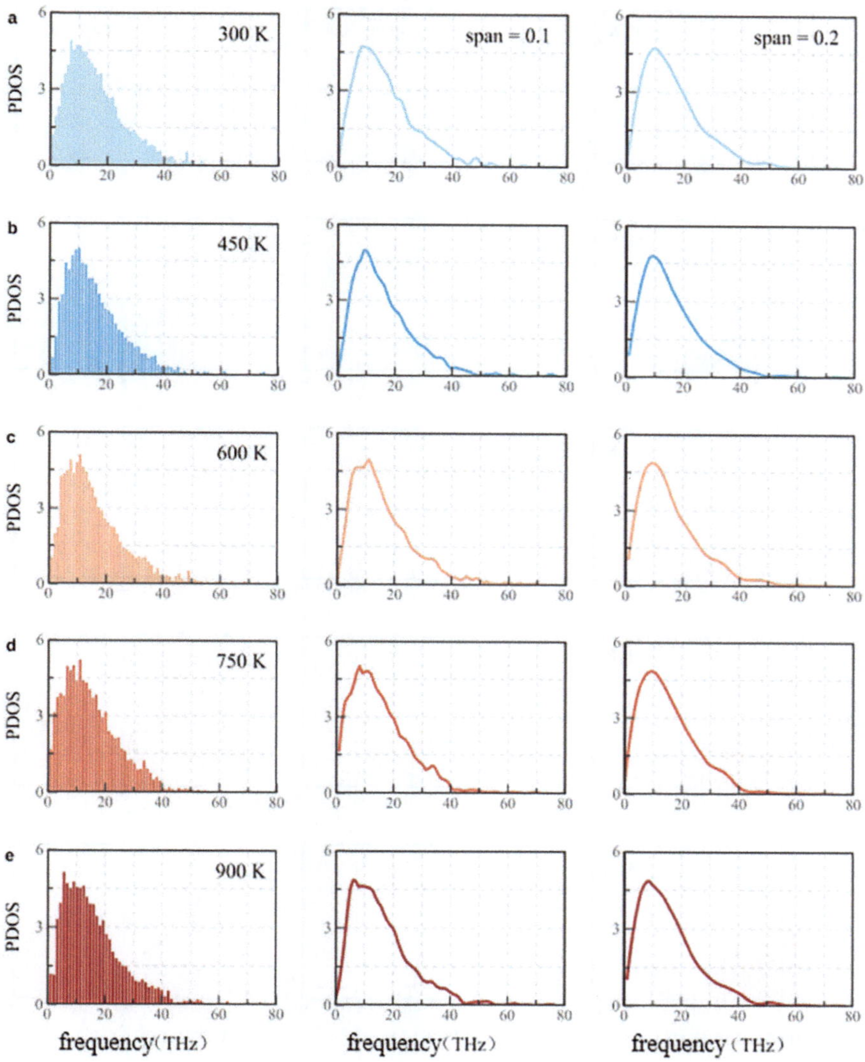

Fig. 10.25 Phonon dividing density of O in coal molecules at different temperatures

temperature growth from 300 to 900 K, and the smoothed curves with parameters of 0.1 and 0.2. It can be seen from the observation chart that the phonon local density of states of C211 is significantly enhanced at 600 K compared with other temperatures, and this position appears at 23.9 THz, which corresponds to the expansion and contraction vibration of naphthalene molecules. The local density of states of C215 and C218 at temperatures of 750 K is 6.7 and 30.2 THz, respectively, corresponding to the telescopic vibration of C–H bond and the bending vibration of C–C bond. The phonon local density of states of C216

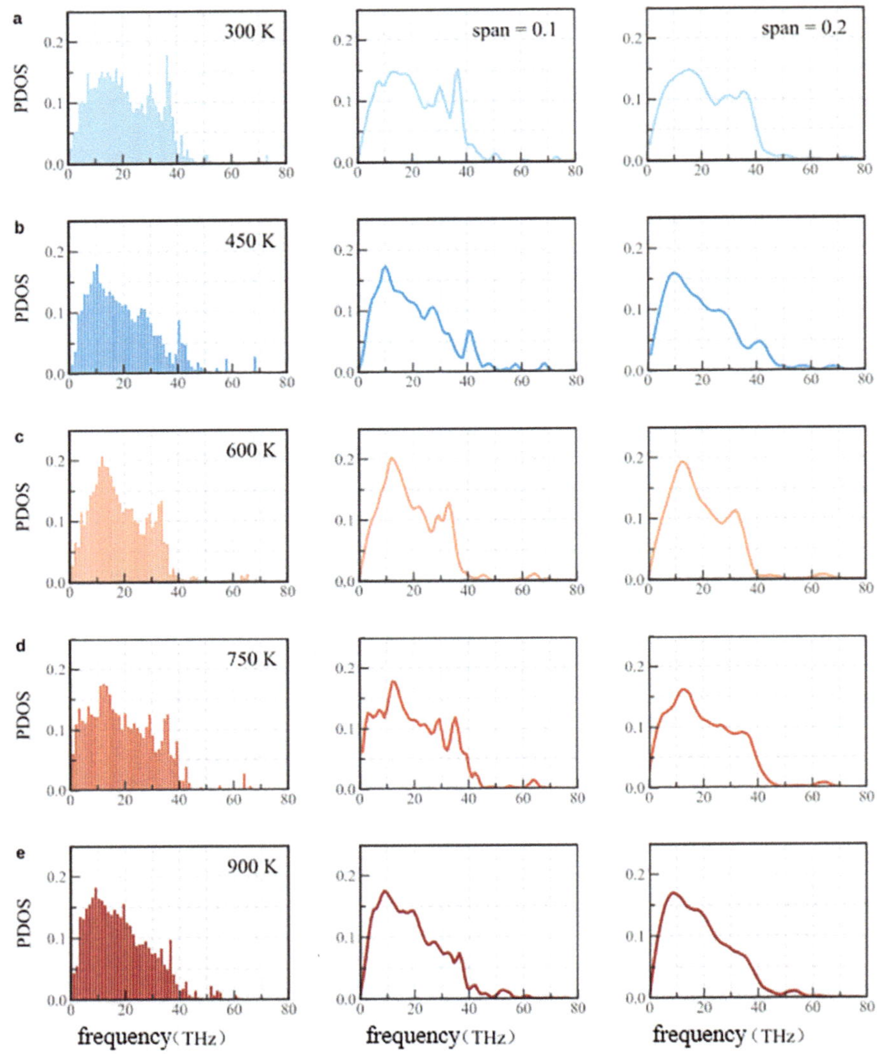

Fig. 10.26 Phonon fractional density of N in coal molecules at different temperatures

and C217 shows that the density of states of these two carbon atoms increases significantly at 20 THz at a temperature of 900 K, similar to C211, this frequency corresponds to the stretching vibration of naphthalene molecules, the difference is that C216 has more phonon state distribution at lower frequencies, because C216 attached to oxygen atoms has a lower vibrational frequency than C217 attached to carbon atoms (Fig. 10.30).

There are three different carbon atoms in the naphthalene molecule, bridge carbon, α carbon and β carbon atom, among which the bridge carbon atom has the highest

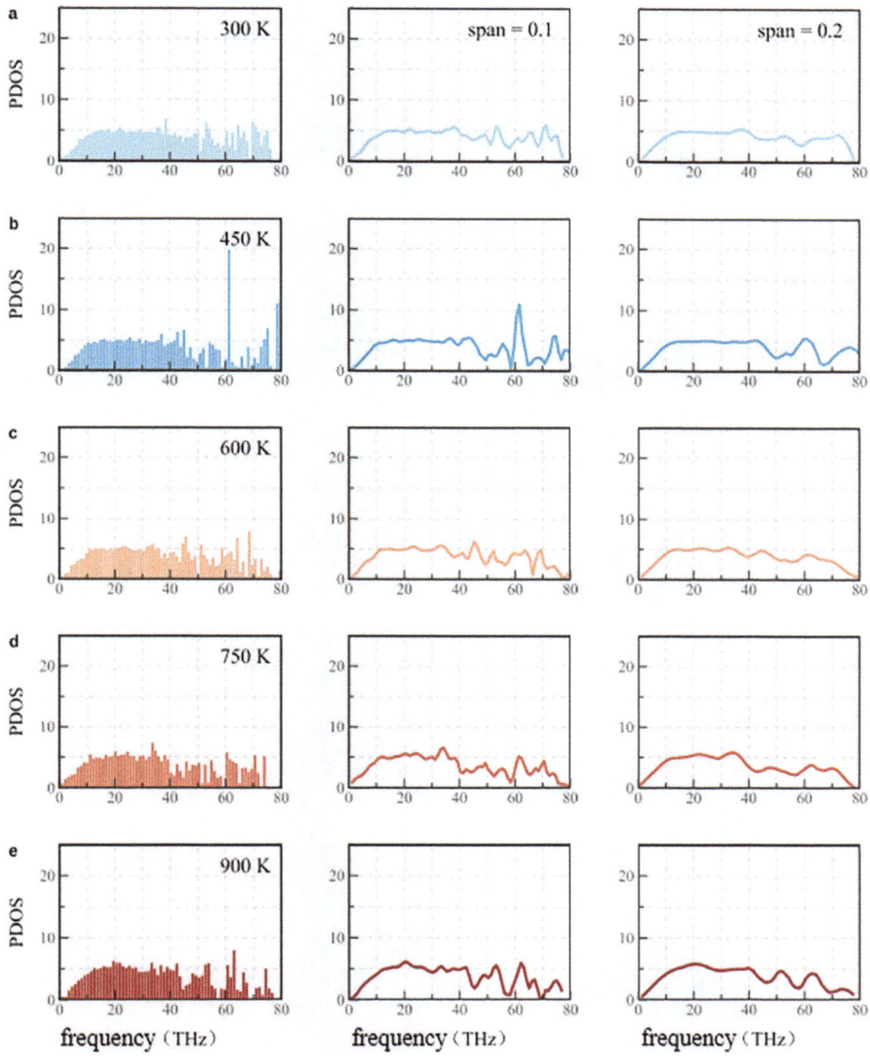

Fig. 10.27 Phonon fractional density of H in coal molecules at different temperatures

temperature sensitivity, and its phonon density of states changes significantly at 600 K. α the temperature sensitivity of carbon atoms is secondary, and the change of phonon density of states frequency and intensity can be clearly observed at 750 K. β carbon atom has the lowest temperature sensitivity, and the change in phononic density of states can be clearly detected at 900 K.

As can be seen in Figs. 10.32, 10.33 and 10.34, the vibrational patterns of the carbon atoms in the benzene molecule are much more complex than those in the naphthalene molecule, and at 450 K, all three atoms have a higher intensity of phonon states in the high-frequency region, and when the temperature increases to 600, 750,

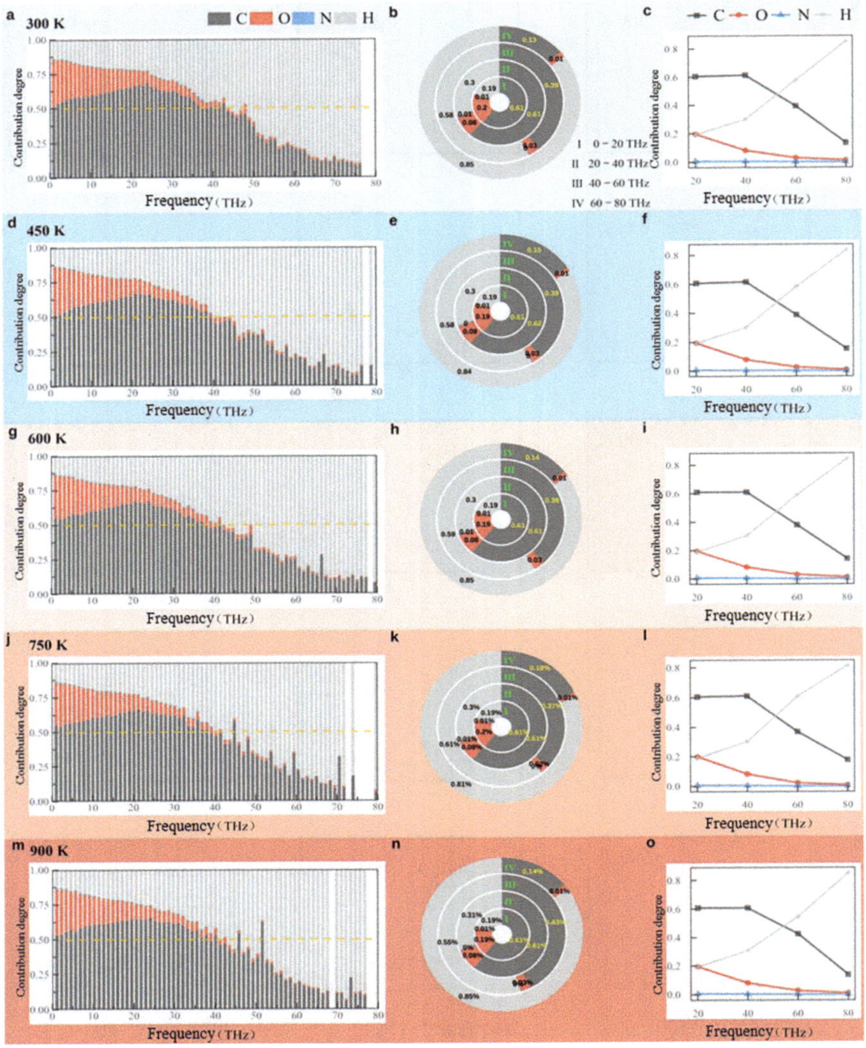

Fig. 10.28 The contribution rate of each fractal density to the total density of states in coal molecules at different temperatures

and 900 K, the higher phonon density of states remains, but the frequency of the higher is gradually reduced. This results in a decrease in the thermal conductivity of its phonons. At 450 K, the high-frequency vibrational pattern in the benzene molecule may result in more energy transfer, resulting in a higher phonon thermal conductivity. However, as the temperature increases, the number of phonon states increases, which may lead to more phonon scattering, which reduces the phonon thermal conductivity.

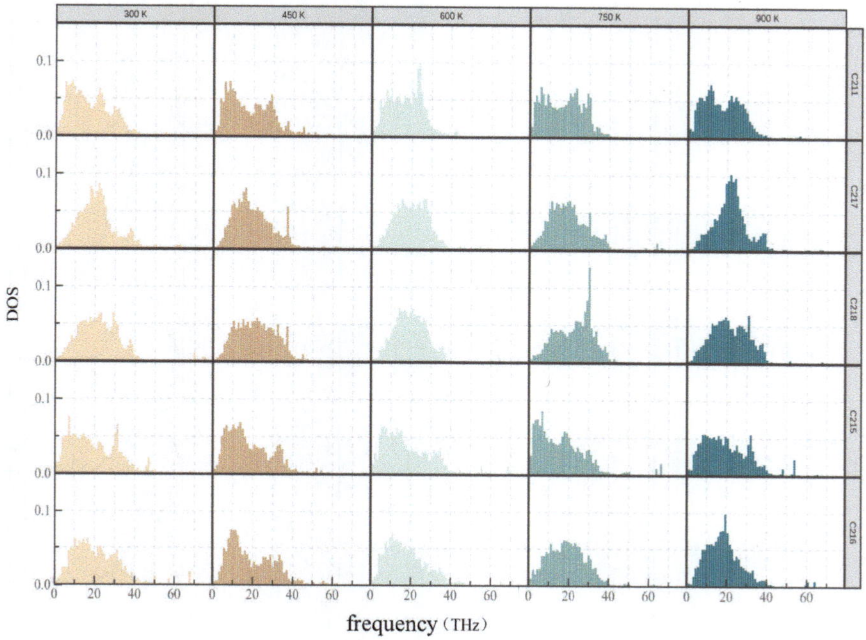

Fig. 10.29 Phonon local density of states of C211, C215, C216, C217, and C218 in coal molecules at different temperatures

With the increase of temperature, the atomic vibration in the lattice becomes more and more intense, the number of phonons excited in the lattice is increasing, and more and more phonons lead to more heat transfer in the low temperature stage, thus further improving the phonon thermal conductivity, but as the temperature gradually increases to the high temperature stage, the number of phonons exceeds a certain value, and the probability of phonon–phonon scattering is greatly increased, resulting in a decrease in the thermal conductivity of the phonons. For the molecular structure of benzene in the coal molecule studied in this paper, this characteristic temperature is 450 K.

In Figs. 10.35, 10.36, 10.37, 10.38 and 10.39, C194 is an aliphatic carbon atom attached to a carbon atom in a benzene molecule, and the number of phonon states at 41.3 THz increases significantly at 600 K. C219 is a carbon-α aliphatic carbon atom attached to naphthalene molecules, and C220 is a carbon-β aliphatic carbon atom attached to naphthalene molecules. C224, a carbon atom covalently bonded to other carbon atoms, at 900 K, its phonon states gather at 19.4 THz, but this is a lower frequency, which means that as the temperature rises to 900 K, the number of phonon states increases, resulting in a large increase in the probability of phonon–phonon scattering, resulting in a decrease in the thermal conductivity of the phonons (Figs. 10.36 and 10.37).

Looking at Figs. 10.38, 10.39 and 10.40, it can be seen that the C188, C225, C228, and C231 carbon atoms are carbon atoms in the carbonyl group, respectively. The

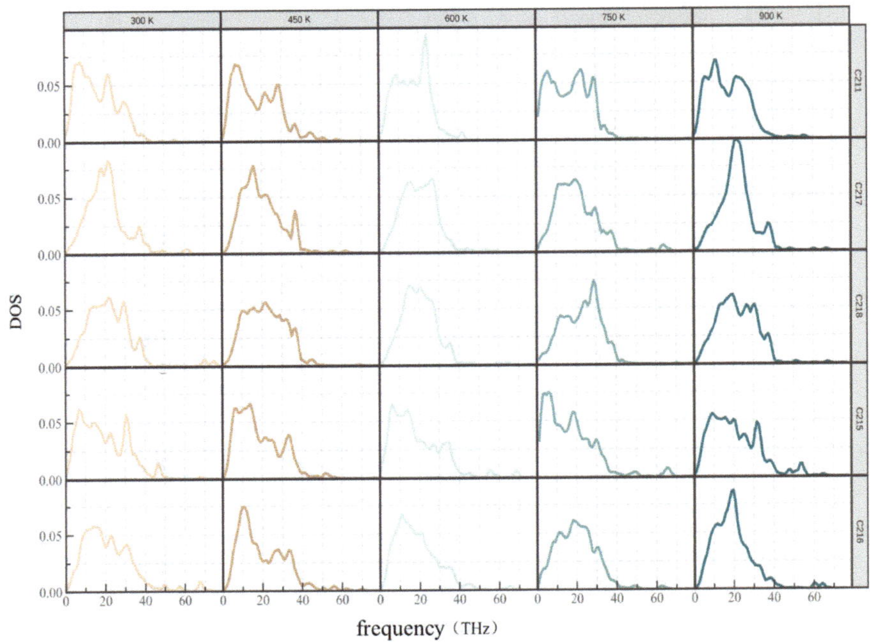

Fig. 10.30 Phonon local densities of states of C211, C215, C216, C217 and C218 in coal molecules at different temperatures (span = 0.1)

change law of the local density of states of the phonons of the carbon atoms on both sides of the carbon – carbon double bond and the carbon atoms in the carboxyl group is similar. From 450 K, the phonon density of states of carbon atoms decreases gradually at 30 THz, and increases at higher frequencies. The lattice vibration frequency usually becomes higher as the temperature increases, usually due to the increasing amplitude of the lattice vibration, which in turn is related to the average position of the atoms in the lattice and the range of vibrations. Thus, when the lattice vibration frequency increases with increasing temperature, the density of states of the phonon also increases, but due to the effect caused by the phonon—honon interaction. The phonon—honon interaction leads to a phenomenon in which the frequency and amplitude of lattice vibrations decrease with increasing temperature, resulting in a phenomenon in which the density of states of sound decreases with increasing temperature. In addition, the phonon scattering mechanism in phonon—honon interaction is more significant at low frequencies, resulting in more frequent scattering of low-frequency phonons, thereby reducing the density of states at low frequencies. At high frequencies, the scattering mechanism in phonon—honon interactions may be more complex due to the higher energy of phonons, resulting in an increase in the density of states of phonons at high frequencies.

(3) Temperature-sensing characteristics of the phonon local density of states of oxygen atoms. Figures 10.41, 10.42 and 10.43 show the phonon local state

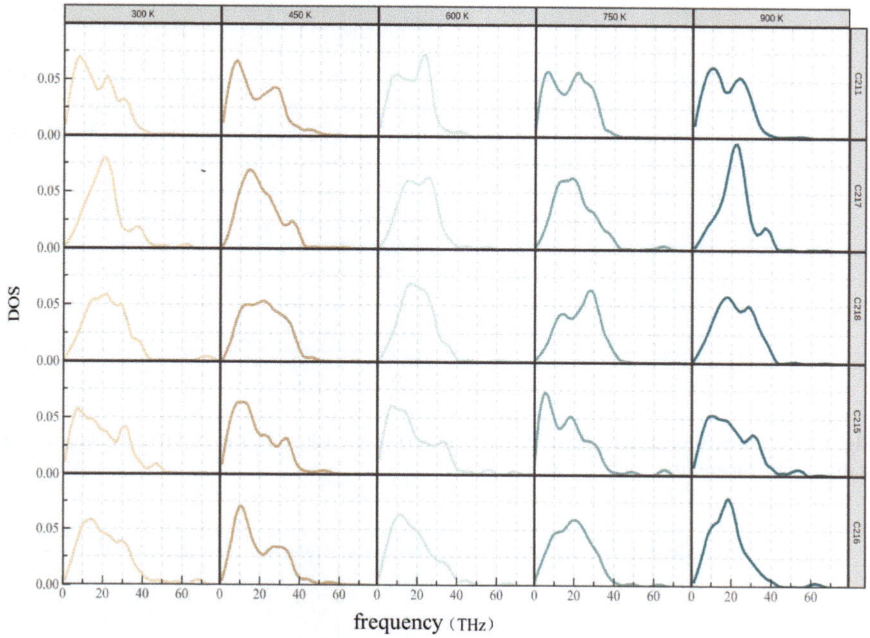

Fig. 10.31 Phonon local densities of states of C211, C215, C216, C217 and C218 in coal molecules at different temperatures (span = 0.2)

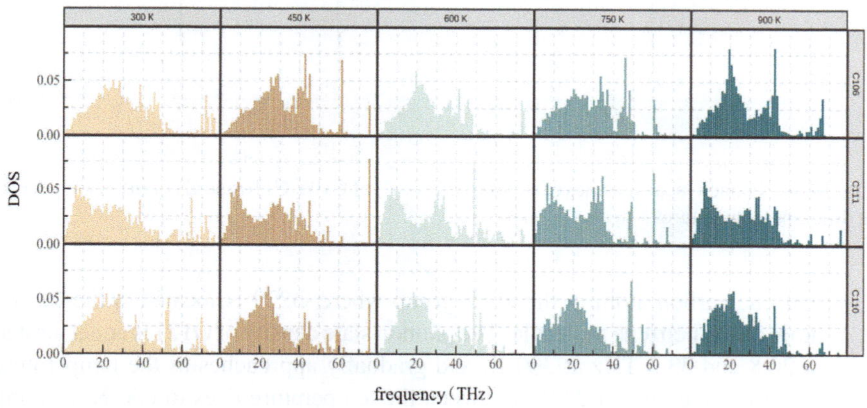

Fig. 10.32 Phonon local densities of states of C106, C110 and C111 in coal molecules at different temperatures

densities of O121, O166 and O229 in coal molecules at different temperatures. These three oxygen atoms are oxygen atoms in the ether group, in which the two carbon atoms attached to O 121 are fatty carbons, one carbon atom in the ether group where O166 is located is the carbon atom on the naphthalene molecule,

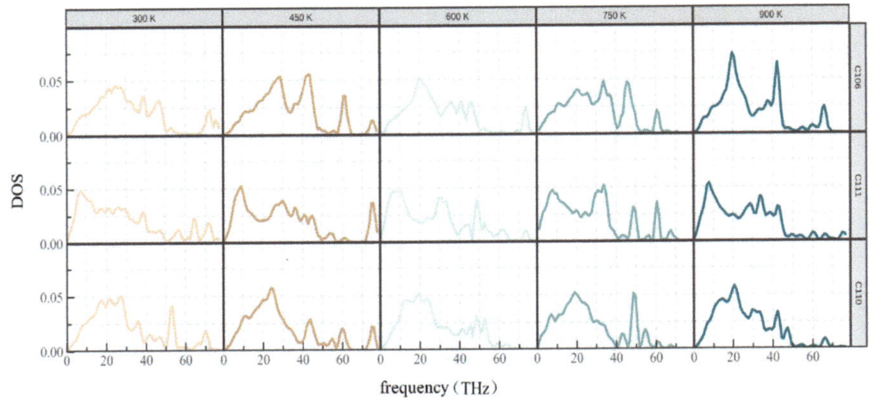

Fig. 10.33 Phonon local densities of states (span = 0.1) of C106, C110 and C111 in coal molecules at different temperatures

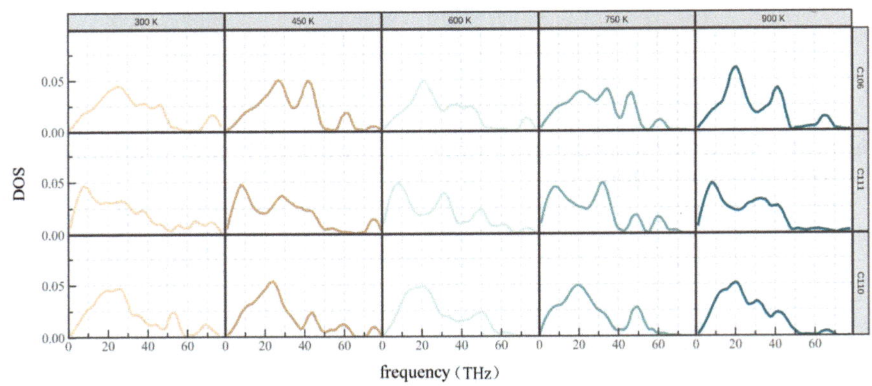

Fig. 10.34 Phonon local densities of states (span = 0.2) of C106, C110 and C111 in coal molecules at different temperatures

and one carbon atom in the ether group where O229 is located is the carbon atom on the benzene molecule. The phonon state density of O121 is concentrated at 20.8 and 33.3 THz at 300 K, and gradually approaches as the temperature increases, merging at 22.9 THz when the temperature rises to 600 K, forming a rather prominent peak, and the height of this peak gradually decreases as the temperature continues to climb to 900 K, and some of the phonon states are distributed at higher frequencies. When the temperature is 450 K, the phonon density of states of these two atoms also has a peak around 11THz, and the height of this peak gradually decreases with the increase of temperature, and the decreasing part shifts to a higher frequency.

Through the comparative analysis of the phonon local density of the three oxygen atoms O121, O166 and O229, it can be seen that the distribution of the phonon

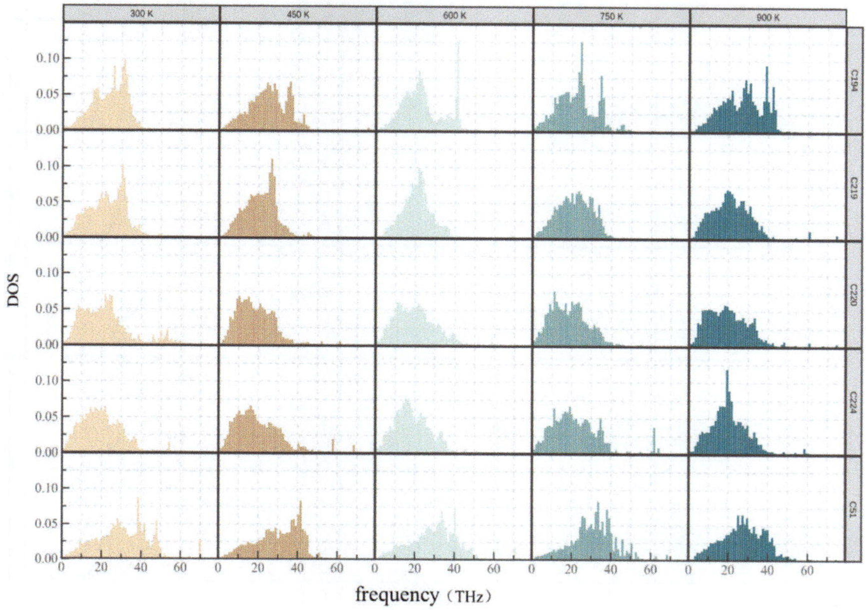

Fig. 10.35 Phonon local densities of states of C51, C194, C219, C220 and C224 in coal molecules at different temperatures

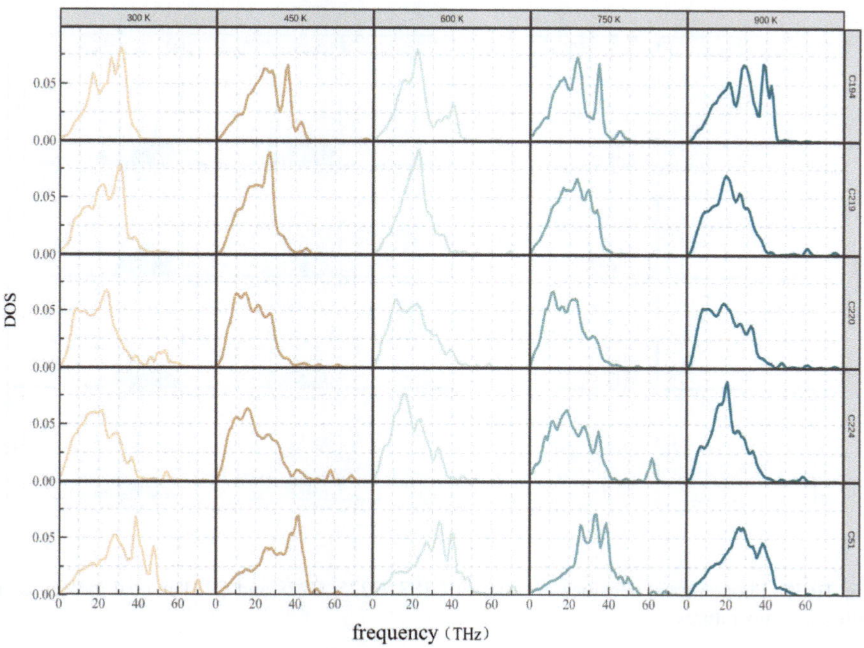

Fig. 10.36 Phonon local state density (span = 0.1) of C51, C194, C219, C220, and C224 in coal molecules at different temperatures

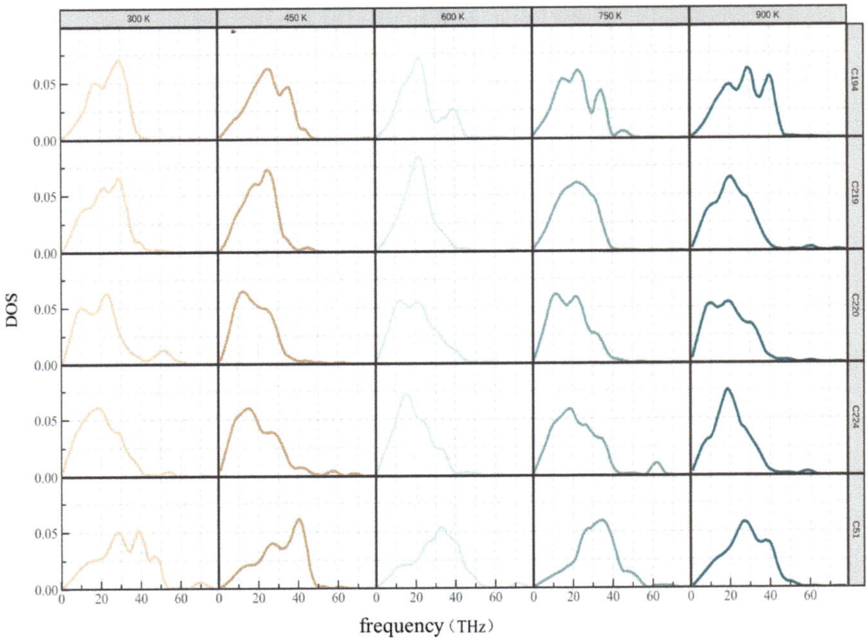

Fig. 10.37 The phonon local state density (span = 0.2) of C51, C194, C219, C220, and C224 in coal molecules at different temperatures

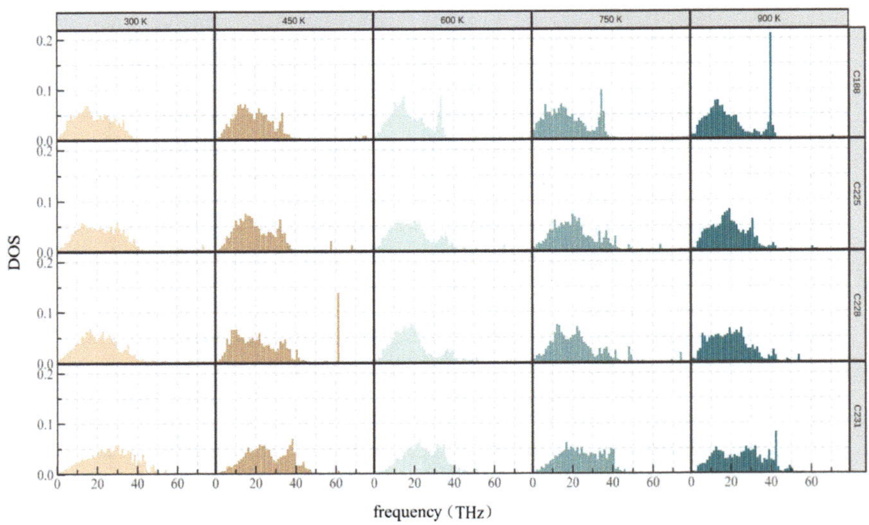

Fig. 10.38 Phonon local densities of states of C188, C225, C228 and C231 in coal molecules at different temperatures

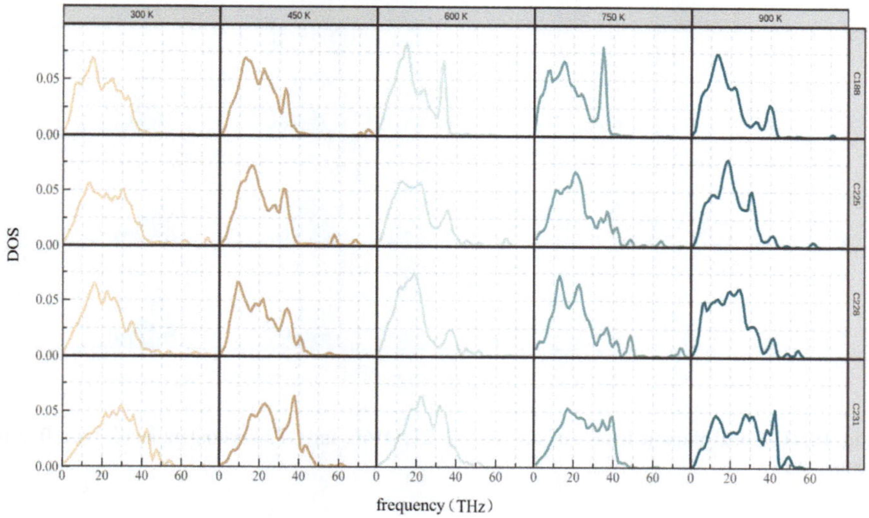

Fig. 10.39 Local density of phonon states of C188, C225, C228 and C231 in coal molecules at different temperatures (span = 0.1)

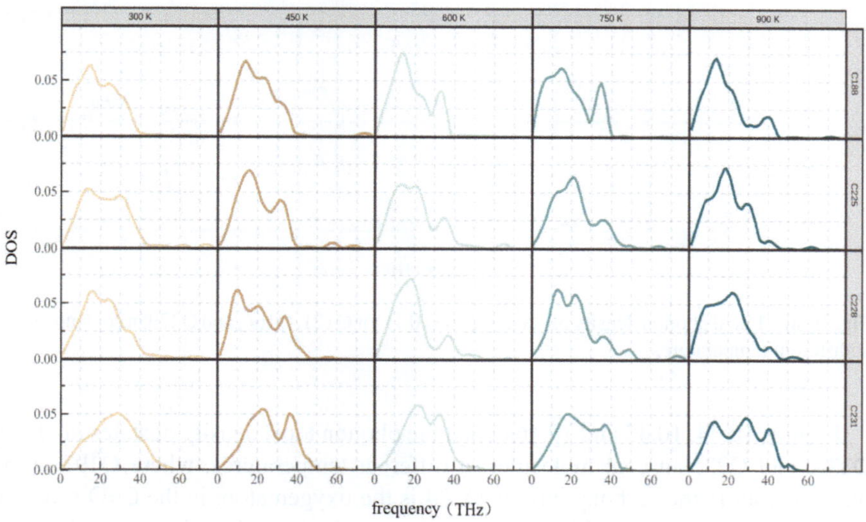

Fig. 10.40 Phonon local state density (span = 0.2) of C188, C225, C228, and C231 in coal molecules at different temperatures

states of the ether-based oxygen atoms at different frequencies in the process of heat accumulation is first concentrated at a lower frequency at a certain temperature, and then transferred to a higher frequency, which is 600 K for O121 and 450 K for O166 and O229 (Fig. 10.42).

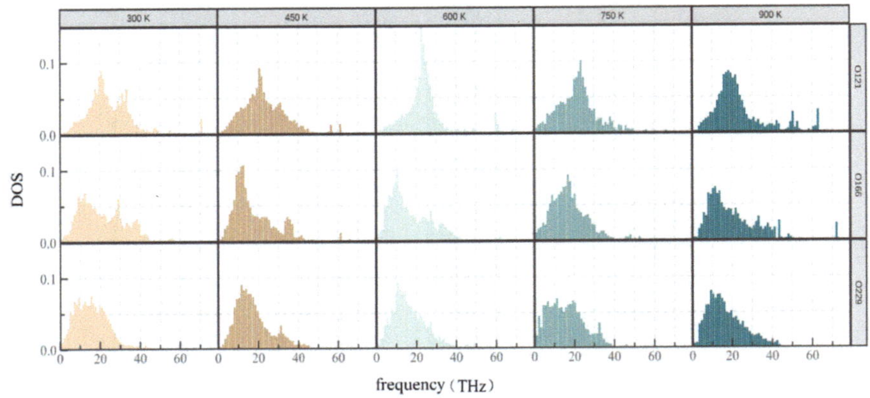

Fig. 10.41 Local phonon density of states for O121, O166, and O229 in coal molecules at different temperatures

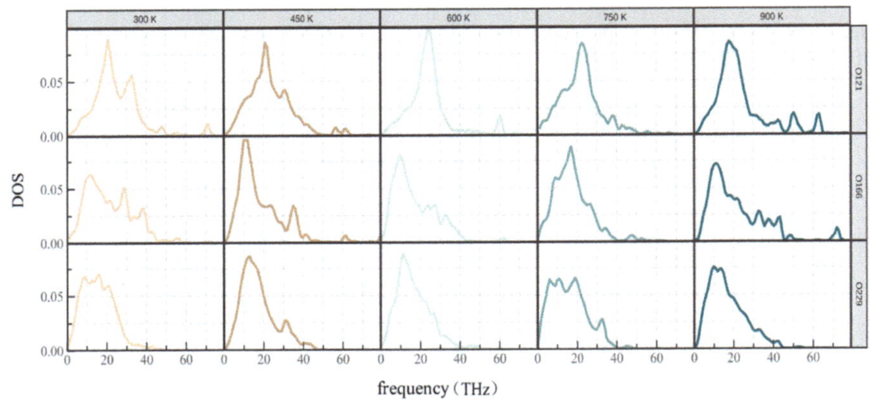

Fig. 10.42 Local phonon density of states (span = 0.1) for O121, O166, and O229 in coal molecules at different temperatures

Figures 10.44, 10.45 and 10.46 depict the phonon local density of states of O193, O234, and O235 in coal molecules at different temperatures, where O193 is the oxygen atom in the carbonyl group, O234 is the oxygen atom in the C=O group in the carboxyl group, and O235 is the oxygen atom in the carboxyl group and C–O.

When the local phonon density of states of O193 rises from 300 to 900 K, the peak and peak of the density of states do not change significantly, but oscillates in a small range, which indicates that the temperature sensitivity of carbonyl oxygen atoms is very low.

O234 and O235 are two oxygen atoms in the same carboxyl group, and their local phonon state densities show a law that the phonon state distribution shifts from low frequency to high frequency with the increase of temperature. Moreover, the distribution of phonon states of O234 at high frequencies is greater than that of

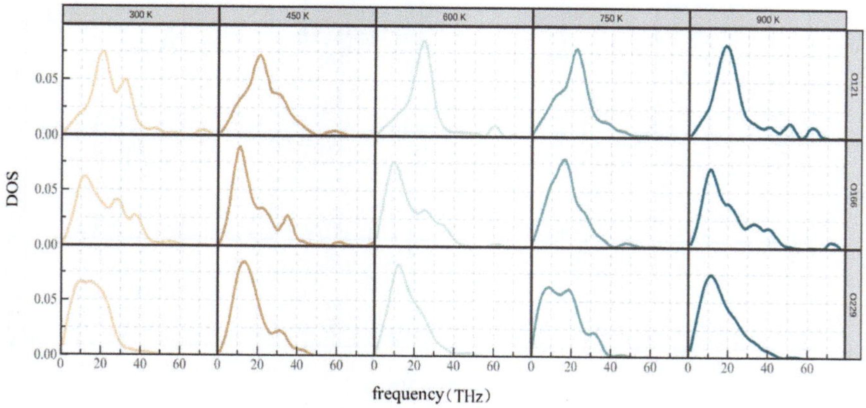

Fig. 10.43 Local phonon density of states (span = 0.2) for O121, O166, and O229 in coal molecules at different temperatures

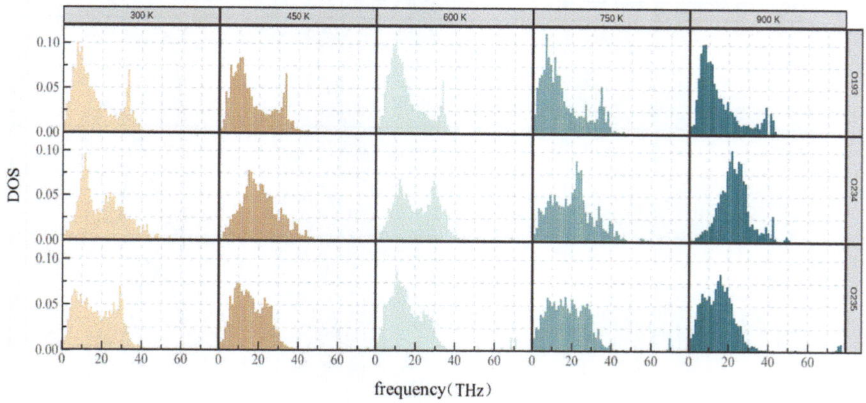

Fig. 10.44 Local phonon density of states for O193, O234, and O235 in coal molecules at different temperatures

O235, indicating that the contribution of the oxygen atom in C=O in the carboxyl group to the phonon thermal conductance is greater than that of the oxygen atom in the carboxyl group and C–O (Fig. 10.45).

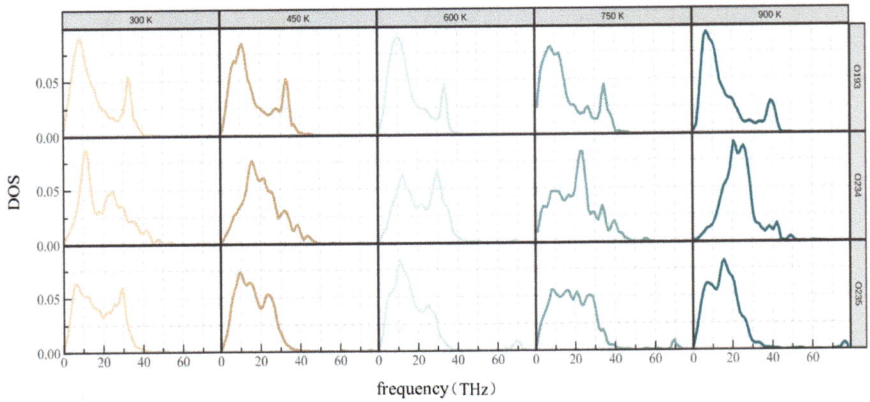

Fig. 10.45 Local phonon density of states (span = 0.1) of O193, O234, and O235 in coal molecules at different temperatures

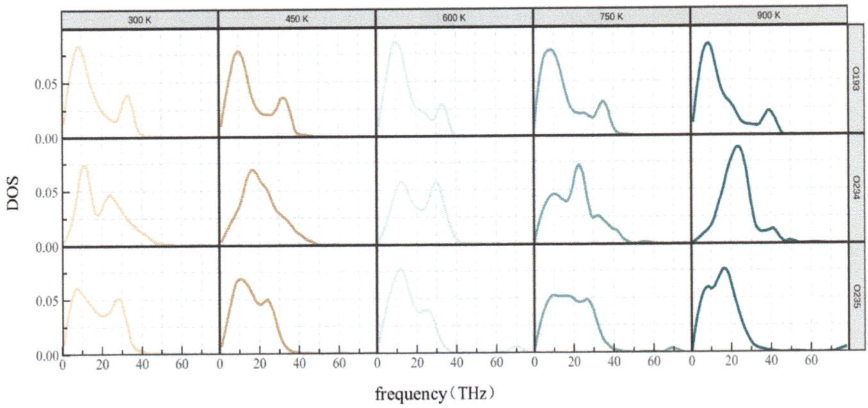

Fig. 10.46 Local phonon density of states (span = 0.2) of O193, O234, and O235 in coal molecules at different temperatures

References

1. National Bureau of Statistics. (2023). *Statistical bulletin on the national economic and social development of the People's Republic of China in 2022*. Peoples Daily. (国家统计局. 中华人民共和国2022年国民经济和社会发展统计公报. 人民日报, 2023–03–01(009).)
2. Heffern, E., & Coates, D. (2003). Geologic history of natural coal-bed fires, Powder River basin, USA. *International Journal of Coal Geology, 59*, 25–47.
3. Kuenzer, C., & Stracher, G. (2011). Geomorphology of coal seam fires. *Geomorphology*, (138), 209–222.
4. Wu, X., & Chen, G. (2018). Coal use embodied in globalized world economy: From source to sink through supply chain. *Renewable and Sustainable Energy Reviews, 81*, 978–993.
5. Deng, J., Li, B., Wang, K., & Wang, C. P. (2016). Research status and outlook on prevention and control technology of coal fire disaster in China. *Coal Science and Technology*, *44*(10),

1–7+101. (邓军, 李贝, 王凯, 王彩萍. 我国煤火灾害防治技术研究现状及展望. 煤炭科学技术,2016,44(10):1–7+101.)

6. Ma, D. M., Chen, Y., Yang, F., Guo, C., Zhang, L., & Zhang, C. Y. (2018). Conductivity changing rules of low-rank coal reservoir during methane absorption and desorption. *Resources & Industries, 20*(04), 1–8. (马东民, 陈跃, 杨甫, 郭晨, 张林, 张晨阳. 低阶煤储层甲烷吸附解吸过程中导电性变化规律. 资源与产业, 2018, 20(04):1–8.)
7. Zeng, Q., Nie, Y., & Pu, Y. (2016). Characteristics of the distribution of typical heavy metals in thesoils of underground coal fire. *Journal of China Coal Society, 41*(08), 1989–1996. (曾强, 聂静, 蒲燕. 地下煤火土壤典型重金属分布特征. 煤炭学报, 2016, 41(08):1989–1996.)
8. Meng, X. G. (2018). Treatment technology and scheme analysis of coal fire in dafeng opencast coal mine's lower group. *Energy Science and Technology, 16*(10), 34–38. (孟兴国. 大峰露天煤矿下组煤火治理技术及方案分析. 神华科技, 2018,16(10):34–38.)
9. Liu, Z. D., Wang, W., Yang, J. R., Shi, M. X., Ma, T. H., & Wang, D. Z., et al. (2020). Review and prospect of study on conductive properties of coal and CBM reservoirs. *Progress in Geophysics, 35*(04), 1415–1423. (刘之的, 王伟, 杨珺茹, 时梦璇, 马廷昊, 王得志, 等. 煤及煤层气储层导电特性研究综述与展望[J].地球物理学进展,2020,35(04):1415–1423.)
10. Wen, H., Tian, Q., Zhao, X. T., & Wang, W. F. (2021). Research status and development trend of coal fire monitoring and control. *Coal Engineering, 53*(09), 101–106. (文虎, 田晴, 赵向涛, 王伟峰. 煤层火灾监测与治理研究现状及发展趋势. 煤炭工程, 2021, 53(09):101–106.)
11. Zhu, H. Q., Yuan, J., Zhao, J. L., & Zhu, X. M. (2019). Research on the status of underground coal fire distributionand development of the detection technology. *Industrial Safety and Environmental Protection, 45*(12), 28–32. (朱红青, 袁杰, 赵金龙, 朱晓曼. 地下煤火分布及探测技术现状研究. 工业安全与环保, 2019, 45(12):28–32.)
12. Liu, S., Wu, D. S., Wang, G., Tian, X. G., & Zhang, G. R. (2019). Application of borehole monitoring in fire area treatment of Wuda coalfield. *China Energy and Environmental Protection, 41*(12), 5–8+23. (刘赛, 吴德山, 王刚, 田新光, 张国荣. 钻孔监测在乌达煤田火区治理中的应用. 能源与环保, 2019, 41(12):5–8+23.)
13. Li, F., Liang, H. D., Zhao, X. P., Bai, J. W., & Cui, Y. K. (2017). Remote sensing monitoring and assessment of fire-fighting effects in Wuda coal field, InnerMongolia. *Remote Sensing for Natural Resources, 29*(03), 217–223. (李峰, 梁汉东, 赵小平, 白江伟, 崔玉坤. 内蒙古乌达煤田煤火治理效果的遥感监测与评估. 国土资源遥感, 2017, 29(03):217–223.)
14. Xiao, Y., Liu, J. W., Zeng, J. F., Lu, X., Tian, Y., & Shu, C. M. (2022). Coupling effect of operational factors on heat extraction from a coal pile using a two-phase closed thermosyphon. *Energy, 239*, Article 122371.
15. Mishra, R. K., Roy, P. N. S., Singh, V. K., & Pandey, J. K. (2018). Detection and delineation of coal mine fire in Jharia coal field, India using geophysical approach: A case study. *Journal of Earth System Science, 127*(08), 1–10.
16. Song, Y. T., & Wang, N. (2019). Exploring temporal and spatial evolution of global coal supply-demand and flow structure. *Energy, 168*, 1073–1080.
17. Huan, X., Tang, Y. G., Xu, J. J., & Xu, M. X. (2020). Nano-level resolution determination of aromatic nucleus in coal. *Fuel, 262*, Article 116532.
18. Liu, K. H., Xiao, Y., Zhang, H., Pang, P., & Shu, C. M. (2022). Inhibiting effects of carbonised and oxidised powders treated with ionic liquids on spontaneous combustion. *Process Safety and Environmental Protection, 157*, 237–245.
19. Liu, J. L. (2020). *Cooperative detection and analysis of multi remote sensing in Miquan fire area, Xinjiang*. Master Thesis, China University of Mining and Technology, Xuzhou China. (刘竞龙. 新疆米泉火区多源遥感协同探测与分析. 中国矿业大学, 2020.)
20. Li, J. L., & Zhang, Y. Y. (2006). Analysis of detection technology of fire origin in colliery. *Journal of Mining and Strata Control Engineering*, (04), 86–87+91. (李金良, 张友谊. 浅谈煤矿火源探测技术. 煤矿开采, 2006, (04):86–87+91.)
21. Qi, F. H., & Zhang, F. Y. (2010). Detection and controlling of underground coal burning. *Coal Geology of China, 22*(S1), 143–146. (齐福辉, 张福英. 地下煤火的探测及防治. 中国煤炭地质, 2010, 22(S1):143–146.)

22. Wu, J. J., Liu, X. C., Jiang, W. G., Gu, L., & Cai, Z. Y. (2010). Spatial analysis of risk for underground coal fire in Xinjiang, China. *Journal of China Coal Society, 35*(07), 1147–1154. (武建军, 刘晓晨, 蒋卫国, 顾磊, 蔡忠勇. 新疆地下煤火风险分布格局探析. 煤炭学报, 2010, 35(07):1147–1154.)
23. Zhu, X. Y., Yu, C. C., Xiong, S. Q., & Chen, B. (2007). The application of the magnetic method to the detection of underground coal fires. *Geophysical and Geochemical Exploration, 31*(02), 115–119. (朱晓颖, 于长春, 熊盛青, 陈斌. 磁法在煤火探测中的应用. 物探与化探, 2007, (02):115–119.)
24. Fuchs, W., & Sandhoff, A. (1942). Theory of coal pyrolysis. *Industrial and Engineering Chemistry, 34*(05), 567–571.
25. Chermin, H., & Krevelen, D. (1957). Chemical structure and properties of coal. *Fuel, 36*(03), 313–320.
26. Given, P. (1959). Structure of bituminous coals: Evidence from distribution of hydrogen. *Nature, 184*(4691), 980–981.
27. Wiser, W. H. (1984). Conversion of bituminous coal to liquids and gases: Chemistry and representative processes. *Journal of Magnetic Resonance, 124*, 325–350.
28. Shinn, J. (1984). From coal to single-stage and two-stage products: A reactive model of coal structure. *Fuel, 63*(09), 1187–1196.
29. Xiang, J. H., Zeng, F. G., Liang, H. Z., Sun, B. L., Zhang, L., & Li, M. F., et al. (2011). Model construction of the macro molecular structure of Yanzhou coal and its molecular simulation. *Journal of Fuel Chemistry and Technology, 39*(07), 481–488. (相建华, 曾凡桂, 梁虎珍, 孙蓓蕾, 张莉, 李美芬, 等. 兖州煤大分子结构模型构建及其分子模拟. 燃料化学学报, 2011, 39(07):481–488.)
30. Zhou, X. Y., Zeng, F. G., Xiang, J. H., Deng, X. P., & Xiang, X. H. (2020). Macromolecular model construction and molecular simulation of organic matter in Majiliang vitrain. *CIESC Journal, 71*(04), 1802–1811. (周星宇, 曾凡桂, 相建华, 邓晓鹏, 项兴华. 马脊梁镜煤有机质大分子模型构建及分子模拟. 化工学报, 2020, 71(04):1802–1811.)
31. Zhang, S., Ma, R. J., Liu, L., Zhang, H., & Liu, Q. F. (2020). Molecular structure characteristics and model construction of anthracite in Jarud. *Coal Geology & Exploration, 48*(01), 62–69. (张帅, 马汝嘉, 刘路, 张浩, 刘钦甫. 扎鲁特地区无烟煤分子结构特征和模型构建. 煤田地质与勘探, 2020, 48(01):62–69.)
32. Yang, J. X., Yang, Y., Jia, Y., & Liu, Z. (2021). The development in preparation and application of thermal conductive composites. *Polymer Bulletin*, (08), 1–8. (杨菊香, 杨莹, 贾园, 刘振. 导热高分子材料的制备及其应用研究进展. 高分子通报, 2021, (08):1–8.)
33. Zhao, J. Q., Meng, F. C., Shen, J. B., & Feng, Y. K. (2020). Research progress of thermal conductive polymer composites. *Chemical Industry and Engineering, 37*(03), 67–73. (赵敬棋, 孟凡成, 申景博, 冯亚凯. 导热高分子复合材料的研究进展. 化学工业与工程, 2020, 37(03):67–73.)
34. Wu, X. S., Tang, W. T., & Xu, X. F. (2020). Recent progresses of thermal conduction intwo-dimensional materials. *Acta Physica Sinica, 69*(19), 37–69. (吴祥水, 汤雯婷, 徐象繁. 二维材料热传导研究进展. 物理学报, 2020, 69(19):37–69.)
35. Hou, S. Y., Yan, H. H., Ren, F., & Di, Y. Y. (2020). Research progress on thermal conducting polymer composites. *Synthetic Materials Aging and Application, 49*(06), 135–138+83. (侯思雨, 闫焕焕, 任芳, 狄莹莹. 高分子复合材料导热性能的研究进展. 合成材料老化与应用, 2020, 49(06):135–138+83.)
36. Cui, Y., Li, S. H., Ying, T., Bao, H., & Zeng, X. Q. (2021). Research on the thermal conductivity of metals based on first principles. *Acta Metallurgica Sinica, 57*(03), 375–384. (崔洋, 李寿航, 应韬, 鲍华, 曾小情. 基于第一性原理的金属导热性能研究. 金属学报, 2021, 57(03):375–384.)
37. Wen, B., & Feng, X. (2015). Thermal conductivity of metal from first principlescalculations and its application in aluminum. *Journal of Yanshan University, 39*(04), 298–305. (温斌, 冯幸. 金属热导率的第一性原理计算方法在铝中的应用. 燕山大学学报, 2015, 39(04):298–305.)
38. Ren, H. Y., Cao, Z. Y., Tong, J. Q., Li, Z. J., & Wang, Z. Y. (2018). Resistivity analysis of primary structural coal and cataclastic coal. *Coal Technology, 37*(09), 139–142. (任华育, 曹志勇, 童继强, 李志军, 王增玉. 原生煤和碎裂煤电阻率对比分析. 煤炭技术, 2018, 37(09):139–142.)

39. Kang, T. H., Dong, D., Wei, J. P., & Xue, Y. L. (2016). An experimental study on the relationship between resistivity and gas content of coal. *Geology and Exploration, 52*(05), 918–923. (康天慧, 董东, 魏建平, 薛玉龙. 煤电阻率与其瓦斯含量关系的实验研究. 地质与勘探, 2016, 52(05):918–923.)
40. Liu, S. Q., Zhang, Y. J., Tuo, K. Y., Wang, L. P., & Chen, G. (2018). Structure, electrical conductivity, and dielectric properties of semi-coke derived from microwave-pyrolyzed low-rank coal. *Fuel Processing Technology, 178*, 139–147.
41. Li, X. C., Zhang, Q., An, X. B., & Zhang, F. (2021). Experimental study of influencing factors of electrical parameters of different kinds of coal mass. *Journal of China University of Mining & Technology, 50*(03), 570−578. (李祥春, 张琪, 安振兴, 张帆. 不同煤体电性参数影响因素实验研究. 中国矿业大学学报, 2021, 50(03):570–578.)
42. Tang, X. Y., Chen, X. J., & Zhang, C. Y. (2019). Experimental study on the change law of resistivity in the primary structure coal with low rank. *Science Technology and Engineering, 19*(26), 294−299. (汤小燕, 陈学健, 张晨阳. 低变质原生结构煤电阻率变化规律实验研究. 科学技术与工程, 2019, 19(26):294–299.)
43. Wang, B. (2019). *Study on response characteristics of resistanceof gas-free coal to moisture content duringwater permeability.* Master Thesis, Henan Polytechnic University, Henan China. (王彬. 不含瓦斯煤渗吸电阻率对含水率的响应特性研究. 河南理工大学, 2019.)
44. Lyu, C., Hao, S. Q., & Sun, Q. (2019). Experiment study on the correlation between the CO_2 adsorption capacity and electrical resistivity of coal with temperature effect. *Greenhouse Gases: Science and Technology, 9*, 924–933.
45. Deng, J., Li, Q. W., Xiao, Y., & Shu, C. M. (2016). Experimental study on the thermal properties of coal during pyrolysis, oxidation, and re-oxidation. *Applied Thermal Engineering, 110*, 1137–1152.
46. Deng, J., Ren, S. J., Xiao, Y., Li, Q. W., & Shu, C. M. (2018). Thermal properties of coals with different metamorphic levels in air atmosphere. *Applied Thermal Engineering, 143*, 542–549.
47. Xiao, Y., Meng, X., Yin, L., Li, Q. W., Shu, C. M., & Tian, Y. (2021). Influence of element composition and microcrystalline structure on thermal properties of bituminous coal under nitrogen atmosphere. *Process Safety and Environmental Protection, 147*, 846–856.
48. Xiao, Y., Yin, L., Deng, J., Wang, C. P., & Shu, C. M. (2019). Thermophysical parameters of coal with various levels of preoxidation. *Journal of Thermal Analysis and Calorimetry, 135*(05), 2819–2829.
49. Deng, J., Ren, S. J., Xiao, Y., Li, Q. W., & Wang, C. P. (2019). Comparative study on heat transfer characteristics of coal during lowtemperature oxidation and pyrolysis. *Journal of China Coal Society, 44*(S1), 171–177. (邓军, 任帅京, 肖旸, 李青蔚, 王彩萍. 煤低温氧化与热解过程的传热特性对比研究. 煤炭学报, 2019, 44(S1):171–177.)
50. Shin, S., Wang, Q., Luo, J., & Chen, R. K. (2020). Advanced materials for high-temperature thermal transport. *Advanced Functional Materials, 30*, 1904815.
51. Tao, L., Yang, Y., & Li, Z. Q. (2017). Low temperature electron scattering mechanismsin three dimensional aluminum doped zinc oxide films. *Low Temperature Physical Letters, 39*(03), 66−71. (陶黎, 杨洋, 李志青. 三维Al掺ZnO薄膜的低温电子散射机制. 低温物理学报,2017,39(03):66–71.)
52. Polshyn, H., Yankowitz, M., Chen, S. W., Zhang, Y. X., Watanabe, K., Taniguchi, T., et al. (2019). Large linear-in-temperature resistivity in twisted bilayer graphene. *Nature Physics, 15*, 1011–1016.
53. Lv, Y. Y., Xu, L., Dong, S. T., Luo, Y. C., Zhang, Y. Y., Chen, Y. B., et al. (2019). Electron-electron scattering dominated electrical and magnetotransport properties in the quasi-two-dimensional Fermi liquid single-crystal Bi_2O_2Se. *Physical Review B, 99*, Article 195143.
54. Jeffries, W. R., Park, K., Vaia, R. A., & Jr, K. L. K. (2020). Resolving electron–electron scattering in plasmonic nanorod ensembles using two-dimensional electronic spectroscopy. *Nano letters, 20*(10), 7722−7727.
55. Jaćimovski, S., Lamovec, J., Šetrajčić, J., & Ilić, D. (2021). Temperature dependence of the relaxation time in scattering of elementary excitations in graphene. *Materials Science and Engineering B, 264*, Article 114933.

56. Huang, X., & Wang, Z. L. (2019). Calculation of the temperature and scale effects for GaAs nanowires' latticethermal conductivity and phonon inelastic scattering. *Journal of Thermal Science and Technology, 18*(06), 444–450. (黄雪, 王照亮. GaAs纳米线晶格热导率温度和尺度效应及声子非弹性散射. 热科学与技术, 2019, 18(06):444–450.)
57. Nshimyimana, E., Hao, S. Q., Su, X. L., Zhang, C., Liu, W., Yan, Y. G., et al. (2020). Discordant nature of Cd in GeTe enhances phonon scattering and improves band convergence for high thermoelectric performance. *Journal of Materials Chemistry A, 8*, 1193–1204.
58. He, X. Y., Lin, F. T., Liu, F., & Zhang, H. (2019). Investigation of phonon scattering on the tunable mechanisms of terahertz graphene metamaterials. *Nanomaterials, 10*(01), 39.
59. Thomas, L. H. (1927). The calculation of atomic fields. *Mathematical Proceedings of the Cambridge Philosophical Society, 23*, 542–548.
60. Fermi, E. (1927). Un metodo statistico per la determinazione di alcune priorieta dell'atome. *Rendiconti Accademia Nazionale dei Lincei, 6*(32), 602–607.
61. Hohenberg, P., & Kohn, W. (1964). Inhomogeneous electron gas. *Physical Review B, 136*(3), 864–871.
62. Kohn, W., & Sham, L. J. (1965). Self-consistent equations including exchange and correlation effects. *Physical Review A, 140*, 133–138.
63. Perdew, J. P., & Zunger, A. (1981). Self-interaction correction to density-functional approximations for many-electron systems. *Physical Review B, 23*, 5048–5079.
64. Perdew, J. P., Burke, K., & Wang, Y. (1996). Generalized gradient approximation for the exchange-correlation hole of a many-electron system. *Physical Review B, 54*(23), 16533–16539.
65. Perdew, J. P., & Wang, Y. (1992). Accurate and simple analytic representation of the electron-gas correlation energy. *Physical review B, 45*(23), 13244–13249.
66. Becke, A. D., & Johnson, E. R. (2007). Exchange-hole dipole moment and the dispersion interaction revisited. *The Journal of chemical physics, 127*(15), Article 154108.
67. Neese, F. (2012). The ORCA program system. *Wiley Interdisciplinary Reviews: Computational Molecular Science, 2*, 73–78.
68. Perdew, J. P., Ruzsinszky, A., Csonka, G. I., Constantin, L. A., & Sun, J. W. (2009). Workhorse semilocal density functional for condensed matter physics and quantum chemistry. *Physical Review Letters, 103*, 26403.
69. Kozuch, S., & Martin, J. (2011). DSD-PBEP86: In search of the best double-hybrid DFT with spin-component scaled MP2 and dispersion corrections. *Physical Chemistry Chemical Physics, 13*, 20104–20107.
70. Kozuch, S., & Martin, J. (2013). Spin-component-scaled double hybrids: An extensive search for the best fifth-rung functionals blending DFT and perturbation theory. *Journal of Computational Chemistry, 34*, 2327–2344.
71. De Oliveira, M. T., Alves, J. M. A., Braga, A. A. C., Wilson, D. J. D., & Barboza, C. A. (2021). Do double-hybrid exchange-correlation functionals provide accurate chemical shifts? A benchmark assessment for proton NMR. *Journal of Chemical Theory and Computation, 17*, 6876–6885.
72. Stoychev, G. L., Auer, A. A., & Neese, F. (2018). Efficient and accurate prediction of nuclear magnetic resonance shielding tensors with double-hybrid density functional theory. *Journal of Chemical Theory and Computation, 14*, 4756–4771.
73. Keal, T., & Tozer, D. (2003). The exchange-correlation potential in Kohn-Sham nuclear magnetic resonance shielding calculations. *The Journal of Chemical Physics, 119*, 3015–3024.
74. Keal, T., & Tozer, D. A. (2004). Semiempirical generalized gradient approximation exchange-correlation functional. *The Journal of Chemical Physics, 121*, 5654–5660.
75. Weigend, F., & Ahlrichs, R. (2005). Balanced basis sets of split valence, triple zeta valence and quadruple zeta valence quality for H to Rn: Design and assessment of accuracy. *Physical Chemistry Chemical Physics, 7*, 3297–3305.
76. Jensen, F. (2015). Segmented contracted basis sets optimized for nuclear magnetic shielding. *Journal of Chemical Theory and Computation, 11*, 132–138.

77. Grimme, S., Ehrlich, S., & Goerigk, L. (2011). Effect of the damping function in dispersion corrected density functional theory. *Journal of Computational Chemistry, 32*, 1456–1465.
78. Stephens, P., Devlin, F., Chabalowski, C., & Frisch, M. J. (1994). Ab initio calculation of vibrational absorption and circular dichroism spectra using density functional force fields. *The Journal of Physical Chemistry, 98*, 11623–11627.
79. Grimme, S., Antony, J., Ehrlich, S., & Krieg, H. (2010). A consistent and accurate ab initio parametrization of density functional dispersion correction (DFT-D) for the 94 elements H-Pu. *The Journal of Chemical Physics, 132*, Article 154104.
80. Lu, T., & Chen, F. (2012). Multiwfn: A multifunctional wavefunction analyzer. *Journal of Computational Chemistry, 33*, 580–592.
81. Kühne, T. D., Iannuzzi, M., Del Ben, M., Rybkin, V. V., Seewald, P., Stein, F., et al. (2020). CP_2K: An electronic structure and molecular dynamics software package-quickstep: Efficient and accurate electronic structure calculations. *The Journal of Chemical Physics, 152*, Article 194103.
82. VandeVondele, J., Krack, M., Mohamed, F., Parrinello, M., Chassaing, T., & Hutter, J. (2005). Quickstep: Fast and accurate density functional calculations using a mixed Gaussian and plane waves approach. *Computer Physics Communications, 167*, 103–128.
83. VandeVondele, J., & Hutter, J. (2007). Gaussian basis sets for accurate calculations on molecular systems in gas and condensed phases. *The Journal of Chemical Physics, 127*, Article 114105.
84. Goedecker, S., Teter, M., & Hutter, J. (1996). Separable dual-space Gaussian pseudopotentials. *Physical Review B, 54*, 1703–1710.
85. Perdew, J. P., Burke, K., & Ernzerhof, M. (1996). Generalized gradient approximation made simple. *Physical Review Letters, 77*, 3865–3868.
86. Togo, A., & Tanaka, I. (2015). First principles phonon calculations in materials science. *Scripta Materialia, 108*, 1–5.

MIX
Papier aus verantwortungsvollen Quellen
Paper from responsible sources
FSC® C105338
FSC
www.fsc.org

If you have any concerns about our products,
you can contact us on
ProductSafety@springernature.com

In case Publisher is established outside the EU,
the EU authorized representative is:
Springer Nature Customer Service Center GmbH
Europaplatz 3, 69115 Heidelberg, Germany

Printed by Libri Plureos GmbH
in Hamburg, Germany